CHILTON'S

TOYOTA / LEXUS
REPAIR MANUAL
1988-1992

Complete service information for all models— cars and light trucks

Sr. Vice President	Ronald A. Hoxter
Publisher and Editor-In-Chief	Kerry A. Freeman, S.A.E.
Managing Editors	Peter M. Conti, Jr. □ W. Calvin Settle, Jr., S.A.E.
Assistant Managing Editor	Nick D'Andrea
Senior Editors	Richard J. Rivele, S.A.E. □ Ron Webb
Director of Manufacturing	Mike D'Imperio
Manager of Manufacturing	John F. Butler

CHILTON BOOK COMPANY

ONE OF THE **DIVERSIFIED PUBLISHING COMPANIES,**
A PART OF **CAPITAL CITIES/ABC, INC.**

Manufactured in USA
© 1992 Chilton Book Company
Chilton Way Radnor, Pa. 19089
ISBN 0-8019-8321-5
ISSN No. 1060-4421

1 2 3 4 5 6 7 8 9 0   1 0 9 8 7 6 5 4 3 2

SAFETY NOTICE

Proper service and repair procedures are vital to the safe, reliable operation of all motor vehicles, as well as the personal safety of those performing service or repairs. This manual outlines procedures for servicing and repairing vehicles using safe, effective methods. The procedures contain many NOTES and CAUTIONS which should be followed along with standard safety procedures to eliminate the possibility of personal injury or improper service which could damage the vehicle or compromise its safety.

It is important to note that repair procedures and techniques, tools and parts for servicing motor vehicles, as well as the skill and experience of the individual performing the work vary widely. It is not possible to anticipate all of the hazards that may result. Standard and accepted safety precautions and equipment should be used when handling toxic or flammable fluids and safety goggles or other protection should be used during cutting, grinding, chiseling, prying or any other process that can cause material removal or projectiles. Similar protection against the high voltages generated in all electronic ignition systems should be employed during service procedures.

Some procedures require the use of tools or test equipment specially designed for a specific purpose. Before substituting another tool or procedure, you must be completely satisfied that neither your personal safety, nor the performance of the vehicle will be endangered.

PART NUMBERS

Part numbers listed in this reference are not recommendations by Chilton for any product by brand name. They are references that can be used with interchange manuals and aftermarket supplier catalogs to locate each brand supplier's discrete part number.

Although information in this manual is based on industry sources and is complete as possible at the time of publication, the possibility exists that some car manufacturers made later changes which could not be included here. While striving for total accuracy, Chilton Book Company cannot assume responsibility for any errors, changes or omissions that may occur in the compilation of this data.

Contents

SERIAL NUMBER IDENTIFICATION

Vehicle Identification Plate

All vehicles have the Vehicle Identification Number (VIN) stamped on a plate which is attached to the left side of the instrument panel. This plate is visible through the windshield.

The serial number consists of a series identification number followed by a 6-digit production number.

Engine Number

Basically, Toyota uses 5 types of engines:

A-Series
3A-C
4A-LC
4A-F, 4A-FE

4A-GE
4A-GEC, 4A-GELC
E-Series
3E, 3E-E
M-Series
5M-GE, 7M-GE, 7M-GTE
S-Series
3S-FE, 3S-GE, 3S-GTE, 5S-FE
Z-Series
2VZ-FE

Engines within each series are similar, as the cylinder block designs are the same. Variances within each series may be due to ignition types, displacements and cylinder head design.

Serial numbers of the engines may be found on the following locations:

A-Series engines—stamped vertically on the left side rear of the engine block.

E-Series engines—stamped on the left side rear of the engine block.

M-Series engines—stamped horizontally on the passenger side of the engine block, behind the alternator.

S-Series engines—the serial number can be found on the rear left side of the block, under the thermostat housing.

Z-Series engines—stamped on the front, right (passenger) side of the cylinder block.

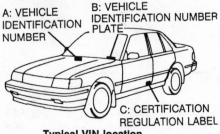

A: VEHICLE IDENTIFICATION NUMBER
B: VEHICLE IDENTIFICATION NUMBER PLATE
C: CERTIFICATION REGULATION LABEL

Typical VIN location

SPECIFICATIONS

ENGINE IDENTIFICATION

Year	Model	Engine Displacement cu. in. (cc/liter)	Engine Series Identification	No. of Cylinders	Engine Type
1988	Tercel	88.6 (1452/1.4)	3A-C	4	SOHC
		88.9 (1456/1.5)	3E	4	SOHC
	Corolla	97.0 (1587/1.6)	4A-LC	4	SOHC
		97.0 (1587/1.6)	4A-GEC, 4A-GELC	4	DOHC
		97.0 (1587/1.6)	4A-F	4	DOHC
	Camry	121.9 (1998/2.0)	3S-FE	4	DOHC
	Celica	121.9 (1998/2.0)	3S-FE, 3S-GE	4	DOHC
		121.9 (1998/2.0)	3S-GTE	4	DOHC, TURBO
	Supra	180.3 (2954/3.0)	7M-GE	6	DOHC
		180.3 (2954/3.0)	7M-GTE	6	DOHC, TURBO
	MR2	97.0 (1587/1.6)	4A-GELC	4	DOHC
		97.0 (1587/1.6)	4A-GZE	4	DOHC, SUPER
	Cressida	168.4 (2759/2.8)	5M-GE	6	DOHC
1989	Tercel	88.9 (1457/1.5)	3E	4	SOHC
	Corolla	97.0 (1587/1.6)	4A-GE	4	DOHC
		97.0 (1587/1.6)	4A-F, 4A-FE	4	DOHC
	Camry	121.9 (1998/2.0)	3S-FE	4	DOHC
		153.0 (2058/2.5)	2VZ-FE	6	DOHC
	Celica	121.9 (1998/2.0)	3S-FE, 3S-GE	4	DOHC
		121.9 (1998/2.0)	3S-GTE	4	DOHC, TURBO
	Supra	180.3 (2954/3.0)	7M-GE	6	DOHC
		180.3 (2954/3.0)	7M-GTE	6	DOHC, TURBO
	MR2	97.0 (1587/1.6)	4A-GELC	4	DOHC
		97.0 (1587/1.6)	4A-GZE	4	DOHC, SUPER

ENGINE IDENTIFICATION

Year	Model	Engine Displacement cu. in. (cc/liter)	Engine Series Identification	No. of Cylinders	Engine Type
1989	Cressida	180.3 (2954/3.0)	7M-GE	6	DOHC
1990	Tercel	88.9 (1457/1.5)	3E	4	SOHC
		88.9 (1457/1.5)	3E-E	4	SOHC
	Corolla	97.0 (1587/1.6)	4A-FE	4	DOHC
		97.0 (1587/1.6)	4A-GE	4	DOHC
	Camry	121.9 (1998/2.0)	3S-FE	4	DOHC
		153.0 (2058/2.5)	2VZ-FE	6	DOHC
	Celica	97.0 (1587/1.6)	4A-FE	4	DOHC
		121.9 (1998/2.0)	3S-GTE	4	DOHC, TURBO
		132.0 (2164/2.2)	5S-FE	4	DOHC
	Supra	180.3 (2954/3.0)	7M-GE	6	DOHC
		180.3 (2954/3.0)	7M-GTE	6	DOHC, TURBO
	MR2	121.9 (1998/2.0)	3S-GTE	4	DOHC, TURBO
		132.0 (2164/2.2)	5S-FE	4	DOHC
	Cressida	180.3 (2954/3.0)	7M-GE	6	DOHC
1991–92	Tercel	88.9 (1457/1.5)	3E-E	4	SOHC
	Corolla	97.0 (1587/1.6)	4A-FE	4	DOHC
		97.0 (1587/1.6)	4A-GE	4	DOHC
	Camry	121.9 (1998/2.0)	3S-FE	4	DOHC
		153.0 (2058/2.5)	2VZ-FE	6	DOHC
	Celica	97.0 (1587/1.6)	4A-FE	4	DOHC
		121.9 (1998/2.0)	3S-GTE	4	DOHC, TURBO
		132.0 (2164/2.2)	5S-FE	4	DOHC
	Supra	180.3 (2954/3.0)	7M-GE	6	DOHC
		180.3 (2954/3.0)	7M-GTE	6	DOHC, TURBO
	MR2	121.9 (1998/2.0)	3S-GTE	4	DOHC, TURBO
		132.0 (2164/2.2)	5S-FE	4	DOHC
	Cressida	180.3 (2954/3.0)	7M-GE	6	DOHC

OHV—Overhead Valves
SOHC—Single Overhead Camshaft
DOHC—Double Overhead Camshaft
TURBO—Turbocharged
SUPER—Supercharged

GENERAL ENGINE SPECIFICATIONS

Year	Model	Engine Displacement cu. in. (cc)	Fuel System Type	Net Horsepower @ rpm	Net Torque @ rpm (ft. lbs.)	Bore × Stroke (in.)	Compression Ratio	Oil Pressure ①
1988	Tercel	88.6 (1452)	2 bbl	62 @ 4800	76 @ 2800	3.05 × 3.03	9.0:1	4.3
		88.9 (1456)	2 bbl	78 @ 6000	87 @ 4000	2.87 × 3.43	9.3:1	4.3
	Corolla	4A-LC 97.0 (1587)	2 bbl	74 @ 5200	86 @ 2800	3.19 × 3.03	9.0:1	4.3
		4A-F 97.0 (1587)	2 bbl	90 @ 6000	95 @ 3600	3.19 × 3.03	9.5:1	4.3
		97.0 (1587)	EFI	116 @ 6600③	110 @ 4800④	3.19 × 3.03	·9.4:1	4.3
	Camry	121.9 (1998)	EFI	115 @ 5200	124 @ 4400	3.39 × 3.39	9.3:1	4.3

GENERAL ENGINE SPECIFICATIONS

Year	Model	Engine Displacement cu. in. (cc)	Fuel System Type	Net Horsepower @ rpm	Net Torque @ rpm (ft. lbs.)	Bore × Stroke (in.)	Compression Ratio	Oil Pressure ①
1988	Celica	3S-FE 121.9 (1998)	EFI	115 @ 5200	124 @ 4400	3.39 × 3.39	9.3:1	4.3
		3S-GE 121.9 (1998)	EFI	135 @ 6000	125 @ 4800	3.39 × 3.39	9.2:1	4.3
		3S-GTE 121.9 (1998)	EFI	190 @ 6000	190 @ 3200	3.39 × 3.39	8.5:1	4.3
	Supra	7M-GE 180.3 (2954)	EFI	200 @ 6000	185 @ 4800	3.27 × 3.58	9.2:1	4.3
		7M-GTE 180.3 (2954)	EFI	230 @ 5600	246 @ 4000	3.27 × 3.58	8.4:1	4.3
	MR2	4A-GELC 97.0 (1587)	EFI	112 @ 6600	100 @ 4800	3.19 × 3.03	9.4:1	4.3
		4A-GZE 97.0 (1587)	EFI	145 @ 6400	140 @ 4000	3.19 × 3.03	8.0:1	4.3
	Cressida	168.4 (2759)	EFI	156 @ 5200	165 @ 4400	3.27 × 3.35	9.2:1	4.3
1989	Tercel	88.6 (1452)	2 bbl	78 @ 6000	87 @ 4000	2.87 × 3.43	9.3:1	4.3
	Corolla	4A-FE 97.0 (1587)	EFI	100 @ 5600	101 @ 4400	3.19 × 3.03	9.5:1	4.3
		4A-F 97.0 (1587)	2 bbl	90 @ 6000	95 @ 3600	3.20 × 3.00	9.5:1	4.3
		97.0 (1587)	EFI	116 @ 6600	100 @ 4800	3.19 × 3.03	9.4:1	4.3
	Camry	121.9 (1998)	EFI	115 @ 5200	124 @ 4400	3.39 × 3.39	9.3:1	4.3
		153.0 (2507)	EFI	153 @ 5600	155 @ 4400	3.44 × 2.74	9.0:1	4.3
	Celica	3S-FE 121.9 (1998)	EFI	115 @ 5200	124 @ 4400	3.39 × 3.39	9.3:1	4.3
		3S-GE 121.9 (1998)	EFI	135 @ 6000	125 @ 4800	3.39 × 3.39	9.2:1	4.3
		3S-GTE 121.9 (1998)	EFI	190 @ 6000	190 @ 3200	3.39 × 3.39	8.5:1	4.3
	Supra	7M-GE 180.3 (2954)	EFI	200 @ 6000	188 @ 3600	3.27 × 3.58	9.2:1	4.3
		7M-GTE 180.3 (2954)	EFI	232 @ 5600	254 @ 3200	3.27 × 3.58	8.4:1	4.3
	MR2	4A-GELC 97.0 (1587)	EFI	115 @ 6600	100 @ 4800	3.19 × 3.03	9.4:1	4.3
		4A-GZE 97.0 (1587)	EFI	145 @ 6400	140 @ 4000	3.19 × 3.03	8.0:1	4.3
	Cressida	180.3 (2954)	EFI	190 @ 5600	185 @ 4400	3.27 × 3.58	9.2:1	4.3
1990	Tercel	3E 88.9 (1457)	1 bbl	78 @ 6000	87 @ 4000	2.87 × 3.43	9.3:1	4.3
		3E-E 88.9 (1457)	EFI	82 @ 5200	89 @ 4400	2.87 × 3.43	9.3:1	4.3
	Corolla	4A-FE 97.0 (1587)	EFI	102 @ 5800	101 @ 4800	3.19 × 3.03	9.5:1	4.3
		4A-GE 97.0 (1587)	EFI	130 @ 6800	102 @ 5800	3.19 × 3.03	9.5:1	4.3

GENERAL ENGINE SPECIFICATIONS

Year	Model	Engine Displacement cu. in. (cc)	Fuel System Type	Net Horsepower @ rpm	Net Torque @ rpm (ft. lbs.)	Bore × Stroke (in.)	Com-pression Ratio	Oil Pressure ①
1990	Camry	3S-FE 121.9 (1998)	EFI	115 @ 5200	124 @ 4400	3.39 × 3.39	9.3:1	4.3
		2VZ-FE 153.0 (2508)	EFI	156 @ 5600	160 @ 4400	3.44 × 2.74	9.0:1	4.3
	Celica	4A-FE 97.0 (1587)	EFI	103 @ 6000 ⑤	102 @ 3200 ⑥	3.19 × 3.03	9.5:1	4.3
		5S-GTE 121.9 (1998)	EFI	200 @ 6000	200 @ 3200	3.39 × 3.39	9.5:1	4.3
		3S-FE 132.0 (2164)	EFI	130 @ 5400	140 @ 4400	3.43 × 3.58	9.5:1	4.3
	Supra	7M-GE 180.3 (2954)	EFI	200 @ 6000	188 @ 3600	3.27 × 3.58	9.2:1	4.3
		7M-GTE 180.3 (2954)	EFI	232 @ 5600	254 @ 3200	3.27 × 3.58	8.4:1	4.3
	MR2	3S-GTE 121.9 (1998)	EFI	200 @ 6000	200 @ 3200	3.39 × 3.39	9.5:1	4.3
		5S-FE 132.0 (2164)	EFI	130 @ 5400	140 @ 4400	3.43 × 3.58	9.5:1	4.3
	Cressida	180.3 (2954)	EFI	190 @ 5600	185 @ 4400	3.27 × 3.35	9.2:1	4.3
1991–92	Tercel	3E-E 88.9 (1457)	EFI	82 @ 5200	90 @ 4400	2.88 × 3.43	9.3:1	4.3
	Corolla	4A-FE 97.0 (1587)	EFI	102 @ 5800	101 @ 4800	3.19 × 3.03	9.5:1	4.3
		4A-GE 97.0 (1587)	EFI	130 @ 6800	105 @ 6000	3.19 × 3.03	10.3:1	4.3
	Camry	3S-FE 121.9 (1998)	EFI	115 @ 5200	124 @ 4400	3.39 × 3.39	9.3:1	4.3
		2VZ-FE 153.0 (2058)	EFI	156 @ 5600	160 @ 4400	3.44 × 2.74	9.0:1	4.3
	Celica	4A-FE 97.0 (1587)	EFI	103 @ 6000	102 @ 3200	3.19 × 3.10	9.5:1	4.3
		3S-GTE 121.9 (1998)	EFI	200 @ 6000	200 @ 3200	3.39 × 3.39	8.8:1	4.3
		5S-FE 132.0 (2164)	EFI	130 @ 5400	140 @ 4400	3.43 × 3.58	9.5:1	4.3
	Supra	7M-GE 180.3 (2954)	EFI	200 @ 6000	188 @ 3600	3.27 × 3.58	9.2:1	4.3
		7M-GTE 180.3 (2954)	EFI	232 @ 5600	254 @ 3200	3.27 × 3.58	8.4:1	4.3
	MR2	3S-GTE 121.9 (1998)	EFI	200 @ 6000	200 @ 3200	3.39 × 3.39	8.8:1	4.3
		5S-FE 132.0 (2164)	EFI	130 @ 5400	140 @ 4400	3.43 × 3.58	9.5:1	4.3
	Cressida	7M-GE 180.3 (2954)	EFI	190 @ 5600	185 @ 4400	3.27 × 3.58	9.2:1	4.3

EFI Electronic Fuel Injection
① At Idle
② FX-16: 108 @ 6600
③ FX-16: 110 @ 6600
④ FX-16: 98 @ 4800
⑤ California: 102 @ 5800
⑥ California: 101 @ 4800

GASOLINE ENGINE TUNE-UP SPECIFICATIONS

Year	Model	Engine Displacement cu. in. (cc)	Spark Plugs Type	Gap (in.)	Ignition Timing (deg.) MT	AT	Compression Pressure (psi)	Fuel Pump (psi)	Idle Speed (rpm) MT	AT	Valve Clearance In.	Ex.
1988	Tercel	88.6 (1452)	BPR5EY-11 ①	0.043	5B	5B	178	2.6–3.5	650	900	0.008	0.012
		88.9 (1456)	BPR5EY-11	0.043	3B	3B	184	2.6–3.5	650	900	0.008	0.008
	Corolla	4A-LC 97.0 (1587)	BPR5EY-11	0.043	5B	5B	163	2.5–3.5	650	750	0.008	0.012
		4A-F 97.0 (1587)	BPR5EY-11	0.043	5B	5B	191	2.5–3.5	650	750	0.008	0.010
		4A-GE 97.0 (1587)	BCPR5EP-11	0.043	10B	10B	179	33–38	800	800	0.008	0.010
	Camry	121.9 (1998)	BCPR5EY-11	0.043	10B	10B	178	38–44	700	750	0.009	0.013
	Celica	3S-FE 121.9 (1998)	BCPR5EY-11	0.043	10B	10B	178	38–44	650	650	0.009	0.013
		3S-GE 121.9 (1998)	BCPR5EP-11	0.043	10B	10B	178	33–38	750	750	0.008	0.010
		3S-GTE 121.9 (1998)	BCPR5EP-8	0.031	10B	—	178	33–38	750	—	0.008	0.010
	Supra	7M-GE 180.3 (2954)	BCPR5EP-11	0.043	10B	10B	156	33–40	700	700	0.008	0.010
		7M-GTE 180.3 (2954)	BCPR6EP-N8	0.031	10B	10B	142	33–40	650	650	0.008	0.010
	MR2	4A-GE 97.0 (1587)	BCPR5EP-11	0.043	10B	10B	179	38–44	800	800	0.008	0.010
		4A-GZE 97.0 (1587)	BCPR6EP-11	0.043	10B	10B	156	33–38	800	800	0.008	0.010
	Cressida	168.4 (2759)	BPR5EP-11	0.043	—	10B	164	35–38	—	650	Hyd.	Hyd.
1989	Tercel	88.9 (1456)	BPR5EY-11	0.043	3B	3B	184	2.6–3.5	700	900	0.008	0.008
	Corolla	4A-FE 97.0 (1587)	BCPR5EY	0.031	10B	10B	191	38–44	800	800	0.008	0.010
		4A-F 97.0 (1587)	BCPR5EY-11	0.043	5B	5B	191	2.5–3.5	650	750	0.008	0.010
		4A-GE 97.0 (1587)	BCPR5EP-11	0.043	10B	10B	179	33–44	800	800	0.008	0.010
	Camry	121.9 (1998)	BCPR5EY-11	0.043	10B	10B	178	38–44	700	700	0.009	0.013
		153.0 (2507)	BCPR6E-11	0.043	10B	10B	142	38–44	700	700	0.007	0.013
	Celica	3S-FE 121.9 (1998)	BCPR5EY-11	0.043	10B	10B	178	38–44	700	700	0.009	0.013
		3S-GE 121.9 (1998)	BCPR5EP-11	0.043	10B	10B	178	33–38	750	750	0.008	0.010
		3S-GTE 121.9 (1998)	BCPR5EP-8	0.031	10B	—	178	33–38	750	—	0.008	0.010
	Supra	7M-GE 180.3 (2954)	BCPR5EP-11	0.043	10B	10B	156	38–44	700	700	0.008	0.010
		7M-GTE 180.3 (2954)	BCPR6EP-N8	0.031	10B	10B	142	33–40	650	650	0.008	0.010

GASOLINE ENGINE TUNE-UP SPECIFICATIONS

Year	Model	Engine Displacement cu. in. (cc)	Spark Plugs Type	Gap (in.)	Ignition Timing (deg.) MT	AT	Compression Pressure (psi)	Fuel Pump (psi)	Idle Speed (rpm) MT	AT	Valve Clearance In.	Ex.
1989	MR2	4A-GE 97.0 (1587)	BCPR5EP-11	0.043	10B	10B	179	38–44	800	800	0.008	0.010
		4A-GZE 97.0 (1587)	BCPR6EP-11	0.043	10B	10B	156	33–38	800	800	0.008	0.010
	Cressida	180.3 (2954)	BCPR5EP-11	0.043	—	10B	156	38–44	—	700	0.008	0.010
1990	Tercel	3E 88.9 (1457)	5PR5EY-11	0.043	3B	3B	184	2.6–3.5	700	900	0.008	0.008
		3E-E 88.9 (1457)	5PR5EY-11	0.043	10B	10B	184	38–44	800	800	0.008	0.008
	Corolla	4A-FE 97.0 (1587)	BCPR5EY	0.031	10B	10B	190	38–44	②	②	0.006–0.010	0.008–0.012
		4A-CE 97.0 (1587)	BCPR5EY	0.031	10B	10B	190	38–44	800	800	0.006–0.010	0.008–0.012
	Camry	3S-FE 121.9 (1998)	BCPR5EY-11	0.043	10B	10B	178	38–44	650–750	650–750	0.007–0.011	0.011–0.015
		2VZ-FE 153.0 (2508)	BCPR6E-11	0.043	10B	10B	178	38–44	650–750	650–750	0.005–0.009	0.011–0.015
	Celica	4A-FE 97.0 (1587)	BCPR5EY	0.031	10B	10B	191	38–44	800	800	0.006–0.010	0.008–0.012
		3S-GTE 121.9 (1998)	BCPR5EP-8	0.031	10B	—	178	38–44	750	—	0.006–0.010	0.008–0.012
		5S-FE 132.0 (2164)	BPR5EYA-11	0.043	10B	10B	178	38–44	650–750③	650–750④	0.007–0.011	0.011–0.015
	Supra	7M-GE 180.3 (2954)	BCPR5EP-11	0.043	10B	10B	156	38–44	700	700	0.006–0.010	0.008–0.012
		7M-GTE 180.3 (2954)	BCPR6EP-N8	0.031	10B	10B	142	33–40	650	650	0.006–0.010	0.006–0.010
	MR2	3S-GTE 121.9 (1998)	BKR6EP-8	0.031	10B	10B	164	33–38	750–850	—	0.006–0.010	0.008–0.012
		5S-FE 132.0 (2164)	BKR5EYA-11	0.043	10B	10B	142	38–44	700–800⑤	650–750④	0.007–0.011	0.011–0.015
	Cressida	180.3 (2954)	BCPR5EP-11	0.043	—	10B	156	38–44	—	700	0.006–0.010	0.008–0.012
1991	Tercel	3E-E 88.9 (1457)	BPR5EY-11	0.043	10B	10B	184	33–37	750	800	0.008	0.008
	Corolla	4A-FE 97.0 (1587)	BCPR5EY	0.031	10B	10B	191	38–44	⑥	⑥	0.006–0.010	0.008–0.012
		4A-GE 97.0 (1587)	BKR6EP-8	0.031	10B	10B	190	38–44	800	800	0.006–0.010	0.008–0.012
	Camry	3S-FE 121.9 (1998)	BCPR5EY-11	0.043	10B	10B	178	38–44	650	650	0.007–0.011	0.011–0.015
		2VZ-FE 153.0 (2058)	BCPR6E-11	0.043	10B	10B	178	38–44	700	700	0.005–0.009	0.011–0.015

GASOLINE ENGINE TUNE-UP SPECIFICATIONS

Year	Model	Engine Displacement cu. in. (cc)	Spark Plugs Type	Gap (in.)	Ignition Timing (deg.) MT	AT	Compression Pressure (psi)	Fuel Pump (psi)	Idle Speed (rpm) MT	AT	Valve Clearance In.	Ex.
1991	Celica	4A-FE 97.0 (1587)	BCPR5EY	0.031	10B	10B	191	38–44	800	800	0.006–0.010	0.008–0.012
		3S-GTE 121.9 (1998)	BKR6EP-8	0.031	10B	10B	164	33–38	800	800	0.006–0.010	0.008–0.012
		5S-FE 132.0 (2164)	BKR5EYA-11	0.043	10B	10B	178	38–44	⑦	750	0.007–0.011	0.011–0.015
	Supra	7M-GE 180.3 (2954)	BCPR5EP-11	0.043	10B	10B	171	38–44	700	700	0.006–0.010	0.008–0.012
		7M-GTE 180.3 (2954)	BCPR6EP-N8	0.031	10B	10B	142	33–40	650	650	0.006–0.010	0.008–0.012
	MR2	3S-GTE 121.9 (1998)	BKR6EP-8	0.031	10B	10B	164	33–38	800	800	0.006–0.010	0.008–0.012
		5S-FE 132.0 (2164)	BKR5EYA-11	0.043	10B	10B	178	38–44	⑧	⑧	0.007–0.011	0.011–0.015
	Cressida	7M-GE 180.3 (2954)	BCPR5EP-11	0.043	10B	10B	171	38–44	700	700	0.006–0.010	0.008–0.011
1992							SEE UNDERHOOD SPECIFICATIONS STICKER					

NOTE: The Underhood Specifications sticker often reflects tune-up specification changes made in production. Sticker figures must be used if they disagree with those in this chart.
MT Manual transmission
AT Automatic transmissino
NA Not adjustable
A After Top Dead Center
B Before Top Dead Center
Hyd. Hydraulic valve lash adjusters
① Can. wagon w/MT: BPR5EY; 0.031 in.
② 2WD: 700
   4WD: 800
③ Canada: 750–850
④ Canada: 700–800
⑤ Canada: 800–900
⑥ 2WD Federal and Canada: 700
   2WD California and 4WD: 800
⑦ USA: 700
⑧ USA M/T, Canada A/T: 750
   USA A/T: 700
   Canada M/T: 850

NOTE: To avoid confusion, always replace spark plug wires one at a time.

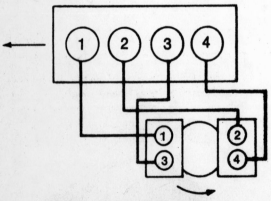

3A-C and 4A-F Engines
Engine Firing Order: 1–3–4–2
Distributor Rotation: Counterclockwise

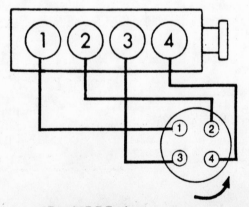

3E and 3E-E Engines
Engine Firing Order: 1–3–4–2
Distributor Rotation: Counterclockwise

3S-FE, 3S-GE, 3S-GTE and 5S-FE Engines
Engine Firing Order: 1–3–4–2
Distributor Rotation: Counterclockwise

4A-F and 4A-FE Engines
Engine Firing Order: 1–3–4–2
Distributor Rotation: Counterclockwise

Front of car

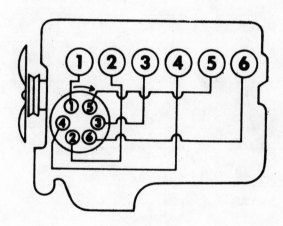

5M-GE Engine
Engine Firing Order: 1–5–3–6–2–4
Distributor Rotation: Clockwise

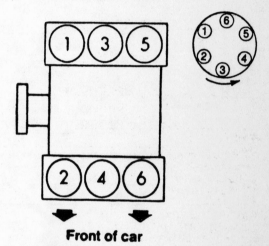

Front of car

2VZ-FE Engine
Engine Firing Order: 1–2–3–4–5–6
Distributor Rotation: Counterclockwise

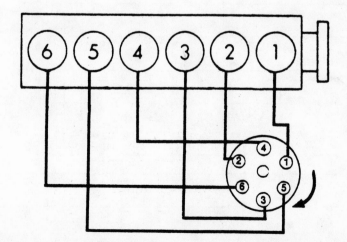

7M-GE Engine
Engine Firing Order: 1–5–3–6–2–4
Distributor Rotation: Clockwise

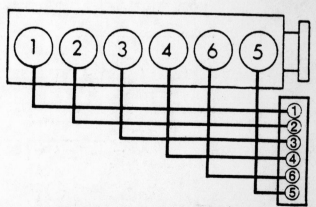

7M-GTE Engine
Engine Firing Order: 1–5–3–6–2–4
Distributorless Ignition System

1 TOYOTA

CAPACITIES

Year	Model	Engine Displacement cu. in. (cc)	Engine Crankcase (qts.) with Filter	Engine Crankcase (qts.) without Filter	Transmission (pts.) 4-spd	Transmission (pts.) 5-Spd	Transmission (pts.) Auto.	Drive Axle (pts.)	Fuel Tank (gal.)	Cooling System (qts.)
1988	Tercel	88.6 (1452)	3.5	3.2	8.2	8.2	8.8	2.2	13.2	5.6
		88.9 (1456)	3.4	3.1	5.0	5.0	4.6	3.0	11.9	5.3
	Corolla	4A-LC 97.0 (1587)	3.5	3.2	5.4	5.4	①	3.0	13.2②	6.4
		4A-F 97.0 (1587)	3.3	3.2	5.4	5.4	①	3.0	13.2	6.3
		4A-GE 97.0 (1587)	3.9	3.5	5.4	5.4	①	3.0	13.2	6.3
	Camry	121.9 (1998)	4.1	3.9	5.4	5.4	5.2	3.4②	15.9	6.8
	Celica	3S-FE 121.9 (1998)	4.1	3.9	5.4	5.4	5.2	3.4	15.9	6.8
		3S-GE 121.9 (1998)	4.1	3.9	5.4	5.4	5.2	3.4	15.9	7.4
		3S-GTE 121.9 (1998)	3.8	3.6	—	10.2	—	—	15.9	8.5
	Supra	7M-GE 180.3 (2954)	4.4	4.1	—	5.0	3.4	2.8	18.5	8.6
		7M-GTE 180.3 (2954)	4.4	4.1	—	6.4	3.4	2.8	18.5	8.7
	MR2	97.0 (1587)	3.5	3.2	—	③	6.6	—	10.8	④
	Cressida	168.4 (2759)	5.2	4.9	—	—	3.4	2.6	18.5	8.7
1989	Tercel	88.9 (1456)	3.4	3.1	5.0	5.0	4.6	3.0	11.9	5.5
	Corolla	4A-F, 4A-FE 97.0 (1587)	3.4	3.2	5.4	5.4	①	3.0	13.2	5.9
		4A-GE 97.0 (1587)	3.9	3.6	5.4	5.4	①	3.0	13.2	6.3
	Camry	121.9 (1998)	4.1	3.9	—	5.4④	5.2	3.4②	15.9	6.8
		153.0 (2507)	4.1	3.9	—	—	5.2	2.2	15.9	9.0
	Celica	3S-FE 121.9 (1998)	4.1	3.9	—	5.4	5.2	3.4	15.9	6.6
		3S-GE 121.9 (1998)	4.1	3.8	—	5.4	5.2	3.4	15.9	6.4
		3S-GTE 121.9 (1998)	3.8	3.5	—	10.2	—	2.4	15.9	6.8
	Supra	7M-GE 180.3 (2954)	4.7	4.3	—	5.0	3.4	2.8	18.5	8.6
		7M-GTE 180.3 (2954)	4.9	4.5	—	6.4	3.4	2.8	18.5	8.7
	MR2	97.0 (1587)	3.5	3.2	—	③	6.6	—	10.8	13.6
	Cressida	180.3 (2954)	4.7	4.3	—	—	3.4	—	18.5	8.8
1990	Tercel	3E 88.9 (1457)	3.4	3.1	—	5.0	4.6	—	11.9	⑤
		3E-E 88.9 (1457)	3.4	3.1	—	5.0	4.6	—	11.9	5.9

CAPACITIES

Year	Model	Engine Displacement cu. in. (cc)	Engine Crankcase (qts.) with Filter	without Filter	Transmission (pts.) 4-spd	5-Spd	Auto.	Drive Axle (pts.)	Fuel Tank (gal.)	Cooling System (qts.)
1990	Corolla	4A-FE 97.0 (1587)	3.3	3.2	—	⑥	⑦	⑧	13.2	⑨
		4A-GE 97.0 (1587)	3.4	3.6	—	⑥	⑦	⑧	13.2	⑨
	Camry	3S-FE 121.9 (1998)	4.1	3.9	—	⑩	⑪	⑫	15.9	⑬
		2VZ-FE 132.0 (2508)	4.1	3.9	—	⑩	⑪	⑫	15.9	10.0
	Celica	4A-FE 97.0 (1587)	3.3	3.1	—	5.4	7.0	—	15.9	⑯
		3S-GTE 121.9 (1998)	3.8	3.5	—	10.2	—	2.4	15.9	6.8
		5S-FE 132.0 (2164)	⑭	⑮	—	5.4	7.0	—	15.9	⑰
	Supra	7M-GE 180.3 (2954)	4.7	4.3	—	5.0	3.4	1.8	18.5	8.6
		7M-GTE 180.3 (2954)	4.7	4.3	—	6.4	3.4	2.8	18.5	8.7
	MR2	3S-GTE 121.9 (1998)	4.1	3.8	—	5.4	7.0	—	10.8	14.4
		5S-FE 132.0 (2164)	4.4	4.0	—	8.8	7.0	—	10.8	13.7
	Cressida	180.3 (2954)	4.3	4.7	—	—	3.4	—	18.5	8.8
1991–92	Tercel	3E-E 88.9 (1457)	3.4	3.1	—	5.0	5.2	3.0	11.9	5.2⑱
	Corolla	4A-FE 97.0 (1587)	3.3	3.2	—	⑲	⑳	㉑	13.2	⑨
		4A-GE 97.0 (1587)	3.9	3.6	—	⑲	⑳	㉑	13.2	⑨
	Camry	3S-FE 121.9 (1998)	4.3	3.9	—	㉒	㉓	㉔	15.9	⑬
		2VZ-FE 153.0 (2058)	4.1	3.9	—	㉒	㉓	㉔	15.9	10.0
	Celica	4A-FE 97.0 (1587)	3.3	3.1	—	㉕	7.0	2.4	15.9	⑯
		3S-GTE 121.9 (1998)	4.1	3.8	—	㉕	7.0	2.4	15.9	6.9
		5S-FE 132.0 (2164)	⑭	⑮	—	㉕	7.0	2.4	15.9	⑰
	Supra	7M-GE 180.3 (2954)	4.7	4.3	—	5.0	3.4	2.8	18.5	8.6
		7M-GTE 180.3 (2954)	5.0	4.7	—	6.4	3.4	2.8	18.5	㉖
	MR2	3S-GTE 121.9 (1998)	4.1	3.8	—	5.4	7.0	—	10.8	14.4
		5S-FE 132.0 (2164)	4.4	4.0	—	8.8	7.0	—	10.8	13.7

CAPACITIES

Year	Model	Engine Displacement cu. in. (cc)	Engine Crankcase (qts.) with Filter	without Filter	Transmission (pts.) 4-spd	5-Spd	Auto.	Drive Axle (pts.)	Fuel Tank (gal.)	Cooling System (qts.)
1991–92	Cressida	7M-GE 180.3 (2954)	4.7	4.3	—	—	3.4	—	18.5	8.8

① A240E, A241H: 6.6
   A131L: 5.2
② 4wd rear diff.: 2.4
③ C52: 5.4; E51: 8.8
④ 42d: 10.4
⑤ M/T: 5.5
   M/T (EL31L-NGKB5A) 5.3
   A/T: 5.4
⑥ C50, C52: 5.4
   All-trac (E5SFS, ES7F5): 10.6
⑦ AL131L: 5.2
   A240L: 6.6
   A241H: 6.6
⑧ Transfer (A241H A/T only): 3.0
   Rear differential: 2.4
⑨ M/T (except AE29L-ACMXRR): 5.9
   M/T (AE29L-ACMXKK): 6.6
   A/T: 6.4
⑩ S51: 5.4
   ES6FS: 10.6
   ES2: 8.8
⑪ A140L, A140E, A540E: 5.2
   A540H: 7.0
⑫ Differential
   SV21 A/T: 3.4
   VZV21 A/T: 2.1
   Transfer (A540H only): 1.48

⑬ M/T: 6.8
   A/T (2WD): 6.7
   A/T (4WD): 7.2
⑭ w/oil cooler: 4.4
   w/out oil cooler: 4.3
⑮ w/oil cooler: 4.0
   w/out oil cooler: 3.9
⑯ M/T: 5.5
   A/T: 5.8
⑰ M/T: 6.6
   A/T: 6.4
⑱ M/T: 5.2
   A/T: 5.7
⑲ M/T (C50, C52): 5.4
   M/T (E55F5, E57F5): 10.6
⑳ A131L: 5.2, A240L: 6.6, A241H:6.6
㉑ A131L only: 3.0
㉒ S51: 5.4, E52: 8.8
㉓ A140L, A140E: 0.8
   A540E: 12.4, A540H: 14.8
㉔ A/T only SV21: 3.4, VZV21: 2.2
㉕ C52, S53: 5.4, E150F: 11.0
㉖ M/T: 8.7, A/T: 8.5

CAMSHAFT SPECIFICATIONS

All measurements given in inches.

Year	Engine Displacement cu. in. (cc)	Journal Diameter 1	2	3	4	5	6	7	Bearing Clearance	Camshaft End Play
1988	3A-C 88.6 (1452)	1.1015–1.1022	1.1015–1.1022	0.1015–1.1022	0.1015–1.1022	—	—	—	0.0015–0.0029	0.0031–0.0071
	3E 88.9 (1456)	1.0622–1.0628	1.0622–1.0628	1.0622–1.0628	1.0622–1.0628	—	—	—	0.0015–0.0029	0.0031–0.0071
	4A-F 97.0 (1587)	0.9035–0.9041③	0.9035–0.9041	0.9035–0.9041	0.9035–0.9041	—	—	—	0.0015–0.0028	②
	4A-LC 97.0 (1587)	1.1015–1.1022	1.1015–1.1022	0.1015–1.1022	0.1015–1.1022	—	—	—	0.0015–0.0029	0.0031–0.0071
	4A-GZE, 4A-GEC, 4A-GELC 97.0 (1587)	1.0610–1.0616	1.0610–1.0616	1.0610–1.0616	1.0610–1.0616	—	—	—	0.0014–0.0028	0.0031–0.0075
	3S-FE 121.9 (1998)	1.0614–1.0620	1.0614–1.0620	1.0614–1.0620	1.0614–1.0620	—	—	—	0.0010–0.0024	0.0018–0.0039
	3S-GE, 3S-GTE 121.9 (1998)	1.0614–1.0620	1.0614–1.0620	1.0614–1.0620	1.0614–1.0620	—	—	—	0.0010–0.0024	0.0047–0.0079④
	5M-GE 168.4 (2759)	1.4944–1.4951	1.6913–1.6919	1.7110–1.7116	1.7307–1.7313	1.7504–1.7510	1.7700–1.7707	1.7897–1.7907	0.0010–0.0026	0.0028–0.0098
	7M-GE, 7M-GTE 180.3 (2954)	1.0610–1.0616	1.0586–1.0620	1.0586–1.0620	1.0586–1.0620	1.0586–1.0620	1.0586–1.0620	1.0586–1.0620	0.0010–0.0037①	0.0031–0.0075

CAMSHAFT SPECIFICATIONS

All measurements given in inches.

Year	Engine Displacement cu. in. (cc)	Journal Diameter							Bearing Clearance	Camshaft End Play
		1	2	3	4	5	6	7		
1989	3E 88.9 (1456)	1.0622–1.0628	1.0622–1.0628	1.0622–1.0628	1.0622–1.0628	—	—	—	0.0015–0.0029	0.0031–0.0071
	4A-F, 4A-FE 97.0 (1587)	0.9035–0.9041 ③	0.9035–0.9041	0.9035–0.9041	0.9035–0.9041	—	—	—	0.0015–0.0028	②
	4A-GZE, 4A-GEC, 4A-GELC 97.0 (1587)	1.0610–1.0616	1.0610–1.0616	1.0610–1.0616	1.0610–1.0616	—	—	—	0.0014–0.0028	0.0031–0.0075
	3S-FE 121.9 (1998)	1.0614–1.0620	1.0614–1.0620	1.0614–1.0620	1.0614–1.0620	—	—	—	0.0010–0.0024	0.0018–0.0039
	3S-GE, 3S-GTE 121.9 (1998)	1.0614–1.0620	1.0614–1.0620	1.0614–1.0620	1.0614–1.0620	—	—	—	0.0010–0.0024	0.0047–0.0114
	2VZ-FE 153.0 (2507)	1.0610–1.0616	1.0610–1.0616	1.0610–1.0616	1.0610–1.0616	1.0610–1.0616	—	—	0.0014–0.0028	0.0012–0.0031
	7M-GE, 7M-GTE 180.3 (2954)	1.0610–1.0616	1.0586–1.0620	1.0586–1.0620	1.0586–1.0620	1.0586–1.0620	1.0586–1.0620	1.0586–1.0620	0.0010–0.0037 ①	0.0031–0.0075
1990	3E, 3E-E 88.9 (1457)	1.0622–1.0628	1.0622–1.0628	1.0622–1.0628	1.0622–1.0628	—	—	—	0.0015–0.0029	0.0031–0.0071
	4A-FE 97.0 (1587)	0.9035–0.9041 ③	0.9035–0.9041	0.9035–0.9041	0.9035–0.9041	—	—	—	0.0014–0.0028	②
	4A-GE 97.0 (1587)	1.0610–1.0616	1.0610–1.0616	1.0610–1.0616	1.0610–1.0616	—	—	—	0.0014–0.0028	0.0031–0.0075
	3S-FE 121.9 (1998)	1.0614–1.0620	1.0614–1.0620	1.0614–1.0620	1.0614–1.0620	—	—	—	0.0010–0.0024	⑤
	3S-GTE 121.9 (1998)	1.0614–1.0620	1.0614–1.0620	1.0614–1.0620	1.0614–1.0620	—	—	—	0.0010–0.0024	0.0047–0.0114
	5S-FE 132.0 (2164)	1.0614–1.0620	1.0614–1.0620	1.0614–1.0620	1.0614–1.0620	—	—	—	0.0010–0.0024	⑤
	2VZ-FE 153.0 (2507)	1.0610–1.0616	1.0610–1.0616	1.0610–1.0616	1.0610–1.0616	1.0610–1.0616	—	—	0.0014–0.0028	0.0012–0.0031
	7M-GE, 7M-GTE 180.3 (2954)	1.0610–1.0616	1.0586–1.0620	1.0586–1.0620	1.0586–1.0620	1.0586–1.0620	1.0586–1.0620	1.0586–1.0620	0.0010–0.0037 ①	0.0031–0.0075
1991–92	3E-E 88.9 (1457)	1.0622–1.0628	1.0622–1.0628	1.0622–1.0628	1.0622–1.0628	—	—	—	0.0015–0.0029	0.0031–0.0071
	4A-FE 97.0 (1587)	0.9035–0.9041 ③	0.9035–0.9041	0.9035–0.9041	0.9035–0.9041	—	—	—	0.0014–0.0028	②
	4A-GE 97.0 (1587)	1.0610–1.0616	1.0610–1.0616	1.0610–1.0616	1.0610–1.0616	—	—	—	0.0014–0.0028	0.0031–0.0071
	3S-FE 121.9 (1998)	1.0614–1.0620	1.0614–1.0620	1.0614–1.0620	1.0614–1.0620	—	—	—	0.0010–0.0024	⑤
	3S-GTE 121.9 (1998)	1.0614–1.0620	1.0614–1.0620	1.0614–1.0620	1.0614–1.0620	—	—	—	0.0010–0.0024	0.0047–0.0114
	5S-FE 132.0 (2164)	1.0614–1.0620	1.0614–1.0620	1.0614–1.0620	1.0614–1.0620	—	—	—	0.0010–0.0024	⑤
	2VZ-FE 153.0 (2507)	1.0610–1.0616	1.0610–1.0616	1.0610–1.0616	1.0610–1.0616	1.0610–1.0616	—	—	0.0014–0.0028	0.0012–0.0031
	7M-GE, 7M-GTE 180.3 (2954)	1.0610–1.0616	1.0586–1.0620	1.0586–1.0620	1.0586–1.0620	1.0586–1.0620	1.0586–1.0620	1.0586–1.0620	0.0010–0.0037 ①	0.0031–0.0075

① No. 1: 0.0014–0.0028
② Intake: 0.0012–0.0033 Exhaust: 0.0014–0.0035
③ Exhaust No. 1: 0.9822–0.9829
④ 35-GTE: 0.0039–0.0094
⑤ Intake: 0.0018–0.0039 Exhaust: 0.0012–0.0033

CRANKSHAFT AND CONNECTING ROD SPECIFICATIONS

All measurements are given in inches.

Year	Engine Displacement cu. in. (cc)	Crankshaft				Connecting Rod		
		Main Brg. Journal Dia.	Main Brg. Oil Clearance	Shaft End-play	Thrust on No.	Journal Diameter	Oil Clearance	Side Clearance
1988	3A-C 88.6 (1452)	1.8891– 1.8898	0.0006– 0.0013	0.0008– 0.0087	3	1.5742– 1.5748	0.0008– 0.0020	0.0059– 0.0098
	3E 88.9 (1456)	1.9683– 1.9685	0.0006– 0.0014	0.0008– 0.0087	3	1.8110– 1.8113	0.0006– 0.0019	0.0059– 0.0138
	3S-FE, 3S-GE, 3S-GTE 121.9 (1998)	2.1648– 2.1653	0.0007– 0.0015 ①	0.0008– 0.0087	3	1.8892– 1.8898	0.0009– 1.0022	0.0063– 0.0123
	4A-F, 4A-LC 97.0 (1587)	1.8891– 1.8898	0.0006– 0.0013	0.0008– 0.0087	3	1.5742– 1.5748	0.0008– 0.0020	0.0059– 0.0098
	4A-GEC, 4A-GELC 97.0 (1587)	1.8891– 1.8898	0.0005– 0.0015	0.0008– 0.0087	3	1.5742– 1.5748	0.0008– 0.0020	0.0059– 0.0098
	4A-GZE 97.0 (1587)	1.8891– 1.8898	0.0006– 0.0013	0.0008– 0.0087	3	1.6529– 1.6535	0.0008– 0.0020	0.0059– 0.0098
	5M-GE 168.4 (2759)	2.3625– 2.3627	0.0012– 0.0048	0.0020– 0.0098	4	2.1659– 2.1663	0.0008– 0.0021	0.0063– 0.0117
	7M-GE, 7M-GTE 180.3 (2954)	2.3625– 2.3627	0.0012– 0.0048	0.0020– 0.0098	4	2.1659– 2.1663	0.0008– 0.0021	0.0063– 0.0117
1989	3E 88.9 (1456)	1.9683– 1.9685	0.0006– 0.0014	0.0008– 0.0087	3	1.8110– 1.8113	0.0006– 0.0019	0.0059– 0.0138
	3S-FE, 3S-GE, 3S-GTE 121.9 (1998)	2.1649– 2.1655	0.0006– 0.0013 ①	0.0008– 0.0087	3	1.8892– 1.8898	0.0009– 0.0022	0.0063– 0.0123
	4A-F, 4A-FE 97.0 (1587)	1.8891– 1.8898	0.0006– 0.0013	0.0008– 0.0087	3	1.5742– 1.5748	0.0008– 0.0020	0.0059– 0.0098
	4A-GEC, 4A-GELC 97.0 (1587)	1.8891– 1.8898	0.0006– 0.0013	0.0008– 0.0087	3	1.6529– 1.6535	0.0008– 0.0020	0.0059– 0.0098
	4A-GZE 97.0 (1587)	1.8891– 1.8898	0.0006– 0.0013	0.0008– 0.0087	3	1.6529– 1.6535	0.0008– 0.0020	0.0059– 0.0098
	2VZ-FE 153.0 (2507)	2.5191– 2.5197	0.0011– 0.0022	0.0008– 0.0087	3	1.8892– 1.8898	0.0011– 0.0026	0.0059– 0.0130
	7M-GE, 7M-GTE 180.3 (2954)	2.3625– 2.3627	0.0012– 0.0019	0.0020– 0.0098	4	2.1659– 2.1663	0.0008– 0.0021	0.0063– 0.0117
1990	3E, 3E-E 88.9 (1457)	1.9683– 1.9685	0.0006– 0.0014	0.0008– 0.0087	3	1.8110– 1.8113	0.0006– 0.0019	0.0059– 0.0138
	4A-FE 97.0 (1587)	1.8891– 1.8898	0.0006– 0.0013	0.0008– 0.0087	3	1.5742– 1.5748	0.0008– 0.0020	0.0059– 0.0098
	4A-GE 97.0 (1587)	1.8891– 1.8898	0.0006– 0.0013	0.0008– 0.0087	3	1.6529– 1.6535	0.0008– 0.0020	0.0059– 0.0098
	3S-FE 121.9 (1998)	2.1649– 2.1655	0.0010– 0.0017	0.0008– 0.0087	3	1.8892– 1.8898	0.0009– 0.0022	0.0063– 0.0123
	3S-GTE 121.9 (1998)	2.1653– 2.1655	0.0006– 0.0013 ①	0.0008– 0.0087	3	1.8892– 1.8898	0.0009– 0.0022	0.0063– 0.0123
	5S-FE 132.0 (2164)	2.1653– 2.1655	0.0006– 0.0013 ①	0.0008– 0.0087	3	1.8892– 1.8898	0.0009– 0.0022	0.0063– 0.0123
	2VZ-FE 153.0 (2507)	2.5191– 2.5197	0.0011– 0.0022	0.0008– 0.0087	3	1.8892– 1.8898	0.0011– 0.0026	0.0059– 0.0123
	7M-GE, 7M-GTE 180.3 (2954)	2.3625– 2.3627	0.0012– 0.0019	0.0020– 0.0098	4	2.0470– 2.0474	0.0008– 0.0021	0.0063– 0.0117

CRANKSHAFT AND CONNECTING ROD SPECIFICATIONS

All measurements are given in inches.

Year	Engine Displacement cu. in. (cc)	Crankshaft Main Brg. Journal Dia.	Crankshaft Main Brg. Oil Clearance	Crankshaft Shaft End-play	Thrust on No.	Connecting Rod Journal Diameter	Connecting Rod Oil Clearance	Connecting Rod Side Clearance
1991–92	3E-E 88.9 (1457)	1.9683–1.9685	0.0006–0.0013	0.0008–0.0087	3	1.8110–1.8113	0.0006–0.0019	0.0059–0.0138
	4A-FE 97.0 (1587)	1.8891–1.8898	0.0006–0.0013	0.0008–0.0087	3	1.5742–1.5748	0.0008–0.0020	0.0059–0.0098
	4A-GE 97.0 (1587)	1.8891–1.8898	0.0006–0.0013	0.0008–0.0087	3	1.6529–1.6535	0.0008–0.0020	0.0059–0.0098
	3S-FE 121.9 (1998)	2.1649–2.1655	0.0010–0.0017	0.0008–0.0087	3	1.8892–1.8898	0.0009–0.0022	0.0063–0.0123
	3S-GTE 121.9 (1998)	2.1653–2.1655	0.0006–0.0013 ①	0.0008–0.0087	3	1.8892–1.8898	0.0009–0.0022	0.0063–0.0123
	5S-FE 132.0 (2164)	2.1653–2.1655	0.0006–0.0013 ①	0.0008–0.0087	3	1.8892–1.8898	0.0009–0.0022	0.0063–0.0123
	2VZ-FE 153.0 (2507)	2.5191–2.5197	0.0011–0.0022	0.0008–0.0087	3	1.8892–1.8898	0.0011–0.0026	0.0059–0.0123
	7M-GE, 7M-GTE 180.3 (2954)	2.3625–2.3627	0.0012–0.0019	0.0020–0.0098	4	2.0470–2.0474	0.0008–0.0021	0.0063–0.0117

① No. 3: 0.0011–0.0019 (1988)
No. 3: 0.0010–0.0017 (1989–91)

VALVE SPECIFICATIONS

Year	Engine Displacement cu. in. (cc)	Seat Angle (deg.)	Face Angle (deg.)	Spring Test Pressure (lbs.)	Spring Installed Height (in.)	Stem-to-Guide Clearance (in.) Intake	Stem-to-Guide Clearance (in.) Exhaust	Stem Diameter (in.) Intake	Stem Diameter (in.) Exhaust
1988	3A-C 88.6 (1452)	45	44.5	52.0	1.520	0.0010–0.0024	0.0012–0.0026	0.2744–0.2750	0.2742–0.2748
	3E 88.9 (1456)	45	44.5	35.1	1.384	0.0010–0.0024	0.0012–0.0026	0.2350–0.2356	0.2348–0.2354
	3S-FE 121.9 (1998)	45.5	44.5	39.6	1.366	0.0010–0.0024	0.0012–0.0026	0.2350–0.2356	0.2348–0.2354
	3S-GE, 3S-GTE 121.9 (1998)	45.5	44.5	38.6 ③	1.366	0.0010–0.0023	0.0012–0.0025	0.2346–0.2352	0.2344–0.2350
	4A-F 97.0 (1587)	45	44.5	34.8	1.366	0.0010–0.0024	0.0012–0.0026	0.2350–0.2356	0.2348–0.2354
	4A-LC 97.0 (1587)	45	44.5	52.0	1.520	0.0010–0.0024	0.0012–0.0026	0.2744–0.2750	0.2742–0.2748
	4A-GEC, 4A-GELC 4A-GZE 97.0 (1587)	45	44.5	35.9	1.366	0.0010–0.0024	0.0012–0.0026	0.2350–0.2356	0.2348–0.2354
	5M-GE 168.4 (2759)	45	44.5	①	②	0.0010–0.0024	0.0012–0.0026	0.3138–0.3144	0.3136–0.3142
	7M-GE, 7M-GTE 180.3 (2954)	45	44.5	35.0	1.378	0.0010–0.0024	0.0012–0.0026	0.2350–0.2356	0.2348–0.2354
1989	3E 88.9 (1456)	45	44.5	35.1	1.384	0.0010–0.0024	0.0012–0.0026	0.2350–0.2356	0.2348–0.2354
	3S-FE 121.9 (1998)	45.5	44.5	39.6	1.366	0.0010–0.0024	0.0012–0.0026	0.2350–0.2356	0.2348–0.2354

VALVE SPECIFICATIONS

Year	Engine Displacement cu. in. (cc)	Seat Angle (deg.)	Face Angle (deg.)	Spring Test Pressure (lbs.)	Spring Installed Height (in.)	Stem-to-Guide Clearance (in.)		Stem Diameter (in.)	
						Intake	Exhaust	Intake	Exhaust
1989	3S-GE, 3S-GTE 121.9 (1998)	45.5	44.5	38.6 ③	1.366	0.0010–0.0023	0.0012–0.0025	0.2346–0.2352	0.2344–0.2350
	4A-F, 4A-FE 97.0 (1587)	45	44.5	34.8	1.366	0.0010–0.0024	0.0012–0.0026	0.2350–0.2356	0.2348–0.2354
	4A-GEC, 4A-GZE 97.0 (1587)	45	44.5	34.7	1.366	0.0010–0.0024	0.0012–0.0026	0.2350–0.2356	0.2348–0.2354
	2VZ-FE 153.0 (2507)	45	44.5	41.0–47.2	1.331	0.0010–0.0024	0.0012–0.0026	0.2350–0.2356	0.2348–0.2354
	7M-GE, 7M-GTE 180.3 (2954)	45	44.5	35.0	1.378	0.0010–0.0024	0.0012–0.0026	0.2350–0.2356	0.2348–0.2354
1990	3E, 3E-E 88.9 (1457)	45	44.5	35.1	1.3842	0.0010–0.0024	0.0012–0.0026	0.2350–0.2356	0.2348–0.2354
	4A-FE 97.0 (1587)	45	45.5	34.8	1.366	0.0010–0.0024	0.0012–0.0026	0.2350–0.2356	0.2348–0.2354
	4A-GE 97.0 (1587)	45	44.5	35.9	1.366	0.0010–0.0024	0.0012–0.0026	0.2350–0.2356	0.2348–0.2354
	3S-FE 121.9 (1998)	45	45.5	42.5	1.366	0.0010–0.0024	0.0012–0.0026	0.2350–0.2356	0.2348–0.2354
	3S-GTE 121.9 (1998)	45	45.5	53.1	1.354	0.0010–0.0023	0.0012–0.0025	0.2346–0.2352	0.2344–0.2350
	5S-FE 132.0 (2164)	45	45.5	42.5	1.366	0.0010–0.0024	0.0012–0.0026	0.2350–0.2356	0.2348–0.2354
	2VZ-FE 153.0 (2507)	45	45.5	47.2	1.331	0.0010–0.0024	0.0012–0.0026	0.2350–0.2356	0.2348–0.2354
	7M-GE, 7M-GTE 180.3 (2954)	45	45.5	35	1.378	0.0010–0.0024	0.0012–0.0026	0.2350–0.2356	0.2348–0.2354
1991–92	3E-E 88.9 (1457)	45	44.5	35.1	1.3842	0.0010–0.0024	0.0012–0.0026	0.2350–0.2356	0.2348–0.2354
	4A-FE 97.0 (1587)	45	45.5	34.8	1.366	0.0010–0.0024	0.0012–0.0026	0.2350–0.2356	0.2348–0.2354
	4A-GE 97.0 (1587)	45	44.5	35.9	1.366	0.0010–0.0024	0.0012–0.0026	0.2350–0.2356	0.2348–0.2354
	3S-FE 121.9 (1998)	45	44.5	42.5	1.366	0.0010–0.0024	0.0012–0.0026	0.2350–0.2356	0.2348–0.2354
	3S-GTE 121.9 (1998)	45	44.5	53.1	1.354	0.0010–0.0023	0.0012–0.0025	0.2346–0.2352	0.2344–0.2350
	5S-FE 132.0 (2164)	45	44.5	42.5	1.366	0.0010–0.0023	0.0012–0.0026	0.2350–0.2356	0.2348–0.2354
	2VZ-FE 153.0 (2507)	45	44.5	47.2	1.331	0.0010–0.0024	0.0012–0.0026	0.2350–0.2356	0.2348–0.2354
	7M-GE, 7M-GTE 180.3 (2954)	45	44.5	35	1.378	0.0010–0.0024	0.0012–0.0026	0.2350–0.2356	0.2348–0.2354

① Intake: 76.5–84.4; Exhaust: 73.4–80.9
② Intake: 1.575; Exhaust: 1.693
③ 3S-GTE: 44.1

PISTON AND RING SPECIFICATIONS

All measurements are given in inches.

Year	Engine Displacement cu. in. (cc)	Piston Clearance	Ring Gap			Ring Side Clearance		
			Top Compression	Bottom Compression	Oil Control	Top Compression	Bottom Compression	Oil Control
1988	3A-C 88.6 (1452)	0.0039–0.0047	0.0079–0.0185	0.0079–0.0204	0.0118–0.0402	0.0016–0.0031	0.0012–0.0028	Snug
	3E 88.9 (1456)	0.0028–0.0035	0.0102–0.0142	0.0118–0.0177	0.0059–0.0157	0.0016–0.0031	0.0012–0.0028	Snug
	3S-FE 121.9 (1998)	0.0018–0.0026	0.0106–0.0205	0.0106–0.0209	0.0079–0.0323	0.0018–0.0028	0.0018–0.0028	Snug
	3S-GE 121.9 (1998)	0.0012–0.0020	0.0130–0.0264	0.0177–0.0323	0.0079–0.0283	0.0012–0.0028	0.0008–0.0024	Snug
	3S-GTE 121.9 (1998)	0.0012–0.0020	0.0130–0.0224	0.0177–0.0272	0.0079–0.0244	0.0015–0.0031	0.0012–0.0028	Snug
	4A-F 97.0 (1587)	0.0024–0.0031	0.0098–0.0138	0.0059–0.0118	0.0039–0.0236	0.0016–0.0031	0.0012–0.0028	Snug
	4A-LC 97.0 (1587)	0.0035–0.0043	0.0098–0.0138	0.0059–0.0165	0.0078–0.0276	0.0016–0.0031	0.0012–0.0028	Snug
	4A-GEC, 4A-GELC, 4A-GZE 97.0 (1587)	0.0039–0.0047 ①	0.0098–0.0185	0.0078–0.0118	②	0.0016–0.0031	0.0012–0.0028	Snug
	5M-GE 168.4 (2759)	0.0024–0.0031	0.0091–0.0150	0.0098–0.0209	0.0040–0.0201	0.0012–0.0028	0.0008–0.0024	Snug
	7M-GE 180.3 (2954)	0.0020–0.0028	0.0091–0.0150	0.0098–0.0209	0.0039–0.0157	0.0012–0.0028	0.0008–0.0024	Snug
	7M-GTE 180.3 (2954)	0.0028–0.0035	0.0114–0.0173	0.0098–0.0209	0.0039–0.0173	0.0012–0.0028	0.0008–0.0024	Snug
1989	3E 88.9 (1456)	0.0028–0.0035	0.0102–0.0142	0.0118–0.0177	0.0059–0.0157	0.0016–0.0031	0.0012–0.0028	Snug
	3S-FE 121.9 (1998)	0.0018–0.0026	0.0106–0.0197	0.0106–0.0201	0.0079–0.0217	0.0012–0.0028	0.0012–0.0028	Snug
	3S-GE 121.9 (1998)	0.0012–0.0020	0.0130–0.0217	0.0177–0.0276	0.0079–0.0236	0.0012–0.0028	0.0008–0.0024	Snug
	3S-GTE 121.9 (1998)	0.0012–0.0020	0.0130–0.0217	0.0177–0.0264	0.0079–0.0236	0.0015–0.0031	0.0012–0.0028	Snug
	4A-F 97.0 (1587)	0.0024–0.0031	0.0098–0.0138	0.0059–0.0118	0.0039–0.0236	0.0016–0.0031	0.0012–0.0028	Snug
	4A-FE 97.0 (1587)	0.0024–0.0031	0.0098–0.0138	0.0059–0.0118	0.0039–0.0236	0.0020–0.0031	0.0012–0.0028	Snug
	4A-GEC, 4A-GELC, 4A-GZE 97.0 (1587)	0.0039–0.0047 ①	0.0098–0.0185	0.0079–0.0165	②	0.0016–0.0031	0.0012–0.0028	Snug
	2VZ-FE 153.0 (2507)	0.0018–0.0026	0.0118–0.0205	0.0138–0.0236	0.0079–0.0217	0.0004–0.0031	0.0012–0.0028	Snug
	7M-GE 180.3 (2954)	0.0020–0.0028 ③	0.0091–0.0150	0.0098–0.0209	0.0039–0.0157	0.0012–0.0028	0.0008–0.0024	Snug
	7M-GTE 180.3 (2954)	0.0028–0.0035	0.0114–0.0173	0.0098–0.0209	0.0039–0.0173	0.0012–0.0028	0.0008–0.0024	Snug

PISTON AND RING SPECIFICATIONS

All measurements are given in inches.

| Year | Engine Displacement cu. in. (cc) | Piston Clearance | Ring Gap | | | Ring Side Clearance | | Oil Control |
			Top Compression	Bottom Compression	Oil Control	Top Compression	Bottom Compression	
1990	3E, 3E-E 88.9 (1457)	0.0028–0.0035	0.0102–0.0189	0.0118–0.0224	0.0059–0.0205	0.0016–0.0031	0.0012–0.0028	Snug
	4A-FE 97.0 (1587)	0.0024–0.0031	0.0098–0.0177	0.0059–0.0157	0.0039–0.0276	0.0016–0.0031	0.0012–0.0028	Snug
	4A-GE 97.0 (1587)	0.0039–0.0047	0.0098–0.0185	0.0079–0.0165	0.0059–0.0205	0.0012–0.0031	0.0012–0.0028	Snug
	3S-FE 121.9 (1998)	0.0018–0.0026	0.0118–0.0205	0.0138–0.0236	0.0079–0.0217	0.0004–0.0031	0.0012–0.0028	Snug
	3S-GTE 121.9 (1998)	0.0028–0.0035	0.0130–0.0217	0.0177–0.0264	0.0079–0.0236	0.0016–0.0031	0.0012–0.0028	Snug
	5S-FE 132.0 (2164)	0.0031–0.0039	0.0106–0.0197	0.0138–0.0234	0.0079–0.0217	0.0012–0.0028	0.0012–0.0028	Snug
	2VZ-FE 153.0 (2507)	0.0018–0.0026	0.0118–0.0205	0.0138–0.0236	0.0079–0.0217	0.0004–0.0031	0.0012–0.0028	Snug
	7M-GE 180.3 (2954)	0.0031–0.0039	0.0091–0.0150	0.0098–0.0209	0.0039–0.0157	0.0012–0.0028	0.0008–0.0024	Snug
	7M-GTE 180.3 (2954)	0.0028–0.0035	0.0114–0.0173	0.0098–0.0209	0.0039–0.0173	0.0008–0.0024	0.0008–0.0024	Snug
1991–92	3E-E 88.9 (1457)	0.0028–0.0035	0.0102–0.0189	0.0118–0.0224	0.0059–0.0205	0.0016–0.0031	0.0012–0.0028	Snug
	4A-FE 97.0 (1587)	0.0024–0.0031	0.0098–0.0177	0.0059–0.0157	0.0039–0.0276	0.0016–0.0031	0.0012–0.0028	Snug
	4A-GE 97.0 (1587)	0.0039–0.0047	0.0098–0.0185	0.0079–0.0165	0.0059–0.0205	0.0012–0.0031	0.0012–0.0028	Snug
	3S-FE 121.9 (1998)	0.0018–0.0026	0.0118–0.0205	0.0138–0.0236	0.0079–0.0217	0.0004–0.0031	0.0012–0.0028	Snug
	3S-GTE 121.9 (1998)	0.0028–0.0035	0.0130–0.0217	0.0177–0.0264	0.0079–0.0236	0.0016–0.0031	0.0012–0.0028	Snug
	5S-FE 132.0 (2164)	0.0031–0.0039	0.0106–0.0197	0.0138–0.0234	0.0079–0.0217	0.0012–0.0028	0.0012–0.0028	Snug
	2VZ-FE 153.0 (2507)	0.0018–0.0026	0.0118–0.0205	0.0138–0.0236	0.0079–0.0217	0.0004–0.0031	0.0012–0.0028	Snug
	7M-GE 180.3 (2954)	0.0031–0.0039	0.0091–0.0150	0.0098–0.0209	0.0039–0.0157	0.0012–0.0028	0.0008–0.0024	Snug
	7M-GTE 180.3 (2954)	0.0028–0.0035	0.0114–0.0173	0.0098–0.0209	0.0039–0.0173	0.0012–0.0024	0.0008–0.0024	Snug

① 4A-GZE: 0.0047–0.0055
② Code T: 0.0059–0.0205
   Code R: 0.0118–0.0402
③ 1990 Supra: 0.0031–0.0039

TORQUE SPECIFICATIONS
All readings in ft. lbs.

Year	Engine Displacement cu. in. (cc)	Cylinder Head Bolts	Main Bearing Bolts	Rod Bearing Bolts	Crankshaft Pulley Bolts	Flywheel Bolts	Manifold Intake	Manifold Exhaust	Spark Plugs
1988	3A-C 88.6 (1452)	40–47	40–47	34–39	80–94	55–61	15–21	15–21	16–20
	3E 88.9 (1456)	②	40–47	27–31	105–117	60–70	11–17	33–42	11–15
	3S-FE 121.9 (1998)	45–50	40–45	33–38	78–82	70–75	11–17	27–33	11–15
	3S-GE, 3S-GTE 121.9 (1998)	38–42	40–45	44–50	78–82	①	12–16	30–34 ③	11–15
	4A-F 97.0 (1587)	40–47	40–47	32–40	80–94	55–61	11–17	15–21	11–15
	4A-LC 97.0 (1587)	40–47	40–47	32–40	80–94	55–61	15–21	15–21	16–20
	4A-GEC, 4A-GELC 4A-GZE 97.0 (1587)	④	40–47	32–40	100–110	50–58	15–21	27–31	11–15
	5M-GE 168.4 (2759)	55–61	72–78	31–34	185–205	51–57	11–15	26–32	11–15
	7M-GE, 7M-GTE 180.2 (2954)	55–61	72–78	45–49	185–205	51–57	11–15	26–32	11–15
1989	3E 88.9 (1456)	②	40–47	27–31	105–117	60–70	11–17	33–42	11–15
	3S-FE 121.9 (1998)	45–50	40–45	33–38	78–82	70–75	11–17	27–33	11–15
	3S-GE, 3S-GTE 121.9 (1998)	38–42	40–45	44–50	78–82	①	12–16	30–34	11–15
	4A-F, 4A-FE 97.0 (1587)	40–47	40–47	32–40	80–94	55–61	11–17	15–21	11–15
	4A-GEC, 4A-GELC 4A-GZE 97.0 (1587)	②	40–47	④	95–105	50–58	18–22	27–31	11–15
	2VZ-FE 153.0 (2507)	②	43–47	16–20	176–186	58–64	11–15	26–32	11–15
	7M-GE, 7M-GTE 180.2 (2954)	55–61	72–78	45–49	185–205	51–57	11–15	26–32	11–15
1990	3E, 3E-E 88.9 (1457)	⑨	42	29	112	88	14	38	13
	4A-FE 97.0 (1587)	44 ⑩	44	36	87	58	14	18	13
	4A-GE 97.0 (1587)	④	44	29	101	54	9	29	13
	3S-FE 121.9 (1998)	⑥	43	36	80	⑤	14	36	13
	3S-GTE 121.9 (1998)	⑥	43	49	80	80	14	38	13
	5S-FE 132.0 (2164)	⑥	43	⑦	80	⑤	14	36	13
	2VZ-FE 153.0 (2507)	⑬	⑧	⑨	181	61	13	13	13

TORQUE SPECIFICATIONS
All readings in ft. lbs.

Year	Engine Displacement cu. in. (cc)	Cylinder Head Bolts	Main Bearing Bolts	Rod Bearing Bolts	Crankshaft Pulley Bolts	Flywheel Bolts	Manifold Intake	Manifold Exhaust	Spark Plugs
1990	7M-GE, 7M-GTE 180.3 (2954)	58	75	47	195	54	13	29	13
1991-92	3E-E 88.9 (1457)	⑨	42	29	112	65	14	35	13
	4A-FE 97.0 (1587)	44 ⑩	44	36	87	58	14	18	13
	4A-GE 97.0 (1587)	⑪	44	⑫	101	54	20	29	13
	3S-FE 121.9 (1998)	⑥	43	36	80	65	14	36	13
	3S-GTE 121.9 (1998)	⑥	43	49	80	80	14	38	13
	5S-FE 132.0 (2164)	⑥	43	⑦	80	65	14	36	13
	2VZ-FE 153.0 (2507)	⑬	⑧	⑦	181	61	13	29	13
	7M-GE, 7M-GTE 180.3 (2954)	58 ⑩	75	47	195	54	13	29	13

① New: 65; Used:63
② See text
③ 3S-GTE: 38
④ 29 ft.lbs. and an additional 90° turn
⑤ M/T: 65
　 A/T: 61
⑥ 36 ft. lbs. and an additional 90° turn
⑦ 18 ft. lbs. and an additional 90° turn
⑧ 45 ft. lbs. and an additional 90° turn
⑨ 22 ft. lbs., 36 ft. lbs. and an additional 90° turn
⑩ Torque in sequence, in 3 steps
⑪ 22 ft. lbs., an additional 90° turn plus an additional 90° turn
⑫ 29 ft. lbs. plus an additional 90° turn
⑬ 25 ft. lbs., an additional 90° turn plus an additional 90° turn

BRAKE SPECIFICATIONS
All measurements in inches unless noted.

Year	Model	Lug Nut Torque (ft. lbs.)	Master Cylinder Bore	Brake Disc Minimum Thickness	Brake Disc Maximum Runout	Standard Brake Drum Diameter	Minimum Lining Thickness Front	Minimum Lining Thickness Rear
1988	Tercel	65–86	①	0.394	0.006	7.126②	0.040	0.040
	Corolla	65–86	①	⑤	0.006	7.874	0.040	0.040
	Camry	65–86	①	0.945⑦	0.003	9.079	0.040	0.040
	Celica	65–86	①	0.827⑥	0.006	7.913	0.040	0.040
	Supra	65–86	①	0.827④	0.005	—	0.040	0.040
	MR2	65–86	①	0.827③	0.006	—	0.040	0.040
	Cressida	65–86	①	0.827④	0.006	—	0.040	0.040

BRAKE SPECIFICATIONS
All measurements in inches unless noted.

Year	Model	Lug Nut Torque (ft. lbs.)	Master Cylinder Bore	Brake Disc		Standard Brake Drum Diameter	Minimum Lining Thickness	
				Minimum Thickness	Maximum Runout		Front	Rear
1989	Tercel	65–86	①	0.394	0.006	7.087②	0.040	0.040
	Corolla	65–86	①	⑤	0.004	7.874	0.040	0.040
	Camry	65–86	①	0.945⑦	0.003⑧	7.874	0.040	0.040
	Celica	65–86	①	0.827⑥	0.003	7.874	0.040	0.040
	Supra	65–86	①	0.827④	0.005	—	0.040	0.040
	MR2	65–86	①	0.827③	0.006	—	0.040	0.040
	Cressida	65–86	①	0.827④	0.003⑧	—	0.040	0.040
1990	Tercel	65–86	①	0.394	0.0059	7.087	0.394	0.039
	Corolla	65–86	①	⑨	⑩	7.874	0.039	0.039
	Camry	65–86	①	⑪	⑫	7.874	0.039	0.039
	Celica	65–86	①	⑬	⑫	7.874	0.039	0.039
	Supra	65–86	①	0.827⑥	0.005	—	0.039	0.039
	MR2	65–86	①	⑭	⑮	—	0.039	0.039
	Cressida	65–86	①	⑯	⑰	—	0.039	0.039
1991–92	Tercel	65–86	①	0.709	0.0035	7.087	0.039	0.039
	Corolla	65–86	①	⑨	⑩	7.874	0.039	0.039
	Camry	65–86	①	⑪	⑫	7.874	0.039	0.039
	Celica	65–86	①	⑲	⑫	7.874	0.039	0.039
	Supra	65–86	①	⑯	0.005	—	0.039	0.039
	MR2	65–86	①	⑭	⑮	—	0.039	0.039
	Cressida	65–86	①	⑯	⑰	—	0.039	0.039

① Not specified by the manufacturer
② Wagon & 4wd: 7.913
③ Rear disc: 0.354
④ Rear disc: 0.669
⑤ 1987 FWD (exc. FX16): 0.492
   1988–89 FWD & FX16: 0.669
   1987 RWD: 0.669
   1987 Rear disc: 0.315
   1988–89 Rear disc: 0.354
⑥ ABS or 4wd: 0.945
⑦ 4wd rear disc: 0.354
⑧ Rear disc: 0.006
⑨ Front:
   4A-FE—0.669
   4A-GE—0.827
   Rear: 0.315
⑩ Front disc: 0.0035
   Rear disc: 0.0039
⑪ Front: 0.945
   Rear: 0.354
⑫ Front: 0.0028
   Rear: 0.0059
⑬ Front: 0.787
   Rear: 0.354
⑭ Front: 0.945
   Rear: 0.591
⑮ Front: 0.0028
   Rear: 0.0039
⑯ Front: 0.827
   Rear: 0.669
⑰ Front: 0.0028
   Rear: 0.0052
⑱ Front: 0.866
   Rear: 0.709
⑲ Front
   2WD: 0.787
   4WD: 0.906
   Rear: 0.354

WHEEL ALIGNMENT

Year	Model	Caster Range (deg.)	Caster Preferred Setting (deg.)	Camber Range (deg.)	Camber Preferred Setting (deg.)	Toe-in (in.)	Steering Axis Inclination (deg.)
1988	Tercel (Sedan)	③	①	3/4N-3/4P	0	0.08 out-0.08 in	11½
	(Wagon)	1 11/16P-3 3/16P	2 1/4P	3/16N-1 5/16P	9/16P	0.12 out-0.04 in	12
	Corolla (FX/FX16)	1/8P-1 5/8P	7/8P	1N-1/2P	1/4N	0.04 out-0.12 in	12½
	(4A-F)	⑤	⑥	15/16N-9/16P	3/16N	0-0.08	12 11/16
	(4A-GE)	9/16P-2 1/16P	1 5/16P	1N-1/2P	1/4N	0-0.08	12 13/16
	Camry (Sedan)	15/16P-2 7/16P	1 11/16P	3/16N-1 5/16P	9/16P	0.04 out-0.12 in	12 3/4
	(Wagon)	1/4P-1 3/4P	1P	1/4N-1 1/4P	1/2P	0.04 out-0.12 in	13
	Celica	7/16P-1 15/16P	1 3/16P	15/16N-9/16P	3/16N	0.08 out-0.08 in	13½
	Supra	6 3/4P-8 1/4P	7 1/2P	13/16N-11/16P	1/16N	0.08 out-0.08 in	11
	MR2	4 15/16P-5 13/16P	5 1/16P	1/2N-1P	1/4P	0-0.08	12
	Cressida	4 1/16P-5 9/16P	4 13/16P	5/16N-1 3/16P	7/16P	0-0.16	10½
1989	Tercel	③	④	3/4N-3/4P	0	0.08 out-0.08 in	11½
	Corolla	③	⑥	15/16N-9/16P	3/16N	0-0.08	12 11/16
	(4A-GE)	9/16P-2 1/16P	1 5/16P	1N-1/2P	1/4N	0-0.08	12 13/16
	Camry (Sedan)	15/16P-2 7/16P	1 11/16P	3/16N-1 5/16P	9/16P	0.04 out-0.12 in	12 3/4
	(Wagon)	1/4P-1 3/4P	1P	1/4N-1 1/4P	1/2P	0.04 out-0.12 in	13
	Celica	7/16P-1 15/16P	1 3/16P	15/16N-9/16P	3/16N	0.08 out-0.08 in	13½
	Supra	6 3/4P-8 1/4P	7 1/2P	13/16N-11/16P	1/16N	0.08 out-0.08 in	11
	MR2	4 15/16P-5 13/16P	5 1/16P	1/2N-1P	1/4P	0-0.08	12 1/16
	Cressida	4 1/16P-5 9/16P	4 13/16P	5/16N-1 3/16P	7/16P	0-0.16	10½
1990	Tercel	③	④	3/4N-3/4P	0	0.08 out-0.08 in	11½
	Corolla	9/16P-2 1/16P	1 5/16P	1N-1/2P	1/4N	0-5/32	12 13/16
	(4WD)	3/4P-1 3/4P	1 1/4P	5/16N-11/16P	3/16P	0-5/32	12
	Camry (Sedan)	1 3/16P-2 3/16P	1 11/16P	1/16N-1 1/16P	9/16P	0.04 out-0.04 in	12 3/4
	(Wagon)	1/2P-1 1/2P	1P	0-1P	1/2P	0.04 out-0.04 in.	12 13/16
	Celica	1/4P-1 3/4P	1P	15/16N-9/16P	3/16N	1/16-1/8	14 3/16
	Supra	7 3/16P-8 3/16P	7 11/16P	11/16N-5/16P	3/16N	3/64 out-3/64 in	10 15/16
	MR2	2P-13 1/4P	2 3/4P	1 13/32N-13/32N	1N	0.04 out-0.04 in	13
	Cressida	6 9/16P-8 1/16P	7 5/16P	0-1P	1/2P	3/32-1/4	13 3/16
1991-92	Tercel	1 3/4P-3 1/4P	2 1/2P	3/4N-3/4P	0	3/64 out-3/64 in	11½
	Corolla (2WD)	9/16P-2 1/16P	1 5/16P	1N-1/2P	1/4N	0-5/64	12 5/8
	(4WD)	3/4P-1 3/4P	1 1/4P	5/16N-11/16P	3/16P	0-5/64	12
	Camry (Sedan)	1 3/16P-2 3/16P	1 11/16P	1/16P-1 1/16P	9/16P	0.04 out-0.04 in	12 3/4
	(Wagon)	1/2P-1 1/2P	1P	0-1P	1/2P	0.04 out-0.04 in.	12 13/16
	Celica	1/4P-1 3/4P	1P	15/16N-9/16P	3/16P	1/32-1/16	14 3/16
	Supra	7 3/16P-8 3/16P	7 11/16P	11/16N-5/16P	3/16N	3/64 out-3/64 in	10 15/16
	MR2	2P-13 1/4P	2 3/4P	1 13/32N-13/32N	1N	0.04 out-0.04 in	13

WHEEL ALIGNMENT

Year	Model	Caster		Camber		Toe-in (in.)	Steering Axis Inclination (deg.)
		Range (deg.)	Preferred Setting (deg.)	Range (deg.)	Preferred Setting (deg.)		
1991-92	Cressida	6⁹/₁₆P–8¹/₁₆P	7⁵/₁₆P	0–1P	½P	³/₃₂–¼	13³/₁₆

N—Negative
P—Positive
① Man. Str.: ¹/₁₆N–1¹/₃P
   Pwr. Str.: 1¼P–3P
② Man. Str.: ⅔P
   Pwr. Str.: 2¼P
③ Man. Str.: ¼P–1¾P
   Pwr. Str.: 1¾–3¼P
④ Man. Str.: 1P
   Pwr. Str.: 2½P
⑤ Exc. Coupe: ⁹/₁₆P–2¹/₁₆P
   Coupe: ¾P–2¼P
⑥ Exc. Coupe: 1⁵/₁₆P
   Coupe: 1½P

ENGINE MECHANICAL

NOTE: Disconnecting the negative battery cable on some vehicles may interfere with the functions of the on board computer systems and may require the computer to undergo a relearning process, once the negative battery cable is reconnected.

Engine Assembly

REMOVAL & INSTALLATION

2VZ-FE Engine

1. Disconnect the negative battery cable.
2. Remove the battery.
3. Drain the cooling system and engine oil.
4. Remove the hood.
5. Remove the ignition coil, igniter and bracket assembly.
6. Remove the radiator.
7. Remove the and coolant reservoir tank.
8. If equipped with automatic transaxle, disconnect the throttle cable from the throttle body.
9. If equipped with cruise control, remove the cruise control actuator and vacuum pump.
10. Remove the air cleaner assembly.
11. If equipped with manual transaxle, remove the clutch release cylinder. Position it aside with the hydraulic line still attached.
12. Disconnect the speedometer and transaxle control cables.
13. Remove the alternator and the belt adjusting bar.
14. Remove the air conditioning compressor and position it aside. Do not disconnect the refrigerant lines.
15. Disconnect the 2 water bypass hoses and fuel lines.
16. Tag and disconnect the brake booster, air conditioning control valve and charcoal canister vacuum hoses.
17. Tag and disconnect any additional wires and lines which may interfere with engine removal.
18. Raise and support the vehicle safely.
19. Remove the engine under covers.
20. Remove the lower suspension crossmember.
21. Remove the halfshafts.
22. Remove the power steering pump and position it aside without disconnecting the hydraulic lines.
23. Remove the front exhaust pipe.
24. Remove the engine mounting center member.
25. Remove the front, center and rear engine mount insulator and bracket assemblies.
26. Lower the vehicle. Remove the glove box and then tag and disconnect the 3 ECU connectors, the circuit opening, cowl wire and instrument wire connectors. Pull the main engine harness out through the firewall.
27. Remove the power steering reservoir tank and position it aside without disconnecting the hydraulic lines.
28. Remove the 2 right side engine mounting stays. Remove the left side engine mounting stay.

To install:

29. When installing the engine pay close attention to the following torque specifications.
   a. Transaxle-to-engine mounting bolts—56 ft. lbs. (77 Nm).
   b. Bracket-to-frame bolt—40 ft. lbs. (56 Nm).
   c. Engine mount bracket-to-block—28 ft. lbs. (39 Nm).
   d. Front and rear engine mounting through bolts and nuts—60 ft. lbs. (84 Nm).
   e. Right hand mounting rubber on body side—30 ft. lbs. (41 Nm). Right hand mounting rubber on engine side—45 ft. lbs. (62 Nm).
   f. Strut-to-body nuts—40 ft. lbs. (56 Nm).
   g. Tie rod end-to-steering knuckle ball joints—29 ft. lbs. (40 Nm).
30. Connect a suitable lifting device to the 2 engine hangers. If with ABS, remove the clamp bolts for the power steering oil cooler pipes. Remove the right and left engine mount insulators and their brackets. Slowly remove the engine/transaxle assembly as a unit.
31. Connect the battery cable, refill all fluids, start the engine and check for leaks.

3A-C Engine

1. Disconnect the negative battery cable.
2. Remove the battery and battery carrier.
3. Remove the hood.
4. Drain the cooling system.
5. Disconnect and plug the transaxle fluid lines, if equipped with automatic transaxle.
6. Remove the radiator.
7. Remove the windshield washer tank.
8. Disconnect the heater hoses.
9. If equipped with air conditioning, remove the condenser fan assembly.

10. If equipped with power steering, remove the power steering pump and position it aside. Plug the lines to prevent fluid leakage.

11. If equipped with air conditioning, remove the air conditioning compressor and position it aside. Leave the refrigerant lines connected.

12. Disconnect and tag the engine ground strap, the oxygen sensor wire, the distributor connector, the ground strap from the dash panel, the oil pressure switch wire, the coolant fan wire, the water temperature gauge wire, the back-up light switch and neutral safety switch wires.

13. Disconnect the accelerator cable. If equipped with automatic transaxle, disconnect the accelerator cable.

14. Disconnect and plug the fuel line hoses. Disconnect and tag all vacuum hoses.

15. Disconnect the air suction filter from the cylinder block.

16. Remove the transaxle upper mount bolts.

17. Raise and support the vehicle safely.

18. Remove the front exhaust pipe.

19. Remove the oil cooler lines, if equipped.

20. If equipped with manual transaxle, disconnect the clutch release cable.

21. Remove the stiffener plates.

22. Disconnect the engine mounting absorber.

23. Remove the engine mount bolts.

24. Remove the torque converter cover and torque converter bolts.

25. Position a lifting device under the transaxle assembly. Remove the lower transaxle retaining bolts. As required, remove the starter.

26. Properly support the engine/transaxle assembly. Connect a suitable lifting device to the engine lifting hooks. Carefully remove the engine from the vehicle.

To install:

27. Properly support the engine/transaxle assembly. Connect a suitable lifting device to the engine lifting hooks. Carefully install the engine into the vehicle.

28. Install the lower transaxle retaining bolts. As required, install the starter.

29. Install the torque converter bolts and cover.

30. Install the engine mount bolts.

31. Connect the engine mounting absorber.

32. Install the stiffener plates.

33. If equipped with manual transaxle, connect the clutch release cable.

34. Install the oil cooler lines, if equipped.

35. Install the front exhaust pipe.

36. Install the transaxle upper mount bolts.

37. Connect the air suction filter to the cylinder block.

38. Connect the fuel line hoses and all vacuum hoses.

39. Connect the accelerator cable. If equipped with automatic transaxle, connect the accelerator cable.

40. Connect the engine ground strap, the oxygen sensor wire, the distributor connector, the ground strap to the dash panel, the oil pressure switch wire, the coolant fan wire, the water temperature gauge wire, the back-up light switch and neutral safety switch wires.

41. If equipped with air conditioning, install the air conditioning compressor.

42. If equipped with power steering, install the power steering pump and position it aside.

43. If equipped with air conditioning, install the condenser fan assembly.

44. Connect the heater hoses.

45. Install the windshield washer tank.

46. Install the radiator.

47. Connect the transaxle fluid lines, if equipped with automatic transaxle.

48. Refill the cooling system.

49. With an assistant, install the hood.

50. Install the battery and battery carrier.

51. Connect the negative battery cable, start the engine and check for leaks.

   a. On manual transaxles, torque the flywheel bolts to 58 ft. lbs. (78 Nm).

   b. On automatic transaxles, torque the driveplate bolts to 47 ft. lbs. (64 Nm).

   c. Engine mount nuts to 29 ft. lbs. (39 Nm).

   d. Stiffener plate bolts to 29 ft. lbs. (34 Nm).

   e. 14mm upper and lower transaxle mount bolts to 29 ft. lbs. (39 Nm); 17mm upper and lower transaxle bolts to 43 ft. lbs. (39 Nm).

3E and 3E-E Engine

1. Disconnect the negative battery cable.

2. Remove the battery.

3. Remove the hood.

4. Remove the engine under covers.

5. Drain the cooling system.

6. If equipped with automatic transaxle, disconnect and plug the transaxle fluid lines.

7. Remove the radiator.

8. Remove the windshield washer tank.

9. Disconnect the heater hoses. On 3E-E engine, disconnect the PCV hoses and remove the air cleaner assembly with the air intake collector.

10. If equipped with cruise control, disconnect and remove the actuator assembly. Disconnect the accelerator cable.

11. If equipped with automatic transaxle, disconnect the accelerator cable.

12. Disconnect and plug the fuel line hoses.

13. Remove the charcoal canister assembly.

14. Disconnect the brake booster hose.

15. Disconnect the transaxle control cables.

16. Disconnect the speedometer cable.

17. If equipped with automatic transaxle, remove the clutch release cylinder and the selecting bell crank.

18. Disconnect the engine ground strap, the oxygen sensor wire, the oil pressure switch wire, the coolant fan wire, the water temperature gauge wire, the back-up light switch and neutral safety switch wires. Tag all wires.

19. Disconnect and tag all vacuum hoses.

20. Disconnect the wiring harness from the intake manifold. Remove the intake manifold ground strap. Disconnect the Cold Mixture Heater (CMH) connector, the alternator electrical connector, starter electrical wires and all other wires necessary to remove the engine.

21. Remove the Vacuum Switching Valve (VSV).

22. If equipped with power steering, remove the power steering pump and position it aside.

23. If equipped with air conditioning, remove the air conditioning compressor and position it aside. Leave the refrigerant lines connected.

24. Disconnect the exhaust pipe at the manifold.

25. Remove the halfshafts.

26. Support the engine/transaxle assembly properly.

27. Connect a suitable lifting device to the engine lifting hooks.

28. If equipped with manual transaxle, remove the rear mounting thru-bolt and the rear mounting assembly. If equipped with automatic transaxle, remove the front mounting thru-bolt and front mounting assembly.

29. Remove the right and left side mounting bolts and brackets.

30. Carefully lift the engine assembly out of the vehicle.

To install:

31. Lower the engine into the vehicle with the help of an assistant.

32. Connect all electrical and hoses fittings.

33. Install all necessary components.

34. During installation, observe the following torque specifications:

   a. On manual transaxles, torque the flywheel bolts to 65 ft. lbs. (88 Nm).

b. On automatic transaxles, torque the driveplate bolts to 13 ft. lbs. (18 Nm).

c. Left bracket and mounting insulator bolts to 35 ft. lbs. (48 Nm).

d. Right mounting insulator and front through bolts to 47 ft. lbs. (64 Nm); rear mounting insulator-to-body bolts to 54 ft. lbs. (73 Nm).

e. If equipped with automatic transaxle, torque the rear mounting insulator bolts to 43 ft. lbs. (58 Nm).

35. Connect the battery cable, refill all fluids, start the engine and check for leaks.

3S-FE Engine

CAMRY (2WD)

1. Disconnect the negative battery cable.
2. Remove the hood.
3. Drain the cooling system.
4. Remove the igniter and bracket assembly.
5. Tag and disconnect all vacuum hoses, electrical wires and cables that are necessary to remove the engine.
6. Remove the radiator and coolant reservoir tank.
7. If equipped with automatic transaxle, disconnect the throttle cable and bracket from the throttle body.
8. Disconnect the accelerator cable from the throttle body.
9. If equipped with cruise control, remove the cruise control actuator and bracket.
10. Disconnect the ground wire from the alternator upper bracket.
11. Remove the air cleaner assembly, air flow meter and air cleaner hose.
12. Remove the heater hoses.
13. Disconnect and plug the fuel lines.
14. Disconnect the speedometer cable.
15. If equipped with manual transaxle, remove the clutch release cylinder and tube bracket. Do not disconnect the tube from the bracket.
16. Disconnect the transaxle control cable.
17. If equipped with air conditioning, remove the air conditioning compressor and position it aside. Do not disconnect the refrigerant lines.
18. If equipped with power steering, remove the power steering pump and position it aside. Do not disconnect the lines.
19. Raise and support the vehicle safely.
20. Drain the engine oil.
21. Remove the engine under covers.
22. Remove the suspension lower crossmember.
23. Remove the halfshafts.
24. Disconnect the exhaust pipe from the catalytic converter.
25. Disconnect the engine mounting center crossmember member.

26. Lower the vehicle.
27. Tag and disconnect the ECU electrical connectors.
28. Connect a suitable lifting device to the engine. Raise the engine slightly and remove the engine retaining brackets and bolts.
29. Carefully remove the engine/transaxle assembly from the vehicle.

NOTE: Be careful not to hit the power steering gear housing or the neutral safety switch.

To install:
30. Lower the engine into the vehicle with the help of an assistant.
31. Connect all electrical and hoses fittings.
32. Install all necessary components.
33. During installation, observe the following torque specifications:

a. On manual transaxles, torque the flywheel bolts to 65 ft. lbs. (88 Nm).

b. On automatic transaxles, torque the driveplate bolts to 13 ft. lbs. (18 Nm).

c. Left bracket and mounting insulator bolts to 35 ft. lbs. (48 Nm).

d. Right mounting insulator and front through bolts to 47 ft. lbs. (64 Nm); rear mounting insulator-to-body bolts to 54 ft. lbs. (73 Nm).

e. If equipped with automatic transaxle, torque the rear mounting insulator bolts to 43 ft. lbs. (58 Nm).

34. Connect the battery cable, refill all fluids, start the engine and check for leaks.

CAMRY (AWD)

1. Disconnect the negative battery cable.
2. Drain the cooling system.
3. Remove the hood.
4. Disconnect the accelerator cable from the throttle body.
5. Remove the radiator and the coolant reservoir tank.
6. Disconnect the heater hoses.
7. Disconnect the inlet hose at the fuel filter. Disconnect the return hose at the fuel return pipe.
8. Disconnect and remove the cruise control actuator.
9. Remove the air cleaner assembly.
10. Remove the clutch slave cylinder and hose bracket without disconnecting the hydraulic line. Position the assembly aside.
11. Disconnect the speedometer cable and the transaxle control cables.
12. If equipped with air conditioning, disconnect and remove the compressor with the refrigerant lines still attached. Move the compressor aside.
13. Tag and disconnect all wires, connecters and vacuum lines necessary to remove engine.
14. Raise and support the vehicle safely.

15. Drain the engine oil and remove the engine under covers.
16. Remove the lower suspension crossmember and the halfshafts.
17. Disconnect and remove the driveshaft.
18. Remove the power steering pump with the hydraulic lines still attached and position it aside.
19. Remove the front exhaust pipe.
20. Remove the engine mounting center member and the stabilizer bar. Lower the vehicle.
21. Tag and disconnect the ECU connectors and pull them out through the firewall.
22. Remove the power steering pump reservoir tank.
23. Connect a suitable lifting device to the eyelets on the engine.
24. Remove the right side engine mount stay and then remove the insulator and bracket.
25. Remove the left side engine mount insulator and bracket.
26. Remove the engine and transaxle as an assembly.

NOTE: Be careful not to hit the power steering gear housing or the neutral safety switch.

To install:
27. Lower the engine into the vehicle with the help of an assistant.
28. Connect all electrical and hoses fittings.
29. Install all necessary components.
30. During installation, observe the following torque specifications:

a. Right and left engine mount bracket bolts and nuts to 38 ft. lbs. (52 Nm).

b. Right side engine mount stay bolt and nut to 54 ft. lbs. (73 Nm).

c. Engine mounting center member: member-to-body bolts—29 ft. lbs. (39 Nm); member-to-other bolts—38 ft. lbs. (52 Nm)

d. Lower crossmember bolts: outer—153 ft. lbs. (206 Nm); inner—29 ft. lbs. (39 Nm).

31. Connect the battery cable, refill all fluids, start the engine and check for leaks.

Celica

1. Disconnect the negative battery cable.
2. Remove the battery.
3. Remove the hood.
4. Drain the cooling system.
5. Tag and disconnect all vacuum hoses, electrical wires and cables that are necessary to remove the engine.
6. Disconnect the ignition coil connector and high tension wire from the coil.
7. Remove the suspension upper brace.
8. Remove the radiator.

9. Remove the coolant reservoir tank.

10. If equipped with automatic transaxle, disconnect the throttle cable and bracket from the throttle body.

11. Disconnect the accelerator cable from the throttle body.

12. If equipped with cruise control, remove the cruise control actuator and bracket.

13. Remove the oxygen sensor.

14. Remove the air cleaner assembly, air flow meter, air cleaner hose and air cleaner bracket.

15. Remove the igniter.

16. Remove the heater hoses.

17. Disconnect and plug the fuel lines.

18. Disconnect the speedometer cable.

19. If equipped with manual transaxle, remove the clutch release cylinder and tube bracket. Do not disconnect the tube from the bracket.

20. Disconnect the transaxle control cable.

21. Remove the air conditioning compressor and position it aside. Do not disconnect the refrigerant lines.

22. Raise and support the vehicle safely.

23. Drain the engine oil and transaxle fluid.

24. Remove the right under cover.

25. Remove the power steering pump and position it aside. Do not disconnect the lines.

26. Remove the suspension lower crossmember.

27. Remove the halfshafts.

28. Disconnect the exhaust pipe from the catalytic converter.

29. Remove the engine rear mounting bolt. Lower the vehicle.

30. Disconnect the ECU electrical connectors.

31. Remove the power steering pump reservoir mounting bolts.

32. Connect a suitable lifting device to the engine. Raise the engine slightly and remove the engine retaining brackets and bolts.

33. Carefully remove the engine/transaxle assembly from the vehicle. Be careful not to hit the power steering gear housing or the neutral safety switch.

To install:

34. Lower the engine into the vehicle with the help of an assistant.

35. Connect all electrical and hoses fittings.

36. Install all necessary components.

37. During installation, observe the following torque specifications:

   a. Right and left engine mount bracket bolts and nuts to 38 ft. lbs. (52 Nm).

   b. Right side engine mount stay bolt and nut to 54 ft. lbs. (73 Nm).

   c. Engine mounting center member: member-to-body bolts—29 ft. lbs. (39 Nm); member-to-other bolts—38 ft. lbs. (52 Nm)

   d. Lower crossmember bolts: outer—153 ft. lbs. (206 Nm); inner—29 ft. lbs. (39 Nm).

38. Connect the battery cable, refill all fluids, start the engine and check for leaks.

3S-GE Engine

1. Disconnect the negative battery cable.

2. Remove the battery.

3. Remove the hood.

4. Drain the cooling system.

5. Tag and disconnect the connector high tension lead at the ignition coil.

6. Remove the 4 bolts and 2 nuts securing the upper suspension brace. Remove the brace.

7. If equipped with automatic transaxle, disconnect the throttle cable and its bracket at the throttle body.

8. If equipped with manual transaxle, disconnect the throttle cable from the throttle body.

9. Remove the coolant overflow tank.

10. Remove the cruise control actuator and its bracket.

11. Remove the oxygen sensor.

12. Tag and disconnect the cooling fan leads at the radiator.

13. Disconnect the heater hoses.

14. If equipped with automatic transaxle, disconnect the fluid cooler lines.

15. Remove the radiator and the 2 supports.

16. Remove the air cleaner assembly and bracket.

17. Remove the igniter.

18. Tag, disconnect and plug the fuel hoses at the filter and fuel return pipe.

19. Disconnect the speedometer cable.

20. If equipped with manual transaxle, disconnect the transaxle control cable at the shift and selector levers. If equipped with automatic transaxle, disconnect the cable at the swivel and at the bracket and then remove it.

21. Unbolt the air conditioning compressor and position it aside with the refrigerant lines still attached.

22. Tag and disconnect any remaining wires or electrical leads. Tag and disconnect any remaining vacuum hoses.

23. Raise and support the vehicle safely.

24. Drain the engine oil.

25. Remove the right side engine under cover.

26. Remove the lower suspension crossmember.

27. Remove both halfshafts.

28. Unbolt the power steering pump. Disconnect the 2 vacuum hoses and remove the drive belt. Position the pump aside with the hydraulic lines still connected to it.

29. Disconnect the exhaust pipe at the manifold.

30. Remove the rear engine mount bolt. Lower the vehicle and then remove the front engine mount bolts.

31. Remove the power steering pump reservoir and position it aside.

32. Connect a suitable lifting device to the engine lifting hooks. Take up the engine's weight with the hoist and remove the right and left engine mounts.

33. Slowly and carefully, remove the engine and transaxle assembly. Be careful not to hit the power steering gear housing or the neutral safety switch.

To install:

34. Lower the engine into the vehicle with the help of an assistant.

35. Connect all electrical and hoses fittings.

36. Install all necessary components.

37. During installation, observe the following torque specifications:

   a. Right and left engine mount bracket bolts and nuts to 38 ft. lbs. (52 Nm).

   b. Right side engine mount stay bolt and nut to 54 ft. lbs. (73 Nm).

   c. Engine mounting center member: member-to-body bolts—29 ft. lbs. (39 Nm); member-to-other bolts—38 ft. lbs. (52 Nm)

   d. Lower crossmember bolts: outer—153 ft. lbs. (206 Nm); inner—29 ft. lbs. (39 Nm).

38. Connect the battery cable, refill all fluids, start the engine and check for leaks.

3S-GTE Engine

CELICA (2WD)

1. Disconnect the negative battery cable.

2. Remove the battery.

3. Drain the coolant from the engine and turbocharger intercooler.

4. Remove the hood.

5. Disconnect the accelerator cable at the throttle body.

6. Remove the radiator.

7. Disconnect the heater and intercooler hoses.

8. Disconnect the fuel inlet line at the fuel filter and the return line at the return pipe.

9. Remove the cruise control actuator and bracket.

10. Remove the air cleaner assembly.

11. Remove the clutch release cylinder and bracket without disconnecting the hydraulic line. Move it aside.

12. Disconnect the speedometer and transaxle control cables.

13. Remove the alternator.

14. Remove the air conditioning compressor and position it aside. Do

not disconnect the refrigerant lines.

15. Tag and disconnect any wires, cables, hoses, connectors and vacuum lines which might interfere with engine removal.

16. Raise and support the vehicle safely.

17. Drain the engine oil and remove the under covers.

18. Remove the lower suspension crossmember.

19. Remove the front halfshafts and the driveshaft.

20. Remove the power steering pump and bracket without disconnecting the hydraulic lines. Position the power steering pump aside.

21. Disconnect the front exhaust pipe at the manifold and tailpipe and remove it.

22. Remove the engine mounting center member and lower the vehicle.

23. Unplug the 3 ECU connectors, remove the 2 screws and pull the connectors out through the firewall.

24. Remove the power steering pump reservoir tank.

25. Connect a suitable lifting device to the lifting brackets on the engine. Remove the 2 bolts holding the right engine mount insulator to the mounting bracket. Remove the 4 bolts holding the left engine mount insulator to the mounting bracket and then lower the engine out of the vehicle.

To install:

26. Lower the engine into the vehicle with the help of an assistant.

27. Connect all electrical and hoses fittings.

28. Install all necessary components.

29. During installation, observe the following torque specifications:

   a. Torque the right and left engine mount bracket bolts to 38 ft. lbs. (52 Nm).

   b. When installing the engine mounting center member, tighten the outer bolts to 29 ft. lbs. (39 Nm), tighten the inner bolts to 38 ft. lbs. (52 Nm).

   c. Tighten the front exhaust pipe bolts to 46 ft. lbs. (62 Nm).

   d. When installing the lower suspension crossmember, tighten the outer bolts to 154 ft. lbs. (208 Nm), tighten the inner bolts to 29 ft. lbs. (39 Nm).

30. Connect the battery cable, refill all fluids, start the engine and check for leaks.

CELICA (AWD)

1. Disconnect the negative battery cable.

2. Remove the hood.

3. Raise and support the vehicle safely.

4. Remove the engine under covers.

5. Drain the cooling system, engine oil and transaxle fluid.

6. Remove the air cleaner assembly.

7. Disconnect the accelerator cable from the throttle body.

8. Remove the relay box from the battery. Disconnect the wires and connectors from the box.

9. Remove the air conditioning relay box from its mounting bracket.

10. Remove the injector solenoid resistor and fuel pump resistor from the engine compartment.

11. Remove the radiator and coolant overflow tank.

12. If equipped with cruise control, disconnect the wiring and remove the cruise control actuator.

13. Remove the wiper arms and outside windshield moulding. Then, remove the upper brace which is retained by 4 nuts and 2 bolts. The brace connects from the struts to the firewall.

14. Remove the ignition coil.

15. From inside the engine compartment, tag and disconnect all electrical wiring and vacuum hoses necessary to remove the engine.

16. Remove the engine wire bracket.

17. Remove the charcoal canister.

18. Disconnect the heater hoses.

19. Disconnect the speedometer cable from the transaxle.

20. Disconnect and plug the fuel hoses.

21. Remove the starter.

22. Remove the clutch release cylinder without disconnecting the hydraulic tube. Move the unit aside.

23. Disconnect the control cables from the transaxle.

24. Remove the turbocharger pressure sensor and air conditioning Air Switching Valve (ASV) from inside the engine compartment.

25. From the passenger compartment, unplug the connectors from the ECU, air conditioning amplifier and cowl wires. Pull the wiring harnesses through the firewall.

26. Remove the suspension lower crossmember.

27. Remove the front halfshafts and the driveshaft.

28. Remove the power steering pump and bracket without disconnecting the hydraulic lines. Position the pump aside.

29. Disconnect the front exhaust pipe at the manifold and tailpipe and remove it.

30. Remove the engine mounting center member and lower the vehicle. Unplug the 3 TCCS ECU connectors, remove the 2 screws and pull the connectors out through the firewall.

31. Remove the power steering pump reservoir tank.

32. Connect a suitable lifting device to the lifting brackets on the engine.

33. Remove the 2 bolts holding the right engine mount insulator to the mounting bracket. Remove the 4 bolts holding the left engine mount insulator to the mounting bracket.

34. Slowly and carefully remove the engine and transaxle assembly for the top of the vehicle.

To install:

35. Lower the engine into the vehicle with the help of an assistant.

36. Connect all electrical and hoses fittings.

37. Install all necessary components.

38. During installation, observe the following torque specifications:

   a. Torque the left mounting bracket-to-transaxle case bolts to 38 ft. lbs. (52 Nm).

   b. Torque the left mounting insulator through bolt to 47 ft. lbs. (63 Nm) and 4 hex head bolts to 64 ft. lbs. (87 Nm).

   c. Torque the right mounting insulator nuts to 38 ft. lbs. (52 Nm) and the thru bolt to 64 ft. lbs. (87 Nm).

   d. Torque the front and rear bolts to 57 ft. lbs. (77 Nm).

   e. Torque the front and rear engine mounting through bolts to 64 ft. lbs. (87 Nm).

   f. Torque the lower crossmember nuts and bolts to 112 ft. lbs. (152 Nm).

   g. Torque the suspension upper brace nuts to 47 ft. lbs. (64 Nm) and bolts to 15 ft. lbs. (21 Nm).

39. Connect the battery cable, refill all fluids, start the engine and check for leaks.

1990–92 MR2

1. Disconnect the negative battery cable.

2. Remove the hood and side panels.

3. Raise and support the vehicle safely.

4. Remove the engine under covers.

5. Drain the cooling system, engine oil and transaxle fluid.

6. Remove the suspension upper brace that criss-crosses from the struts to the firewall.

7. Remove the air cleaner assembly.

8. Remove both air connector tubes.

9. Disconnect the accelerator cable from the throttle body.

10. If equipped with cruise control, disconnect the wiring and remove the cruise control actuator and accelerator linkage assemblies.

11. Disconnect the brake booster vacuum hose.

12. Disconnect the ground strap connector.

13. Remove the check connector and turbocharger pressure sensor.

14. Remove the injector solenoid resistor, fuel pump relay, fuel pump resistor and the air conditioning vacuum switching valve.

15. Disconnect the filler and over-

flow hoses from the water filler connection. Remove the water filler from the engine.

16. Remove the engine relay box. Disconnect the wires and connectors from the box.

17. Remove the ignition coil and igniter.

18. From inside the luggage compartment, disconnect the wiring harnesses for the ECU, starter relay, cooling fan and engine wires.

19. Disconnect the starter wiring.

20. Disconnect the radiator hose from the water inlet.

21. Disconnect and plug the fuel inlet and return hoses.

22. Disconnect the radiator hoses from the water outlet housing.

23. Disconnect the heater hoses.

24. Disconnect the control cables from the transaxle.

25. Remove the tailpipe and front exhaust pipe.

26. Remove the engine compartment cooling fan.

27. Remove the idler pulley bracket and unbolt the air conditioning compressor. Move the compressor aside. Leave the refrigerant lines connected.

28. Remove the intercooler.

29. Remove the rear engine mounting insulator.

30. Disconnect the speedometer cable from the transaxle.

31. Disconnect the stabilizer link from the shock absorber.

32. Remove the wire clamp bolt and remove the ABS speed sensor.

33. Remove the lower suspension arms.

34. Remove the driveshafts.

35. Remove the 4 bolts and remove the lower crossmember.

36. Remove the front engine mounting insulator.

37. Remove the nut and bolt attaching the clutch release cylinder to the transaxle. Remove the mounting bracket bolts and remove the clutch release cylinder without disconnecting the hydraulic tube.

38. Remove the right and left engine mounting stays.

39. Remove the lateral control rod and air cleaner case bracket.

40. Connect a suitable lifting device to the engine hanger brackets. Tension the lifting device to support the weight of the engine, then remove the left and right mounting insulator fasteners; 2 bolts and 3 nuts for each insulator.

41. Carefully lower then raise the engine from the vehicle.

To install:

42. Lower the engine into the vehicle with the help of an assistant.

43. Connect all electrical and hoses fittings.

44. Install all necessary components.

45. During installation, observe the following torque specifications:

a. Torque the rear engine mounting bracket-to-transaxle case bolts to 38 ft. lbs. (52 Nm) for the 14mm bolts and 57 ft. lbs. (77 Nm) for the 17mm bolts.

b. Torque the left mounting insulator through bolt to 47 ft. lbs. (63 Nm) and 4 hex head bolts to 54 ft. lbs. (73 Nm).

c. Torque the right mounting insulator nuts to 54 ft. lbs. (73 Nm) and the thru bolt to 64 ft. lbs. (87 Nm).

d. Torque the front and rear bolts to 57 ft. lbs. (77 Nm).

e. Torque the front and rear engine mounting through bolts to 64 ft. lbs. (87 Nm).

f. Torque the lower crossmember nuts and bolts to 112 ft. lbs. (152 Nm).

g. Torque the suspension upper brace nuts to 47 ft. lbs. (64 Nm) and bolts to 15 ft. lbs. (21 Nm).

46. Connect the battery cable, refill all fluids, start the engine and check for leaks.

1988 4A-LC Engine

1. Disconnect the negative battery cable.

2. Remove the battery.

3. Remove the hood.

4. Drain the cooling system and the oil.

5. Remove the air cleaner hose and air cleaner.

6. Remove the coolant reservoir tank, radiator and fan shroud.

7. Disconnect the actuator, accelerator and throttle cables at the carburetor.

8. Tag and disconnect all wires, hoses and lines which might interfere with engine removal.

9. Disconnect the fuel lines at the fuel pump.

10. Remove the heater hoses.

11. Remove the power steering pump and air conditioning compressor and position them aside. Do not disconnect the hydraulic or refrigerant lines.

12. Disconnect the speedometer cable at the transaxle.

13. Remove the clutch release cylinder without disconnecting the hydraulic line and position it aside.

14. Disconnect the shift control cable and then raise the vehicle and support it with safety stands.

15. Disconnect the exhaust pipe at the manifold.

16. Disconnect the front and rear engine mounts at the center member. Remove the center member.

17. Disconnect the halfshafts at the transaxle and lower the vehicle.

18. Connect a suitable lifting device to the lift brackets on the engine.

19. Remove the right engine mount thru-bolt. Remove the left engine mount and bracket.

20. Lift the engine out of the vehicle.

To install:

21. Lower the engine into the vehicle with the help of an assistant.

22. Connect all electrical and hoses fittings.

23. Install all necessary components.

24. During installation, observe the following torque specifications:

a. Tighten the engine mount center member to 29 ft. lbs. (39 Nm).

b. Tighten the front and rear engine mounts to 29 ft. lbs. (39 Nm).

c. Tighten the exhaust pipe-to-manifold bolts to 46 ft. lbs. (62 Nm).

25. Connect the battery cable, refill all fluids, start the engine and check for leaks.

4A-F Engine

1. Disconnect the negative battery cable.

2. Remove the battery.

3. Remove the hood.

4. Remove the engine under covers.

5. Drain the cooling system, engine oil and transaxle oil.

6. Remove the air cleaner and air cleaner flexible hose.

7. Remove the coolant reservoir tank, radiator and cooling fan.

8. If equipped with automatic transaxle, disconnect the accelerator and throttle cables at the carburetor.

9. Disconnect all electrical wires and vacuum lines necessary to remove the engine.

10. Disconnect the fuel lines at the fuel pump. Plug the lines.

11. Disconnect the heater hoses at the water inlet housing.

12. Remove the power steering pump and set it aside with the hydraulic lines still attached.

13. Remove the air conditioning compressor and set it aside with the refrigerant lines still attached.

14. Disconnect the speedometer cable from the transaxle.

15. If equipped with manual transaxle, remove the clutch release cylinder and position it aside with the hydraulic lines still attached.

16. Disconnect the shift control cables.

17. Raise and support the vehicle safely.

18. Remove the 2 nuts from the flange and then disconnect the exhaust pipe at the manifold.

19. Disconnect the halfshafts at the transaxle.

20. Remove the 2 hole covers and then remove the front, center and rear engine mounts from the center member. Remove the 5 bolts and insulators and remove the center member.

21. Connect a suitable lifting device

to the lifting brackets on the engine.

22. Remove the 3 bolts and mounting stay. Remove the bolt, 2 nuts and the thru-bolt and pull out the right side engine mount. Remove the 2 bolts and the left mounting stay. Remove the 3 bolts and disconnect the left engine mount bracket from the transaxle.

23. Lift the engine/transaxle assembly out of the vehicle.

To install:

24. Lower the engine into the vehicle with the help of an assistant.

25. Connect all electrical and hoses fittings.

26. Install all necessary components.

27. During installation, observe the following torque specifications:

    a. Torque the right engine mount insulator bolt to 47 ft. lbs. (64 Nm); tighten the nut to 38 ft. lbs. (52 Nm). Align the insulator with the bracket on the body and tighten the bolt to 64 ft. lbs. (87 Nm).

    b. Align the left engine mount insulator bracket with the transaxle bracket and tighten the bolt to 35 ft. lbs. (48 Nm).

    c. Install the right mounting stay and tighten the 3 bolts to 31 ft. lbs. (42 Nm). Install the left stay and tighten the 2 bolts to 15 ft. lbs. (21 Nm).

    d. Install the engine center member and tighten the 5 bolts to 45 ft. lbs. (61 Nm).

    e. Install the front and rear engine mounts and bolts. Align the bolts holes in the brackets with the center member and tighten the front mount bolts to 35 ft. lbs. (48 Nm); tighten the center and rear mounts to 38 ft. lbs. (52 Nm). Install the 2 hole covers and tighten the rear mounting bolt to 58 ft. lbs. (78 Nm).

28. Connect the battery cable, refill all fluids, start the engine and check for leaks.

4A-FE Engine

COROLLA

1. Disconnect the negative battery cable.

2. Remove the battery.

3. Remove the hood.

4. Remove the engine under covers.

5. Drain the cooling system, engine and transaxle oil.

6. Remove the air cleaner and air cleaner flexible hose.

7. Remove the coolant reservoir tank, radiator and cooling fan.

8. If equipped with automatic transaxle, disconnect the accelerator and throttle cables.

9. If equipped with cruise control, remove the cruise control actuator.

10. Disconnect the No. 2 junction block, the ground strap connector and the ground strap.

11. Disconnect the check, vacuum sensor and oxygen sensor connectors. Disconnect the air conditioning compressor wire.

12. Disconnect the vacuum hoses at the brake booster, power steering pump, vacuum sensor, charcoal canister and vacuum switch.

13. Disconnect the fuel lines at the fuel pump.

14. Disconnect the heater hoses at the water inlet housing.

15. Remove the power steering pump and set it aside with the hydraulic lines still attached.

16. Remove the air conditioning compressor and set it aside with the refrigerant lines still attached.

17. Disconnect the speedometer cable at the transaxle.

18. If equipped with manual transaxle, remove the clutch release cylinder and position it aside with the hydraulic lines still attached.

19. Disconnect the shift control cables.

20. Raise and support the vehicle safely.

21. Disconnect the oil cooler lines and the exhaust pipe (at the manifold).

22. Disconnect the halfshafts and the driveshaft at the transaxle.

23. Connect a suitable lifting device to the lifting brackets on the engine and raise it just enough to relieve pressure on the mounts.

24. Pull out the hole covers and remove the 5 bolts on the front and rear engine mounts. Remove the mounts from the center crossmember. Remove the 4 center crossmember bolts and the 8 bolts from the sub-frame. Remove the front and rear mounting bolts and then remove the member.

25. Remove the engine mount stay and the mount.

26. Remove the air cleaner bracket. Disconnect the left side mounting bracket from the transaxle bracket and then lift out the engine/transaxle assembly slowly and carefully.

To install:

27. During installation, observe the following torque specifications:

    a. Torque the right engine mount insulator bolt and nuts to 38 ft. lbs. (52 Nm). Align the insulator with the bracket on the body and tighten the bolt to 64 ft. lbs. (87 Nm).

    b. Align the left engine mount insulator bracket with the transaxle bracket and tighten the bolt to 35 ft. lbs. (48 Nm).

    c. Install the left stay and tighten the 2 bolts to 15 ft. lbs. (21 Nm).

    d. Install the engine center member and tighten the 5 bolts to 45 ft. lbs. (61 Nm).

    e. Install the front and rear engine mounts and bolts. Align the bolts holes in the brackets with the center member and tighten the front mount bolts to 35 ft. lbs. (48 Nm); tighten the center and rear mounts to 42 ft. lbs. (57 Nm). Install the 8 sub-frame bolts and tighten the lower arm bolt to 152 ft. lbs. (206 Nm) and the rear bolt to 94 ft. lbs. (127 Nm).

1990–92 CELICA

1. Disconnect the negative battery cable.

2. Remove the battery.

3. Raise the vehicle and support safely.

4. Remove the engine under covers.

5. Drain the cooling system and engine oil.

6. Remove the air cleaner assembly along with its hose and any attachments.

7. Disconnect the accelerator and throttle cables at the bracket.

8. Remove the lower cover from the relay box. Disconnect the fusible link cassette and connectors. Remove the engine relay box.

9. Remove the air conditioning relay box from the bracket.

10. Remove the coolant reservoir tank, radiator and cooling fan.

11. Disconnect the check connector, vacuum sensor connector and ground strap from the left front fender apron. Remove the engine wiring bracket. Disconnect the noise filter assembly.

12. Remove the charcoal canister.

13. Disconnect the heater hose from the water inlet.

14. Disconnect the speedometer cable at the transaxle.

15. Disconnect the fuel hose.

16. If equipped with manual transaxle, remove the clutch release cylinder and position it aside with the hydraulic lines still attached.

17. Disconnect the shift control cables from the transaxle.

18. Tag and disconnect all vacuum hoses and electrical wires necessary to remove the engine.

19. Remove the suspension lower crossmember.

20. Disconnect the oxygen sensor connector.

21. Remove the front exhaust pipe assembly.

22. If equipped with automatic transaxle, disconnect control cable from engine mounting center member.

23. Remove the front halfshafts.

24. Unbolt the air conditioning compressor and wire it aside with the refrigerant lines still attached.

25. Remove the power steering pump assembly without disconnecting the hydraulic lines.

26. Connect a suitable lifting device to the engine lifting brackets. Tension the lifting device slightly to take the pressure off the mounts.

27. Remove the engine mounting center member.

28. Remove the front engine mounting insulator and bracket.

29. Remove the rear mounting insulator and bracket.

30. Disconnect the ground wire from the fender apron. Remove the ground strap from the transaxle.

31. Remove the right and left engine mounting stay.

32. Slowly and carefully, remove the engine and transaxle assembly from the top of the vehicle.

NOTE: Be careful not to hit the power steering gear housing or the neutral safety switch.

To install:

33. Lower the engine into the vehicle with the help of an assistant.

34. Connect all electrical and hoses fittings.

35. Install all necessary components.

36. During installation, observe the following torque specifications:

   a. Torque the left mounting bracket-to-transaxle case bolts to 38 ft. lbs. (52 Nm).

   b. Torque the left mounting insulator-to-bracket bolts to 35 ft. lbs. (48 Nm) and the thru bolt to 64 ft. lbs. (87 Nm).

   c. Torque the right engine stay bolts to 31 ft. lbs. and the left engine stay bolts to 15 ft. lbs. (21 Nm).

   d. Torque the front and rear engine mounting bracket and insulator fasteners to 57 ft. lbs. (77 Nm).

   e. Torque the engine center member bolts to 38 ft. lbs. (52 Nm) and the center member-to-insulator bolts to 47 ft. lbs. (64 Nm).

   f. Torque the front and rear engine mounting through bolts to 64 ft. lbs. (87 Nm).

37. Connect the battery cable, refill all fluids, start the engine and check for leaks.

4A-GEC and 4A-GELC Engine

1. Disconnect the negative battery cable.

2. Remove the battery.

3. Remove the air cleaner assembly.

4. Drain the cooling system and engine oil.

5. Remove the fuel tank protectors and the engine under cover.

6. Disconnect the accelerator cable.

7. If equipped with cruise control, disconnect the cruise control at the cable actuator. If equipped with automatic transaxle, disconnect the throttle cable.

8. Disconnect the heater hoses at the water inlet housing on the rear of the cylinder head cover. Disconnect the radiator hose and the air bleeder hose at the water inlet housing.

9. Disconnect and plug the fuel line at the fuel filter. Disconnect the fuel return hose. Tag and disconnect the vacuum hose at the charcoal canister.

10. Tag and disconnect the engine ground strap and the main wiring harness connector at the engine. Disconnect the back-up light switch connector as required.

11. Disconnect the speedometer cable.

12. Remove the transaxle gravel shield.

13. Remove the ground strap from the water inlet housing.

14. Remove the radiator overflow tank.

15. Remove the air conditioning and alternator drive belts. Remove the alternator.

16. Disconnect the radiator hose at the water outlet housing.

17. Tag and disconnect the 2 connectors at the igniter, the noise filter connector, the cooling fan electrical connector, the cylinder head ground strap, the air condition compressor connector and the high tension leads at the ignition coil.

18. Remove the rear luggage compartment trim.

19. Tag and disconnect the circuit opening relay connector, the ball connections at the electronic control unit and the electrical lead for the cooling fan computer.

20. Pull the main wiring harness out and through the engine compartment.

21. Remove the mounting bolts and remove the air conditioning compressor. Position it aside without disconnecting the refrigerant lines.

22. If equipped with a manual transaxle, disconnect the control cables from the outer shift lever and gear shift selector lever. If equipped with automatic transaxle, disconnect the control cable at the gear shift lever.

23. If equipped with a manual transaxle, remove the control cable bracket on the transaxle. Remove the clutch release cylinder.

24. Disconnect the engine oil cooler lines, if equipped. Disconnect the automatic transaxle fluid lines if equipped.

25. Remove the exhaust pipe assembly. Remove the oxygen sensor at the exhaust manifold.

26. If equipped with automatic transaxle, remove the mounting bolts and remove the stiffener plate at the transaxle. Remove the flywheel shield.

27. Remove the right halfshaft. Disconnect the left halfshaft from the side gear shaft and position it aside.

28. Remove the front and rear engine mount bolts. Place a block of wood on an hydraulic floor jack and carefully position the jack under the engine. Raise the jack just enough to ease the engine's weight on the mounts. Remove the right and left engine mounts.

29. Make sure there are no remaining wires or hoses connected to the engine and then slowly and carefully raise the vehicle while lowering the jack supporting the engine/transaxle assembly.

To install:

30. Lower the engine into the vehicle with the help of an assistant.

31. Connect all electrical and hoses fittings.

32. Install all necessary components.

33. During installation, observe the following torque specifications:

   a. Torque the right engine mount insulator bolt and nuts to 38 ft. lbs. (52 Nm). Align the insulator with the bracket on the body and tighten the bolt to 64 ft. lbs. (87 Nm).

   b. Align the left engine mount insulator bracket with the transaxle bracket and tighten the bolt to 35 ft. lbs. (48 Nm).

   c. Install the left stay and tighten the 2 bolts to 15 ft. lbs. (21 Nm).

   d. Install the engine center member and tighten the 5 bolts to 45 ft. lbs. (61 Nm).

   e. Install the front, center and rear engine mounts and bolts. Align the bolts holes in the brackets with the center member and tighten the front mount bolts to 35 ft. lbs. (48 Nm); tighten the center mounts to 38 ft. lbs. (52 Nm); and the rear mount bolts to 42 ft. lbs. (57 Nm).

   f. Bounce the engine several times to unload the front and rear mounts, for automatic transmission only, and then tighten the rear bolt to 64 ft. lbs. (87 Nm). Install the front bolt and tighten it to 64 ft. lbs. (87 Nm).

34. Connect the battery cable, refill all fluids, start the engine and check for leaks.

4A-GZE Engine

1. Disconnect the negative battery cable.

2. Raise and support the vehicle safely.

3. Remove the fuel tank protectors and the engine under cover.

4. Drain the cooling system and engine oil.

5. Remove the supercharger intercooler.

6. Remove the battery.

7. Disconnect the air conditioning idle-up, charcoal canister and cruise control vacuum hoses. Disconnect the air bleeder hose at the water inlet housing.

8. If equipped with automatic transmission, disconnect the throttle cable.

9. Disconnect the cruise control cable, the heater hoses and radiator hose.

10. Disconnect the fuel inlet and return lines.

11. Disconnect the speedometer cable and the brake booster vacuum line.

12. Remove the radiator reservoir tank. Disconnect the radiator hose at the water outlet housing.

13. Tag and disconnect any remaining hoses, wires or lines which may interfere with engine removal.

14. Remove the 5 clip fasteners and pull out the rear luggage compartment trim. Disconnect the connectors and pull the wiring harness through the engine compartment.

15. Remove the air conditioning compressor and position it aside with the refrigerant lines still connected.

16. Disconnect the shifter control cables.

17. Remove the clutch release cylinder and bracket and position it aside with the hydraulic lines still attached.

18. Disconnect the coolant lines at the engine and transmission oil coolers.

19. Remove the front exhaust pipe.

20. Remove the rear halfshafts.

21. Remove the front and rear engine mount insulators. Lower the engine slightly and then remove the right and left engine mounts. Carefully support the engine and raise the vehicle, over the engine.

To install:

22. Lower the engine into the vehicle with the help of an assistant.

23. Connect all electrical and hoses fittings.

24. Install all necessary components.

25. During installation, observe the following torque specifications:

   a. When installing the rear engine mount insulator, tighten the 10mm bolts to 38 ft. lbs. (52 Nm); tighten the 12mm bolts to 58 ft. lbs. (78 Nm).

   b. Install the front engine mount insulator to the body and tighten the inner bolt to 38 ft. lbs. (52 Nm), tighten the outer bolts to 54 ft. lbs. (73 Nm).

   c. Connect the mounting bracket to the insulator and install the thru-bolt. Bounce the engine several times and tighten the thru-bolt to 58 ft. lbs. (78 Nm).

26. Connect the battery cable, refill all fluids, start the engine and check for leaks.

5M-GE Engine

1. Disconnect the negative battery cable.

2. Remove the battery.

3. Remove the hood.

4. Remove the air cleaner assembly.

5. Remove the fan shroud.

6. Drain the cooling system.

7. Disconnect the upper and lower radiator hoses.

8. If equipped with automatic transmission, disconnect and plug the oil lines from the oil cooler.

9. Detach the hose which runs to the thermal expansion tank and remove the expansion tank from its mounting bracket.

10. Remove the radiator.

11. If equipped with automatic transmission, remove the throttle cable bracket from the cylinder head. Remove the accelerator and actuator cable bracket from the cylinder head.

12. Tag and disconnect the cylinder head ground cable, the oxygen sensor wire, the oil pressure sending unit, alternator wires, the high tension coil wire, the water temperature sending, the thermo-switch wires, the starter wires, the ECT connectors, the solenoid resistor wire connector and the knock sensor wire.

13. Tag and disconnect the brake booster vacuum hose from the air intake chamber, along with the EGR valve vacuum hose. Disconnect the actuator vacuum hose from the air intake chamber, if equipped with cruise control. Disconnect the heater bypass hoses from the engine.

14. Remove the glove box, and remove the ECU computer module. Disconnect the 3 connectors, and pull out the EFI wiring harness from the engine compartment side of the firewall.

15. Remove the 4 shroud and 4 fluid coupling screws, and the shroud and coupling as a unit.

16. Remove the engine undercover protector.

17. Disconnect the coolant reservoir hose. Remove the radiator and the coolant expansion tank.

18. Remove the air conditioning compressor drive belt, and remove the compressor mounting bolts. Without disconnecting the refrigerant hoses, lay the compressor aside and secure it.

19. Disconnect the power steering pump drive belt and remove the pump stay. Unbolt the pump and lay it aside without disconnecting the fluid hoses.

20. Remove the engine mounting bolts from each side of the engine. Remove the engine ground cable.

21. If equipped with manual transmission, remove the shift lever from the inside of the vehicle.

22. Raise and support the vehicle safely.

23. Drain the engine oil.

24. Disconnect the exhaust pipe from the exhaust manifold. Remove the exhaust pipe clamp from the transmission housing.

25. If equipped with manual transmission, remove the clutch slave cylinder.

26. Disconnect the speedometer cable at the transmission.

27. If equipped with automatic transmission, disconnect the shift linkage from the shift lever. If equipped with manual transmission, disconnect the wire from the back-up light switch.

28. Remove the stiffener plate from the ground cable.

29. Disconnect and plug the fuel line from the fuel filter and the return hose from the fuel hose support.

30. Remove the 2 bolts from the top and bottom of the steering universal, and remove the sliding yoke.

31. Disconnect the tie rod ends. Disconnect the pressure line mounting bolts from the front crossmember.

32. Remove the intermediate shaft from the driveshaft.

33. Position a jack under the transmission, with a wooden block between the 2 to prevent damage to the transmission case. Place a wooden block between the cowl panel and the cylinder head rear end to prevent damage to the heater hoses.

34. Unbolt the engine rear support member from the frame, along with the ground cable.

35. Make sure all wiring is disconnected, all hoses disconnected, and everything clear of the engine and transmission. Connect a suitable lifting device to the lift brackets on the engine, and carefully lift the engine and transmission up and out of the vehicle.

To install:

36. Lower the engine into the vehicle with the help of an assistant.

37. Connect all electrical and hoses fittings.

38. Install all necessary components.

39. During installation, observe the following torque specifications:

   a. When installing the rear engine mount insulator, tighten the 10mm bolts to 38 ft. lbs. (52 Nm); tighten the 12mm bolts to 58 ft. lbs. (78 Nm).

   b. Install the front engine mount insulator to the body and tighten the inner bolt to 38 ft. lbs. (52 Nm), tighten the outer bolts to 54 ft. lbs. (73 Nm).

   c. Connect the mounting bracket to the insulator and install the thru-bolt. Bounce the engine several times and tighten the thru-bolt to 58 ft. lbs. (78 Nm).

40. Connect the battery cable, refill all fluids, start the engine and check for leaks.

5S-FE Engine

1990–92 CELICA

1. Disconnect the negative battery cable.

2. Remove the battery.

3. Remove the hood.

4. Raise the vehicle and support safely.

5. Remove the engine under covers.

6. Drain the cooling system and engine oil.

7. Remove the air cleaner assembly along with hoses and any attachments.

8. Disconnect the accelerator and throttle cables at the bracket.

9. Remove the lower cover from the relay box. Disconnect the fusible link cassette and connectors. Remove the engine relay box.

10. Remove the air conditioning relay box from the bracket.

11. Remove the cruise control actuator assembly.

12. Remove the coolant reservoir tank, radiator and cooling fan.

13. Remove the 2 wiper arms and outside lower windshield moulding. Remove the suspension upper brace where it attaches to the struts and the firewall.

14. Remove the ignition coil assembly. Disconnect the check connector, igniter connector, vacuum sensor connector and ground strap from the left front fender apron. Remove the engine wiring bracket. Disconnect the noise filter assembly.

15. Remove the charcoal canister.

16. Disconnect the heater hose from the water inlet.

17. Disconnect the speedometer cable.

18. Disconnect the fuel hose.

19. If equipped with a manual transaxle, remove the clutch release cylinder and position it aside with the hydraulic lines still attached.

20. Disconnect the shift control cables from the transaxle.

21. Tag and disconnect the vacuum sensor hose from the gas filter on the air intake chamber, brake booster vacuum hose and air conditioning vacuum hoses on air intake chamber.

22. Disconnect 2 cowl wire connectors and engine wire clamp from engine fender apron.

23. Tag and disconnect: engine ECU connector, cowl wire connectors and air conditioning amplifier connector.

24. Remove the suspension lower crossmember.

25. Disconnect the oxygen sensor connector.

26. Remove all necessary brackets and retaining bolts.

27. Remove the front exhaust pipe assembly.

28. If equipped with automatic transaxle, disconnect control cable from engine mounting centermember. Remove the front halfshafts.

29. Unbolt the air conditioning compressor and then wire it aside with the refrigerant lines still attached.

30. Remove the power steering pump assembly without disconnecting the hydraulic lines.

31. Connect a suitable lifting device to the engine lifting brackets.

32. Remove the engine mounting center member.

33. Remove the front engine mounting insulator and bracket.

34. Remove the rear mounting insulator and bracket.

35. Disconnect the ground wire from the fender apron. Remove the ground strap from the transaxle.

36. Remove the right and left engine mounting stay.

37. Slowly and carefully, remove the engine and transaxle assembly from the top of the vehicle.

NOTE: Be careful not to hit the power steering gear housing or the neutral safety switch.

To install:

38. Lower the engine into the vehicle with the help of an assistant.

39. Connect all electrical and hoses fittings.

40. Install all necessary components.

41. During installation, observe the following torque specifications:

　a. Torque the left mounting bracket-to-transaxle case bolts to 38 ft. lbs. (52 Nm).

　b. Torque the left mounting insulator-to-bracket bolts to 35 ft. lbs. (48 Nm) and the thru bolt to 64 ft. lbs. (87 Nm).

　c. Torque the right engine stay bolts to 31 ft. lbs. and the left engine stay bolts to 15 ft. lbs. (21 Nm).

　d. Torque the front and rear engine mounting bracket and insulator fasteners to 57 ft. lbs. (77 Nm).

　e. Torque the engine center member bolts to 38 ft. lbs. (52 Nm) and the centermember-to-insulator bolts to 47 ft. lbs. (64 Nm).

　f. Torque the front and rear engine mounting through bolts to 64 ft. lbs. (87 Nm).

42. Connect the battery cable, refill all fluids, start the engine and check for leaks.

1990–92 MR2

1. Disconnect the negative battery cable.

2. Remove the hood and engine side panels.

3. Raise and support the vehicle safely.

4. Remove the engine under covers.

5. Drain the cooling system, engine oil and transaxle fluid.

6. Remove the suspension upper brace that criss-crosses from the struts to the firewall.

7. Remove the air cleaner assembly.

8. Disconnect the accelerator cable from the throttle body.

9. If equipped with cruise control, disconnect the wiring and remove the cruise control actuator and accelerator linkage assemblies.

10. Disconnect the brake booster vacuum hose.

11. Disconnect the ground strap connector.

12. Remove the check connector and vacuum sensor.

13. Remove the the air conditioning vacuum switching valve.

14. Disconnect the filler and overflow hoses from the water filler connection. Remove the water filler from the engine.

15. Remove the engine relay box. Disconnect the wires and connectors from the box.

16. Remove the ignition coil and igniter.

17. From inside the luggage compartment, disconnect the wiring harnesses for the ECU, starter relay, cooling fan and engine wires.

18. Disconnect the starter wiring.

19. Disconnect the radiator hose from the water inlet.

20. Disconnect and plug the fuel inlet and return hoses.

21. Disconnect the radiator hoses from the water outlet housing.

22. Disconnect the heater hoses.

23. Disconnect the control cables from the transaxle.

24. If equipped with automatic transaxle, disconnect and plug the oil cooler hoses.

25. Remove the front exhaust pipe.

26. Remove the driveshafts.

27. Remove the idler pulley bracket and unbolt the air conditioning compressor. Move the compressor aside. Leave the refrigerant lines connected.

28. Remove the front and rear engine mounting insulator.

29. Disconnect the speedometer cable from the transaxle.

30. If equipped with manual transaxle, remove the nut and bolt attaching the clutch release cylinder to the transaxle. Remove the mounting bracket bolts and remove the clutch release cylinder without disconnecting the hydraulic tube. If equipped with automatic transaxle, unbolt and remove the control cable bracket from the transaxle.

31. Remove the rear engine mounting bracket.

32. Remove the right and left engine mounting stays.

33. If equipped with manual transaxle, remove the lateral control rod and air cleaner case bracket. If equipped with automatic transaxle, unbolt the air cleaner case bracket, disconnect the charcoal canister tube and the ground strap from the transaxle.

34. Connect a suitable lifting device to the engine hanger brackets. Tension the lifting device to support the weight of the engine, then remove the left and right mounting insulator fasteners.

35. Carefully lower then raise the engine from the vehicle.

To install:

36. Lower the engine into the vehicle with the help of an assistant.
37. Connect all electrical and hoses fittings.
38. Install all necessary components.
39. During installation, observe the following torque specifications:

    a. Torque the left mounting bracket-to-transaxle case bolts to 38 ft. lbs. (52 Nm).

    b. Torque the left mounting insulator-to-bracket bolts to 35 ft. lbs. (48 Nm) and the thru bolt to 64 ft. lbs. (87 Nm).

    c. Torque the right engine stay bolts to 31 ft. lbs. and the left engine stay bolts to 15 ft. lbs. (21 Nm).

    d. Torque the front and rear engine mounting bracket and insulator fasteners to 57 ft. lbs. (77 Nm).

    e. Torque the engine center member bolts to 38 ft. lbs. (52 Nm) and the center member-to-insulator bolts to 47 ft. lbs. (64 Nm).

    f. Torque the front and rear engine mounting through bolts to 64 ft. lbs. (87 Nm).

40. Connect the battery cable, refill all fluids, start the engine and check for leaks.

7M-GE and 7M-GTE Engine
SUPRA

1. Disconnect the negative battery cable.
2. Remove the hood.
3. Raise and support the vehicle safely.
4. Remove the engine under cover.
5. Drain the cooling system and engine oil.
6. Remove the radiator.
7. On 7M-GE, remove the air cleaner assembly. On 7M-GTE, remove the No. 4 air cleaner pipe along with the No. 1 and 2 air cleaner hose.
8. Remove the No. 7 air cleaner hose with the air flow meter and air cleaner cap.
9. Remove the air conditioning belt. Remove the alternator drive belt, water pump pulley and fan assembly. Remove the power steering belt.
10. Disconnect the brake booster hose, the heater valve hose, the cruise control hose and the charcoal canister hose.
11. Remove the heater hoses.
12. Tag and disconnect all electrical wire and vacuum hoses necessary to remove the engine.
13. If equipped with cruise control, disconnect the cruise control cable.
14. Disconnect the accelerator cable.
15. If equipped with automatic transmission, disconnect the throttle cable.
16. Remove the air conditioning compressor. Position the unit aside.

Do not disconnect the refrigerant lines.

17. On the 7M-GTE, remove the No. 6 air cleaner hose and the upper radiator outlet hose.
18. Remove the power steering pump. Position the unit aside; do not disconnect the hydraulic lines.
19. If equipped with manual transmission, remove the shift lever.
20. Disconnect the ground strap from the fuel hose clamp. On the 7M-GTE, remove the engine mounting absorber.
21. Disconnect and plug the fuel lines.
22. Raise the vehicle and support safely.
23. Remove the exhaust pipe.
24. Remove the driveshaft.
25. Disconnect the speedometer cable.
26. If equipped with automatic transmission, remove the shift linkage. If equipped with manual transmission, remove the clutch release cylinder.
27. Properly support the engine and transmission assembly. Remove the No. 1 front crossmember. Remove the engine retaining mounts.
28. Position a piece of wood between the engine firewall and the rear of the cylinder head to prevent damage to the heater hose.
29. Make sure there are no remaining wires or hoses connected to the engine and then slowly and carefully remove the engine and transmission from the vehicle.

To install:

30. Lower the engine into the vehicle with the help of an assistant.
31. Connect all electrical and hoses fittings.
32. Install all necessary components.
33. During installation, observe the following torque specifications:

    a. Torque the left mounting bracket-to-transaxle case bolts to 38 ft. lbs. (52 Nm).

    b. Torque the left mounting insulator-to-bracket bolts to 35 ft. lbs. (48 Nm) and the thru bolt to 64 ft. lbs. (87 Nm).

    c. Torque the right engine stay bolts to 31 ft. lbs. and the left engine stay bolts to 15 ft. lbs. (21 Nm).

    d. Torque the front and rear engine mounting bracket and insulator fasteners to 57 ft. lbs. (77 Nm).

    e. Torque the engine center member bolts to 38 ft. lbs. (52 Nm) and the center member-to-insulator bolts to 47 ft. lbs. (64 Nm).

    f. Torque the front and rear engine mounting through bolts to 64 ft. lbs. (87 Nm).

34. Connect the battery cable, refill all fluids, start the engine and check for leaks.

CRESSIDA

1. Disconnect the negative battery cable.
2. Drain the cooling system.
3. Remove the hood.
4. Remove the battery and tray.
5. Disconnect the accelerator, throttle and cruise control cables.
6. Remove the air cleaner assembly complete with the air flow meter, hoses and connector pipe.
7. Tag and disconnect all electrical wires and vacuum hoses necessary to remove the engine.
8. Remove the radiator.
9. Remove the drive belt and unbolt the air conditioning compressor. Position it aside and suspend it with wire. Do not disconnect the refrigerant lines.
10. Unbolt the power steering pump. Position it aside and suspend it with wire. Do not disconnect the hydraulic lines.
11. Remove the windshield washer fluid reservoir.
12. Remove the glove box and disconnect the 6 connectors from the main wiring harnerss and then pull the main wiring harness through the firewall and into the engine compartment.
13. Disconnect the heater hoses.
14. Raise the vehicle and support it safely.
15. Remove the engine under cover and drain the oil.
16. Disconnect the exhaust pipe at the manifold.
17. Disconnect the driveshaft at the transmission flange and position it aside.
18. Disconnect the speedometer cable and the transmission linkage.
19. Disconnect the starter lead and the ground lines at the stiffener plate and left side engine mount.
20. Disconnect and plug the fuel lines.
21. Remove the front wheels and then disconnect the power steering rack. Leave the hydraulic lines attached and lay the rack aside.
22. Loosen the 8 bolts and the ground strap and then remove the rear engine support.
23. Lower the vehicle and remove the 4 engine mount-to-suspension bolts. Attach an engine hoist to the 2 engine hangers and then slowly and carefully lift the engine out of the vehicle.

To install:

24. Lower the engine into the vehicle with the help of an assistant.
25. Connect all electrical and hoses fittings.
26. Install all necessary components.
27. During installation, observe the following torque specifications:

    a. Torque the left mounting

bracket-to-transaxle case bolts to 38 ft. lbs. (52 Nm).

b. Torque the left mounting insulator-to-bracket bolts to 35 ft. lbs. (48 Nm) and the thru bolt to 64 ft. lbs. (87 Nm).

c. Torque the right engine stay bolts to 31 ft. lbs. and the left engine stay bolts to 15 ft. lbs. (21 Nm).

d. Torque the front and rear engine mounting bracket and insulator fasteners to 57 ft. lbs. (77 Nm).

e. Torque the engine center member bolts to 38 ft. lbs. (52 Nm) and the center member-to-insulator bolts to 47 ft. lbs. (64 Nm).

f. Torque the front and rear engine mounting through bolts to 64 ft. lbs. (87 Nm).

28. Connect the battery cable, refill all fluids, start the engine and check for leaks.

Cylinder Head

REMOVAL & INSTALLATION

2VZ-FE Engine

1. Disconnect the negative battery cable and drain the cooling system.

2. Disconnect the throttle cable at the throttle body. If equipped with cruise control, remove the cruise control actuator and vacuum pump.

3. Remove the air cleaner hose.

4. Raise the vehicle and support safely. Remove the engine undercovers.

5. Remove the lower suspension crossmember and the front exhaust pipe.

6. Remove the alternator. Remove the ISC valve.

7. Remove the throttle body, EGR pipe, EGR valve and vacuum modulator.

8. Remove the vacuum pipe and the distributor.

9. Remove the exhaust crossover pipe. Disconnect the cold start injector and then remove the injector tube.

10. Tag and disconnect all hoses leading to the air intake chamber and then remove the chamber.

11. Remove the fuel delivery pipes and the injectors.

12. Disconnect the water temperature sensor and remove the upper radiator hose. Remove the water outlet. Remove the water bypass outlet.

13. Loosen the 2 bolts and remove the cylinder head rear plate.

14. Remove the intake and exhaust manifolds.

15. Remove the timing belt, camshaft pulleys and the No. 2 idler pulley.

16. Remove the No. 3 timing belt cover.

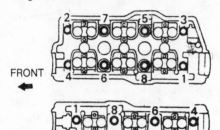

Remove the 2 hex head bolts—2VZ-FE engine

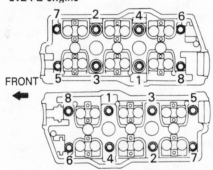

Cylinder head bolt loosening sequence—2VZ-FE engine

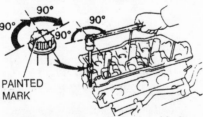

Cylinder head bolt tightening sequence—2VZ-FE engine

Angle torquing the cylinder head bolts—2VZ-FE engine

17. Remove the cylinder head covers and then remove the camshafts.

18. Remove the 2 hex head cylinder head bolts. Loosen and remove the remaining cylinder head bolts, in several stages, in the proper sequence. Remove the cylinder heads from the block.

To install:

19. Install new cylinder head gaskets on the block and then position the cylinder heads.

20. Torque the cylinder head bolts as follows:

a. Install the regular (12-sided) cylinder head bolts and tighten them to 25 ft. lbs. (34 Nm) in the the proper sequence.

b. Mark the front of each bolt with a dab of paint and then tighten each bolt, in order, an additional 90 degrees (the dab of paint will be at the 3 o'clock position.

c. Retighten all bolts, in sequence, an additional 90 degrees so the paint dab is now at 6 o'clock.

d. Coat the threads of the 2 hex head bolts with engine oil and install them. Tighten each bolt to 13 ft. lbs. (18 Nm).

21. Installation of the remaining components. Start the engine and check for leaks.

3A-C, 4A-C and 4A-LC Engines

EXCEPT 1988 COROLLA FX

1. Disconnect the negative battery cable. Remove the exhaust pipe from the manifold. Drain the cooling system.

2. Remove the air cleaner. Tag and disconnect all hoses necessary to remove the cylinder head.

3. Remove all linkages from the carburetor. Disconnect and plug the fuel lines at the cylinder head and manifold.

4. Remove the fuel pump, carburetor and intake manifold.

5. Remove the cylinder head cover. Note the position of the spark plug wires and remove them. Remove the spark plugs.

6. Set the No. 1 cylinder to TDC of its compression stroke. This is accomplished by removing the No. 1 spark plug, placing finger over the hole and then turning the crankshaft pulley until pressure is exerted against finger.

7. Remove the crankshaft pulley with the proper tool. Remove the water pump pulley. Remove the top and bottom timing cover.

8. Matchmark the camshaft pulley and timing belt for reassembly. Loosen the belt tensioner. Remove the water pump.

9. Remove the timing belt. Do not bend, twist, or turn the belt inside out.

10. Remove the rocker arm bolts and remove the rocker arms. Remove the camshaft pulley by holding the camshaft with a suitable tool and removing the belt in the pulley end of the shaft. Do not hold the cam on the lobes, as damage will result.

11. Remove the camshaft seal. Remove the camshaft bearing caps and set them down in the order they appear on the engine. Remove the camshaft.

12. Loosen the head bolts in the re-

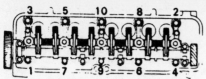

Cylinder head bolt loosening sequence—
3A-C, 4A-C and 4A-LC engines

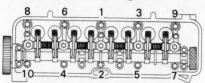

Cylinder head bolt tightening sequence—
3A-C, 4A-C and 4A-LC engines

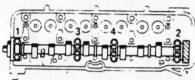

Camshaft bearing cap loosening
sequence—3A-C, 4A-C and 4A-LC
engines

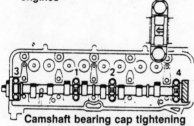

Camshaft bearing cap tightening
sequence—3A-C, 4A-C and 4A-LC
engines

verse order of the torque sequence. Lift the head directly up. Do not attempt to slide it off.

To install:

13. Clean the gasket mating surfaces. Use care not to damage the aluminum components. Lower the cylinder head onto the engine with the help of an assistant. Make sure the dowel pins are aligned and no hoses or wires are between the head and cylinder block.

14. During installation, observe the following torques:

Cam bearing caps 8–10 ft. lbs. (11–14 Nm).

Cam sprocket 29–39 ft. lbs. (39–53 Nm).

Crankshaft pulley 55–61 ft. lbs. (75–83 Nm)

Manifold bolts 15–21 ft. lbs. (20–29 Nm).

Rocker arm bolts 17–19 ft. lbs. (23–26 Nm).

Timing gear idler bolt 22–32 ft. lbs. (30–44 Nm).

Adjust belt tension 0.24–0.28 in. (6–7mm).

Adjust the valves.

15. Connect all electrical and hoses fittings.

16. Install all necessary components.

17. Connect the battery cable, refill

all fluids, start the engine and check for leaks.

1988 COROLLA FX

1. Disconnect the negative battery cable. Raise the vehicle and support safely. Remove the engine under cover. Drain the cooling system and engine oil.

2. Disconnect the exhaust pipe at the manifold. Remove the air cleaner assembly.

3. If equipped with automatic transaxle, disconnect the accelerator and throttle cables. Disconnect and plug the fuel lines at the fuel pump.

4. Disconnect the water hose and remove the water outlet housing. Disconnect the water hoses at the cylinder head. Disconnect the water pump pulley.

5. If equipped with power steering, remove the pump stay. Remove the distributor and spark plugs.

6. Remove the cylinder head cover and gasket. Lift out the half circle plug.

7. Remove the No. 1 and No. 2 timing belt covers. With the engine at TDC of the compression stroke, matchmark the timing belt to the camshaft timing pulley and then slide it off the pulley. Be sure to secure the bottom of the belt so as not to lose valve timing.

8. Loosen each rocker arm support bolt a little at a time and in sequence. Remove the rocker arms.

9. Secure the camshaft, using the proper tool, and remove the camshaft timing pulley. Measure the camshaft thrust clearance. With the camshaft still secure, loosen the distributor drive gear bolt. Loosen the camshaft bearing cap bolts gradually and in sequence. Remove the camshaft.

10. Loosen and remove the cylinder head bolts gradually, in several stages. Remove the cylinder head.

To install:

11. Position the cylinder head on the block with a new gasket and tighten the head bolts in several stages, in the order shown, to a final torque of 43 ft. lbs. (59 Nm).

12. Install the distributor drive gear and plate washer to the camshaft, coat the bearing journals with clean engine oil and position the cam into the head. Position the bearing caps over each journal with the arrow pointing forward. Install a new oil seal and then tighten the cap bolts to 9 ft. lbs. (13 Nm) in the correct order.

13. Install the camshaft timing pulley and tighten the bolt to 34 ft. lbs. (47 Nm).

14. Install and tighten the rocker arm support bolts gradually in 3 passes and in the proper sequence. Tighten to 18 ft. lbs. (25 Nm).

NOTE: Loosen the rocker arm adjusting screw before installation.

15. Align the mark on the No. 1 camshaft bearing cap with the small hole in the timing pulley and install the timing belt so the marks made earlier are in alignment.

16. Connect all electrical and hoses fittings.

17. Install all necessary components.

18. Connect the battery cable, refill all fluids, start the engine and check for leaks.

3E Engine

1. Disconnect the negative battery cable.

2. Drain the cooling system.

3. On 1990–92 3E engine, remove the air cleaner assembly.

4. Remove the right engine under cover.

5. If equipped with power steering, remove the power steering pump and bracket. If equipped with air conditioning and without power steering, remove the idler pulley bracket.

6. Disconnect the radiator hoses. Disconnect the accelerator cable. If equipped with automatic transaxle, disconnect the throttle cable from the bracket mounted to the transaxle case.

7. Remove the timing belt and camshaft timing pulley. Disconnect the heater inlet hose. Disconnect and plug the fuel lines.

8. Remove the air suction hose and valve assembly. Disconnect the brake booster hose from the intake manifold. Disconnect the water inlet hose. Disconnect the intake manifold water hose from the intake manifold.

9. Tag and disconnect all electrical wires, vacuum lines and cables that will interfere with cylinder head removal.

10. Remove the EVAP, VSV and the No. 2 cold enrichment breaker valves. Disconnect the water bypass hoses from the carburetor. Remove the valve cover.

11. Disconnect the exhaust pipe. Remove the intake manifold stay and ground strap. Remove the wire harness clamp bolt from the intake manifold.

12. Measure the cylinder head camshaft thrust clearance using a dial indicator gauge. Standard clearance should be 0.0031–0.071 in. Maximum clearance should be 0.0098 in. If not within specification replace defective parts as required.

13. Loosen then remove the cylinder head bolts in 3 phases and in the proper sequence. Remove the cylinder head from the engine.

14. Installation is the reverse of the removal procedure. During installa-

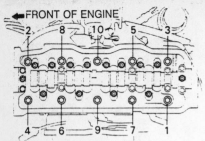

Cylinder head bolt loosening sequence— 3E and 3E-E engines

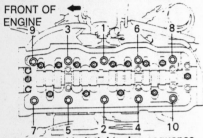

Cylinder head bolt tightening sequence— 3E and 3E-E engines

tion, use a new head gasket. Torque the cylinder head bolts as follows:

a. Tighten the cylinder head bolts in sequence to 22 ft. lbs. (29 Nm).

b. Tighten the bolts is sequence again to 36 ft. lbs. (49 Nm).

c. Retighten each bolt an additional 90 degree turn.

3E-E Engine

1. Disconnect the negative battery cable.
2. Remove the right engine under cover.
3. Drain the cooling system.
4. Disconnect the accelerator and throttle cables.
5. Remove the PCV hoses.
6. Remove the air cleaner and air intake collector assembly.
7. If equipped with power steering, remove the power steering pump and bracket. If equipped with air conditioning and without power steering remove the idler pulley bracket.
8. Remove the pulsation damper; disconnect the fuel inlet and return hoses from the delivery pipe. Plug the hoses to prevent fuel leakage.
9. Disconnect the radiator hoses.
10. Disconnect the heater and water inlet hoses.
11. From the water inlet pipe, disconnect the water bypass hose that connects to the auxiliary air valve.
12. Tag and disconnect all electrical wire and vacuum hoses that interfere with removal of the cylinder head.
13. Remove the exhaust pipe stay and disconnect the exhaust pipe from the manifold.
14. Remove the intake manifold stay.
15. Remove the timing belt and the camshaft timing pulley.

16. Remove the valve cover.
17. Loosen then remove the cylinder head bolts in 3 phases and in the proper sequence. Remove the cylinder head from the engine.

To install:

18. Clean the gasket mating surfaces. Use care not to damage the aluminum components. Lower the cylinder head onto the engine with the help of an assistant. Make sure the dowel pins are aligned and no hoses or wires are between the head and cylinder block.
19. During installation, observe the following torques:

a. Tighten the cylinder head bolts in sequence to 22 ft. lbs. (29 Nm).

b. Tighten the bolts is sequence again to 36 ft. lbs. (49 Nm).

c. Retighten each bolt an additional 90 degree turn each.

20. Connect all electrical and hoses fittings.
21. Install all necessary components.
22. Connect the battery cable, refill all fluids, start the engine and check for leaks.

3S-FE Engine

1. Disconnect the negative battery cable.
2. Drain the cooling system.
3. If equipped with automatic transaxle, disconnect the throttle cable and bracket from the throttle body.
4. Disconnect the accelerator cable and bracket from the throttle body and intake chamber. If equipped with cruise control, remove the actuator and bracket.
5. Remove the air cleaner hose and the alternator.
6. Remove the oil pressure gauge, engine hangers and alternator upper bracket. Raise the vehicle and support safely. Remove the right wheel and tire assembly.
7. Remove the right under cover. Remove the suspension lower crossmember. Disconnect the exhaust pipe from the catalytic converter. Separate the exhaust pipe from the catalytic converter.
8. Disconnect the water temperature sender gauge connector, water temperature sensor connector, cold start injector time switch connector, upper radiator hose, water hoses, and the emission control vacuum hoses.
9. Remove the water outlet and gaskets. Remove the distributor. Remove the water bypass pipe. Remove the EGR valve and modulator.
10. Remove the throttle body assembly. Remove the cold start injector pipe. Remove the air intake chamber air hose, the throttle body air hose, and the power steering pump hoses, if equipped. Remove the air tube.
11. Remove the intake manifold re-

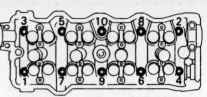

Cylinder head bolt LOOSENING sequence – 3S-FE engines

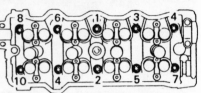

Cylinder head bolt tightening sequence – 3S-FE engines

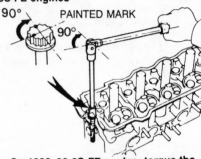

On 1990–92 3S-FE engine, torque the cylinder head bolts an additional 90 degrees in sequence.

taining bolts. Remove the intake manifold. Remove the fuel delivery pipe and the injectors. Remove the spark plugs.

12. Remove the camshaft timing pulley. Remove the No. 1 idler pulley and tension spring. Remove the No. 3 timing belt cover. Properly support the timing belt so meshing of the crankshaft timing pulley does not occur and the timing belt does not shift.
13. Remove the cylinder head cover. Arrange the grommets in order so they can be reinstalled in the correct order.
14. Remove the camshafts.
15. Loosen, then remove the cylinder head bolts in 3 phases and in the proper sequence. Remove the cylinder head from the engine.

To install:

16. Clean the gasket mating surfaces. Use care not to damage the aluminum components. Lower the cylinder head onto the engine with the help of an assistant. Make sure the dowel pins are aligned and no hoses or wires are between the head and cylinder block.
17. During installation, observe the following torques:

a. Apply a light coat of clean engine oil to the threads of the head bolts prior to installation.

b. On 1988–89 vehicles, torque the cylinder head to specification and in 3 phases to 47 ft. lbs. (64 Nm).

c. On 1990–92 vehicles, torque the cylinder head to specification and in 3 phases to 47 ft. lbs. (64 Nm). Then, mark the front of each cylinder head bolt with a dab of paint. Finally, re-torque the cylinder head bolts an additional 90 degrees in the proper sequence.

18. Connect all electrical and hoses fittings.

19. Install all necessary components.

20. Connect the battery cable, refill all fluids, start the engine and check for leaks.

3S-GE Engine

1. Disconnect the negative battery cable. Drain the cooling system.

2. Tag and disconnect the ignition coil connector and the spark plug wire at the ignition coil. Remove the 4 nuts and 2 bolts and lift out the upper suspension brace.

3. If equipped with automatic transaxle, disconnect the throttle cable with its bracket from the throttle body. Disconnect the accelerator cable from the throttle body. Remove the radiator overflow tank.

4. If equipped with cruise control, remove the cruise control actuator and its bracket.

5. Disconnect the air flow meter connector. Remove the air cleaner cap clips. Loosen the hose clamp and remove the air cleaner hose and the air flow meter along with the air cleaner top. Lift out the filter element and then remove the air cleaner case.

6. Tag and disconnect the oxygen sensor lead. Remove the 4 mounting bolts and remove the exhaust manifold heat insulator. Remove the alternator and bracket.

7. Raise the vehicle and support safely. Remove the right front wheel and tire assembly.

8. Remove the right side engine under cover and remove the lower suspension crossmember.

9. Disconnect the exhaust pipe at the manifold. Remove the exhaust manifold stay and the EGR pipe. Unbolt the manifold and remove it along with the lower heat insulator.

10. Remove the distributor. Disconnect the oil pressure switch connector.

11. Tag and disconnect all electrical leads and vacuum hoses at the water outlet. Remove the upper radiator hoses, the heater outlet hose and the water bypass hose. Remove the water outlet.

12. Disconnect the heater inlet hose and the water bypass hose and then remove the water bypass pipe.

13. Disconnect the throttle position sensor lead, the ventilation hose, the air valve hose and any emission control vacuum hoses at the throttle body.

Remove the 4 bolts and lift out the throttle body.

14. Remove the forward engine hanger and the No. 2 intake manifold stay. Remove the EGR vacuum modulator.

15. Tag and disconnect any remaining vacuum hoses which may interfere with cylinder head removal. Tag and disconnect the fuel injector electrical leads at the injector.

16. Disconnect the fuel inlet hose at the fuel filter. Disconnect the fuel return hose at the return pipe.

17. Remove the No. 1 and No. 3 intake manifold stays. Tag and disconnect the 2 VSV connectors. Disconnect the 2 power steering vacuum hoses. Remove the intake manifold and the air control valve.

18. Remove the fuel delivery pipe with the injectors attached. Pull the 4 injector insulators out of the injector holes in the cylinder head.

19. Remove the cylinder head cover. Remove the spark plugs. Remove the No. 1 engine hanger.

20. Remove the power steering reservoir and position it aside with the hydraulic lines still attached.

21. Remove the camshaft timing pulleys. Remove the No. 1 idler pulley and tension spring.

22. Remove the bolt holding the No. 2 and No. 3 timing covers. Remove the 4 mounting bolts and remove the No. 3 timing cover.

23. Loosen and remove the camshaft bearing caps, in several stages, and in the proper sequence. Lift out the camshafts and the oil seal. When removing the camshaft bearing caps, keep them in the proper order.

24. Loosen and remove the cylinder head bolts, in several stages, and in the proper sequence. Remove the cylinder head.

To install:

25. Position the cylinder head onto the cylinder block with a new gasket. Lightly coat the cylinder head bolts with engine oil, install them into the head and tighten them in several passes, in the proper sequence, to 40 ft. lbs. (53 Nm).

26. Position the camshafts into the cylinder head so the No. 1 cam lobes are facing outward.

27. Apply silicone sealant to the outer edge of the mating surface on the No. 1 bearing cap only. Position the bearing caps over each journal with the arrows pointing forward and in numerical order from the front to the rear.

28. Lightly coat the cap bolt threads with engine oil. Tighten them in several stages, and in the proper sequence, to 14 ft. lbs. (19 Nm).

29. Check the camshaft thrust clearance. Coat the inside of a new oil seal

with grease and carefully tap it onto the camshaft with a suitable drift. Install the No. 3 timing belt cover.

30. Connect the idler pulley tension spring to the pulley and the pin on the cylinder head. Install the idler pulley onto the pivot pin, force it to the left as far as it will go and tighten it. Make sure the tension spring is not out of the groove in the pin. Install the camshaft timing pulleys and the timing belt.

31. Installation of the remaining components is in the reverse order of removal. Tighten the lower suspension crossmember end bolts to 154 ft.

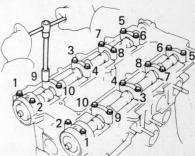
Remove the camshaft bearing caps in the order – 3S-GE and 3S-GTE engines

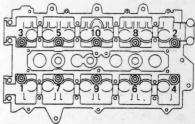
Cylinder head bolt loosening sequence – 3S-GE and 3S-GTE engines

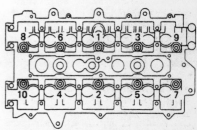
Cylinder head bolt tightening sequence – 3S-GE and 3S-GTE engines

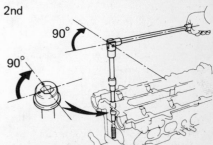
Mark the head bolts prior to angle-torquing – 3S-GE and 3S-GTE engines

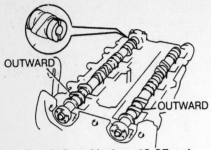

Camshaft positioning—3S-GE and 3S-GTE engines

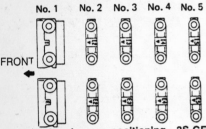

Camshaft bearing cap positioning—3S-GE and 3S-GTE engines

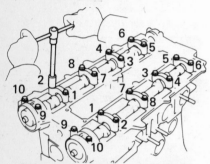

Tighten the bearing cap bolts in this order—3S-GE and 3S-GTE engines

lbs. (209 Nm) and the center bolt to 29 ft. lbs. (39 Nm). Tighten the upper suspension brace bolts to 15 ft. lbs. (20 Nm) and the nuts to 47 ft. lbs. (64 Nm). Refill the engine with coolant. Check the idle speed and ignition timing.

3S-GTE Engine

CELICA

1. Disconnect the negative battery cable. Drain the coolant from the engine and intercooler.

2. Remove the upper suspension brace that runs between the strut towers. Disconnect the accelerator cable at the throttle body. Remove the radiator reservoir tank.

3. Remove the air cleaner assembly. Remove the alternator. Raise the vehicle and support it safely.

4. Remove the right wheel and tire assembly and engine undercovers.

5. Remove the front exhaust pipe and catalytic converter. Remove the alternator brackets.

6. Remove the turbocharger, exhaust manifold and distributor.

7. Disconnect the air hose and remove the No. 2 air pipe.

8. Remove the left engine hanger along with the reservoir tank. Remove the oil pressure switch.

9. Remove the water outlet housing and the water bypass pipe.

10. Remove the throttle body and disconnect the cold start injector lead.

11. Remove the EGR valve, vacuum modulator and EGR control Vacuum Switching Valve (VSV). Remove the delivery pipe and all injectors. Remove the vacuum pipe.

12. Remove the intake manifold stays and the No. 1 air pipe.

13. Disconnect the vacuum hose and remove the fuel pressure VSV. Remove the T-VIS Vacuum Switching (VSV), vacuum tank and the turbo pressure VSV.

14. Remove the intake manifold with the air control valve. Remove the power steering reservoir tank without disconnecting the hydraulic hoses. Position the pump aside.

15. Remove the spark plugs and the No. 2 front cover. Remove the timing belt and the PCV pipe.

16. Remove the cylinder head cover. Remove the camshaft timing pulleys and the No. 1 idler pulley. Remove the No. 3 timing belt cover.

NOTE: Secure the timing belt so the belt does not unmesh from the cranskahsft pulley.

17. Gradually loosen and remove the camshaft bearing cap bolts, in several passes, in the proper sequence. Remove the bearing caps, oil seals and lift out the camshafts.

18. Remove the right rear engine hanger. Remove the cylinder head bolts, in several stages, in the sequence and lift off the cylinder head.

To install:

19. Position the cylinder head and a new gasket on the block and torque the bolts as follows:

   a. Coat the head bolts with engine oil and tighten them in several passes, in sequence, to 40 ft. lbs. (54 Nm).

   b. Mark the front of each bolt with a dab of paint.

   c. Retighten the bolts an additional 90 degrees turn. The paint dabs should now all be at a 90 degree angle to the front of the head.

20. Install the right rear engine hanger and tighten it to 14 ft. lbs. (19 Nm).

21. Position the camshafts in the cylinder head with the No. 1 lobes facing outward. Coat the No. 1 bearing cap with seal packing and then install all the caps over the bearing journals. Coat the bearing cap bolts with engine oil and then tighten them to 14 ft. lbs. (19 Nm) in several stages, in the order

shown. Grease 2 new oil seals and install them into the camshafts.

22. Install the No. 3 timing belt cover and the No. 1 idler pulley. Install the camshaft timing pulleys.

23. Install the cylinder head cover and the timing belt.

24. Install the remaining components, start the engine and check for leaks.

1990–92 MR2

1. Disconnect the negative battery cable.

2. Drain the coolant from the engine and intercooler.

3. Tag and disconnect all hoses, lines and wiring that interfere with removal of the turbocharger, exhaust manifold, intake manifold and cylinder head.

4. Remove the engine hood side panels.

4. Remove the upper suspension brace that runs between the strut towers.

5. Disconnect the accelerator cable at the throttle body.

6. If equipped with cruise control, remove the cruise control actuator and disconnect the accelerator linkage.

7. Remove the air cleaner cap.

8. Remove the right front engine hanger.

9. Remove the intercooler.

10. Remove the front exhaust pipe, catalytic converter and turbocharger.

11. Remove the throttle body and cold start injector.

12. Remove the exhaust manifold and distributor.

13. Remove the No. 2 air tube.

14. Remove the left engine hanger.

15. Remove the EGR vacuum modulator and Vacuum Switching Valve (VSV).

16. Remove the vacuum pipe, EGR valve and EGR pipe.

17. Remove the water outlet and housing.

18. Remove the oil pressure switch.

19. Remove the oil cooler.

20. Remove the water bypass pipe.

21. Remove the intake manifold stays and the No. 1 air pipe.

22. Remove the T-VIS Vacuum Switching (VSV), vacuum tank and the turbocharger pressure VSV.

23. Remove the intake manifold with the air control valve.

24. Remove the delivery pipe and fuel injectors.

25. Remove the cylinder head cover.

26. Remove the camshaft timing pulleys and the No. 1 idler pulley.

27. Remove the No. 3 timing belt cover.

NOTE: Secure the timing belt so the belt does not unmesh from the cranskahsft pulley.

28. Gradually loosen and remove the camshaft bearing cap bolts in several passes, in the proper sequence. Remove the bearing caps, oil seals and lift out the camshafts.

29. Remove the cylinder head bolts in several stages, in the the proper sequence and lift off the cylinder head from the alignment dowels on the block.

To install:

30. Position the cylinder head and a new gasket on the block.

31. Torque the cylinder head bolts as follows:

    a. Coat the head bolts with engine oil and tighten them in several passes, in sequence to 36 ft. lbs. (49 Nm).

    b. Mark the front of each bolt with a dab of paint.

    c. Retighten the bolts an additional 90 degrees turn. The paint dabs should now all be at a 90 degree angle to the front of the head.

32. Position the camshafts in the cylinder head with the No. 1 lobes facing outward. Coat the No. 1 bearing cap with seal packing and then install all the caps over the bearing journals. Coat the bearing cap bolts with engine oil and then tighten them to 14 ft. lbs. (19 Nm) in several stages, in the order shown. Grease 2 new oil seals and install them into the camshafts.

33. Check and adjust the valve clearance, as necessary.

34. Install the No. 3 timing belt cover, No. 1 idler pulley and camshaft timing pulleys.

35. Install the cylinder head cover with 2 new gaskets and 12 new bolt seal washers. Torque the cover bolts to 21 inch lbs. (2.5 Nm).

36. Install the remaining components, start the engine and check for leaks.

4A-F Engine

1. Disconnect the negative battery cable. Drain the cooling system.

2. Remove the engine undercover and then disconnect the exhaust pipe at the manifold.

3. Remove the air cleaner and hoses. If equipped with automatic transaxle, disconnect the accelerator and throttle cables at the bracket.

4. Tag and disconnect all wires, lines and hoses that may interfere with removal of the exhaust manifold, intake manifold and cylinder head.

5. Disconnect the fuel lines at the fuel pump. Disconnect the heater hoses at the engine.

6. Disconnect the water hose and the bypass hose at the rear of the cylinder head. Remove the 2 bolts and pull off the water outlet pipe.

7. Remove the 2 mounting bolts and lift out the exhaust manifold stay. Remove the upper manifold insulator and then remove the exhaust manifold.

8. Remove the distributor.

9. Disconnect the 2 water hoses at the water inlet (front of head) and then remove the inlet housing.

10. Remove the fuel pump.

11. Disconnect the PCV and water hoses at the intake manifold. Remove the manifold stay and then remove the intake manifold and wire clamp.

12. Remove the drive belts and the power steering pump support.

13. Remove the spark plugs and the cylinder head cover.

14. Remove the No. 3 and No. 2 front covers. Turn the crankshaft pulley and align its groove with the **0** mark on the No. 1 front cover. Check that the camshaft pulley hole aligns with the mark on the No. 1 camshaft bearing cap (exhaust side).

15. Remove the plug from the No. 1 front cover and matchmark the timing belt to the camshaft pulley. Loosen the idler pulley mounting bolt and push the pulley to the left as far as it will go; tighten the bolt. Slide the timing belt off the camshaft pulley and support it so it won't fall into the case.

16. Remove the camshaft pulley and check the camshaft thrust clearance. Remove the camshafts.

17. Gradually loosen the cylinder head mounting bolts in several passes, in the sequence. Remove the cylinder head.

To install:

18. Clean the gasket mating surfaces. Use care not to damage the aluminum components. Lower the cylinder head onto the engine with the help of an assistant. Make sure the dowel pins are aligned and no hoses or wires are between the head and cylinder block.

19. Lightly coat the cylinder head bolts with engine oil and then install them. Tighten the bolts in 3 stages, in the proper sequence. On the final pass, torque the bolt to 44 ft. lbs. (60 Nm).

19. Position the camshafts into the cylinder head. Position the bearing caps over each journal with the arrows pointing forward.

20. Tighten each bearing cap a little at a time and in the reverse of the removal sequence. Tighten to 9 ft. lbs. (13 Nm) Recheck the camshaft endplay.

21. Install the camshaft timing pulleys making sure the camshaft knock pins and the matchmarks are in alignment. Lock each camshaft and tighten the pulley bolts to 43 ft. lbs. (59 Nm).

22. Align the matchmarks made during removal and then install the timing belt on the camshaft pulley. Loosen the idler pulley set bolt. Make sure the timing belt meshing at the crankshaft pulley does not shift.

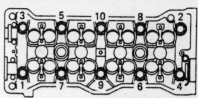

Cylinder head bolt loosening sequence—4A-F and 4A-FE engines (Corolla and Celica)

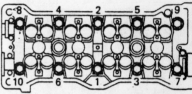

Cylinder head bolt tightening sequence—4A-F and 4A-FE engines (Corolla)

23. Rotate the crankshaft clockwise 2 revolutions from TDC to TDC. Make sure each pulley aligns with the marks made previously. If the marks are not in alignment, the valve timing is wrong. Shift the timing belt meshing slightly and then repeat Steps 21–23.

24. Tighten the set bolt on the timing belt idler pulley to 27 ft. lbs. (37 Nm). Measure the timing belt deflection at the top span between the 2 camshaft pulleys. It should deflect no more than 0.16 in. at 4.4 lbs. of pressure. If deflection is greater, readjust by using the idler pulley.

25. Install the remaining components, start the engine and check for leaks.

4A-FE Engine

COROLLA

1. Disconnect the negative battery cable at the battery. Drain the cooling system.

2. Remove the engine undercover and then disconnect the exhaust pipe at the manifold.

3. Remove the air cleaner and hoses; disconnect the intake air temperature sensor. Disconnect the accelerator and throttle cables at the bracket on vehicles with automatic transaxle.

4. Remove the cruise control actuator cable.

5. Tag and disconnect all wires, lines and hoses that may interfere with exhaust manifold, intake manifold and cylinder head removal.

6. Disconnect the fuel lines at the fuel pump. Disconnect the heater hoses at the engine.

7. Disconnect the water hose and the bypass hose at the rear of the cylinder head. Remove the 2 bolts and pull off the water outlet pipe.

8. Remove the 2 mounting bolts and lift out the exhaust manifold stay. Remove the upper manifold insulator and then remove the exhaust manifold.

9. Remove the distributor.

10. Disconnect the 2 water hoses at the water inlet (front of head) and then remove the inlet housing.

11. Disconnect the PCV, fuel return and vacuum sensing hoses.

12. Remove the fuel inlet pipe and the cold start injector pipe. Disconnect the 4 vacuum hoses and then remove the EGR vacuum modulator.

13. Remove the fuel delivery pipe along with the injectors, spacers and insulators.

14. Unbolt the engine wire cover at the intake manifold and then disconnect the wire at the cylinder head.

15. Remove the intake manifold.

16. Remove the drive belts and then remove the water pump.

17. Remove the spark plugs, cylinder head cover and semi-circular plug.

18. Remove the No. 3 and No. 2 front covers. Turn the crankshaft pulley and align its groove with the **0** mark on the No. 1 front cover. Check that the camshaft pulley hole aligns with the mark on the No. 1 camshaft bearing cap (exhaust side). If not, rotate the crankshaft 360 degrees until the marks are aligned.

19. Remove the plug from the No. 1 front cover and matchmark the timing belt to the camshaft pulley. Loosen the idler pulley mounting bolt and push the pulley to the left as far as it will go; tighten the bolt. Slide the timing belt off the camshaft pulley and support it so it won't fall into the case.

20. Remove the camshaft pulley and check the camshaft thrust clearance. Remove the camshafts.

21. Gradually loosen the cylinder head mounting bolts in several passes, in the the proper sequence. Remove the cylinder head.

NOTE: On 1990–92 vehicles, the cylinder head bolts on the right (intake) side of the cylinder head are 3.54 in. (90mm) and the bolts on the left (exhaust) side of the head are 4.25 in. (108mm). Label the bolts to ensure proper installation.

To install:

22. Position the cylinder head on the block with a new gasket. Lightly coat the cylinder head bolts with engine oil and then install them. Tighten the bolts in 3 stages, in the proper sequence. On the final pass, torque the bolt to 44 ft. lbs. (60 Nm).

23. Position the camshafts into the cylinder head. Position the bearing caps over each journal with the arrows pointing forward.

24. Tighten each bearing cap a little at a time and in the reverse of the removal sequence. Tighten to 9 ft. lbs. (13 Nm) and recheck the camshaft endplay.

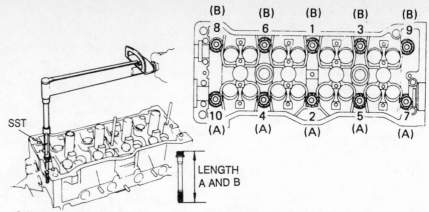

Cylinder head bolt tightening sequence and positioning on 1990–92 4A-FE engine (Corolla and Celica). The "A" bolts are 3.54 in. (90mm) and the "B" bolts are 4.25 in. (108mm)

25. Install the camshaft timing pulleys making sure the camshaft knock pins and the matchmarks are in alignment. Lock each camshaft and tighten the pulley bolts to 43 ft. lbs. (59 Nm).

26. Align the matchmarks made during removal and then install the timing belt on the camshaft pulley. Loosen the idler pulley set bolt. Make sure the timing belt meshing at the crankshaft pulley does not shift.

27. Rotate the crankshaft clockwise 2 revolutions from TDC to TDC. Make sure each pulley aligns with the marks made previously. If the marks are not in alignment, the valve timing is wrong. Shift the timing belt meshing slightly and then repeat Steps 24–26.

28. Tighten the set bolt on the timing belt idler pulley to 27 ft. lbs. (37 Nm). Measure the timing belt deflection at the top span between the 2 camshaft pulleys. It should deflect no more than 0.24 in. (6mm) at 4.4 lbs. of pressure. If deflection is greater, readjust by using the idler pulley.

29. Install the remaining components, start the engine and check for leaks.

1990–92 CELICA

1. Disconnect the negative battery cable. Drain the cooling system.

2. If equipped with automatic transaxle, disconnect the throttle cable and bracket from the throttle body.

3. Disconnect the accelerator cable and bracket from the throttle body.

4. Remove the air cleaner cap and hose.

5. Remove the engine under covers. Remove the suspension lower crossmember.

6. Disconnect all lines, hoses and electrical wires that interfere with exhaust manifold, intake manifold and cylinder head removal.

7. Disconnect the front exhaust pipe, distributor and exhaust manifold.

8. Remove the water outlet and gaskets. Remove the water inlet and inlet housing.

9. Unbolt and remove the power steering pump without disconnecting hoses.

10. Remove the throttle body, cold start injector pipe, cold start injector, delivery pipe and fuel injectors.

11. Remove the Air Control Valve (ACV) assembly. Disconnect engine wiring from the timing belt cover and from the intake manifold.

12. Remove the vacuum pipe, EGR vacuum modulator and EGR Vacuum Switching Valve (VSV) assembly.

13. Remove the EGR valve and gasket. Remove the water inlet pipe and fuel return hose from the fuel filter.

14. Remove the intake manifold with retaining manifold stay (bracket).

15. Remove the valve cover.

16. Remove the camshaft timing pulley, No. 1 idler pulley and tension spring and No. 3 timing belt cover. Properly support the timing belt so meshing of the crankshaft timing pulley does not occur and the timing belt does not shift.

17. Remove the fan belt adjusting bar, engine hangers and power steering drive belt adjusting strut or bracket.

18. Remove the camshafts. Make sure to uniformly loosen and remove bearing cap bolts in several phases and in the proper sequence when removing the camshafts.

19. Loosen then remove the cylinder head bolts in 3 phases and in the proper sequence. Remove the cylinder head from the engine.

NOTE: The cylinder head bolts on the right (intake) side of the cylinder head are 3.54 in. (90mm) and the bolts on the left (exhaust) side of the head are 4.25 in. (108mm). Label the bolts to ensure proper installation.

To install:

20. Install the cylinder head on the

cylinder head block. Place the cylinder head in position on the cylinder head gasket.

21. Apply a light coat of clean engine oil to the threads of the head bolts before installation. Tighten the bolts in 3 stages, in the proper sequence. On the final pass, torque the bolt to 44 ft. lbs. (60 Nm).

22. Installation of the camshafts and remaining components. Start the engine and check for leaks.

4A-GE, 4A-GEC and 4A-GELC Engines

1. Disconnect the negative battery cable. Remove the engine undercover. Drain the cooling system and engine oil.

2. Loosen the clamp and then disconnect the No. 1 air cleaner hose from the throttle body. Disconnect the actuator and accelerator cables from the bracket on the throttle body.

3. If equipped with power steering, Remove the power steering pump and its bracket. Position the pump aside with the hydraulic lines connected.

4. Loosen the water pump pulley set nuts. Remove the drive belt adjusting bolt and then remove the belt. Remove the set nuts and then remove the fluid coupling along with the fan and the water pump pulley.

5. Disconnect the upper radiator hose at the water outlet on the cylinder head. Disconnect the 2 heater hoses at the water bypass pipe and the cylinder head rear plate.

6. Remove the distributor. Remove the cold start injector pipe and the PCV hose from the cylinder head.

7. Remove the pulsation damper from the delivery pipe. Disconnect the fuel return hose from the pressure regulator.

8. Tag and disconnect all vacuum hoses which may interfere with cylinder head remvoval. Remove the wiring harness and the vacuum pipe from the No. 3 timing cover. Tag and disconnect all wires which might interfere with exhaust manifold, intake manifold and cylinder head removal.

9. Disconnect the exhaust bracket from the exhaust pipe. Disconnect the exhaust manifold from the exhaust pipe.

10. Remove the vacuum tank and the VCV valve. Remove the exhaust manifold.

11. Remove the 2 mounting bolts and remove the water outlet housing from the cylinder head with the No. 1 bypass pipe and gasket. Pull the No. 1 bypass pipe out of the housing.

12. Remove the fuel delivery pipe along with the fuel injectors.

13. Remove the intake manifold stay. Remove the intake manifold along with the air control valve.

14. Remove the cylinder head covers and their gaskets. Remove the spark plugs. Remove the No. 1 and No. 2 timing belt covers and their gaskets.

15. Rotate the crankshaft pulley until its groove is in alignment with the **0** mark on the No. 1 timing belt cover. Check that the valve lifters on the No. 1 cylinder are loose. If not, rotate the crankshaft 1 complete revolution (360 degrees).

16. Place matchmarks on the timing belt and 2 timing pulleys. Loosen the idler pulley bolts and move the pulley to the left as far as it will go and then retighten the bolt.

17. Remove the timing belt from the camshaft pulleys. When removing the timing belt, support the belt so the meshing of the crankshaft timing pulley and the timing belt does not shift. Never drop anything inside the timing case cover. Be sure the timing belt does not come in contact with dust or oil.

18. Lock the camshafts and remove the timing pulleys. Remove the No. 4 timing belt cover.

19. Using a dial indicator, measure the endplay of each camshaft. If not within specification, replace the thrust bearing.

20. Loosen each camshaft bearing cap bolt a little at a time and in the correct sequence. Remove the bearing caps, camshaft and oil seal.

21. Loosen the cylinder head bolts gradually in 3 stages, and in the proper order using the proper tool.

22. Remove the cylinder head.

NOTE: On 1990–92 engines, the cylinder head bolts on the right (intake) side of the cylinder head are 3.54 in. (90mm) and the bolts on the left (exhaust) side of the head are 4.25 in. (108mm). Label the bolts to ensure proper installation.

To install:

23. Position the cylinder head on the block with a new gasket. Lightly coat the cylinder head bolts with engine oil and then install the short head bolts on the intake side and the long ones on the exhaust side. Tighten the bolts in 3 stages, in the proper sequence. On the final pass, torque the bolt to 43 ft. lbs.

24. On 1988–92 engines, coat the head bolts with engine oil and tighten them in several passes, in the sequence shown to 22 ft. lbs. (29 Nm). Mark the front of each bolt with a dab of paint and then retighten the bolts a further 90 degree turn. The paint dabs should now all be at a 90 degree angle to the front of the head. Retighten the bolts one more time a further 90 degree turn. The paint dabs should now all be pointing toward the rear of the head.

25. Position the camshafts into the cylinder head. Position the bearing caps over each journal with the arrows pointing forward.

26. Tighten each bearing cap a little at a time and in the reverse of the removal sequence. Tighten to 9 ft. lbs. (13 Nm). Recheck the camshaft endplay.

27. Drive the camshaft oil seals onto the end of the camshafts using a suitable seal installer. Be careful not to install the oil seals crooked. Install the No. 4 timing belt cover.

28. Install the camshaft timing pulleys making sure the camshaft knock pins and the matchmarks are in alignment. Lock each camshaft and tighten the pulley bolts to 34 ft. lbs. (47 Nm).

29. Align the matchmarks made during removal and then install the timing belt on the camshaft pulley. Loosen the idler pulley set bolt. Make sure the timing belt meshing at the crankshaft pulley does not shift.

30. Rotate the crankshaft clockwise 2 revolutions from TDC to TDC. Make sure each pulley aligns with the marks made previously. If the marks are not in alignment, the valve timing is wrong. Shift the timing belt meshing slightly and then repeat Steps 28–30.

31. Tighten the set bolt on the timing belt idler pulley to 27 ft. lbs. (37 Nm). Measure the timing belt deflection at the top span between the 2 camshaft pulleys. It should deflect no more than 0.16 in. at 4.4 lbs. of pressure. If deflection is greater, readjust by using the idler pulley.

32. Install the remaining components, start the engine and check for leaks.

4A-GZE Engine

1. Disconnect the negative battery cable. Remove the hood and engine undercovers.

2. Drain the engine coolant and remove the intercooler. Remove the battery.

3. Disconnect the air bleeder hose at the water inlet housing. Disconnect the cruise control vacuum hose and the throttle cable.

4. Remove the air flow meter with the No. 3 air cleaner hose. Disconnect the accelerator cable.

5. Remove the accelerator link and disconnect the air conditioning idle-up vacuum hoses.

6. Disconnect the heater hose at the rear of the cylinder head. Disconnect the brake booster vacuum hose and remove the radiator reservoir tank. Disconnect the No. 1 radiator hose at the water outlet housing.

7. Tag and disconnect any remaining hoses, wires or connections which may interfere with exhaust manifold, intake manifold and cylinder head removal.

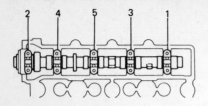

Loosen the camshaft bearing cap bolts in this order—4A-GE (all) and 4A-GZE engines

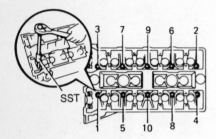

Cylinder head bolt loosening sequence—4A-GE and 4A-GZE engines

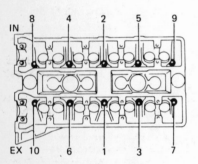

Cylinder head bolt tightening sequence—4A-GE and 4A-GZE engines

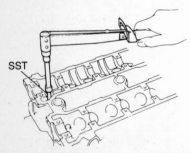

Cylinder head bolt tightening sequence—4A-GE (all) and 4A-GZE engines

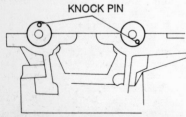

Position the camshafts into the cylinder head as shown—4A-GE (all) and 4A-GZE engines

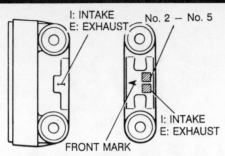

Camshaft bearing cap positioning (the arrows must always point forward)—4A-GE (all) and 4A-GZE engines

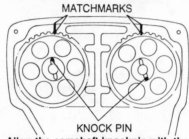

Align the camshaft knockpin with the camshaft timing pulley—4A-GE (all) and 4A-GZE engines

8. Remove all drive belts and then remove the water pump pulley. Remove the supercharger.

9. Remove the No. 2 air outlet duct and the No. 3 fuel pipe. Remove the No. 1 vacuum transmitting pipe.

10. Disconnect the No. 2 fuel hose and remove the cylinder head rear cover.

11. Remove the fuel delivery pipe and the fuel injectors.

12. Remove the water outlet with the bypass pipe. Remove the EGR valve with the pipe still attached.

13. Remove the intake manifold. Remove the front exhaust pipe.

14. Remove the air conditioning compressor and bracket. Remove the distributor and alternator with its bracket.

15. Remove the exhaust manifold.

16. Rotate the crankshaft pulley until its groove is in alignment with the 0 mark on the No. 1 timing belt cover. Check that the valve lifters on the No. 1 cylinder are loose. If not, rotate the crankshaft 1 complete revolution (360 degrees).

17. Place matchmarks on the timing belt and 2 timing pulleys. Loosen the idler pulley bolts and move the pulley to the left as far as it will go and then retighten the bolt.

18. Remove the timing belt from the camshaft pulleys. When removing the timing belt, support the belt so the meshing of the crankshaft timing pulley and the timing belt does not shift. Never drop anything inside the timing case cover. Be sure the timing belt does not come in contact with dust or oil.

19. Lock the camshafts and remove the timing pulleys. Remove the No. 4 timing belt cover.

20. Using a dial indicator, measure the endplay of each camshaft. If not within specification, replace the thrust bearing.

21. Loosen each camshaft bearing cap bolt a little at a time and in the sequence shown. Remove the bearing caps, camshaft and oil seal.

22. Loosen the cylinder head bolts gradually in 3 stages, and in the proper order, using the proper tool.

23. Remove the cylinder head from the vehicle.

To install:

24. Position the cylinder head on the block with a new gasket.

25. Torque the cylinder head bolts as follows:

    a. Lightly coat the cylinder head bolts with engine oil and then install the short head bolts on the intake side and the long ones on the exhaust side.

    b. Tighten them in several passes, in the proper sequence to 22 ft. lbs. (29 Nm).

    c. Mark the front of each bolt with a dab of paint and then retighten the bolts a further 90 degree turn. The paint dabs should now all be at a 90 degree angle to the front of the head.

    d. Retighten the bolts one more time a further 90 degree turn. The paint dabs should now all be pointing toward the rear of the head.

26. Position the camshafts into the cylinder head. Position the bearing caps over each journal with the arrows pointing forward.

27. Tighten each bearing cap a little at a time and in the reverse of the removal sequence. Tighten to 9 ft. lbs. (13 Nm). Recheck the camshaft endplay.

28. Drive the camshaft oil seals onto the end of the camshafts using a suitable seal installer. Be careful not to install the oil seals crooked. Install the No. 4 timing belt cover.

29. Install the camshaft timing pulleys making sure the camshaft knock pins and the matchmarks are in alignment. Lock each camshaft and tighten the pulley bolts to 34 ft. lbs. (47 Nm).

30. Align the matchmarks made during removal and then install the timing belt on the camshaft pulley. Loosen the idler pulley set bolt. Make sure the timing belt meshing at the crankshaft pulley does not shift.

31. Rotate the crankshaft clockwise 2 revolutions from TDC to TDC. Make sure each pulley aligns with the marks made previously. If the marks are not in alignment, the valve timing is wrong. Shift the timing belt meshing slightly and then repeat Steps 28–30.

32. Tighten the set bolt on the timing belt idler pulley to 27 ft. lbs. (37 Nm). Measure the timing belt deflection at the top span between the 2 camshaft pulleys. It should deflect no more than 0.16 in. (4mm). at 4.4 lbs. of pressure. If deflection is greater, readjust by using the idler pulley.

33. Install the remaining components, start the engine and check for leaks.

5M-GE Engine

1. Disconnect the negative battery cable. Drain the cooling system.

2. Disconnect the exhaust pipe from the exhaust manifold.

3. Remove the throttle cable bracket from the cylinder head if equipped with automatic transmission, and remove the accelerator and actuator cable bracket.

4. Tag and disconnect the ground cable, oxygen sensor wire, high tension coil wire, distributor connector, solenoid resistor wire connector and thermo-switch wire, if equipped with automatic transmission.

5. Tag and disconnect the brake booster vacuum hose, EGR valve vacuum hose, fuel hose from the intake manifold and actuator vacuum hose, if equipped with cruise control.

6. Disconnect the upper radiator hose from the thermostat housing, and disconnect the 2 heater hoses.

7. Disconnect the No. 1 air hose from the air intake connector. Remove the 2 clamp bolts, loosen the throttle body hose clamp and remove the air intake connector and the connector pipe.

8. Tag and disconnect all emission control hoses from the throttle body and air intake chamber, the 2 PCV hoses from the cam cover and the fuel hose from the fuel hose support.

9. Remove the air intake chamber stay and the vacuum pipe and ground cable. Remove the bolt that attaches the spark plug wire clip, leaving the wires attached to the clip. Remove the distributor from the cylinder head with the cap and wires attached, by removing the distributor holding bolt.

10. Tag and disconnect the cold start injector wire and disconnect the cold start injector fuel hose from the delivery pipe.

11. Loosen the nut of the EGR pipe, remove the 5 bolts and 2 nuts and remove the air intake chamber and gasket.

12. Remove the glove box and remove the ECU module. Disconnect the 3 connectors and pull the EFI (fuel injection) wire harness out through the engine side of the firewall.

13. Remove the pulsation damper and the No. 1 fuel pipe. Remove the water outlet housing by first loosening the clamp and disconnecting the water bypass hose.

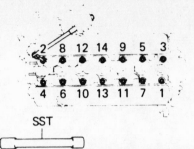

Cylinder head bolt loosening sequence – 5M-GE engine

14. Remove the intake manifold.

15. Disconnect the power steering pump drive belt and remove the power steering pump without disconnecting the fluid hoses. Position the pump aside.

16. Disconnect the oxygen sensor connector and remove the exhaust manifold.

17. Remove the timing belt and camshaft timing gears. Remove the timing belt cover stay, and remove the oil pressure regulator and gasket. Remove the No. 2 timing belt cover and gasket.

18. Tag and disconnect any other wires, linkage and/or hoses still attached to the cylinder head or may interfere with its removal.

19. Carefully remove the 14 head bolts gradually in 2–3 passes and in the proper sequence.

20. Carefully lift the cylinder head from the dowels on the cylinder block and remove it.

21. Installation is the reverse of the removal procedure. Torque the cylinder head bolts, in several stages, to 58 ft. lbs. (78 Nm).

5S-FE Engine

1. Disconnect the negative battery cable. Drain the cooling system.

2. Tag and disconnect all lines, hoses and electrical wires that interfere with exhaust manifold, intake manifold and cylinder head removal.

3. MR2, remove the engine under covers, engine hood side panels and the brace that runs across the struts.

4. If equipped with automatic transaxle, disconnect the throttle cable and bracket from the throttle body.

5. Disconnect the accelerator cable and bracket from the throttle body and intake chamber.

6. If equipped with cruise control, remove the actuator and bracket.

7. Remove the air cleaner cap.

8. On Celica, remove the alternator and unbolt the air conditioning compressor from its mounting bracket. Leave the refrigerant lines connected and wire the compressor aside.

9. Remove the distributor.

10. On Celica, raise and support the

vehicle safely. Remove the right tire and wheel assembly and engine under covers.

11. Remove the suspension lower crossmember.

12. Disconnect the exhaust pipe from the catalytic converter.

13. Remove the exhaust pipe and catalytic converter.

14. Remove the water outlet and the water bypass pipe.

15. Remove the throttle body and cold start injector. On Celica, remove the cold start injector pipe.

16. Remove the EGR valve and modulator.

17. On MR2, remove the fuel pressure Vacuum Switching Valve (VSV). On Celica and MR2, remove the EGR vacuum switching valve.

18. Remove the air intake chamber air hose, the throttle body air hose, and the power steering pump hoses, if equipped. On Celica, remove the air tube.

19. Remove the intake manifold.

20. Remove the fuel delivery pipe and the injectors.

21. Remove the camshaft timing pulley, No. 1 idler pulley and tension spring and No. 3 timing belt cover. Properly support the timing belt so meshing of the crankshaft timing pulley does not occur and the timing belt does not shift.

22. Remove the engine hangers and oil pressure switch. On Celica, remove the alternator bracket.

23. Remove the valve cover.

24. Remove the camshafts.

25. Loosen then remove the cylinder head bolts in 3 phases and in the proper sequence. Remove the cylinder head from the engine.

To install:

26. Install the cylinder head on the cylinder head block. Place the cylinder head in position on the cylinder head gasket.

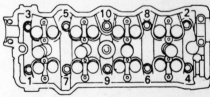

Cylinder head bolt loosening sequence – 5S-FE engine

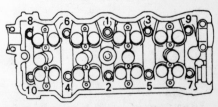

Cylinder head bolt tightening sequence – 5S-FE engine

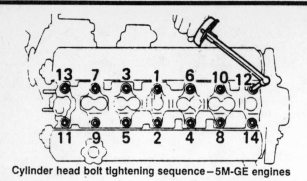

Cylinder head bolt tightening sequence—5M-GE engines

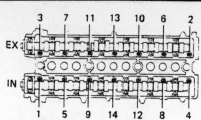

Cylinder head bolt loosening sequence—
7M-GE and 7M-GTE engines

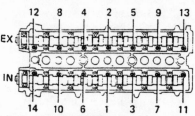

Cylinder head bolt tightening sequence—
7M-GE and 7M-GTE engines

27. Torque the cylinder head bolts as follows:

    a. Apply a light coat of clean engine oil to the threads of the head bolts before installation.

    b. Tighten the bolts in several passes, in the proper sequence to 36 ft. lbs. (49 Nm).

    c. Mark the front of each bolt with a dab of paint.

    d. Retighten the bolts a further 90 degree turn. The paint dabs should now all be at a 90 degree angle to the front of the head.

28. Installation of the remaining components is the reverse of the removal precedure.

7M-GE and 7M-GTE Engine

1. Disconnect the negative battery cable. Drain the cooling system.

2. Disconnect the exhaust pipe from the exhaust manifold. Disconnect the cruise control cable, if equipped.

3. Disconnect the accelerator cable. Disconnect the throttle cable, if equipped with automatic transmission. Disconnect the engine ground strap.

4. On the 7M-GE, remove the No. 1 air cleaner hose along with the intake air pipe assembly. On the 7M-GTE, remove the No. 4 air cleaner pipe along with the No. 1 and No. 2 air cleaner hose.

5. Disconnect the cruise control vacuum hose, the charcoal canister hose and the brake booster hose.

6. Remove the radiator and heater inlet hoses. Remove the alternator.

7. On the 7M-GTE, remove the power steering reservoir tank. On the 7M-GTE, remove the cam position sensor.

8. Remove the air intake chamber with the connector. Remove the PCV pipe. Disconnect and tag all lines, hoses and electrical wires that interfere with exhaust manifold, intake manifold and cylinder head removal.

9. Remove the EGR pipe mounting bolts. Remove the manifold stay retaining bolts. On the 7M-GE, remove the throttle body bracket. On the 7M-GTE, remove the ISC pipe.

10. Remove the air intake connector

mounting bolt (7M-GTE). On the 7M-GE, remove the cold start injector tube. On the 7M-GTE, disconnect the cold start injector. Disconnect the EGR vacuum modulator from the bracket.

11. Disconnect the engine wire from the clamps of the intake chamber. Remove the nuts and bolts, vacuum pipes and intake chamber with the connector and gasket.

12. On the 7M-GTE, remove the ignition coil and bracket.

13. Remove the pulsation damper, the VSV and the No. 1 fuel pipe. Remove the No. 2 and No. 3 fuel pipes. On the 7M-GTE, remove the auxiliary air pipe.

14. On the 7M-GE, remove the high tension wires and the distributor. Remove the oil dipstick. On the 7M-GTE, remove the turbocharger assembly.

15. Remove the exhaust manifold. Remove the water outlet housing. Remove the cylinder head covers. Remove the spark plugs.

16. Remove the timing belt and the camshaft timing pulleys. Remove the cylinder head retaining bolts gradually and in the proper sequence. carefully remove the cylinder head from the engine. As the cylinder head is lifted, separate the No. 5 water bypass line from its union.

To install:

17. Clean the gasket mating surfaces. Use care not to damage the aluminum components. Lower the cylinder head onto the engine with the help of an assistant. Make sure the dowel pins are aligned and no hoses or wires are between the head and cylinder block.

18. During installation, observe the following torques:

    Cylinder head bolts; 1st step to 20 ft. lbs. (27 Nm), 2nd step to 40 ft. lbs. (54 Nm) and 3rd step to 58 ft. lbs. (78 Nm).

    Cam bearing caps 8–10 ft. lbs. (11–14 Nm).

    Cam sprocket 29–39 ft. lbs. (39–53 Nm).

    Crankshaft pulley 55–61 ft. lbs. (75–83 Nm).

    Manifold bolts 15–21 ft. lbs. (20–29 Nm).

    Rocker arm bolts 17–19 ft. lbs. (23–26 Nm).

    Timing gear idler bolt 22–32 ft. lbs. (30–44 Nm).

    Adjust belt tension 0.24–0.28 in. (6–7mm).

    Adjust the valves.

19. Connect all electrical and hoses fittings.

20. Install all necessary components.

21. Connect the battery cable, refill all fluids, start the engine and check for leaks.

Valve Lifters

REMOVAL & INSTALLATION

1. Disconnect the negative battery cable.

2. Drain the cooling system.

3. Remove the valve cover(s).

4. Remove the camshafts.

5. Remove the valve lifters and shims from the cylinder head.

6. Label each lifter and shim with the respective cylinder head bore.

7. Inspect the lifters and shims for excessive wear. Replace worn lifters and shims as required.

To install:

8. Install the lifters and shims into the cylinder head. Make sure the lifter can be rotated freely by hand.

9. Install the camshafts.

10. Install the valve cover(s) using new gaskets and sealant, as required.

11. Fill the cooling system to the proper level.

12. Connect the negative battery cable.

Valve Lash

ADJUSTMENT

2VZ-FE Engine

1. Remove the air intake chamber and the cylinder head covers.

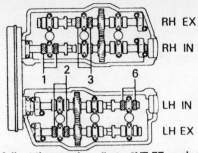

Adjust these valves first – 2VZ-FE engine

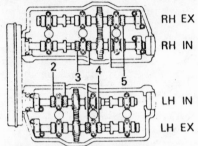

Adjust these valves second – 2VZ-FE engine

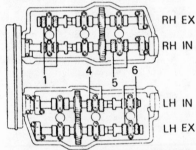

Adjust these valves third – 2VZ-FE engine

2. Use a wrench and turn the crankshaft until the notch in the pulley aligns with the timing mark **0** of the No. 1 timing belt cover. This will insure that No. 1 piston is at TDC of the compression stroke.

NOTE: Check that the valve lifters on the No. 1 (intake) cylinder are loose and those on No. 1 cylinder (exhaust) are tight. If not, turn the crankshaft 1 complete revolution (360 degrees) and then re-align the marks.

3. Using a flat feeler gauge measure the clearance between the camshaft lobe and the valve lifter on the first set of valves shown. This measurement should correspond to specification.

NOTE: If the measurement is within specifications, go on to the next step. If not, record the measurement taken for each individual valve.

4. Turn the crankshaft ⅔ revolution (240 degrees).

5. Measure the clearance of the second set of valves shown.

NOTE: If the measurement is within specifications, go on to the next step. If not, record the measurement taken for each individual valve.

6. Turn the crankshaft ⅔ revolution (240 degrees).

7. Measure the clearance of the third set of valves shown.

NOTE: If the measurement for this set of valves (and also the previous ones) is within specifications, go no further, the procedure is finished. If not, record the measurements and then proceed to Step 8.

8. Turn the crankshaft to position the intake camshaft lobe of the cylinder to be adjusted, upward.

9. Using a suitable tool, turn the valve lifter so the notch is easily accessible; it should be toward the spark plug.

10. Install tool 09248-55010 or equivalent, between the 2 camshafts lobes and then turn the handle so the tool presses down the valve lifter evenly.

11. Using a suitable tool and a magnet, remove the valve shims.

12. Measure the thickness of the old shim with a micrometer. Using this measurement and the clearance ones made earlier (from Step 3, 5 or 7), determine what size replacement shim will be required in order to bring the valve clearance into specification.

NOTE: Replacement shims are available in 17 sizes, in increments of 0.0020 in. (0.05mm), from 0.0984 in. to 0.1299 in. (2.50mm to 3.300mm)

13. Install the new shim, remove the special tool and then recheck the valve clearances.

3A-C, 3E, 3E-E, 4A-C and 4A-LC Engines

1. Start the engine and run it until it reaches normal operating temperature.

2. Stop the engine. Remove the air cleaner assembly and the cylinder head cover.

3. Turn the crankshaft until the point or notch on the pulley aligns with the **0** or **T** mark on the timing scale. This will set the engine to TDC of the compression stroke.

NOTE: Check that the rocker arms on the No. 1 cylinder are loose. If not, turn the crankshaft one complete revolution (360 degrees).

4. Retighten the cylinder head bolts to specifications. Also, retighten the valve rocker support bolts to the proper specifications.

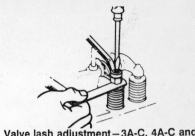

Valve lash adjustment – 3A-C, 4A-C and 4A-LC engines

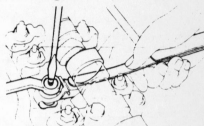

Valve lash adjustment – 3E and 3E-E engines

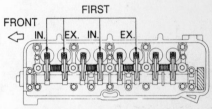

Adjust these valves first – 3A-C, 4A-C and 4A-LC engines

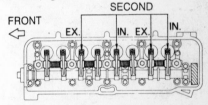

Adjust these valves second – 3A-C, 4A-C and 4A-LC engines

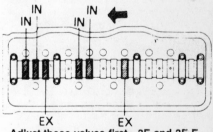

Adjust these valves first – 3E and 3E-E engines

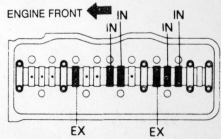

Adjust these valves second – 3E and 3E-E engines

5. Using a flat feeler gauge, check the clearance between the bottom of the rocker arm; top for 3E and 3E-E engines, and the top of the valve stem; bottom of cam lobe for 3E and 3E-E engines.

6. If the clearance is not within specification, the valves will require adjustment. Loosen the locknut on the end of the rocker arm and, still holding the nut with an open end wrench, turn the adjustments screw to achieve the correct clearance.

7. Once the correct valve clearance is achieved, keep the adjustment screw from turning with a suitable tool and then tighten the locknut. Recheck the valve clearances.

8. Turn the engine 1 complete revolution (360 degrees) and adjust the remaining valves.

9. Use a new gasket and then install the cylinder head cover. Install the air cleaner assembly.

3S-FE, 3S-GE, 3S-GTE and 5S-FE Engines

1. Remove the cylinder head covers.

2. Use a wrench and turn the crankshaft until the notch in the pulley aligns with the timing mark **0** of the No. 1 timing belt cover. This will insure that No. 1 piston is at TDC of the compression stroke.

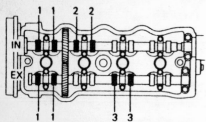

Adjust these valves first — 3S-FE engine

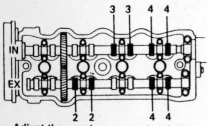

Adjust these valves second — 3S-FE engine

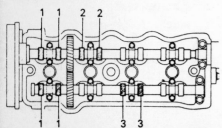

Adjust these valves first — 3S-GE, 3S-GTE and 5S-FE engines

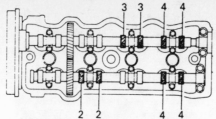

Adjust these valves second — 3S-GE, 3S-GTE and 5S-FE engines

NOTE: Check that the valve lifters on the No. 1 cylinder are loose and those on No. 4 cylinder are tight. If not, turn the crankshaft 1 complete revolution (360 degrees) and then realign the marks.

3. Using a flat feeler gauge measure the clearance between the camshaft lobe and the valve lifter on the first set of valves shown. This measurement should correspond to specification.

NOTE: If the measurement is within specifications, go on to the next step. If not, record the measurement taken for each individual valve.

4. Turn the crankshaft 1 complete revolution and realign the timing marks.

5. Measure the clearance of the second set of valves.

NOTE: If the measurement for this set of valves (and also the previous one) is within specifications, go no further, the procedure is finished. If not, record the measurements and then proceed to Step 6.

6. Turn the crankshaft to position the intake camshaft lobe of the cylinder to be adjusted, upward.

NOTE: Both intake and exhaust valve clearance may be adjusted at the same time, if required.

7. Using a suitable tool, turn the valve lifter so the notch is easily accessible.

8. Install tool 09248–70012 for 3S-GE or 09248–55010 for 3S-FE, 3S-GTE, 5S-FE, between the 2 camshaft lobes and then turn the handle so the tool presses down both (intake and exhaust) valve lifters evenly.

9. Using a suitable tool and a magnet, remove the valve shims.

10. Measure the thickness of the old shim with a micrometer. Using this measurement and the clearance ones made earlier (from Step 3 or 5), determine what size replacement shim will be required in order to bring the valve clearance into specification.

NOTE: Replacement shims are available in 27 sizes, in increments of 0.0020 in. (0.05mm), from 0.0787 in. to 0.1299 in. (2.00mm to 3.3mm).

11. Install the new shim, remove the special tool and then recheck the valve clearances.

4A-F, 4A-FE, 4A-GE, 4A-GEC, 4A-GELC and 4A-GZE Engines

1. Start the engine and run it until it reaches normal operating temperature.

2. Stop the engine. Remove the air cleaner assembly and the valve cover.

3. Use a wrench and turn the crankshaft until the notch in the pulley aligns with the timing pointer in the front cover. This will insure that engine is at TDC.

NOTE: Check that the valve lifters on the No. 1 cylinder are loose and those on No. 4 cylinder are tight. If not, turn the crankshaft one complete revolution (360 degrees) and then re-align the marks.

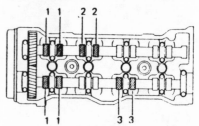

Adjust these valves first — 4A-F and 4A-FE engines

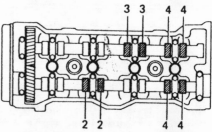

Adjust these valves second — 4A-F and 4A-FE engines

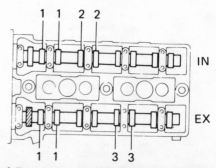

Adjust these valves first — 4A-GE, 4A-GEC and 4A-GELC (Corolla) engines

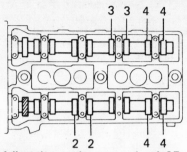

Adjust these valves second—4A-GE, 4A-GEC and 4A-GELC (Corolla) engines

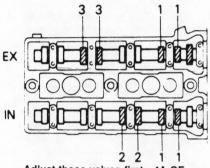

Adjust these valves first—4A-GE, 4A-GELC and 4A-GZE (MR2) engines

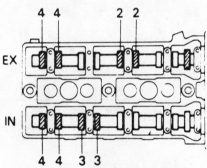

Adjust these valves second—4A-GE, 4A-GELC and 4A-GZE (MR2) engines

4. Using a flat feeler gauge measure the clearance between the camshaft lobe and the valve lifter. Check the first set of valves shown.

NOTE: If the measurement is within specifications, go on to the next step. If not, record the measurement taken for each individual valve.

5. Turn the crankshaft 1 complete revolution and realign the timing marks. Measure the clearance of the second set of valves shown.

NOTE: If the measurement for this set of valves (and also the previous one) is within specification, go no further, the procedure is finished. If not, record the measurements and then proceed to Step 6.

6. Turn the crankshaft to position

the intake camshaft lobe of the cylinder to be adjusted, upward. Both intake and exhaust valve clearance may be adjusted at the same time, if required.

7. Using a suitable tool, turn the valve lifter so the notch is easily accessible.

8. Install tool 09248–70011 for 4A-GE or 09248–55010 4A-F, 4A-FE, 4A-GEC, 4A-GELC, 4A-GZE or equivalent, between the 2 camshafts lobes and then turn the handle so the tool presses down both (intake and exhaust) valve lifters evenly. On the 4A-GE, the tool will work on only one valve lifter at a time.

NOTE: On the 4A-FE, 4A-GEC, 4A-GELC and 4A-GZE, position the notch toward the spark plug before pressing down the valve lifter.

9. Using a suitable tool and a magnet, remove the valve shims.

10. Measure the thickness of the old shim with a micrometer. Using this measurement and the clearance of ones made earlier, determine what size replacement shim will be required in order to bring the valve clearance into specification.

11. Install the new shim, remove the special tool and then recheck the valve clearance.

5M-GE Engines

These engine is equipped with hydraulic lash adjusters in the valve train. The adjusters maintain a 0 clearance between the rocker arm and valve stem, no adjustment is possible or necessary.

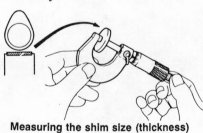

Measuring the shim size (thickness)

Depressing the valve lifter to remove the shim—3S-GE and 4A-GE engines

7M-GE and 7M-GTE Engines

1. Remove the cylinder head covers.
2. Use a wrench and turn the crank-

shaft until the notch in the pulley aligns with the timing mark **0** of the No. 1 timing belt cover. This will insure that engine is at TDC.

NOTE: Check that the valve lifters on the No. 1 cylinder are loose and those on No. 6 cylinder are tight. If not, turn the crankshaft 1 complete revolution (360 degrees) and then realign the marks.

3. Using a flat feeler gauge measure the clearance between the camshaft lobe and the valve lifter. This measurement should correspond to specification. Check the first set of valves shown.

NOTE: If the measurement is within specifications, go on to the next step. If not, record the measurement taken for each individual valve.

4. Turn the crankshaft ⅔ revolution (240 degrees).
5. Measure the clearance of the second set of valves shown.

NOTE: If the measurement is within specifications, go on to the next step. If not, record the measurement taken for each individual valve.

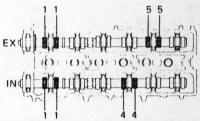

Adjust these valves first—7M-GE and 7M-GTE engines

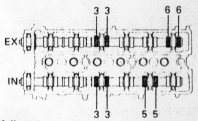

Adjust these valves second—7M-GE and 7M-GTE engines

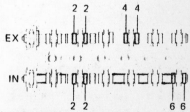

Adjust these valves third—7M-GE and 7M-GTE engines

6. Turn the crankshaft ⅔ revolution (240 degrees).

7. Measure the clearance of the third set of valves shown.

NOTE: If the measurement for this set of valves (and also the previous ones) is within specifications, go no further, the procedure is finished. If not, record the measurements and then proceed to Step 8.

8. Turn the crankshaft to position the intake camshaft lobe of the cylinder to be adjusted, upward.

NOTE: Both intake and exhaust valve clearance may be adjusted at the same time, if required.

9. Using a suitable tool, turn the valve lifter so the notch is easily accessible.

10. Install tool 09248–55010 or equivalent, between the 2 camshafts lobes and then turn the handle so the tool presses down both (intake and exhaust) valve lifters evenly.

11. Using a suitable tool and a magnet, remove the valve shims.

12. Measure the thickness of the old shim with a micrometer. Using this measurement and the clearance ones made earlier (from Step 3, 5 or 7), determine what size replacement shim will be required in order to bring the valve clearance into specification.

NOTE: Replacement shims are available in 17 sizes, in increments of 0.0020 in. (0.05mm), from 0.0787 in. to 0.1299 in. (2.00mm to 3.300mm).

13. Install the new shim, remove the special tool and then recheck the valve clearance.

5M-GE

The 1988 5M-GE engine is equipped with hydraulic valve lash adjusters. No adjustment is necessary. After servicing, the lash adjuster has to be bleed by immersing in engine oil and inserting special tool SST 09276–70010 or equivalent into the plunger hole and slide the plunger up and down several times while pushing down lightly on the check ball. Replace the adjuster if the plunger stroke exceeds 0.020 in. (0.5mm) even after repeating the first 2 steps.

Rocker Arms/Shafts

REMOVAL & INSTALLATION

3A-C, 4A-C and 4A-LC Engines

1. Disconnect the negative battery terminal.

2. Remove the air cleaner and all necessary hoses.

3. Remove all linkage from the carburetor.

4. Remove the valve cover and gasket.

5. Remove the rocker arm bolts.

6. Installation is the reverse of the removal procedure. Install a new valve cover gasket before replacing the valve cover. Tighten the rocker arm bolts to 17–19 ft. lbs. (23–26 Nm).

3E and 3E-E Engines

1. Disconnect the negative battery cable.

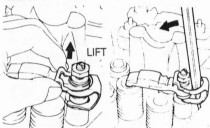

Rocker arm spring clip removal – 3E and 3E-E engines

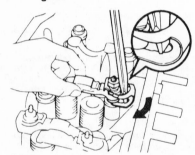

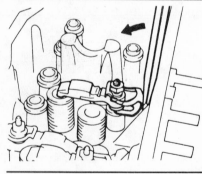

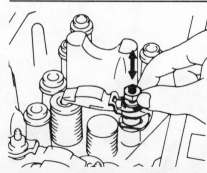

Rocker arm spring clip installation – 3E and 3E-E engines

2. Remove the camshaft.

3. Loosen the rocker arm adjusting screw locknuts.

4. Pull up on the top of the spring while prying the spring with a suitable tool.

5. Remove the rocker arms and arrange them in order. Check the contact surface for any signs of pitting or wear.

6. Check that the adjusting screw is as shown and install a new spring to the rocker arm.

7. Press the bottom lip of the spring until it fits into the groove on the rocker arm pivot.

NOTE: Put the valve adjusting screw in the rocker arm pivot.

8. Pry the rocker spring clip onto the pivot. Pull the rocker arm up and down to check that there is spring tension and that the rocker does not rattle.

All Other Engines

All other engines do not utilize rocker arms shafts. The valves are activated directly by the camshaft via valve adjusting shims.

Intake Manifold

REMOVAL & INSTALLATION

2VZ-FE Engine

1. Disconnect the negative battery cable and drain the cooling system.

2. Disconnect the throttle cable at the throttle body. If equipped, remove the cruise control actuator and vacuum pump.

3. Remove the air cleaner hose.

4. Raise the vehicle and support safely. Remove the engine undercovers.

5. Remove the lower suspension crossmember and the front exhaust pipe.

6. Remove the alternator. Remove the ISC valve.

7. Remove the throttle body. Remove the EGR pipe, valve and vacuum modulator.

8. Remove the vacuum pipe and the distributor.

9. Remove the exhaust crossover pipe. Disconnect the cold start injector and then remove the injector tube.

10. Tag and disconnect all hoses leading to the air intake chamber and then remove the chamber.

11. Remove the fuel delivery pipes and the injectors.

12. Disconnect the water temperature sensor and remove the upper radiator hose. Remove the water outlet. Remove the water bypass outlet.

13. Loosen the 2 bolts and remove the cylinder head rear plate.

14. Remove the No. 2 idler pulley bracket stay.

15. Remove the 8 bolts and 4 nuts and lift out the intake manifold with its 2 gaskets.

To install:

16. Clean the gasket mating surfaces being careful not to damage the aluminum surfaces. Check the mating surfaces for warpage with a straight edge.

17. Match the old gasket with the new for an exact match. Use new gaskets when installing the manifold and tighten the bolts, from the center outward, to 13 ft. lbs. (18 Nm).

18. Install the remaining components, start the engine and check for leaks.

3E and 3E-E Engines

1. Disconnect the negative battery cable. Drain the coolant. Remove the air cleaner assembly.

2. Tag and disconnect all wires, hoses or cables that interfere with intake manifold removal.

3. Remove the necessary components in order to gain access to the intake manifold retaining bolts.

4. Remove the carburetor from the throttle body.

5. Disconnect the intake manifold water hoses.

6. Remove the intake manifold retaining bolts. Remove the intake manifold from the vehicle.

To install:

7. Clean the gasket mating surfaces being careful not to damage the aluminum surfaces. Check the mating surfaces for warpage with a straight edge.

8. Match the old gasket with the new for an exact match. Use new gaskets when installing the manifold and tighten the bolts, from the center outward, to 13 ft. lbs. (18 Nm).

9. Install the remaining components, start the engine and check for leaks.

3S-FE and 5S-FE Engines

1. Disconnect the negative battery cable. Drain the cooling system. Remove the air cleaner assembly.

2. Tag and disconnect all wires, hoses or cables that interfere with intake manifold removal.

3. Remove the necessary components in order to gain access to the intake manifold retaining bolts.

4. Remove the throttle body and cold start injector pipe.

5. Remove the air tube assembly. If equipped with power steering, remove the hoses before removing the air tube assembly.

6. Remove the intake manifold retaining bolts. Remove the intake manifold from the vehicle.

To install:

7. Clean the gasket mating surfaces being careful not to damage the alumi-

num surfaces. Check the mating surfaces for warpage with a straight edge.

8. Match the old gasket with the new for an exact match. Use new gaskets when installing the manifold and tighten the bolts, from the center outward. Tighten the intake manifold mounting bolts to 14 ft. lbs. (19 Nm). Tighten the 12mm manifold stay bolt to 14 ft. lbs. (19 Nm); tighten the 14mm bolts to 31 ft. lbs. (42 Nm).

9. Install the remaining components, start the engine and check for leaks.

3S-GE and 3S-GTE Engines

1. Disconnect the negative battery cable. Drain the cooling system. Remove the air cleaner assembly.

2. Tag and disconnect all wires, hoses or cables that interfere with intake manifold removal.

3. Remove the necessary components in order to gain access to the intake manifold retaining bolts.

4. Remove the intake manifold retaining bolts and nuts. Remove the intake manifold from the vehicle.

To install:

5. Clean the gasket mating surfaces being careful not to damage the aluminum surfaces. Check the mating surfaces for warpage with a straight edge.

6. Match the old gasket with the new for an exact match. Use new gaskets when installing the manifold and tighten the bolts, from the center outward, to 13 ft. lbs. (18 Nm).

7. Install the remaining components, start the engine and check for leaks.

4A-F, 4A-FE, 4A-GE, 4A-GEC, 4A-GELC and 4A-GZE Engines

1. Disconnect the negative battery cable. Drain the coolant. Remove the air cleaner assembly.

2. Tag wires, hoses or cables in the way of manifold removal.

3. Remove the necessary components in order to gain access to the intake manifold retaining bolts.

4. Remove the intake manifold retaining bolts. Remove the intake manifold from the vehicle.

To install:

5. Clean the gasket mating surfaces being careful not to damage the aluminum surfaces. Check the mating surfaces for warpage with a straight edge.

6. Match the old gasket with the new for an exact match. Use new gaskets when installing the manifold and tighten the bolts, from the center outward, to the proper specification.

7. Install the remaining components, start the engine and check for leaks.

5M-GE Engine

1. Disconnect the negative battery cable. Drain the engine coolant.

2. Tag and disconnect all wires, hoses or cables that interfere with intake manifold removal.

3. Remove the air intake chamber.

4. Disconnect and move the wiring away from the fuel delivery and injector pipe.

4. Remove the fuel injector and delivery pipe.

5. Remove the fuel pressure regulator, which is mounted on the center of the intake manifold.

6. Remove the EGR valve from the rear of the manifold.

7. Disconnect the radiator hoses, heater hoses, and vacuum lines from the intake manifold.

8. Remove the distributor cap and position it aside.

9. Remove the intake manifold retaining bolts. Remove the intake manifold and gasket from the engine.

To install:

10. Clean the gasket mating surfaces being careful not to damage the aluminum surfaces. Check the mating surfaces for warpage with a straight edge.

11. Match the old gasket with the new for an exact match. Use new gaskets when installing the manifold and tighten the bolts, from the center outward and torque the manifold fasteners to 10–15 ft. lbs. (13.6–20 Nm).

12. Install the remaining components, start the engine and check for leaks.

7M-GE and 7M-GTE Engines

1. Disconnect the negative battery cable. Drain the cooling system.

2. Remove the air cleaner assembly.

3. Tag and disconnect all wires, hoses or cables that interfere with intake manifold removal.

4. Remove the necessary components in order to gain access to the intake manifold retaining bolts.

5. Remove the air intake connector along with the air intake chamber assembly.

6. Remove the fuel delivery pipe with the injectors still attached.

7. Remove the intake manifold retaining bolts. Remove the intake manifold from the vehicle.

To install:

8. Clean the gasket mating surfaces being careful not to damage the aluminum surfaces. Check the mating surfaces for warpage with a straight edge.

9. Match the old gasket with the new for an exact match. Use new gaskets when installing the manifold and tighten the bolts, from the center outward, to 13 ft. lbs. (18 Nm).

10. Install the remaining components, start the engine and check for leaks.

Exhaust Manifold
REMOVAL & INSTALLATION

2VZ-FE Engine

1. Disconnect the negative battery cable and drain the cooling system.
2. Disconnect the throttle cable at the throttle body. If equipped with cruise control, remove the cruise control actuator and vacuum pump.
3. Remove the air cleaner hose.
4. Raise the vehicle and support safely. Remove the engine undercovers.
5. Remove the lower suspension crossmember and the front exhaust pipe.
6. Remove the alternator and the Idle Speed Control (ISC) valve.
7. Remove the throttle body, EGR pipe, EGR valve and vacuum modulator.
8. Remove the vacuum pipe and the distributor.
9. Remove the exhaust crossover pipe. Disconnect the cold start injector and then remove the injector tube.
10. Tag and disconnect all hoses leading to the air intake chamber and then remove the chamber.
11. Remove the fuel delivery pipes and the injectors.
12. Disconnect the water temperature sensor and remove the upper radiator hose. Remove the water outlet. Remove the water bypass outlet.
13. Loosen the 2 bolts and remove the cylinder head rear plate.
14. Remove the intake manifold.
15. Disconnect the oxygen sensor and then remove the outside heat insulator for the righ manifold. Remove the manifold and gasket and then remove the inner heat shield.
16. Remove the left side heat shield and then remove the manifold.
To install:
17. Clean the gasket mating surfaces being careful not to damage the aluminum surfaces. Check the mating surfaces for warpage with a straight edge.
18. Match the old gasket with the new for an exact match. Use new gaskets when installing the manifold and tighten the bolts, from the center outward, to 29 ft. lbs. (39 Nm).
19. Install the remaining components, start the engine and check for leaks.

3E and 3E-E Engines

1. Disconnect the negative battery cable. Remove the exhaust manifold heat insulator shield assembly.
2. Remove the necessary components in order to gain access to the exhaust manifold retaining bolts.
3. Disconnect the exhaust manifold bolts at the exhaust pipe. Disconnect

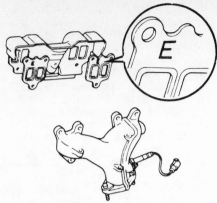

Exhaust manifold gasket installation—3E and 3E-E engines

the oxygen sensor electrical wire. It may be necessary to raise and support the vehicle safely before removing these bolts.
4. Remove the exhaust manifold retaining bolts. Remove the exhaust manifold from the vehicle.
To install:
5. Clean the gasket mating surfaces being careful not to damage the aluminum surfaces. Check the mating surfaces for warpage with a straight edge.
6. Match the old gasket with the new for an exact match. During installation, the **E** mark on the gasket must face outward. Use new gaskets when installing the manifold and tighten the bolts, from the center outward, to 38 ft. lbs. (51 Nm).
7. Install the remaining components, start the engine and check for leaks.

3S-FE and 5S-FE Engines

1. Disconnect the negative battery cable.
2. Raise and support the vehicle safely.
3. Remove the exhaust manifold heat insulator shield assembly.
4. Remove the necessary components in order to gain access to the exhaust manifold retaining bolts.
5. Disconnect the exhaust manifold bolts at the exhaust pipe or catalytic converter. It may be necessary to raise and support the vehicle safely before removing these bolts.
6. Remove the exhaust manifold retaining bolts. Remove the exhaust manifold from the vehicle. On 5S-FE engine, the exhaust manifold and catalytic converter are removed as one unit.
To install:
7. Be sure to use new gaskets. On 5S-FE, the **R** mark should be toward the rear. On 3S-FE, tighten the exhaust manifold bolts to 31 ft. lbs. (41 Nm). On 5S-FE, torque the exhaust manifold retaining bolts to 36 ft. lbs. (49 Nm).

5M-GE, 3S-GE, 3S-GTE, 4A-F, 4A-FE, 4A-GE, 4A-GEC, 4A-GELC and 4A-GZE Engines

1. Disconnect the negative battery cable. Raise the vehicle and support safely. Remove the right gravel shield from under the vehicle.
2. Remove the throttle body and turbocharger.
3. Remove the exhaust pipe support stay. Unbolt the exhaust pipe from the exhaust manifold flange.
4. Disconnect the oxygen sensor connector. On the 3S-GE and 4A-F, remove the upper heat insulator.
5. Remove the manifold retaining nuts. Remove the exhaust manifold from the vehicle.
To install:
6. Clean the gasket mating surfaces being careful not to damage the aluminum surfaces. Check the mating surfaces for warpage with a straight edge.
7. Match the old gasket with the new for an exact match. Use new gaskets when installing the manifold and tighten the bolts, from the center outward, to the proper specification.
8. Install the remaining components, start the engine and check for leaks.

7M-GE and 7M-GTE Engines

1. Disconnect the negative battery cable. Remove the exhaust manifold heat insulator shield assembly, if equipped.
2. Remove the necessary components in order to gain access to the exhaust manifold retaining bolts.
3. Disconnect the exhaust manifold bolts at the exhaust pipe. It may be necessary to raise and support the vehicle safely before removing these bolts.
4. On 7M-GTE, remove the turbocharger.
5. Remove the exhaust manifold retaining bolts. Remove the exhaust manifold from the vehicle.
To install:
6. Clean the gasket mating surfaces being careful not to damage the aluminum surfaces. Check the mating surfaces for warpage with a straight edge.
7. Match the old gasket with the new for an exact match. Use new gaskets when installing the manifold and tighten the bolts, from the center outward, to 29 ft. lbs. (39 Nm).
8. Install the remaining components, start the engine and check for leaks.

Combination Manifold
REMOVAL & INSTALLATION

3A-C, 4A-C and 4A-LC Engines

1. Disconnect the negative battery cable.

2. Remove the air cleaner and all necessary hoses.

3. Remove all the carburetor linkages.

4. Remove the carburetor.

5. Remove the intake/exhaust manifold pipe.

6. Remove the intake/exhaust manifold.

7. Installation is the reverse of the removal procedure. Tighten the manifold bolts to 15–21 ft. lbs. (20–29 Nm).

Supercharger

REMOVAL & INSTALLATION

4A-GZE Engine

1. Disconnect the negative cable at the battery. Drain the cooling system and remove the radiator reservoir tank.

2. Remove the Vacuum Switching Valve (VSV) and the intercooler.

3. Remove the air flow meter with the No. 3 air cleaner hose. Disconnect the accelerator cable (rod) and throttle cable.

4. Disconnect the PCV, brake booster, ACV, air conditioning idle-up and emission control vacuum hoses.

5. Remove the No. 1 intake air connector pipe and its air hose.

6. Loosen the idler pulley locknut and adjusting bolt and remove the supercharger drive belt.

7. Disconnect the No. 2 and 3 water

bypass hoses. Loosen the air hose clamp.

8. Remove the air inlet duct stay. Remove the throttle body.

9. Disconnect the ACV and supercharger connectors and the 2 ACV hoses. Remove the 2 nuts and the ACV. Remove the pivot bolt and nut, remove the 2 stud bolts and then rotate the assembly so the hub is facing upward; remove the supercharger.

10. Installation is the reverse of the removal procedure. Tighten the 2 stud bolts to 25 ft. lbs. (34 Nm).

Turbocharger

REMOVAL & INSTALLATION

3S-GTE Engine
CELICA

1. Disconnect the negative battery cable. Drain the coolant from the engine and intercooler.

2. Remove the air cleaner assembly.

3. Remove the catalytic converter and the oxygen sensor.

4. Disconnect the 2 intercooler water lines and the reservoir tank line. Loosen the clamps, disconnect the air hose and remove the intercooler.

5. Remove the alternator duct and the No. 2 alternator bracket.

6. Remove the turbocharger heat insulator and the turbocharger outlet elbow. Remove the turbocharger stay.

7. Remove the turbocharger.

8. Installation is in the reverse order of removal. Pour about 20cc of new oil into the turbocharger oil inlet and then spin the impeller to lubricate the bearing. Tighten the turbo-to-manifold bolts to 47 ft. lbs. (64 Nm).

1990–92 MR2

1. Disconnect the negative battery cable.

2. Drain the coolant from the engine and the intercooler.

3. Raise and support the vehicle safely.

4. Remove the engine under covers and engine hood side panels.

5. Tag and disconnect all water hoses, vacuum lines, air tubes, engine control cables, transaxle control cables and electrical wires that interfere with turbocharger removal.

6. Remove the brace that runs across the struts.

7. Remove the air cleaner assembly.

8. Remove the front exhaust pipe.

9. Discharge the air conditioning system. Disconnect the refrigerant hoses and electrical wiring from the compressor. Remove the compressor and idler pulley bracket from the engine.

10. Remove the front engine mounting insulator.

11. Remove the front mounting bracket and clutch release cylinder. Leave the hydraulic lines connected and position the release cylinder aside.

12. Remove the engine cooling fan.

13. Remove the catalytic converter.

14. Remove the Vacuum Transmitting Valve (VTV).

15. Remove the air bypass valve.

16. Remove the heat insulator from the turbocharger.

17. Remove the oxygen sensor.

18. Remove the heat insulators from the turbocharger outlet elbow.

19. Disconnect the oil hose from the turbocharger oil pipe.

20. Remove the turbocharger mounting stay.

21. Unbolt and remove the turbocharger oil pipe from the block.

22. Remove the 4 nuts and separate the turbocharger and gasket from the exhaust manifold.

To install:

23. Clean the turbocharger and exhaust manifold gasket surfaces.

24. Prior to installing the turbocharger, pour approximately 1.2 cu. in. (20cc) of new oil into the oil inlet and then turn the impeller wheel by hand a few times in order to lubricate the bearing.

25. Install a new gasket onto the exhaust manifold.

26. Mount the turbocharger onto the gasket and install the 4 nuts. Torque the nuts in a criss-cross pattern to 47 ft. lbs. (64 Nm).

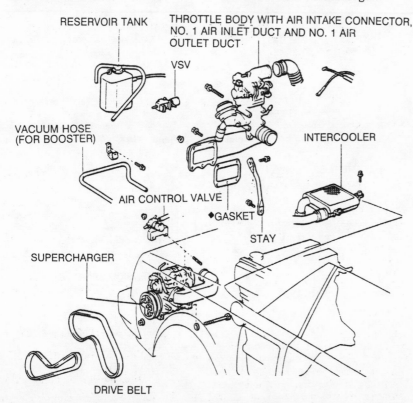

Supercharger and related components

27. Installation of the remaining components is the reverse of the removal procedure. Torque the oil pipe union bolt to the block to 38 ft. lbs. (51 Nm); stay-to-turbocharger bolts to 51 ft. lbs.; stay-to-block bolts to 43 ft. lbs. (59 Nm) and oxygen sensor to 33 ft. lbs. (44 Nm).

7M-GTE Engine

1. Disconnect the negative battery cable and drain the cooling system.
2. Remove the No. 4 air cleaner pipe with the No. 1 and No. 2 air cleaner hoses still attached.
3. Disconnect the 3 air hoses, the PCV hose and the electrical lead at the air flow meter. Disconnect the power steering idle-up air hose and then remove the No. 7 air cleaner hose with the air flow meter and cap still attached.
4. Disconnect the oxygen sensor and remove the turbocharger heat insulator.
5. Remove the oil dipstick guide.
6. Remove the No. 1 air cleaner pipe with the No. 6 air cleaner hose.
7. Disconnect the front exhaust pipe.
8. Remove the mounting nuts and union bolt for the turbocharger oil line. Remove the turbocharger stay.
9. Remove the No. 2 turbocharger stay. Disconnect the No. 1 turbocharger water hose at the water outlet housing. Disconnect the union pipe.
10. Remove the turbocharger and its gasket.

To install:

11. Prior to installing the turbocharger, pour approximately 1.2 cu. in. (20cc) of new oil into the oil inlet and then turn the impeller wheel by hand a few times in order to lubricate the bearing.
12. Position a new gasket with the protrusion pointing toward the rear

and then install the turbocharger unit. Tighten the mounting bolts to 33 ft. lbs. (44 Nm). Tighten the union bolt to 25 ft. lbs. (34 Nm) and the nut to 9 ft. lbs. (13 Nm). The remainder of the installation is in the reverse order of removal.

Timing Belt Front Cover

REMOVAL & INSTALLATION

3A-C, 4A-LC and 4A-C Engines

1. Disconnect the negative battery cable.
2. Remove all drive belts.
3. Bring the No. 1 cylinder to TDC on the compression stroke.
4. Remove the crankshaft pulley with a suitable puller.
5. Remove the water pump pulley.
6. Remove the upper and lower timing case covers.
7. Installation is the reverse of removal. Tighten the timing belt cover to 5–8 ft. lbs. (7–11 Nm).

3E, 3E-E and 3S-FE Engines

1. Disconnect the negative battery cable.
2. On 3E and 3E-E engines, remove the air cleaner assembly. On 3E-E engine, disconnect the accelerator and throttle cables.
3. Remove all drive belts.
4. On the 3S-FE engine remove the alternator, alternator bracket and right engine mounting stay (2WD). If equipped with cruise control remove the actuator and bracket assembly.
5. Raise and support the vehicle safely.
6. Remove the right tire and wheel assembly. Remove the right side engine under cover. Remove the right side engine mount insulator.
7. On the 3E and 3E-E engines, remove the cylinder head cover.
8. Set the No. 1 piston to TDC of the compression stroke and remove the crankshaft pulley using the proper tools.
9. Remove the engine front cover retaining bolts. Remove both front covers from the engine.

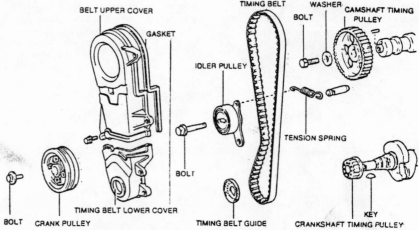

Timing belt and related components—3A-C engine

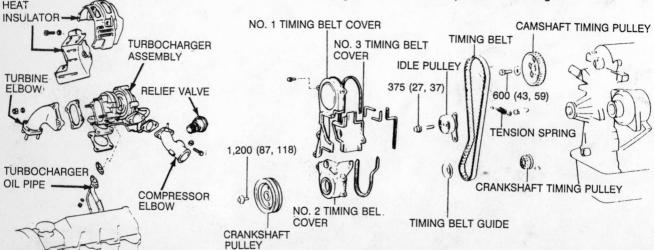

Timing belt and related components—4A-C and 4A-LC engines

Typical turbocharger assembly

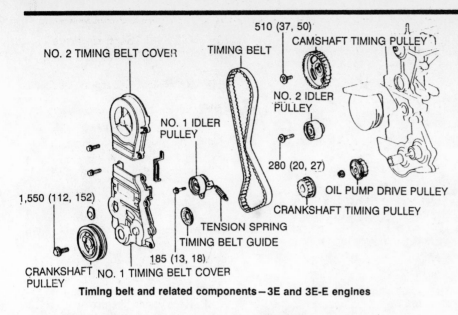

Timing belt and related components—3E and 3E-E engines

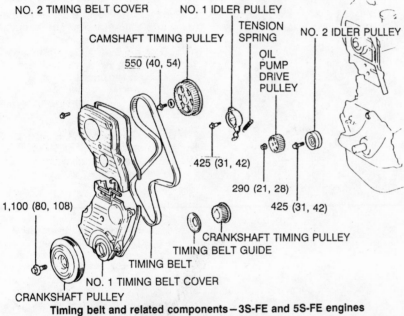

Timing belt and related components—3S-FE and 5S-FE engines

10. Installation is the reverse of the removal procedure. Use new cover seals. On 3E and 3E-E engines, torque the crankshaft pulley bolt to 112 ft. lbs. (154 Nm). On 3S-FE engine, torque the crankshaft pulley bolt to 80 ft. lbs. (108 Nm).

7M-GE and 7M-GTE Engines

1. Disconnect the negative battery cable.
2. Drain the cooling system.
3. Remove the radiator and water outlet.
4. Remove the spark plugs, drive belts and alternator.
5. Remove the upper timing belt cover and seal.
6. Set the No. 1 piston to TDC of the compression stroke.
7. Remove the timing belt from the camshaft sprockets.
8. Remove the crankshaft pulley using the proper tool.
9. Remove the lower timing belt cover and seal.
10. Installation is the reverse of the removal procedure. Use new cover seals. Torque the crankshaft pulley bolt to 195 ft. lbs. (265 Nm).

OIL SEAL REPLACEMENT

1. Remove the front cover.
2. Inspect the oil seal for signs of wear, leakage, or damage.
3. If worn, pry the old seal out. Remove it toward the front of the cover. Once the seal has been removed, it must be replaced.
4. Wipe the seal bore with a clean rag.
5. Drive the oil seal into place using a suitable seal installer. Work from the front of the cover. Be extremely careful not to damage the seal.
6. Install the front cover.

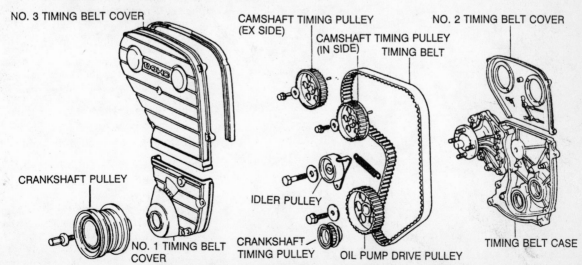

Timing belt and related components—5M-GE, 7M-GE and 7M-GTE engines

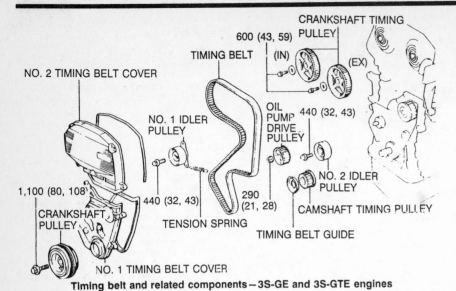

NO. 2 TIMING BELT COVER

CRANKSHAFT TIMING PULLEY

600 (43, 59)

(IN) **(EX)**

TIMING BELT

NO. 1 IDLER PULLEY

OIL PUMP DRIVE PULLEY **440 (32, 43)**

440 (32, 43)

NO. 2 IDLER PULLEY

1,100 (80, 108)

290 (21, 28)

CAMSHAFT TIMING PULLEY

CRANKSHAFT PULLEY

TENSION SPRING

TIMING BELT GUIDE

NO. 1 TIMING BELT COVER

Timing belt and related components — 3S-GE and 3S-GTE engines

NO. 3 TIMING BELT COVER

CRANKSHAFT TIMING PULLEY

600 (43, 59)

NO. 2 TIMING BELT COVER

IDLE PULLEY

PLUG

1,200 (87, 118)

TENSION SPRING

375 (27, 37)

CAMSHAFT TIMING PULLEY

CRANKSHAFT PULLEY

NO. 1 TIMING BELT COVER

TIMING BELT GUIDE

Timing belt and related components — 4A-F and 4A-FE engines

KG-CM (FT. LBS., NM) TIGHTENING TORQUE

NO. 3 TIMING BELT COVER

TENSION SPRING

CAMSHAFT TIMING PULLEY (IN SIDE)

NO. 2 TIMING BELT COVER

GASKET

TIMING BELT

(EX SIDE)

FAN WITH FLUID COUPLING

WATER PUMP PULLEY

475 (34, 47)

1,200 (87, 118) PULLEY BOLT

DRIVE BELT

CRANKSHAFT PULLEY

IDLER PULLEY MOUNTING BOLT

CRANKSHAFT TIMING PULLEY

TIMING BELT GUIDE

IDLER PULLEY

NO. 1 TIMING BELT COVER

Timing belt and related components — 4A-GE and 4A-GZE engines

Timing Belt and Tensioner

REMOVAL & INSTALLATION

2VZ-FE Engine

1. Disconnect the negative battery cable. If equipped, remove the cruise control actuator and vacuum pump.

2. Remove the power steering oil reservoir tank and position it aside without disconnecting the hydraulic lines.

3. Raise the vehicle and support it safely. Remove the right wheel and tire assembly.

4. Remove the alternator and power steering pump drive belts.

5. Remove the right side fender apron seal.

6. Remove the right side engine mounting stays, position a floor jack under the engine and raise it just enough to release the pressure on the mount and then remove it. If with ABS, first remove the clamp bolts for the power steering oil cooler lines.

7. Remove the spark plugs.

8. Remove the upper timing belt cover and then remove the right side engine mounting bracket.

NOTE: If re-using the old timing belt, matchmark it to each of the timing pulleys and to mark it for the direction of rotation.

9. Rotate the engine until the groove in the crankshaft pulley is aligned with the **0** mark on the lower timing belt cover. Check that the marks on the camshaft timing pulleys are aligned with the ones on the inner timing belt cover; if not, rotate the engine one complete revolution (360 degrees).

10. Remove the timing belt tensioner and dust cover.

11. Turn the left side camshaft timing pulley clockwise slightly to release the tension on the timing belt and then slide the belt off both pulleys.

12. Remove the camshaft sprocket retaining bolts and knock pins and pull off the sprockets. Do not mix them up.

13. Remove the No. 2 idler pulley.

14. Remove the crankshaft pulley and then remove the lower timing belt cover.

15. Remove the timing belt guide and then remove the timing belt.

To install:

16. Inspect the timing belt for any cracks, tears or other defects. Replace as required. Inspect the idler pulleys and timing sprockets; replace defective components as necessary.

17. Install the timing belt over the

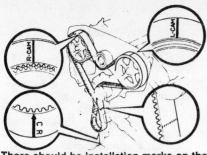

There should be installation marks on the timing belt—2VZ-FE engine

Check that the marks on the sprockets and the No. 3 cover are aligned—2VZ-FE engine

Installing the belt on the crankshaft—2VZ-FE engine

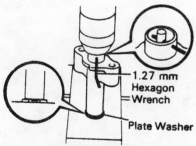

1.27 mm Hexagon Wrench

Plate Washer

Set the timing belt tensioner—2VZ-FE engine

Install the hex wrench—2VZ-FE engine

crankshaft sprocket so the mark on the belt aligns with the drilled mark on the sprocket. Install the belt over the No. 1 idler and water pump pulleys.

18. Install the timing belt guide so the cupped side faces outward. Install the No. 1 timing belt cover with a new gasket.

19. Install the crankshaft pulley so the set key is aligned with the groove in the shaft and then tighten the retaining bolt to 181 ft. lbs. (245 Nm).

20. Install the No. 2 idler pulley and tighten the bolt to 29 ft. lbs. (39 Nm). Check the pulley for smooth operation.

21. Install the left camshaft sprocket, flange side out, so the camshaft knock pin hole aligns with the groove in the sprocket. Install the knock pin and tighten the sprocket bolt to 80 ft. lbs. (108 Nm).

22. Set the No. 1 piston to TDC of the compression stroke again. Rotate the right camshaft until the knock pin hole is aligned with the timing mark on the No. 3 timing belt cover. Rotate the left camshaft sprocket until the timing mark on the sprocket is aligned with the one on the No. 3 cover.

23. Check that the mark on the timing belt aligns with the edge of the No. 1 timing belt cover. Rotate the left camshaft sprocket clockwise slightly so the mark on the timing belt will align with the timing mark on the sprocket, slide the belt over the sprocket and then align the mark on the sprocket with the one on the No. 3 belt cover. Check for tension between the crankshaft and camshaft sprockets.

24. Align the mark on the timing belt with the timing mark on the right camshaft sprocket.

25. Hang the belt over the sprocket, flange side inward, and then align the marks on the sprocket with the one on the No. 3 timing belt cover.

26. Slide the sprocket onto the camshaft so the knock pin hole and groove align. Install the knock pin and then tighten the retaining bolt to 80 ft. lbs. (108 Nm).

27. Install a plate washer between the timing belt tensioner and the cylinder block and then slowly press in the pushrod. When the holes in the pushrod and the housing align, insert a 1.27mm Allen wrench to retain the pushrod and then release the pressure. Install the tensioner and tighten the mounting bolts to 20 ft. lbs. (26 Nm); remove the Allen wrench.

28. Rotate the crankshaft pulley 2 complete revolutions clockwise and check that each pulley is still aligned with the timing marks. If not, remove the belt and start over again.

29. Installation of the remaining components is in the reverse order of

removal. Make all necessary engine adjustments.

3A-C, 4A-C and 4A-LC Engines

1. Remove the timing belt upper and lower covers.

2. If the timing belt is to be reused, mark an arrow in the direction of engine revolution on its surface. Matchmark the belt to the pulleys at the proper locations.

3. Loosen the idler pulley bolt, push it to the left as far as it will go and then temporarily tighten it.

4. Remove the timing belt, idler pulley bolt, idler pulley and the return spring Do not bend, twist, or turn the belt inside out. Do not allow grease or water to come in contact with it.

5. Inspect the timing belt for cracks, missing teeth or overall wear. Replace as necessary. Install the return spring and idler pulley.

6. Install the timing belt. Align the marks made earlier, if reusing the old belt.

7. Adjust the idler pulley so the belt deflection is 0.24–0.28 in. (6–7mm) at 4.5 lbs. of pressure. Check the valve timing.

8. Installation of the remaining components is in the reverse order of removal.

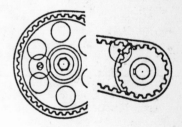

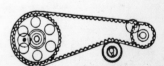

When checking the valve timing, turn the crankshaft 2 complete revolutions clockwise from TDC to TDC and make sure each pulley aligns with the marks shown

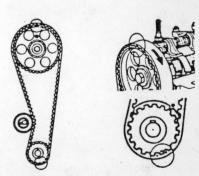

Mark the timing belt before removal

3E and 3E-E Engines

1. Disconnect the negative battery cable. Remove the right side engine under cover. On 3E-E engine, disconnect the accelerator and throttle cables.

2. Remove the drive belts, alternator and alternator bracket. Remove the air cleaner and air intake collector (3E-E) assemblies and spark plugs.

3. Raise the engine and remove the right side engine mounting insulator assembly.

4. Remove the cylinder head cover. Set the engine to TDC on the compression stroke. Remove the crankshaft pulley using the proper removal tool.

5. Remove both timing belt covers. Remove the timing belt guide. Remove the timing belt and the No. 1 idler pulley. If using the old belt matchmark it in the direction of engine rotation. Matchmark the pulleys.

6. Remove the tension spring. Remove the No. 2 idler pulley. Remove the crankshaft pulley, camshaft pulley and oil pump pulley using the proper tools.

To install:

7. Inspect the belt for defects. Replace as required. Inspect the idler pulleys and springs. Replace defective components as required.

8. Align and install the oil pump pulley. Torque the retaining bolt to 20 ft. lbs. (26 Nm).

9. To install the camshaft timing pulley, align the camshaft knock pin with the No. 1 bearing cap mark. Align the knock pin hole on the 3E mark side with the camshaft knock pin hole. Torque the retaining bolt to 37 ft. lbs. (50 Nm).

10. Install the crankshaft timing pulley and align the TDC marks on the oil pump body and the crankshaft timing

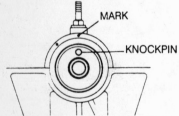

MARK

KNOCKPIN

Camshaft alignment—3E and 3E-E engines

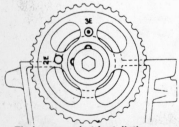

Timing sprocket installation— 3E and 3E-E engines

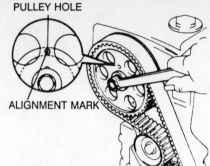

PULLEY HOLE

ALIGNMENT MARK

Timing sprocket alignment—3E and 3E-E engines

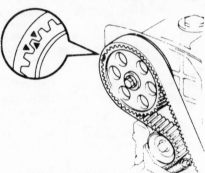

Timing belt alignment—3E and 3E-E engines

pulley. Install the No. 1 idler pulley. Pry the idler pulley toward the left as far as it will go and temporarily tighten the retaining bolt.

11. Install the No. 2 idler pulley and torque the retaining bolt to 20 ft. lbs. (27 Nm). Install the timing belt. If reusing the old belt align it with the marks made during the removal procedure.

12. Inspect the valve timing and the belt tension by loosening the No. 1 idler pulley set bolt. Temporarily install the crankshaft pulley bolt and turn the crankshaft 2 complete revolutions in the clockwise direction.

13. Check that each pulley aligns with the proper markings. Torque the No. 1 idler pulley bolt to 13 ft. lbs. (18 Nm). Check for proper belt tension. Install the belt guide.

14. Install the timing belt covers. Align and install the crankshaft pulley. Torque the retaining bolt to 112 ft. lbs. (152 Nm).

15. Installation of the remaining components is the reverse of the removal procedure.

3S-FE and 5S-FE Engines

1. Disconnect the negative battery cable. Raise and support the vehicle safely. Remove the right tire and wheel assembly.

2. If equipped, remove the cruise control actuator and bracket. Remove the drive belts.

3. Remove the alternator and alternator bracket and right engine mount-

ing stay. Raise the engine enough to remove the right side engine mounting insulator and brackets.

4. Remove the spark plugs. Remove the upper timing cover. Position the No. 1 cylinder to TDC on the compression stroke so the groove in the crankshaft pulley is aligned with the **0** mark in the No. 1 front cover. If the hole in the camshaft pulley is not aligned with the mark on the bearing cap, turn the crankshaft 1 complete revolution (360 degrees).

5. If reusing the belt place matchmarks on the timing belt and the camshaft pulley. Loosen the mount bolt of the No. 1 idler pulley and position the pulley toward the left as far as it will go. Tighten the bolt. Remove the belt from the camshaft pulley.

6. Remove the camshaft pulley. Remove the crankshaft pulley using the proper removal tool. Remove the lower timing cover.

7. Remove the timing belt and the belt guide. If reusing the belt mark the belt and the crankshaft pulley in the direction of engine rotation.

8. Remove the No. 1 idler pulley and the tension spring. Remove the No. 2 idler pulley. Remove the crankshaft timing pulley. Remove the oil pump pulley.

To install:

9. Inspect the belt for defects. Replace as required. Inspect the idler pulleys and springs. Replace defective components as required.

10. Align the cutouts of the oil pump pulley and shaft. Install the oil pump pulley and torque the retaining nut to 21 ft. lbs. (28 Nm).

11. To install the crankshaft pulley, align the pulley set key with the key groove of the pulley and slide it in position. Install the No. 2 idler pulley and torque the bolt to 31 ft. lbs. (42 Nm). Be sure the pulley moves freely.

12. Temporarily install the No. 1 idler pulley and tension spring. Pry the pulley toward the left as far as it will go. Tighten the bolt.

13. Temporarily install the timing belt. If reusing the old belt align the marks made during removal. Install the timing belt guide.

14. Install the No. 1 timing belt cover. Install the crankshaft pulley and tighten the bolt to 80 ft. lbs. (108 Nm).

15. Install the camshaft pulley by aligning the camshaft knock pin with the knock pin groove in the pulley. Install the washer and torque the retaining bolt to 40 ft. lbs. (54 Nm).

16. With the engine set at TDC on the compression stroke install the timing belt. If reusing the belt align with the marks made during the removal procedure.

17. Once the belt is installed be sure

there is tension between the crankshaft pulley, water pump pulley and camshaft pulley. Loosen the No. 1 idler pulley mount bolt ½ turn. Turn the crankshaft pulley 2 revolutions from TDC to TDC, in the clockwise direction. Torque the No. 1 idler pulley mount bolt to 31 ft. lbs. (42 Nm).

18. Installation the remaining components is the reverse of the removal procedure.

3S-GE and 3S-GTE Engines

CELICA (2WD)

1. Disconnect the negative battery cable. Raise the vehicle and support safely. Remove the right front tire and wheel assembly. Remove the right side fender liner.

2. Remove the windshield washer and radiator reservoir tanks. Remove the cruise control actuator, if equipped.

3. Remove the power steering belt. Remove the power steering pump and position it aside with the hydraulic lines still attached.

4. Remove the alternator and support bracket. Remove the upper timing belt cover.

5. Set the No. 1 piston to TDC of the compression stroke by aligning the groove on the crankshaft pulley with the **0** mark on the lower timing belt cover. Check that the matchmarks on the 2 camshaft timing pulleys and the rear timing belt cover are aligned, if not, turn the crankshaft 1 complete revolution clockwise (360 degrees).

6. If the timing belt is to be reused, draw a directional arrow on it and matchmark the belt to the 2 camshaft pulleys. Loosen the No. 1 idler pulley bolt and shift the pulley as far left as possible; tighten the set bolt. Remove the timing belt from the 2 camshaft pulleys. Support the belt so the meshing of the belt with the remaining pulleys does not shift.

7. Carefully hold the camshafts with an adjustable wrench and remove the camshaft pulley set bolts. Remove the pulleys and their set pins.

8. Remove the crankshaft pulley. Remove the lower timing belt.

9. Remove the timing belt guide and then remove the timing belt from the remaining pulleys. Be sure to matchmark the belt to the pulleys if it is to be reused.

10. Remove the No. 1 idler pulley and the tension spring. Remove the No. 2 idler pulley, the crankshaft timing pulley and the oil pump pulley.

To install:

11. Install the oil pump pulley and tighten it to 21 ft. lbs. (28 Nm). Install the crankshaft timing pulley by sliding it onto the crankshaft over the Woodruff key. Install the No. 2 idler pulley and tighten it to 32 ft. lbs. (43 Nm).

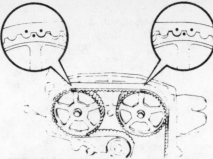

If the timing belt is to be reused on the 3S-GE engine, place matchmarks on the belt and pulleys

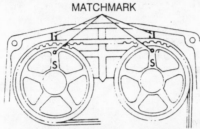

Align the matchmarks on the camshaft timing pulleys with those on the rear timing belt cover—3S-GE engine

Hold the camshaft with an adjustable wrench when removing the camshaft timing pulley—3S-GE engine

12. Install the No. 1 idler pulley and the tension spring. Move the pulley as far to the left as it will go and then tighten it.

13. Install the timing belt on all pulleys except the 2 camshaft pulleys. Make sure the matchmarks made earlier are in alignment.

14. Install the timing belt guide with the cup side out. Install the lower timing belt cover and then install the crankshaft pulley. Tighten it to 80 ft. lbs. (108 Nm).

15. Check that the No. 1 cylinder is at TDC of the compression stroke for the crankshaft. The crankshaft pulley groove should be aligned with the **0** mark on the lower timing belt cover.

16. Check that the No. 1 cylinder is at TDC of the compression stroke for the camshaft.

NOTE: There are 2 types of camshafts, one with 2 holes on the

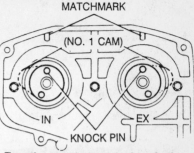

Turn the camshaft so the knock pins align with the matchmark on the rear timing cover and the No. 1 lobes are facing outward—3S-GA engine (2 hole type)

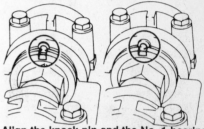

Align the knock pin and the No. 1 bearing cap mark—3S-GE engine (5 hole type)

HOLE TYPE

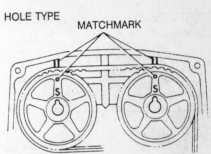

Align the camshaft knock pin with the hole in the camshaft timing pulley—3S-GE engine (1 hole type pulley only)

FIVE HOLE TYPE

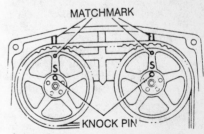

Insert the knock pin into whichever camshaft timing pulley and camshaft holes are aligned—3S-GE engine (5 hole type pulley only)

timing pulley contact surface and one with 5 holes on the timing pulley contact surface. All replacement camshaft have 5 holes.

2 Hole: Using a wrench, turn the camshafts so the camshaft knock pin aligns with the matchmark on the rear

timing belt cover. And the No. 1 cam lobe is pointing outward.

5 Hole: Using a wrench, turn the camshaft so the knock pin aligns with the notch in the No. 1 camshaft bearing cap.

17. Hang the timing belt on the 2 camshaft timing pulleys. Align all matchmarks made during removal. The **S** mark on the pulley should face outward.

NOTE: There are 2 types of camshaft pulleys. One has 5 holes on the camshaft contact surface and one has 1 hole on the contact surface. All replacement pulleys have 5 holes.

18. Align the timing pulley matchmark with the rear timing belt cover matchmark and install the pulleys with the belt.

NOTE: On 1 hole pulleys, match the camshaft knock pin with the camshaft pulley hole. On 5 hole pulleys, insert the knock pin into whichever pulley and camshaft holes are aligned.

19. Hold the camshaft with an adjustable wrench and tighten the pulley set bolt to 43 ft. lbs. (59 Nm).

20. Loosen the No. 1 idler pulley set bolt just enough to move the pad earlier are in alignment.

21. Install the timing belt guide with the cup side out. Install the lower timing belt cover and then install the crankshaft pulley. Tighten it to 80 ft. lbs. (108 Nm).

22. Using a wrench, turn the camshaft so the knock pin aligns with the notch in the No. 1 camshaft bearing cap. Install the camshaft pulleys so the **S** mark faces up. Then, align the holes in the pulley and camshaft and insert the knockpin. Hold the camshaft with an adjustable wrench and tighten the pulley set bolt to 43 ft. lbs. (59 Nm).

23. Check that the No. 1 cylinder is at TDC of the compression stroke for the crankshaft. The crankshaft pulley groove should be aligned with the **0** mark on the lower timing belt cover.

24. Align the timing marks on the camshaft pulleys with the cut-outs in the inner timing belt cover.

25. Install the timing belt. Make sure there is tension between the intake camshaft pulley and the crankshaft pulley.

26. Set the timing belt tensioner as follows:

a. Mount the tensioner in a vise and align the holes in the tensioner pushrod with the holes in the housing.

b. Using a press, slowly press the pushrod into the housing until the holes are exactly aligned.

c. Insert a 1.27mm hex wrench

through the holes to retain the position of the pushrod.

d. Release the press.

27. Using a torque wrench, turn the No. 1 idler pulley bolt counterclockwise to 13 ft. lbs. (18 Nm), then temporarily install the timing belt tensioner with the 2 mounting bolts.

28. Slowly turn the crankshaft pulley $^5/_6$ turn and align the groove in the pulley with the 60 degrees ATDC mark on the timing belt cover. Always turn the crankshaft in the clockwise direction.

29. Insert a 0.75 in. (1.90mm) feeler gauge between the tensioner body and the No. 1 idler pulley stopper.

30. Using a suitable torque wrench, turn the No. 1 idler pulley bolt counterclockwise to 13 ft. lbs. (18 Nm). Push on the tensioner and torque the tensioner retaining bolts to 15 ft. lbs. (21 Nm).

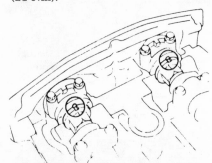

Aligning the camshaft grooves with the drilled marks in the No. 1 camshaft bearing caps — 3S-GTE engine

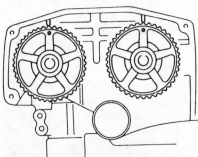

Aligning the camshaft and timing belt cover marks — 3S-GTE engine

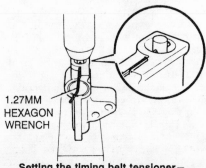

1.27MM HEXAGON WRENCH

Setting the timing belt tensioner — 3S-GTE engine

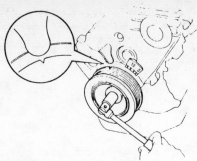

Aligning crankshaft pulley groove with 60 degree ATDC mark on the timing belt cover

1.9MM FEELER GAUGE

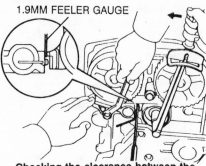

Checking the clearance between the tensioner body and the No. 1 idler pulley stopper

31. Remove the 1.27mm hex wrench from the tensioner.

32. Slowly turn the crankshaft pulley in the clockwise direction $^5/_6$ turn and align the groove in the pulley with the 60 degrees ATDC mark on the timing belt cover. Always turn the crankshaft in the clockwise direction.

33. Turn the No. 1 idler pulley bolt to 18 ft. lbs. (25 Nm).

34. Using a feeler gauge, check the clearance between the tensioner body and the No. 1 idler pulley stop. The clearance should be 0.071–0.087 in. (1.8–2.2mm). If the clearance is not as specified, remove the tensioner and reinstall it.

35. Installation of the remaining components is the reverse of the removal procedure.

4A-F and 4A-FE Engines

1. Raise and support the vehicle safely. Remove the right wheel and undercover. Remove the air cleaner.

2. Remove the drive belts. Remove the power steering pump and the air conditioning compressor, with brackets, and position them aside. Leave the hydraulic and refrigerant lines connected.

3. Remove the spark plugs and the cylinder head cover. Rotate the crankshaft pulley so the **0** mark is in alignment with the groove in the No. 1 front cover. Check that the lifters on the No. 1 cylinder are loose; if not, turn the crankshaft 1 complete revolution (360 degrees).

4. Position a floor jack under the engine and remove the right side engine mounting insulator.

5. Remove the water pump and crankshaft pulleys. The crankshaft pulley will require a 2-armed puller.

6. Loosen the 9 bolts and remove the No. 1, No. 2 and No. 3 front covers. Remove the timing belt guide.

7. Loosen the bolt on the idler pulley, push it to the left as far as it will go and then retighten it. If reusing the timing belt, draw an arrow on it in the direction of engine revolution (clockwise) and then matchmark the belt to the pulleys.

8. Remove the timing belt. Remove the idler pulley bolt, the pulley and the tension spring.

9. Remove the crankshaft timing pulley.

10. Lock the camshaft and remove the camshaft timing pulleys.

To install:

11. Install the camshaft timing pulley so it aligns with the knockpin on the exhaust camshaft. Tighten the pulley to 34 ft. lbs. (47 Nm) – 1988; 43 ft. lbs. (59 Nm) – 1989–92. Align the mark on the No. 1 camshaft bearing cap with the center of the small hole in the pulley.

12. Install the crankshaft timing pulley so the marks on the pulley and the oil pump body are in alignment.

13. Install the idler pulley and its tension spring, move it to the left as far as it will go and tighten it temporarily.

14. Align the matchmarks made during removal and then install the timing belt on the camshaft pulley. Loosen the idler pulley set bolt. Make sure the timing belt meshing at the crankshaft pulley does not shift.

15. Rotate the crankshaft clockwise 2 revolutions from TDC to TDC. Make sure each pulley aligns with the marks made previously. If the marks are not in alignment, the valve timing is wrong. Shift the timing belt meshing slightly and then repeat Steps 14–15.

16. Tighten the set bolt on the timing belt idler pulley to 27 ft. lbs. (37 Nm). Measure the timing belt deflection at the top span between the 2 camshaft pulleys. It should deflect no more than 0.16 in. (4mm) at 4.4 lbs. of pressure – 1988; 0.24 in. (7mm) at 4.4 lbs. of pressure – 1989–92. If deflection is greater, readjust by using the idler pulley.

17. Installation of the remaining components is the reverse of the removal procedure.

4A-GE, 4A-GEC, 4A-GELC and 4A-GZE Engines

1. Disconnect the negative battery cable. Disconnect the No. 2 air cleaner hose from the air cleaner.

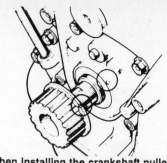

When installing the crankshaft pulley, make sure the TDC marks on the oil pump body and the pulley are in alignment—4A-GE engine

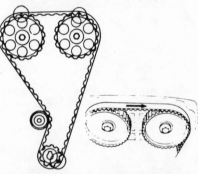

If the timing belt is to be reused, draw a directional arrow and matchmark the belt to the pulleys as shown—4A-GE engine

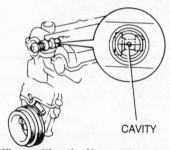

When setting the No. 1 cylinder at TDC on the 4A-GE engine, remove the oil filler cap and check that the cavity in the camshaft is visible

2. If equipped with power steering, remove the power steering pump and position it aside. Do not disconnect the pump hydraulic lines.

3. Loosen the water pump pulley set nuts, remove the drive belt adjusting bolt and then remove the drive belt. Remove the set nuts and then remove the fluid coupling along with the fan and the water pump pulley.

4. Remove the spark plugs. Rotate the crankshaft pulley so the groove on it is in alignment with the **0** mark on the No. 1 timing belt cover. Remove the oil filler cap and check that the cavity in the camshaft is visible. If not, turn the camshaft 1 complete revolution (360 degrees).

5. On the MR2 and 1988–92 Corol-

la, remove the right side engine mount insulator.

6. Lock the crankshaft pulley and remove the pulley bolt. Using a gear puller, remove the crankshaft pulley. Remove the three timing belt covers and their gaskets. Remove the timing belt guide.

7. Loosen the bolt on the idler pulley, push it to the left as far as it will go and then retighten it. If reusing the timing belt, draw an arrow on it in the direction of engine revolution (clockwise) and then matchmark the belt to the pulleys.

8. Remove the timing belt. Remove the idler pulley bolt, the pulley and the tension spring.

9. Remove the cylinder head covers, lock the camshaft and remove the camshaft timing pulleys.

To install:

10. Install the camshaft timing pulleys and cylinder head covers. Tighten the pulley to 34 ft. lbs. (47 Nm).

11. Install the crankshaft timing pulley so the marks on the pulley and the oil pump body are in alignment.

12. Install the idler pulley and its tension spring, move it to the left as far as it will go and tighten it temporarily.

13. Install the timing belt. If the old one is being used, align all the marks made during removal.

14. Installation of the remaining components is the reverse order of the removal procedure.

5M-GE Engine

1. Disconnect the negative battery cable.

2. Loosen the mounting bolts of each of the crankshaft-driven components at the front of the engine and remove the drive belts.

3. Rotate the crankshaft in order to set the No. 1 cylinder to TDC of its compression stroke (both valves of the No. 1 cylinder closed, and TDC marks aligned).

4. Remove the upper and front timing belt cover and gaskets.

5. Loosen the idler pulley bolt and lever the idler pulley toward the alternator side of the engine in order to relieve the tension on the timing belt. Hand tighten the idler pulley bolt.

6. Remove the timing belt from the camshaft pulleys.

7. Remove the camshaft timing pulleys as follows. Hold the pulleys stationary with a spanner wrench. Remove the center pulley bolt. Do not attempt to use timing belt tension as a tool to remove the center pulley bolts, as the belt could become damaged.

NOTE: Do not interchange the intake and exhaust timing pulleys, as they differ for use with each camshaft.

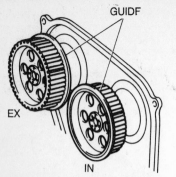

5M-GE engines—When installing the camshaft sprockets, be sure that the guides are positioned as shown. (IN—intake camshaft sprocket; EX—exhaust camshaft sprocket)

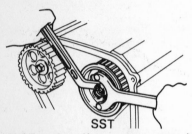

5M-GE engine—Use a spanner wrench 9SST, as shown, to hold the camshaft sprocket while loosening the camshaft sprocket bolt. Do not attempt to use belt tension to hold the sprocket in place while removing the camshaft sprocket bolt

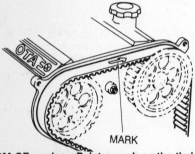

5M-GE engine—Paint a mark on the timing belt prior to belt removal to indicate the belts direction of normal rotation. Point the mark in the same direction if the belt is to be reinstalled

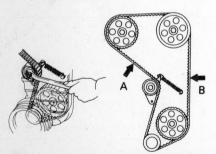

5M-GE engine—When adjusting the timing belt tension, be sure the tension at "A" is the same as that at "B"

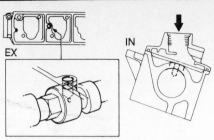

5M-GE engine—Proper alignment of the camshaft matchmarks with the match holes of the camshaft housings (IN—intake; EX—exhaust)

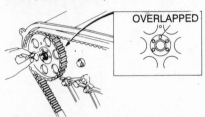

5M-GE engine—Locating the overlapped holes of the camshaft and the camshaft sprocket. Install the match pin into the aligned set of holes (typical of either the intake or exhaust camshaft)

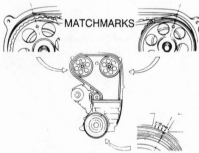

5M-GE engine—Alignment of the camshaft sprocket marks with the No. 2 timing cover marks. Note the position of the crankshaft pulley (TDC)

8. Remove the center crankshaft pulley bolt. Using a puller, remove the crankshaft pulley.

9. Using chalk or crayon, mark the timing belt to indicate its direction of rotation. This mark must face the same direction during installation of the belt.

10. Remove the lower timing belt cover, then the belt.

11. If damaged, the crankshaft pulley can be removed using a puller; the oil pump driveshaft pulley can be removed in the same manner as the camshaft pulleys.

12. Inspect the timing belt for damage, such as cuts, cracks, missing teeth, abrasions, nicks, etc. If the belt teeth are damaged, check that the camshafts rotate freely and correct as necessary.

13. Should damage be evident on the belt face, check the idler pulley belt surface for damage. If damage is present on one side of the belt only, check the belt guide and the alignment of each pulley. If the belt teeth are excessively worn, check the timing belt cover gasket for damage and/or proper installation.

14. Check the idler pulley for damage and smoothness of rotation. Also check the free length of the tension spring, which should be 2.8 in. (71mm), measured between the inside of each end "clip". Replace the spring if the length exceeds this limit.

To install:

15. Install the crankshaft and oil pump driveshaft if these items were removed previously. Torque the oil pump driveshaft center pulley bolt to 16 ft. lbs. (22 Nm). The crankshaft pulley must be evenly driven into place.

16. Install the idler pulley and the tension spring. Lever the pulley towards the alternator side of the engine and tighten the bolt.

17. Check the mark made during Step 9 of removal and temporarily install the timing belt on the crankshaft pulley. The mark must face in the same direction as it did originally.

18. Install the lower timing belt cover. Install the crankshaft pulley and torque the center pulley bolt to 195 ft. lbs. (265 Nm).

19. Remove the oil filter cap of the intake camshaft cover, and the complete camshaft cover on the exhaust side.

20. Check that the match holes of both No. 2 camshaft journals are visible through the camshaft housing match holes. If necessary, temporarily install the camshaft pulley and guide pin, and rotate the camshaft(s) until the holes are aligned.

21. Install the timing pulleys. Note that the belt guide of the exhaust camshaft pulley should be positioned towards the engine; the belt guide of the intake camshaft pulley should be positioned away from the engine. Do not yet install the pulley retaining bolts.

22. Align the following marks. Each camshaft pulley mark must be aligned with its respective mark on the rear, upper timing belt cover. Align the crankshaft pulley notch with the TDC (0) mark of the timing tab.

NOTE: The No. 1 cylinder must be positioned at TDC on its compression stroke.

23. Install the timing belt.

24. Loosen the idler pulley bolt and tension the timing belt. The timing belt tension must be the same between the exhaust camshaft pulley and the crankshaft pulley, as it is between the intake camshaft pulley and the oil pump driveshaft pulley.

25. There are 5 pin holes on each camshaft and each timing pulley. On

the exhaust side: Install the match pin into the one hole of the pulley which is aligned with one of the camshaft pin holes. Repeat this on the intake side. Only one of the holes of each side should be aligned to allow insertion of the match pins.

26. Using a spanner wrench to hold the camshaft pulleys, install and tighten the camshaft pulley bolts. These bolts should be torqued to 51 ft. lbs. (69 Nm).

27. Install the exhaust camshaft cover, using a new gasket. Install the oil filler cap. Install the timing belt cover and gasket.

28. Install and adjust the drive belts at the front of the engine. Reconnect the battery cable.

7M-GE and 7M-GTE Engines

1. Disconnect the negative battery cable. Drain the cooling system. Remove the radiator. Remove the water outlet.

2. Remove the spark plugs. Remove the drive belts. Remove the No. 3 timing belt cover.

3. Position the engine at TDC on the compression stroke. Remove the timing belt from the camshaft sprockets. If reusing the belt, matchmark the belt and the sprockets in the direction of engine rotation.

4. Remove the camshaft pulleys. Remove the crankshaft pulley using the proper removal tools. Remove the power steering air pipe, if equipped.

5. If equipped with air conditioning remove the compressor and position it aside. Do not disconnect the refrigerant lines.

6. Remove the No. 1 timing belt cover. Remove the timing belt. Remove the idler pulley and the tension spring. Remove the oil pump drive pulley.

To install:

7. Inspect the belt for defects. Replace as required. Inspect the idler pulleys and springs. Replace defective components as required.

8. Install the oil pump drive pulley and retaining bolt. Tighten the bolt to 16 ft. lbs. (22 Nm).

9. Install the crankshaft timing pulley. Temporarily install the idler pul-

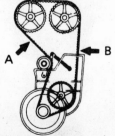

7M-GE and 7M-GTE engines timing belt tension check

ley and tension spring. Tighten the assembly to 36 ft. lbs. (49 Nm). Pry the idler pulley toward the left as far as it will go and temporarily tighten the bolt.

10. Temporarily install the timing belt. If reusing the old belt install it using the marks made during the removal procedure. Install the No. 1 timing belt cover.

11. If equipped with air conditioning, install the compressor assembly. If equipped, install the power steering air pipe.

12. Align the set key with the key groove and install the crankshaft pulley and torque the retaining bolt to 195 ft. lbs. (265 Nm).

13. Install the camshaft timing pulleys. Torque the retaining bolts to 36 ft. lbs. (49 Nm).

14. Loosen the idler pulley bolt. Install the timing belt to the INTAKE side and the EXHAUST side. Tighten the idler pulley bolt to 36 ft. lbs. (49 Nm).

15. Make sure the timing belt tension **A** is equal to the timing belt tension **B**. If not adjust the idler pulley. Turn the engine 2 complete revolutions in the clockwise direction and check to see that everything is aligned properly.

16. Turn both the intake and exhaust camshaft pulleys inward at the same time to loosen the timing belt between the 2 sprockets. Belt deflection should be 4.4–6.6 lbs. If not adjust the idler pulley.

17. Install the remaining components, start the engine and check for leaks and proper operation.

Timing Sprockets

REMOVAL & INSTALLATION

Timing sprocket/pulley removal and installation procedures are detailed within the individual Timing Belt sections.

Camshaft

REMOVAL & INSTALLATION

2VZ-FE Engine

NOTE: Due to a nominal thrust clearance, the camshafts must be held absolutely level during removal. If not, the section of the cylinder head receiving the thrust may crack or be damaged, thus causing the camshaft to break.

1. Remove the cylinder head(s).

2. Rotate the exhaust camshaft in the right cylinder head until the 2 pointed marks on the camshaft drive and driven gears are aligned.

3. Secure the exhaust camshaft sub-gear to the driven gear with bolt. This is important as it will eliminate the torsional spring force of the sub-gear.

4. Loosen the bearing cap bolts in the proper sequence and then remove the 4 bearing caps and the right side exhaust camshaft.

5. Loosen the bearing cap bolts in the proper sequence and then remove the 5 bearing caps and the right side intake camshaft.

NOTE: Be sure to arrange all the bearing caps in their proper order.

6. Rotate the exhaust camshaft in the left cylinder head until the pointed mark on the camshaft drive and driven gears are aligned.

7. Secure the exhaust camshaft sub-gear to the driven gear with bolt. This is important as it will eliminate the torsional spring force of the sub-gear.

8. Loosen the bearing cap bolts in the proper sequence and then remove the 4 bearing caps and the left side exhaust camshaft.

9. Loosen the bearing cap bolts in the proper sequence and then remove the 5 bearing caps and the left side intake camshaft.

NOTE: Be sure to arrange all the bearing caps in their proper order.

To install:

10. Coat the thrust portion of the right side intake camshaft with suitable grease and then position the camshaft into the head so the 2 timing marks are at a 90 degree angle to the head.

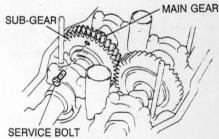

Install a service bolt in the exhaust camshaft sub-gear (right)—2VZ-FE engine

Exhaust camshaft bearing cap bolt loosening sequence (right)—2VZ-FE engine

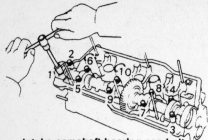

Intake camshaft bearing cap bolt loosening sequence (right) – 2VZ-Fe engine

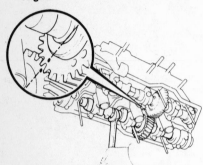

Align the single mark on the exhaust camshaft (left) – 2VZ-FE engine

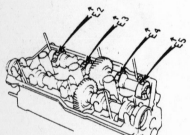

Exhaust camshaft bearing cap installation (right) – 2VZ-FE engine

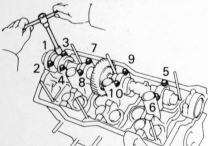

Intake camshaft bearing cap bolt loosening sequence (left) – 2VZ-FE engine

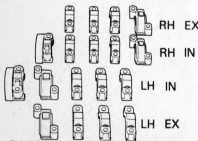

Camshaft bearing cap identification – 2VZ-FE engine

RH EX
RH IN
LH IN
LH EX

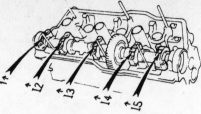

Intake camshaft bearing cap installation (right) – 2VZ-FE engine

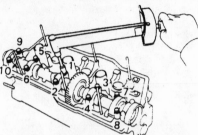

Intake camshaft bearing cap bolt tightening sequence (right) – 2VZ-FE engine

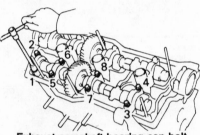

Exhaust camshaft bearing cap bolt loosening sequence (left) – 2VZ-FE engine

SERVICE BOLT (B)

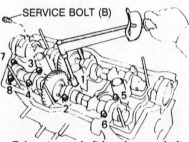

Exhaust camshaft bearing cap bolt tightening sequence (right) – 2VZ-FE engine

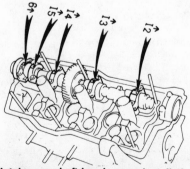

Intake camshaft bearing cap installation (left) – 2VZ-FE engine

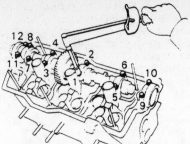

Intake camshaft bearing cap bolt tightening sequence (left) – 2VZ-FE engine

Exhaust camshaft bearing cap installation (left) – 2VZ-FE engine

SERVICE BOLT (B)

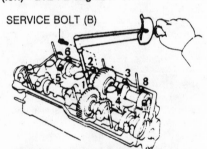

Exhaust camshaft bearing cap bolt tightening sequence (left) – 2VZ-FE engine

11. Coat the edges of the No. 1 bearing cap with sealant and then install all 5 caps in their proper locations. Coat the bolts with engine oil and then tighten them, in sequence, in several stages, to 12 ft. lbs. (16 Nm).

12. Coat the thrust portion of the right side exhaust camshaft with grease and then position the camshaft into the head so the 2 timing marks align with those on the intake shaft.

13. Install all 4 bearing caps in their proper locations. Coat the bolts with engine oil and then tighten them, in sequence, in several stages, to 12 ft. lbs. (16 Nm).

14. Remove the service bolt.

15. Coat the thrust portion of the left side intake camshaft with grease and then position the camshaft into the head so the timing mark is at a 90 degree angle to the head.

16. Coat the edges of the No. 1 bearing cap with sealant and then install all 5 caps in their proper locations. Coat the bolts with engine oil and then tighten them, in sequence, in several stages, to 12 ft. lbs. (16 Nm).

17. Coat the thrust portion of the left side exhaust camshaft with grease and then position the camshaft into the head so the timing mark aligns with the one on the intake shaft.

18. Install all 4 bearing caps in their proper locations. Coat the bolts with engine oil and then tighten them, in sequence, in several stages, to 12 ft. lbs. (16 Nm).

19. Remove the service bolt.

20. Installation of the remaining components is in the reverse order of removal.

3S-FE and 5S-FE Engines

1. Remove the cylinder head.

2. To remove the exhaust camshaft, set the knock pin of the exhaust camshaft at 10–45 degree BTDC of camshaft angle. This angle will help to lift the exhaust camshaft level and evenly by pushing No. 2 and No. 4 cylinder camshaft lobes of the exhaust camshaft toward their valve lifters.

3. Secure the exhaust camshaft sub-gear to the main gear using a service bolt. When removing the exhaust camshaft be sure the torsional spring force of the sub-gear has been eliminated.

4. Remove the No. 1 and No. 2 rear bearing cap bolts and remove the cap. Uniformly loosen and remove bearing cap bolts No. 3 to No. 8 in several passes and in the proper sequence. Do not remove bearing cap bolts No. 9 and 10 at this time. Remove the No. 1, 2 and 4 bearing caps.

5. Alternately loosen and remove bearing cap bolts No. 9 and 10. As these bolts are loosened check to see that the camshaft is being lifted out straight and level.

NOTE: If the camshaft is not lifted out straight and level retighten No. 9 and 10 bearing cap bolts. Reverse Steps 4 through 1, than start over from Step 3. Do not attempt to pry the camshaft from its mounting.

6. Remove the exhaust camshaft from the engine.

7. To remove the intake camshaft, set the knock pin of the intake camshaft at 80–115 degrees BTDC of camshaft angle. This angle will help to lift the intake camshaft level and evenly by pushing No. 1 and No. 3 cylinder camshaft lobes of the intake camshaft toward their valve lifters.

8. Remove the No. 1 and No. 2 front bearing cap bolts and remove the front bearing cap and oil seal. If the cap will not come apart easily, leave it in place without the bolts.

9. Uniformly loosen and remove bearing cap bolts No. 3 to No. 8 in several phases and in the proper sequence. Do not remove bearing cap

bolts No. 9 and 10 at this time. Remove No. 1, 3 and 4 bearing caps.

10. Alternately loosen and remove bearing cap bolts No. 9 and 10. As these bolts are loosened and after breaking the adhesion on the front bearing cap, check to see that the camshaft is being lifted out straight and level.

NOTE: If the camshaft is not lifted out straight and level retighten No. 9 and 10 bearing cap bolts. Reverse Steps 10 through 7, than start over from Step 8. Do not attempt to pry the camshaft from its mounting.

11. Remove the intake camshaft from the engine.
To install:

12. Before installing the intake camshaft, apply multi-purpose grease to the thrust portion of the camshaft. Position the camshaft at 80 degrees BTDC of camshaft angle on the cylinder head. Apply seal packing kit 08826–00080 or equivalent, and apply it to the front bearing cap. Coat the bearing cap bolts with clean engine oil. Uniformly and in several phases tighten the camshaft bearing caps to 14 ft. lbs. (19 Nm).

13. To install the exhaust camshaft, set the knock pin of the camshaft at 10 degrees BTDC of camshaft angle. Apply multi-purpose grease to the thrust portion of the camshaft. Position the exhaust camshaft gear with the intake camshaft gear so the timing marks are in alignment with one another. Be sure to use the proper alignment marks on the gears. Do not use the assembly reference marks.

14. Turn the intake camshaft clockwise or counterclockwise little by little until the exhaust camshaft sits in the bearing journals evenly without rocking the camshaft on the bearing journals.

15. Coat the bearing cap bolts with clean engine oil. Uniformly and in several phases tighten the camshaft bearing caps to 14 ft. lbs. (19 Nm). Remove the service bolt from the assembly.

16. Installation of the remaining components is the reverse of the removal procedure.

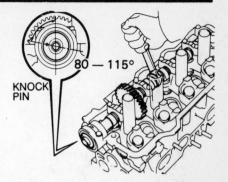

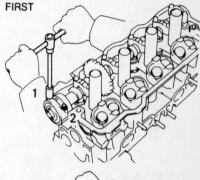

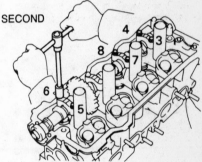

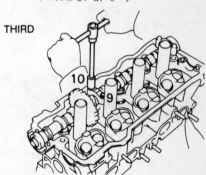

Intake camshaft removal procedure – 3S-FE and 5S-FE engines

FRONT   NO. 1   NO. 2   NO. 3   NO. 4

Intake camshaft bearing positioning – 3S-FE and 5S-FE engines

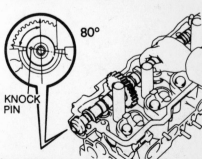

Installing the intake camshaft – 3S-FE and 5S-FE engines

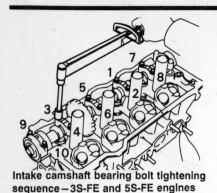

Intake camshaft bearing bolt tightening sequence—3S-FE and 5S-FE engines

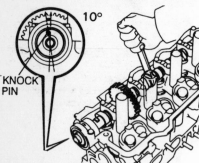

Exhaust camshaft installation—3S-FE and 5S-FE engines

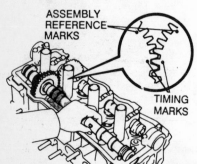

Intake and exhaust camshaft engagement—3S-FE and 5S-FE engines

No. 1   No. 2   No. 3   No. 4   Rear

Exhaust camshaft bearing cap positioning—3S-FE and 5S-FE engines

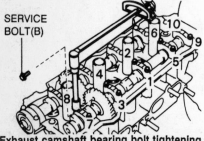

Exhaust camshaft bearing bolt tightening sequence—3S-FE and 5S-FE engines

4A-F and 4A-FE Engines

1. Disconnect the negative battery cable. Drain the cooling system.

2. Remove the spark plugs and the cylinder head cover.

3. Remove the No. 3 and No. 2 front covers. Turn the crankshaft pulley and align its groove with the **0** mark on the No. 1 front cover. Check that the camshaft pulley hole aligns with the mark on the No. 1 camshaft bearing cap (exhaust side).

4. Remove the plug from the No. 1 front cover and matchmark the timing belt to the camshaft pulley. Loosen the idler pulley mounting bolt and push the pulley to the left as far as it will go; tighten the bolt. Slide the timing belt off the camshaft pulley and support it so it won't fall into the case.

5. Remove the camshaft pulley and check the camshaft thrust clearance. Remove the camshafts.

NOTE: Due to the relatively small amount of camshaft thrust clearance, the camshaft must be held level during removal. If the camshaft is not level on removal, the portion of the head receiving the thrust may crack or be damaged.

6. Set the service bolt hole on the intake camshaft gear (the one not attached to the timing pulley) at the 12 o'clock position so the Nos. 1 and 3 cylinder camshaft lobed can push their lifters evenly. Loosen the No. 1 bearing caps on each camshaft a little at a time and remove them.

7. Secure the intake camshaft subgear to the main gear with a service bolt to eliminate any torsional spring force. Loosen the remaining bearing caps a little at a time, in the proper sequence and remove the intake camshaft. If the camshaft cannot be lifted out straight and level, retighten the bolts in the No. 3 bearing cap and loosen them a little at a time with the gear pulled up.

8. Turn the exhaust camshaft approximately 105 degrees so the knock pin is about 5 minutes before the 6:30 o'clock position. Loosen the remaining bearing caps a little at a time, in the proper sequence and remove the exhaust camshaft. If the camshaft can not be lifted out straight and level, retighten the bolts in the No. 3 bearing cap and loosen them a little at a time with the gear pulled up.

To install:

9. Position the exhaust camshaft into the cylinder head as it was removed. Position the bearing caps over each journal so the arrows point forward and then tighten the bolts gradually, in the proper sequence to 9 ft. lbs. (13 Nm).

10. Coat the lip of a new oil seal with MP grease and drive it into the camshaft.

11. Set the knock pin on the exhaust camshaft so it is just above the edge of the cylinder head and engage the intake camshaft gear to the exhaust gear so the mark on each gear is in alignment. Roll the intake camshaft down onto the bearing journals while engaging the gears with each other.

12. Position the bearing caps over each journal on the intake camshaft so the arrows point forward and then tighten the bolts gradually, in the proper sequence to 9 ft. lbs. (13 Nm).

13. Remove the service bolt and install the No. 1 intake bearing cap. If it does not fit properly, pry the camshaft gear backwards until it does. Tighten the bolts to 9 ft. lbs. (13 Nm).

14. Rotate the camshafts 1 revolution (360 degrees) from TDC to TDC and check that the marks on the 2 gears are still aligned.

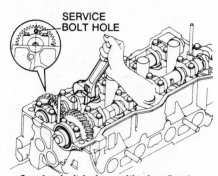

Service bolt hole positioning (intake camshaft)—4A-F and 4A-FE engines

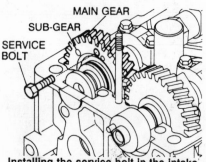

Installing the service bolt in the intake camshaft—4A-F and 4A-FE engines

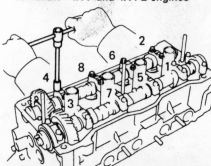

Intake camshaft bearing cap loosening sequence—4A-FE engine (1990-92)

15. Install the camshaft timing pulley making sure the camshaft knock pins and the matchmarks are in alignment. Lock each camshaft and tighten the pulley bolts to 43 ft. lbs. (59 Nm).

16. Align the matchmarks made during removal and then install the timing belt on the camshaft pulley. Loosen the idler pulley set bolt. Make sure the timing belt meshing at the crankshaft pulley does not shift.

17. Rotate the crankshaft clockwise 2 revolutions from TDC to TDC. Make sure each pulley aligns with the marks made previously.

18. Tighten the set bolt on the timing belt idler pulley to 27 ft. lbs. (37 Nm). Measure the timing belt deflection at the top span between the 2 camshaft pulleys. It should deflect no more than 0.24 in. (7mm) at 4.4 lbs. of pressure. If deflection is greater, readjust by using the idler pulley.

19. Installation of the remaining components is the reverse of the removal procedure.

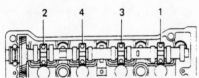

Camshaft bearing cap bolt loosening sequence—4A-F and 4A-FE engines (1988–89)

Exhaust camshaft bearing cap loosening sequence—4A-FE engine (1990–92)

Knockpin positioning on the exhaust camshaft—4A-F and 4A-FE engines

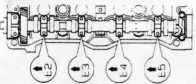

Exhaust camshaft bearing cap positioning—4A-F and 4A-FE engines

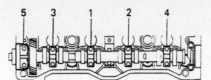

Camshaft bearing cap bolt tightening sequence—4A-F and 4A-FE engines (1988–89)

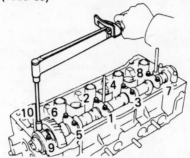

Exhaust camshaft bearing cap tightening sequence—4A-FE engine (1990–92)

Intake camshaft bearing cap tightening sequence—4A-FE engine (1990–92)

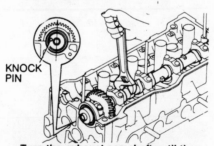

KNOCK PIN

Turn the exhaust camshaft until the knockpin is here—4A-F and 4A-FE engines

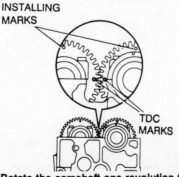

INSTALLING MARKS

TDC MARKS

Rotate the camshaft one revolution from TDC to TDC and check that the marks are lined up—4A-F and 4A-FE engines

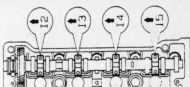

Intake camshaft bearing cap positioning—4A-F and 4A-FE engines

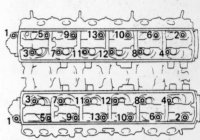

5M-FE engine camshaft housing bolt removal sequence. Loosen bolts gradually on three passes

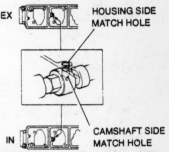

EX

HOUSING SIDE MATCH HOLE

IN

CAMSHAFT SIDE MATCH HOLE

5M-GE engine camshaft housing torque sequence

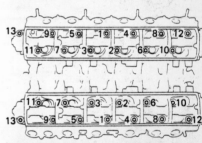

Before installing the camshaft housings, align the match hole on each No. 2 cam journal with the hole in the housing

5M-GE Engine

1. Remove the 2 camshaft covers.

2. Remove the timing belt assembly.

3. Following the sequence shown, loosen the camshaft housing nuts and bolts in 3 passes. Remove the housings (with camshafts) from the cylinder head.

4. Remove the camshaft housing rear covers. Squirt clean oil down around the cam journals in the housing, to lubricate the lobes, oil seals and bearings as the cam is removed. Begin to pull the camshaft out of the back of the housing slowly, turning and pulling. Remove the cam completely.

5. To install, lubricate the entire camshaft with clean oil. Insert the cam into the housing from the back, and slowly turn it and push it into the housing. Install new O-rings and the housing end covers.

6. Installation of the remaining components is in the reverse order of removal. Tighten camshaft housing bolts to 15–17 ft. lbs. (20–23 Nm) in the proper sequence.

All Other Engines

The procedure for removing the camshaft is given as part of the cylinder head removal procedure.

NOTE: It will not be necessary to completely remove the cylinder head in order to remove the camshaft(s). Therefore, proceed only as far as necessary, to remove the camshaft, with the cylinder head removal procedure.

Piston and Connecting Rod

POSITIONING

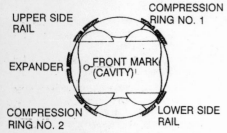

Piston ring gap positioning—2VZ-FE engine

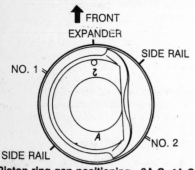

Piston ring gap positioning—3A-C, 4A-C and 4A-LC engines

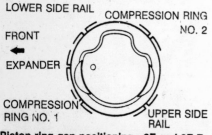

Piston ring gap positioning—3E and 3E-E engines

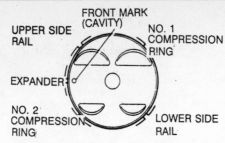

Piston ring gap positioning—3S-GE and 3S-GTE engines

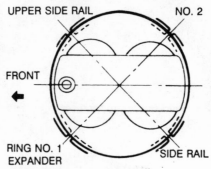

Piston ring gap positioning—4A-F, 4A-FE and 4A-GE (all) engines

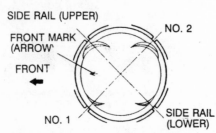

Piston ring gap positioning—4A-GZE engine

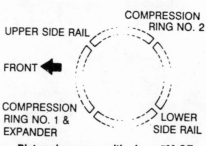

Piston ring gap positioning—5M-GE engine

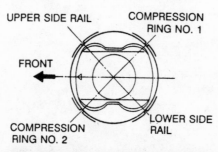

Piston ring gap positioning—7M-GE and 7M-GTE engines

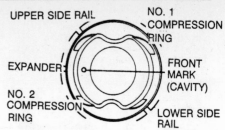

Piston ring gap positioning—3S-FE and 5S-FE engines

ENGINE LUBRICATION

Oil Pan

REMOVAL & INSTALLATION

Corolla

1. Disconnect the negative battery cable. Raise and support the vehicle safely. Drain the oil.

2. Remove the splash shield from under the engine.

3. Place a jack under the transaxle to support it.

4. Remove the bolts which secure the engine rear supporting crossmember to the chassis. On the 4A-GE, remove the center mounting and stiffener plate.

5. Raise the jack under the transaxle, slightly.

6. Remove the front exhaust pipe.

7. Remove the oil pan retaining bolts. Remove the oil pan from the vehicle. If the oil pan does not come out easily, it may be necessary to unbolt the rear engine mounts from the crossmember.

To install:

8. Clean all gasket mating surfaces. Take care of any rust before installation.

9. Install the oil pan and torque the oil pan bolts to 43 inch lbs. (4.9 Nm).

10. Install the remaining components, start the engine and check for leaks.

Camry

1. Disconnect the negative battery cable. Raise the vehicle and support it safely. Drain the oil.

2. Remove the engine undercover. Remove the dipstick.

3. On the 3S-FE, disconnect the exhaust pipe. Remove the suspension lower crossmember. Remove the engine mounting center member. Remove the front engine mount insulator and bracket; 2VZ-FE only. Remove the stiffener plate.

4. Remove the oil pan retaining bolts. Remove the oil pan.

To install:

5. Clean the gasket mating surfaces. Always use a new pan gasket. Some engines were assembled using RTV gasket material in place of a conventional gasket. In that case, apply a thin (5mm) bead of RTV material to the groove around the pan mating surface.

6. Assemble the pan within 15 minutes. Torque pan bolts to 48 inch lbs. (5.4 Nm). On the 2VZ-FE, tighten the pan bolts to 52 inch lbs. (5.9 Nm).

Cressida, Celica and Supra

3S-FE, 5S-FE, 3S-GE AND 3S-GTE ENGINES

1. Disconnect the negative battery cable. Raise the vehicle and support it safely. Drain the engine oil.

2. Remove the engine undercovers.

3. On the 3S-GE and 5S-FE, disconnect the exhaust pipe from the exhaust manifold.

4. Remove the lower suspension crossmember. Remove the engine mounting center member.

5. Remove the engine stiffener plate and the oil level gauge.

6. Remove the oil pan retaining bolts. Remove the oil pan.

To install:

7. Clean all gasket mating surfaces. Take care of any rust before installation..

8. Apply a 5mm bead of RTV gasket material to the groove around the pan flange. Apply the oil within 3 minutes of application.

9. Install the oil pan and torque the oil pan bolts to 48 inch lbs. (5.4 Nm).

10. Install the remaining components, start the engine and check for leaks.

5M-GE ENGINE

1. Disconnect the negative battery cable. Raise the vehicle and support it safely. Drain the oil and cooling system.

2. Remove the air cleaner assembly. Mark any disconnected lines and/or hoses for easy reassembly. Remove the oil level gauge.

3. Disconnect the upper radiator hose at the radiator. Loosen the drive belts.

4. Remove the fan shroud bolts. Remove the 4 fluid coupling flange attaching nuts, then remove the fluid coupling along with the fan and the fan shroud.

5. Remove the engine undercover. Remove the exhaust pipe clamp bolt from the exhaust pipe stay.

6. Remove the 2 stiffener plates from the exhaust pipe. If equipped with manual transaxle, remove the clutch housing undercover.

7. Remove the 4 engine mount bolts from each side of the engine.

8. Place a jack under the transaxle and raise the engine about 1¾ in.

9. Remove the oil pan retaining bolts. Remove the oil pan from the engine.

To install:

10. Clean all gasket mating surfaces. Take care of any rust before installation.

11. Use a new oil pan gasket during installation. Apply a small amount of sealer to the oil pan gasket at each of the 4 corners of the oil pan.

12. Torque the oil pan fasteners to 57–82 inch lbs. (8–11 Nm).

13. Install the remaining components, start the engine and check for leaks.

Supra

7M-GE AND 7M-GTE ENGINES

1. Disconnect the negative battery cable. Remove the hood.

2. Raise the vehicle and support it safely. Remove the engine under cover. Drain the engine oil.

3. If equipped with automatic transmission, remove the fluid cooler hose clamp.

4. Remove the No. 1 front suspension crossmember. Remove the front exhaust pipe bracket and stiffener plates.

5. On the 7M-GTE disconnect the engine oil cooler hose from the engine oil pan.

6. Remove the brake hose brackets and clips. Disconnect the intermediate shaft. Disconnect the stabilizer bar links from the lower control arms.

7. Properly support the engine assembly. Remove the engine mounting bolts. Remove the TEMS actuator assembly.

8. Remove the shock absorbers from the body. Disconnect the front suspension member.

9. Remove the oil pan retaining bolts. Remove the oil pan from the engine.

To install:

10. Clean all gasket mating surfaces. Take care of any rust before installation.

11. Install the oil pan and torque the oil pan bolts to 9 ft. lbs. (13 Nm).

12. Install the remaining components, start the engine and check for leaks.

Cressida

7M-GE ENGINE

1. Disconnect the negative battery cable and drain the cooling system.

2. Raise and safely support the vehicle. Remove the engine under cover and drain the oil.

3. Disconnect the front exhaust pipe at the manifold and at the main tube and remove it.

4. Disconnect the automatic transmission oil cooler pipe.

5. Remove the 9 bolts, ground strap, exhaust pipe stay and the engine rear end plate and then remove the stiffener plates.

6. Loosen the bolt and disconnect the intermediate shaft.

7. Disconnect the front suspension crossmember at the front engine mounts. Position a floor jack under the crossmember, remove the remaining mounting bolts and then lower the crossmember.

8. Remove the pan retaining bolts and then carefully pry the pan from the cylinder block.

To install:

9. Clean all gasket mating surfaces. Take care of any rust before installation.

10. Install the oil pan and torque the oil pan bolts to 9 ft. lbs. (13 Nm).

11. Install the remaining components, start the engine and check for leaks.

Tercel

3A-C ENGINE

1. Disconnect the negative battery cable. Drain the cooling system. Remove the radiator.

2. Raise the vehicle and support it safely. Drain the engine oil.

3. Remove the engine under cover. Remove the stabilizer bracket bolts and lower the stabilizer assembly. Remove the right and left stiffener plates.

4. Remove the oil pan retaining bolts. Remove the oil pan from the vehicle.

To install:

5. Clean all gasket mating surfaces. Take care of any rust before installation.

6. Install the oil pan and torque the oil pan bolts to 43 inch lbs. (4.9 Nm).

7. Install the remaining components, start the engine and check for leaks.

3E AND 3E-E ENGINES

1. Disconnect the negative battery terminal. Raise the vehicle and support it safely. Drain the oil.

2. Remove the right engine under cover. Remove the sway bar and any other necessary steering linkage parts.

3. Disconnect the exhaust pipe from the manifold. Raise the engine enough to take the weight off it.

4. Remove the timing belt.

5. Continue to raise the engine enough to remove the oil pan. Remove the oil pan retaining bolts. Remove the oil pan.

To install:

6. Clean all gasket mating surfaces.

Take care of any rust before installation.

7. Install the oil pan and torque the oil pan bolts to 43 inch lbs. (4.9 Nm).

8. Install the remaining components, start the engine and check for leaks.

MR2
4A-GELC AND 4A-GZE ENGINES

1. Disconnect the negative battery cable. Raise and support the vehicle safely. Drain the engine oil.

2. Remove the exhaust manifold pipe. Remove the timing belt. Remove the crankshaft timing pulley.

3. Support the weight of the engine with a floor jack and then remove the right side engine mount.

4. Remove the oil pan retaining bolts. Remove the oil pan.

To install:

5. Clean all gasket mating surfaces. Take care of any rust before installation.

6. Apply a 5mm bead of RTV gasket material to the groove around the pan flange. Apply the oil pan within 3 minutes of application and tighten the mounting bolts and nuts to 43 inch lbs. (6 Nm).

7. Install the remaining components, start the engine and check for leaks.

3S-GTE AND 5S-FE ENGINES

1. Disconnect the negative battery cable.

2. Drain the engine oil and remove the engine under covers.

3. Remove the right engine hood side panel.

4. Remove the brace that runs across the struts.

5. If equipped with cruise control, remove the cruise control actuator assembly and disconnect the accelerator linkage.

6. Remove the front exhaust pipe.

7. On 3S-GTE with air conditioning, unbolt the compressor and move it aside. Leave the refrigerant lines connected. On 5S-FE, remove the air conditioner idler pulley.

8. On 3S-GTE, remove the catalytic converter and the intercooler.

9. Remove the stiffener plate.

10. On 3S-GTE, disconnect the turbocharger outlet hose where it connects to the oil pan.

11. Remove the dipstick.

12. Remove the oil pan 17 bolts and 2 nuts that attach the oil pan to the block.

13. Insert a suitable seal cutting tool between the oil pan and the block. Work the tool around the pan to break the sealant. Remove the oil pan.

To install:

14. Clean all gasket mating surfaces. Take care of any rust before installation.

15. Apply a 5mm bead of RTV gasket material to the groove around the pan flange. Apply the oil pan within 5 minutes of application and tighten the mounting bolts and nuts to 43 inch lbs. (6 Nm).

16. Install the remaining components, start the engine and check for leaks.

Oil Pump

REMOVAL & INSTALLATION
2VZ-FE, 3E, 3E-E and 3S-FE Engines

1. Remove the oil pan. Remove the oil strainer. On the 3E, remove the dipstick.

2. Raise the engine using a chain hoist. Remove the timing belt and pulleys.

3. On the 2VZ-FE, remove the alternator and the air conditioning compressor and bracket. Do not disconnect the refrigerant lines.

4. Remove the oil pump from the engine.

5. Installation is the reverse of the removal procedure. Clean the gasket mating surfaces. Pack the oil pump cavities with petroleum jelly. Start the engine and check for oil pressure.

3A-C, 3S-GE, 3S-GTE (Except MR2), 4A-F and 4A-GE Engines

1. Remove the fan shroud. Raise and support the vehicle safely.

2. Drain the oil. On the Tercel, drain the coolant and remove the radiator.

3. Remove the oil pan and the oil strainer. Remove the oil pan baffle plate on the 4A-GE. Remove the crankshaft pulley and the timing belt. Remove the oil dipstick guide and then the dipstick.

4. Remove the mounting bolts and then use a rubber mallet to carefully tap the oil pump body from the cylinder block.

To install:

5. To install, position a new gasket on the cylinder block.

6. Position the oil pump on the block so the teeth on the pump drive gear are engaged with the teeth of the crankshaft gear.

7. Clean the gasket mating surfaces. Pack the oil pump cavities with petroleum jelly. Start the engine and check for oil pressure.

8. Installation of the remaining components is the reverse of the removal procedure.

3S-GTE (1990–92 MR2) and 5S-FE (1990–92 Celica and MR2) Engines

1. Disconnect the negative battery cable.

2. Drain the engine oil.

3. Remove the oil pan.

4. Remove the oil pump strainer and baffle plate.

5. Connect a suitable lifting device to the engine and raise the engine a small amount.

6. Remove the timing belt.

7. Remove the No. 2 idler pulley, crankshaft timing pulley and oil pump pulley.

8. Remove the oil pump retaining bolts.

9. Remove the oil pump and gasket by carefully tapping on the outside of the oil pump body. Discard the gasket.

NOTE: One of the oil pump bolts is longer than the rest. Make sure this bolt is identified so it may be installed in the original location.

To install:

10. Clean the gasket mating surfaces. Pack the oil pump cavities with petroleum jelly.

11. Use a new oil pump gasket. On 3S-GTE, torque the oil pump bolts to 69 inch lbs. (8 Nm). On 5S-FE, torque the bolts to 82 inch lbs. (9.3 Nm).

12. Installation of the remaining components, start the engine and check for leaks.

All Other Engines

1. Remove the oil pan.

2. Unbolt the oil pump retaining bolts. Remove the oil pump from the engine.

To install:

3. Clean the gasket mating surfaces. Pack the oil pump cavities with petroleum jelly. Start the engine and check for oil pressure.

Rear Main Bearing Oil Seal

REMOVAL & INSTALLATION

NOTE: The 3A-C engine must be removed from the vehicle before this procedure can be attempted.

1. Remove the transmission or transaxle.

2. Remove the clutch cover assembly and flywheel.

3. Remove the oil seal retaining plate, complete with the oil seal.

4. Using a suitable tool pry the old seal from the retaining plate. Be careful not to damage the plate.

5. Install the new seal, carefully, by using a block of wood to drift it into place. Do not damage the seal as a leak will result.

6. Lubricate the lips of the seal with multipurpose grease. Installation is the reverse of removal.

ENGINE COOLING

Radiator

REMOVAL & INSTALLATION

1. Disconnect the negative battery cable.
2. Drain the cooling system.
3. On 1990–92 MR2, remove the front under covers.
4. Remove the radiator hoses.
5. If equipped with an automatic transmission or transaxle, disconnect and plug the oil cooler lines.
6. Remove the ignition coil, ignitor and bracket assembly on the 2VZ-FE.
7. Remove the hood lock from the radiator upper support, as required. It may be necessary to remove the grille in order to gain access to the hood lock/radiator support assembly.
8. Remove the fan shroud, as required. If equipped with an electric fan (2 on the MR2), disconnect the wiring harness and thermo-switch connectors.
9. Disconnect the overflow hose from the thermal expansion tank and remove the tank from its bracket.
10. Unbolt and remove the radiator upper support.
11. Remove the radiator retaining bolts. Raise the radiator and cooling fan(s) from the lower supports and remove from vehicle.

To install:

12. Lower the radiator and cooling fan(s) onto the lower supports and install the retaining bolts.
13. Install the radiator upper support.
14. Mount the thermal expansion tank and connect the overflow hose.
15. Connect the cooling fan wiring harnesses and thermo-switch connectors. Install the fan shroud, if removed.
16. Install the hood lock and grille, if removed.
17. On 2VZ-FE, install the ignition coil, igniter and bracket assembly.
18. If equipped automatic transmission or transaxle, connect the oil cooler lines.
19. Install the radiator hoses.
20. On 1990–92 MR2, install the front under covers.
21. Fill the cooling system to the proper level.
22. Connect the negative battery cable. Start the engine and check for leaks.

Heater Core

NOTE: On some vehicles, the air conditioning assembly is integral with the heater assembly (including the heater core) and therefore the heater core removal may differ from the procedures detailed below. In some case it may be necessary to remove the air conditioning/heater housing and assembly to remove the heater core. Due to the lack of information available at the time of this publication, a general heater core removal and installation procedure is outlined for each vehicle. The removal steps can be altered as required.

REMOVAL & INSTALLATION

Tercel

1988–90

1. Disconnect the negative battery terminal.
2. Drain the radiator.
3. Remove the ash tray and retainer.
4. Remove the rear heater duct (optional).
5. Remove the left and right side defroster ducts.
6. Remove the under tray (optional).
7. Remove the glove box.
8. Remove the main air duct.
9. Disconnect the radio and remove it.
10. Disconnect the heater control cables and remove them. Mark each cable with the control lever that it connects to.
11. Disconnect the heater hoses.
12. Remove the front and rear air ducts.
13. Disconnect the electrical connectors and vacuum hoses going to the heater unit.
14. Remove the heater bolts and remove the heater. Slide the heater to the right side of vehicle to remove it.
15. Remove the heater core.

To install:

16. Install the heater core into the housing.
17. Fill the cooling system to the proper level. Operate the heater and check for leaks.
18. Install the ramaining components, start the engine and check for proper operation.

1991–92

1. Disconnect the negative battery cable and drain the cooling system.
2. Remove the safety pad from the instrument panel.
3. If equipped with air conditioning, recover the refrigerant from the air conditioning system.
4. Disconnect and cap the evaporator hoses.
5. Remove the instrument lower finish panel and disconnect the harness from the heater and air conditioning assemblies.
6. Remove the air conditioning amplifier.
7. Remove the evaporator housing by removing the 3 screws.
8. Disconnect the heater hoses from the heater core.
9. Remove the heater/air conditioning control assembly.
10. Remove the heater register center duct, instrument panel reinforcements and remove the heater assembly.
11. Remove the screws and clips from the heater assembly case halves. Remove the core from the case.

To install:

12. Install the heater core into the case and install the retainers.
13. Install the heater case assembly and instrument panel components.
14. Install the air conditioning assembly.
15. Connect all hoses and harnesses.
16. Evacuate, recharge and leak test the air conditioning system.
17. Refill the cooling system, start the engine and check for leaks.

Corolla

1. Disconnect the negative battery cable.
2. Drain the cooling system.
3. Remove the gear shift knob and console as necessary.
4. Tag and disconnect the vacuum hoses from heater housing assembly.
5. Remove the under tray or package tray from the right side of the vehicle.
6. Release the 2 clamps and remove the blower duct from the right side of the heater housing.
7. Remove any interfering air ducts.
8. Disconnect the 2 water (heater) hoses from the rear of the heater housing.
9. Tag and disconnect all wires and cables leading from the heater housing and position them aside.
10. Remove all mounting bolts and then remove the heater housing carefully toward the rear of the vehicle.
11. Remove the heater housing assembly from the vehicle. Remove any retaining brackets or hardware that may retain the heater core to the heater housing. Grasp the heater core by the end plate and carefully pull it out of the heater housing.

To install:

12. Install the heater core into the heater housing, make sure to clean heater housing of all dirt, leaves, etc. before heater core installation.
13. Fill the cooling system to the proper level. Operate the heater and check for leaks.
14. Install the remaining compo-

nents, start the engine and check for leaks after the cooling system has pressurized.

Camry

1. Drain the cooling system.
2. Remove the console, if equipped, by removing the shift knob (manual), wiring connector, and console attaching screws.
3. Remove the carpeting from the tunnel.
4. If necessary, remove the cigarette lighter and ash tray.
5. Remove the package tray, if it makes access to the heater core difficult.
6. Remove the bottom cover/intake assembly screws and withdraw the assembly.
7. Remove the cover from the water valve.
8. Remove the water valve.
9. Remove the hose clamps and remove the hoses from the core.
10. Remove the heater core.

To install:

11. Install the heater core into the heater housing, make sure to clean heater housing of all dirt, leaves, etc. before heater core installation.
12. Fill the cooling system to the proper level. Operate the heater and check for leaks.
13. Install the remaining components, start the engine and check for leaks after the cooling system has pressurized.

Cressida

1. Disconnect the negative battery cable.
2. Drain the cooling system.
3. Remove the hood release and the fuel lid release levers.
4. Remove the left instrument panel undercover and lower center pad. Remove the finish plate, then remove the radio assembly.
5. Remove the heater control knobs, heater control panel and ashtray.
6. Remove the right side instrument panel undercover, glove box door and glove box.
7. Remove the front pillar garnish, cluster finish panel and instrument cluster gauge assembly.
8. Remove the safety pad and side defroster hose. Remove the heater assembly air ducts.
9. Remove the lower pad reinforcement and remove the front seats. Remove the center console assembly and the cowl side trim panel.
10. Remove the scuff plate, then position the floor carpeting aside. Remove the rear heater duct, if equipped, and heater control assembly.
11. Disconnect the heater hoses from

the heater core assembly, remove the heater core grommet.
12. Remove the blower motor duct, center duct and instrument panel brace. Remove the heater core assembly from the vehicle.
13. Remove the nuts securing the heater core to the heater core assembly and remove the heater core.

To install:

14. Install the heater core into the heater housing, make sure to clean heater housing of all dirt, leaves, etc. before heater core installation.
15. Fill the cooling system to the proper level. Operate the heater and check for leaks.
16. Install the remaining components, start the engine and check for leaks after the cooling system has pressurized.

Corolla

1. Disconnect the negative battery cable.
2. Drain the cooling system.
3. Remove the center console, scuff plate and front seats.
4. Position the floor carpet aside and remove the heater duct, if equipped.
5. Remove the under tray, glove box and blower duct.
6. On the Corolla station wagon and sedan vehicles, remove the following components:
   a. Remove the heater control knobs and lens. Remove the cluster lower center panel finish, ashtray and heater control assembly.
   b. Remove the instrument cluster finish panel, radio and air ducts.
7. On the Corolla coupe and liftback vehicles, remove the following components:
   a. The instrument cluster finish panel, instrument cluster, radio trim panel and radio.
   b. Ashtray, heater control knobs, heater control panel, heater control assembly and air duct.
8. Disconnect the heater hoses from the heater core assembly and remove the heater hose grommet.
9. Remove the heater core assembly retaining screws and remove the heater core assembly from the vehicle.
10. Remove the heater core from the heater core assembly.

To install:

11. Install the heater core into the heater housing, make sure to clean heater housing of all dirt, leaves, etc. before heater core installation.
12. Fill the cooling system to the proper level. Operate the heater and check for leaks.
13. Install the remaining components, start the engine and check for leaks after the cooling system has pressurized.

MR2

1988–89

1. Disconnect the negative battery cable.
2. Drain the cooling system.
3. Disconnect the heater hose at the engine compartment.
4. Remove the clips retaining the lower part of the heater unit case, then remove the lower part of the case.
5. Using a suitable tool, carefully pry open the lower part of the heater unit case.
6. Remove the heater core assembly from the heater unit case.
7. Installation is the reverse of the removal procedure. Fill the cooling system to the proper level. Operate the heater and check for leaks.

1990–92

1. Disconnect the negative battery cable.
2. Drain the engine cooling system and discharge the air conditioning system. Plug the open refrigerant lines to prevent contamination.
3. Remove the door scuff plate and kick panels.

CAUTION

Use extreme caution when working around the SRS system. Accidental air bag deployment may occur and cause personal injury. Work must be started after approximately 20 seconds or longer from the time the ignition switch is turned to the Lock position and the negative battery terminal cable is disconnected from the battery. The air bag system is equipped with a backup power source so that if work is started within 20 seconds of disconnecting the battery cable , the air bag may be deployed.

---

5. Remove the steering wheels and column cover.
6. Remove the rear console box, console upper panel and console box.
7. Remove the instrument panel lower finish panel and backing plate.
8. Remove the heater duct.
9. Remove the combination switch from the steering column and remove the turn signal bracket.
10. Remove the glove box under cover and glove box assembly.
11. Remove the center cluster finish panel.
12. Remove the instrument cluster finish panel by inserting a taped suitable tool under the panel and pry outward.
13. Remove the instrument panel, pull out and disconnect any harnesses.
14. Remove the radio, heater control, ash tray retainer and clock.
15. Remove the side defroster nozzle and bracket.
16. Disconnect the steering column and remove.
17. Remove the instrument panel re-

tainers and carefully remove the panel from the vehicle. That was't so bad, was it?

18. Disconnect the heater hoses.

19. Remove the heater case retainers and remove the assembly from the vehicle.

20. Remove the case half screws and clips. Separate the 2 halves and remove the heater core.

To install:

21. Install the heater core. Make sure the gaskets are in place. Assemble the case.

22. Install the heater assembly and case retainers.

23. Connect the heater hoses. Fill the cooling system and check for leaks before installing the instrument panel.

24. Install the instrument panel and retainers. Be careful not to damage the panel.

25. Connect the steering column.

26. Install the side defroster nozzle and bracket.

27. Install the radio, heater control, ash tray retainer and clock.

28. Install the instrument cluster and connect harnesses.

29. Install the instrument cluster finish panel by snapping into place.

30. Install the center cluster finish panel.

31. Install the glove box under cover and glove box assembly.

32. Install the turn signal bracket and combination switch to the steering column.

33. Install the heater duct.

34. Install the instrument panel lower finish panel and backing plate.

35. Install the rear console box, console upper panel and console box.

36. Install the steering wheel and column cover.

37. Install the door scuff plate and kick panels.

38. Evacuate, recharge and leak test the air conditioning system.

39. Connect the negative battery cable.

Water Pump
REMOVAL & INSTALLATION

1. Disconnect the negative battery cable.

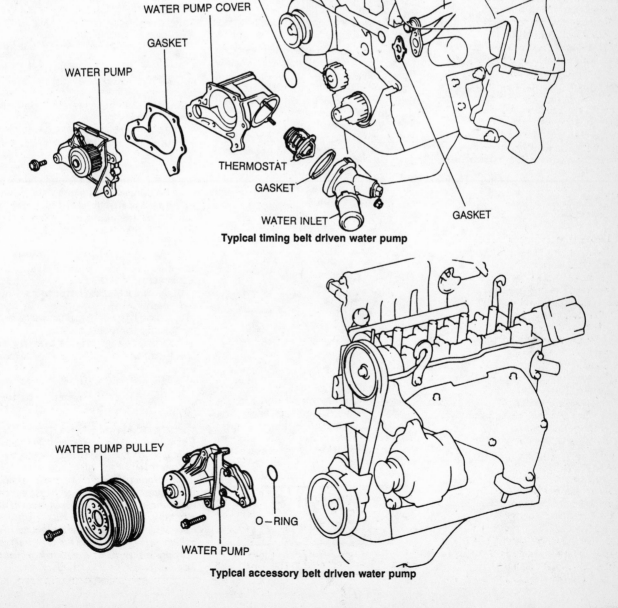

Typical timing belt driven water pump

Typical accessory belt driven water pump

2. Drain the cooling system.

2. Remove the fan shroud retaining bolts and remove the fan shroud, if equipped (RWD vehicles). Loosen and remove all drive belts.

3. Remove all necessary components in order to gain access to the water pump retaining bolts.

4. On some vehicles, it will be necessary to remove the timing covers. On the 1988–92 Camry and 1990–92 MR2 and 1990–92 Celica (5S-FE), remove the timing belt and pulleys.

5. Remove the oil dipstick for 4A series engines.

6. As required, remove the complete air cleaner assembly.

7. Remove all hoses and inlet pipe from the water pump assembly.

8. Remove the water pump retaining bolts. Remove the water pump (and fan) assembly.

To install:

9. Always use a new gasket, O-ring or sealant between the pump body and its mounting.

10. Check for leaks after water pump installation is completed.

Thermostat

REMOVAL & INSTALLATION

1. Disconnect the negative battery cable. Drain the cooling system.

2. Remove the upper radiator or water inlet hose from the thermostat housing.

3. Disconnect the electrical wire from the thermo-switch on the thermostat housing, if equipped.

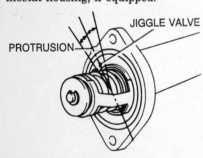

Thermostat jiggle valve alignment

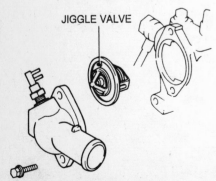

Typical thermostat assembly

4. Remove the thermostat housing retaining bolts. Remove the thermostat housing from the engine.

5. Remove the thermostat.

To install:

6. Clean the gasket mating surfaces. Be sure to use a new thermostat gasket. Be sure the thermostat is installed with the spring pointing down and the jiggle valve up. On the 3S-FE, 5S-FE, 4A-FE (type A thermostat) and 3S-GTE engines, align the jiggle valve with the protrusion on the thermostat housing. The jiggle valve may be aligned within 5–10 degrees on either side of the protrusion. On the 2VZ-FE and 4A-FE (type B thermostat), align the jiggle valve with the upper stud bolt in the housing.

Cooling System Bleeding

1. Fill the radiator with the proper type of coolant.

2. Loosen a fitting in a coolant passage located near the highest point on the engine. A sending unit of vacuum switching valve for example. Apply thread sealing tape or equivalent to the fitting threads. Fill the radiator until coolant comes from the hole and tighten the fitting.

3. With the radiator cap off, start the engine and allow it to run and reach normal operating temperature.

4. Run the heater at full force and with the temperature lever in the hot position. Be sure the heater control valve is functioning.

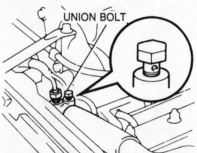

Cooling system air relief location— 2VZ-FE engine

5. Shut the engine off and recheck the coolant level, refill as necessary. On 2VZ-FE engine, release the air from the cooling system by backing off on air relief union bolt about 4–5 turns.

ENGINE ELECTRICAL

—— CAUTION ——

If equipped with an air bag, wait 20 seconds

or longer after disconnecting the negative battery cable before attempting to remove or service any electrical component. The air bag control system is equipped with a back-up power source that remains charged for a minimum of 20 seconds after the negative battery cable is disconnected. Attempting to remove or service an electrical component without allowing the time interval to elapse may result in deployment of the air bag and possible personal injury.

NOTE: Disconnecting the negative battery cable on some vehicles may interfere with the functions of the on board computer systems and may require the computer to undergo a relearning process, once the negative battery cable is reconnected.

Distributor

REMOVAL

5M-GE Engine

1988 Cressida

1. Disconnect the negative battery cable.

2. Disconnect the cables from the spark plugs, after marking the wiring order.

3. Disconnect the high tension cable from the coil.

4. Remove the primary wire and the vacuum line from the distributor.

5. Remove the distributor cap.

6. Matchmark the distributor housing and the engine block and the rotor to the distributor housing; this will aid in correct positioning of the distributor during installation.

7. Remove the clamp from the distributor.

8. Withdraw the distributor from the block.

To install:

9. Install the distributor to its original position.

10. Install the spark plug wires and remaining components.

11. Set the ignition timing to specifications.

Except 5M-GE Engine

1. Disconnect the negative battery cable.

—— CAUTION ——

If equipped with an air bag, wait 20 seconds or longer after disconnecting the negative battery cable before attempting to remove the distributor. The air bag control system is equipped with a back-up power source that remains charged for a minimum of 20 seconds after the negative battery cable is disconnected. Attempting to remove the distributor without allowing the time interval to elapse may result in deployment of the air bag and possible personal injury.

2. Disconnect the electrical leads and spark plug wires from the distributor.

3. Remove the water-proof cover, if installed.

4. On the Supra, with the 7M-GE engine, remove the oil filler cap and rotate the crankshaft clockwise until the nose of the camshaft is visible through the hole. Turn the crankshaft counterclockwise 120 degrees. Now, turn it clockwise 10–40 degrees until the TDC marks on the front cover and the crankshaft pulley are aligned.

5. Remove the intercooler on the 3S-GTE engine.

6. Matchmark the distributor housing and the engine block and the rotor to the distributor housing; this will aid in correct positioning of the distributor during installation.

7. Remove the hold-down bolts and pull the distributor from the engine.

To install:

8. Install the distributor to its original position.

9. Install the spark plug wires and remaining components.

10. Set the ignition timing to specifications.

INSTALLATION

Timing Not Disturbed

5M-GE ENGINE (1988 CRESSIDA)

1. Insert the distributor in the block and align the matchmarks.

2. Engage the distributor drive with the oil pump driveshaft; make sure the gear teeth mesh properly.

3. Install the distributor clamp, cap, high tension wire, primary wire and vacuum line.

4. Install the wires on the spark plugs.

5. Start the engine. Check the timing and adjust, as necessary.

EXCEPT 5M-GE ENGINE

1. Install a new distributor housing O-ring. Apply a thin coat of clean engine oil to the new O-ring before installation.

2. Insert the distributor in the block and align the matchmarks on the housing and the rotor made during removal.

3. Install the distributor hold-down bolts.

4. On 3S-GTE engine, install the intercooler.

4. Install the water-proof cover, if removed.

5. Connect the electrical leads and spark plug wires to the distributor.

6. Connect the negative battery cable.

7. Set the timing.

Timing Disturbed

5M-GE ENGINE (1988 CRESSIDA)

1. Determine Top Dead Center

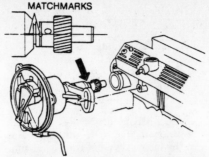

Align the matchmarks on the distributor gear and housing—5M-GE engine

(TDC) of the No. 1 cylinder's compression stroke by removing the spark plug from the No. 1 cylinder and placing a finger or a compression gauge over the spark plug hole. Crank the engine slowly until compression pressure starts to build up. Continue cranking the engine until the timing marks indicate TDC or 0 degrees.

2. Remove the oil filler cap. Looking into the camshaft housing with the aid of a flashlight, check to make sure the match hole on the second (No. 2) journal of the camshaft housing is aligned with the hole in the No. 2 journal of the camshaft. If the holes are not aligned, rotate the camshaft 1 full turn.

3. Install a new O-ring on the distributor cap shaft; make sure the distributor cap is still removed at this time. Align the matchmark on the distributor drive gear with that of the distributor housing.

4. Insert the distributor into the camshaft housing, align the center of the mounting flange with the bolt hole in the side of the housing.

5. Align the rotor tooth in the distributor with the pickup coil. Temporarily install the distributor pinch bolt. Install the distributor cap and install the oil filler cap.

6. Connect the cables to the spark plugs in the proper order by using the marks made during removal. Install the high tension wire on the coil.

7. Start the engine. Adjust the ignition timing.

EXCEPT 5M-GE ENGINE

1. Set the engine at TDC of the No. 1 cylinder's firing stroke. This can be accomplished by removing the No. 1 spark plug and turning the engine by hand with a finger over the spark plug hole. As No. 1 is coming up on its firing stroke, pressure will be felt. Make sure the timing marks are set as follows:

a. On 4A-GE, 4A-GEC and 4A-GELC, align the groove on the crankshaft pulley with the 0 mark on the No. 1 timing cover.

b. For all, except the Supra (7M-GE), 1989–92 Cressida, MR2, Corolla (4A-GE) and Camry, coat the spi-

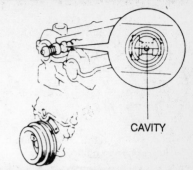

Setting the No. 1 cylinder to TDC of the compression stroke—4A-GE engine

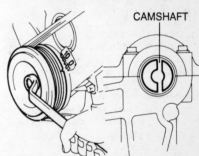

Positioning the No. 1 camshaft—3S-FE, 3S-GE and 3S-GTE engines

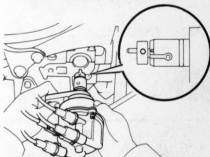

Distributor alignment—3S-FE and 3S-GE engines

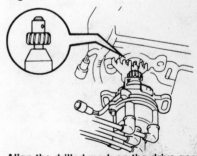

Align the drilled mark on the drive gear with the cavity of the housing—4A-GE engine

Position the camshaft slit as shown—4A-F engine

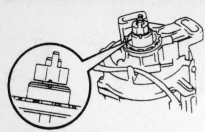

Align the marks on the coupling and the housing—2VZ-FE engine

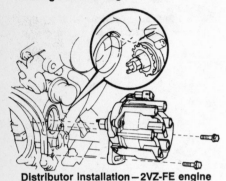

Distributor Installation—2VZ-FE engine

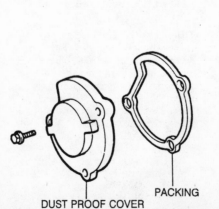

Positioning the No. 1 camshaft—5S-FE engine

ral gear and governor shaft tip with clean engine oil. Align the protrusion on the distributor housing with the pin on the spiral gear drill mark side. Insert the distributor and align the center of the flange with the bolt hole on the cylinder head. Tighten the bolts.

c. On the Supra (7M-GE) and 1989–92 Cressida, align the drilled mark on the driven gear with the groove on the distributor housing. Insert the distributor and align the stationary flange center with bolt hole in the head. Tighten the bolts.

d. On Celica and Camry, turn the crankshaft clockwise until the slot in the forward end of the No. 1 camshaft (front of vehicle) is positioned in the vertical position. Lightly coat a new O-ring with the engine oil and slide it into position. Align the drilled mark or cutout, on the coupling with the notch of the shaft

housing. Insert the distributor into the cylinder head so the center of the flange is aligned with the bolt hole on the cylinder head.

e. On the MR2, except 5S-FE, and the Corolla GTS with 4A-GE, install a new O-ring. Align the drilled mark on the distributor driven gear with the cavity of the housing. On MR2 with 5S-FE, turn the crankshaft clockwise until the slot in the forward end of the No. 1 camshaft, front of vehicle, is positioned in the vertical position. Then, align the cutout portion of the coupling with the groove in the housing. Insert the distributor and align the center of the flange with the bolt hole on the cylinder head. Tighten the hold-down bolts.

f. On the Corolla (4A-F, 4A-FE), install a new O-ring. Align the protrusion on the distributor housing with the groove of the coupling side. On the 4A-FE, align the center of the flange with the bolt hole on the cylinder head. Tighten the hold-down bolts.

g. On Camry with 2VZ-FE, align the cut-out marks of the coupling and the housing and then insert the distributor so the line on the housing and the cut-out on the distributor attachment cap are aligned. Tighten the hold-down bolts.

2. Connect the spark plug wires; check the idle speed and the ignition timing.

Distributorless Ignition

REMOVAL & INSTALLATION

Camshaft Position Sensor

SUPRA (7M-GTE)

1. Disconnect the negative battery cable.
2. Disconnect the cam position sensor connector.
3. Remove the oil filler cap.

4. Look into the oil filler opening with a flashlight and rotate the crankshaft clockwise until the the nose of the cam can be seen.

5. Once the nose of the cam comes into view, rotate the crankshaft approximately 120 degrees counterclockwise.

6. Turn the crankshaft approximately 10–40 degrees clockwise until the TDC mark on the timing belt cover is aligned with the TDC mark on the crankshaft pulley; the engine is now at TDC. Don't move it from this position.

7. Remove the No. 4 air cleaner pipe with the No. 1 and No. 2 air cleaner hoses.

8. Disconnect the 3 air hoses and the PCV hose.

9. Disconnect the air flow meter connector.

10. Disconnect the power steering idle up air hose.

11. Remove the air flow meter mounting bolt and attendant hose clamps. Remove the No. 7 air cleaner hose, air flow meter and air cleaner cap as a unit.

12. Unbolt and remove the power steering reservoir tank. Leave the hoses connected and move the tank aside.

13. Remove the cam position sensor hold-down bolt.

14. Withdraw the cam position sensor from the cylinder head.

15. Remove the cam position sensor O-ring. Discard the O-ring and replace with new.

To install:

16. Install a new O-ring.

17. Align the drilled mark on the driven gear with the groove of the housing.

18. Insert the cam position sensor into the cylinder head so the center of the sensor flange is aligned with the bolt hole in the head.

19. Lightly tighten the hold-down bolt.

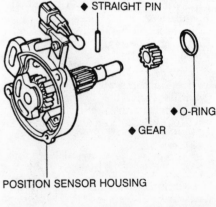

Cam position sensor exploded view—7M-GTE engine

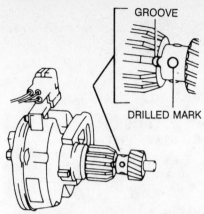

Align the mark on the gear with the groove in the housing—7M-GTE engine

20. Install the power steering reservoir tank.

21. Install the air cleaner cap, air flow meter and No. 7 air cleaner hose. Install the mounting bolt and tighten the clamps.

22. Connect the power steering idle up air hose.

23. Connect the air flow meter connector.

24. Connect the PCV hose and 3 air hoses.

25. Install the air cleaner pip and No. 1 and No. 2 air cleaner hoses.

26. Connect the cam position sensor connector.

27. Start and warm up the engine. Adjust the timing.

Ignition Timing

ADJUSTMENT

3A-C, 4A-LC and 4A-F Engine

1. Warm up the engine and set the parking brake. Connect a tachometer and check the engine speed to make sure it is within specifications. Adjust as required.

2. Connect the dwell meter or tachometer to the negative side of the coil, not to the distributor primary lead, damage to the ignition control will result.

3. All engines require a special type of tachometer which hooks up to the service connector wire coming out of the distributor.

4. Connect a timing light to the engine, as outlined in the instructions supplied by the manufacturer of the light.

5. Disconnect and plug the vacuum line from the distributor vacuum unit. If a vacuum advance/retard distributor is used, disconnect and plug both vacuum lines from the distributor.

6. Allow the engine to run at the specified idle speed with the gear shift in **N** for a manual transmission or **D**

for an automatic transmission. Be sure the parking brake is firmly set and the wheels are chocked.

7. Point the timing light at the timing marks. With the engine at idle, timing should be at the specification. If not, loosen the pinch bolt at the base and rotate the distributor to advance or retard the timing, as required.

8. Stop the engine and tighten the pinch bolt. Start the engine and recheck the timing. Stop the engine and disconnect the timing light and the tachometer. Connect the vacuum line(s) to the vacuum and advance unit.

Except 3E and 3E-E Engine

1. Connect a timing light to the engine following the manufacturer's instructions. On the 7M-GTE, connect the timing light pick-up to the No. 6 spark plug wire.

2. The engines require a special type of tachometer which hooks up to the service connector wire coming from the distributor.

3. Start the engine and run it at idle. Remove the rubber cap from the check connector or open the lid; short the connector at terminals **T** and **E₁** on 1988 California vehicles. On all 1989–92 vehicles, with the 4A-GE, 4A-FE, 3S-FE, 3S-GTE, 5S-FE, 2VZ-FE and 7M-GE, short the **TE₁** and **E₁** terminals.

4. Loosen the distributor pinch bolt so the distributor can be turned. Aim the timing light at the marks on the crankshaft pulley and slowly turn the distributor until the timing mark is aligned. Tighten the distributor pinch bolt. Unshort the connector.

NOTE: The 7M-GTE utilizes a cam position sensor in place of a distributor. Turn this the same as a distributor.

3E Engine

1. Remove the cap. Using the proper tachometer, connect the test probe of the tachometer to the service probe connector at the Integrated Ignition Assembly (IIA).

2. Disconnect the vacuum hose from the IIA sub-diaphragm and plug it.

3. With the engine idling and the electric fan off, check the timing.

4. Loosen the hold-down bolt and adjust the timing, as required.

5. Retighten the hold-down bolt and recheck the ignition timing.

3E-E Engine

1. Warm up the engine to normal operating temperature.

2. Connect a tachometer and timing light to the engine. The tachometer may be connected either to the service

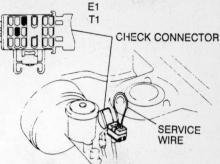

Shorting the test connector—1989 2VZ-FE engine

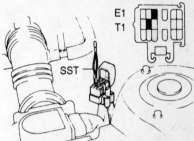

Shorting the test connector—1990–92 2VZ-FE engine

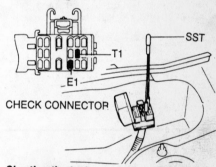

Shorting the test connector—3E-E engine

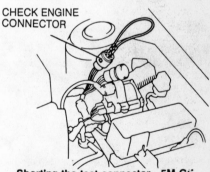

Shorting the test connector—5M-GE (Supra) engine

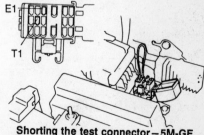

Shorting the test connector—5M-GE (Cressida) engine

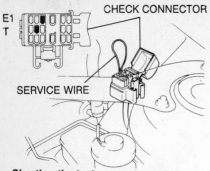

Shorting the test connector—3S-FE, 3S-GE and 3S-GTE engines

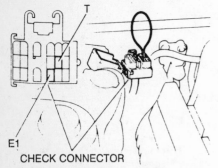

Shorting the test connector—4A-GEC, 4A-GELC and 4A-GZE engines

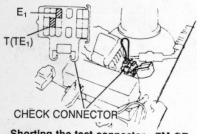

Shorting the test connector—7M-GE (Supra) engine

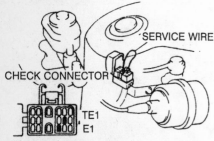

Shorting the test connector—7M-GE engine (Cressida)

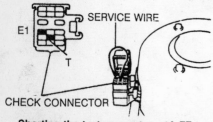

Shorting the test connector—4A-FE engine

Shorting the test connector—5S-FE engine

connector of the distributor or to the **IG** terminal of the check connector.

NOTE: The engines require a special type of tachometer which hooks up to the service connector wire coming from the distributor.

3. Open the lid on the check connector and short the connector at terminals **T** and **E**$_1$.

4. Loosen the distributor pinch bolt so the distributor can be turned. Aim the timing light at the marks on the crankshaft pulley and slowly turn the distributor until the timing mark is aligned. Tighten the distributor pinch bolt. Unshort the connector.

Alternator

PRECAUTIONS

Several precautions must be observed with alternator equipped vehicles to avoid damage to the unit.

• If the battery is removed for any reason, make sure it is reconnected with the correct polarity. Reversing the battery connections may result in damage to the one-way rectifiers.

• When utilizing a booster battery as a starting aid, always connect the positive to positive terminals and the negative terminal from the booster battery to a good engine ground on the vehicle being started.

• Never use a fast charger as a booster to start vehicles.

• Disconnect the battery cables when charging the battery with a fast charger.

• Never attempt to polarize the alternator.

• Do not use test lamps of more than 12 volts when checking diode continuity.

• Do not short across or ground any of the alternator terminals.

• The polarity of the battery, alternator and regulator must be matched and considered before making any electrical connections within the system.

• Never separate the alternator on

an open circuit. Make sure all connections within the circuit are clean and tight.

• Disconnect the battery ground terminal when performing any service on electrical components.

• Disconnect the battery if arc welding is to be done on the vehicle.

BELT TENSION ADJUSTMENT

Inspection and adjustment to the alternator drive belt should be performed every 3000 miles or if the alternator has been removed.

1. Inspect the drive belt to see if it is cracked or worn; be sure it's surfaces are free of grease or oil.

2. Push down on the belt halfway between the fan and the alternator pulleys or crankshaft pulley with thumb pressure; belt deflection should be ⅜–½ in.

3. If the belt tension requires adjustment, loosen the adjusting link bolt and move the alternator until the proper belt tension is obtained.

4. Do not over-tighten the belt, as damage to the alternator bearings could result. Tighten the adjusting link bolt.

5. Drive the vehicle and re-check the belt tension; adjust, as necessary.

REMOVAL & INSTALLATION

Except Celica AWD

NOTE: On some vehicles, the alternator is mounted very low on the engine. On these vehicles it may be necessary to remove the gravel shield and work from under the vehicle in order to gain access to the alternator.

1. Disconnect the negative battery cable.

2. Remove the air cleaner, if necessary, to gain access to the alternator.

3. Remove the power steering or air conditioning drive belts, as required.

4. Unfasten the bolts which attach the adjusting link to the alternator. Remove the alternator drive belt.

NOTE: On the 2VZ-FE, remove the No. 2 right side mounting stay.

5. Unfasten and tag the alternator bolt and withdraw the alternator from it's bracket.

6. Installation is the reverse of the removal procedure. After installing the alternator, adjust the belt tension.

Celica AWD

1. Disconnect the negative battery cable.

2. Remove the lower alternator duct.

3. Loosen the idler pulley bolt.

4. Loosen the adjusting bolt and remove the drive belt.

5. Disconnect the alternator connectors, alternator lead wire, air conditioning compressor connector, water temperature switch connector and oxygen sensor connector.

6. Unbolt and disconnect the ground strap and engine wire from the brackets.

7. Unbolt and remove the alternator bracket.

8. Unbolt and remove the alternator.

9. Remove the upper alternator duct and disconnect the lead wire.

To install:

10. Connect the lead wire and attach the alternator duct.

11. Install the alternator. Torque the 12mm bolt to 14 ft. lbs. (19 Nm) and the 14mm bolt to 38 ft. lbs. (52 Nm).

12. Install the alternator bracket. Torque the turbine outlet elbow bolt to 32 ft. lbs. (43 Nm) and the bracket bolt to 39 ft. lbs. (29 Nm).

13. Install the engine wire and ground strap.

14. Connect the alternator and engine wiring.

15. Install the drive belt and adjust the drive belt tension.

16. Install the lower alternator duct.

17. Connect the negative battery cable.

Starter

REMOVAL & INSTALLATION

1. Disconnect the negative battery cable. Disconnect the cable which runs from the starter to the battery, at the battery end.

2. Remove the air cleaner assembly, if necessary, to gain access to the starter.

3. If equipped with an automatic transmission/transaxle, it may be necessary to disconnect the throttle linkage connecting rod or the transmission/tranasaxle oil filler tube.

4. On the 3S-GE, disconnect the exhaust pipe at the manifold. On the Camry with 2VZ-FE, remove the ignitor bracket. On 1990–92 Celica with 3S-GTE, remove the engine compartment relay box and the battery. On 1990–92 Celica with 4A-FE, remove the lower suspension crossmember and the air cleaner cap. On 1990–92 Celica with 5S-FE, cruise control and ABS, remove the engine compartment relay box and the cruise control actuator. On 1990–92 Corolla with 4A-GE, remove both engine undercovers, front exhaust pipe and electric cooling fan.

5. Disconnect all of the wiring at the starter. Remove the starter retaining bolts. Remove the starter from the vehicle.

6. Installation is the reverse of the removal procedure.

FUEL SYSTEM

Fuel System Service Precautions

Safety is the most important factor when performing not only fuel system maintenance but any type of maintenance. Failure to conduct maintenance and repairs in a safe manner may result in serious personal injury or death. Maintenance and testing of the vehicle's fuel system components can be accomplished safely and effectively by adhering to the following rules and guidelines.

• To avoid the possibility of fire and personal injury, always disconnect the negative battery cable unless the repair or test procedure requires that battery voltage be applied.

• Always relieve the fuel system pressure prior to disconnecting any fuel system component (injector, fuel rail, pressure regulator, etc.), fitting or fuel line connection. Exercise extreme caution whenever relieving fuel system pressure to avoid exposing skin, face and eyes to fuel spray. Please be advised that fuel under pressure may penetrate the skin or any part of the body that it contacts.

• Always place a shop towel or cloth around the fitting or connection prior to loosening to absorb any excess fuel due to spillage. Ensure that all fuel spillage (should it occur) is quickly removed from engine surfaces. Ensure that all fuel soaked cloths or towels are deposited into a suitable waste container.

• Always keep a dry chemical (Class B) fire extinguisher near the work area.

• Do not allow fuel spray or fuel vapors to come into contact with a spark or open flame.

• Always use a backup wrench when loosening and tightening fuel line connection fittings. This will prevent unnecessary stress and torsion to fuel line piping. Always follow the proper torque specifications.

• Always replace worn fuel fitting O-rings with new. Do not substitute fuel hose or equivalent where fuel pipe is installed.

RELIEVING FUEL SYSTEM PRESSURE

1. Remove the fuel pump fuse from the fuse block, fuel pump relay or disconnect the harness connector at the tank while engine is running.

2. It should run and then stall when the fuel in the lines is exhausted. When the engine stops, crank the starter for about 3 seconds to make sure all pressure in the fuel lines is released.

3. Install the fuel pump fuse, relay or harness connector after repair is made.

Fuel Tank
REMOVAL & INSTALLATION

Carbureted Engines

1. Disconnect the negative battery cable. Relieve the fuel pressure.

2. Drain the fuel using an approved pump and container.

3. Label and disconnect the fuel filler, breather, tank-to-evaporator, filler hose and tank-to-return pipe hoses.

4. Raise the vehicle and support safely. Disconnect the tank wiring and remove the under cover, if equipped.

5. Place a floor jack under the tank, remove the retaining bolts and lower the tank far enough to disconnect any hoses and electrical connectors still connected.

To install:

6. Raise the tank with the jack and connect the hoses and electrical connectors. Install the retaining bolts. Torque the retainer to 15 ft. lbs. (20 Nm).

7. Connect the wiring and fuel hoses.

8. Connect the breather hose and fuel filler hose.

9. Connect the battery cable, fill the tank with fuel and check for leaks.

Fuel Injected Engines

1. Disconnect the negative battery cable. Relieve the fuel system pressure.

2. Drain the tank with an approved pump and container.

3. Raise the vehicle and support safely. Remove the rear and center exhaust pipes.

4. Remove the driveshaft for AWD and RWD vehicles. Disconnect the fuel filler and air breather hose.

5. Disconnect the feed, return and evaporative hoses.

6. Disconnect the parking brake cable bracket and return spring.

7. Disconnect the tank harness connectors.

8. Place a floor jack under the tank and remove the tank retainers. Lower the tank and disconnect any wiring or hoses.

To install:

9. Raise the tank into position and

install the tank retainers. Torque the retainers to 15 ft. lbs. (20 Nm).

10. Connect the tank harnesses and hoses.

11. Connect the parking brake cable bracket and return spring.

12. Connect the feed, return and evaporative hoses.

13. Install the driveshaft, if removed. Connect the fuel filler and air breather hoses.

14. Install the rear and center exhaust pipes.

15. Refill the tank and check for leaks.

16. Connect the battery cable.

Fuel Filter

REMOVAL & INSTALLATION

1. Disconnect the negative battery cable. Unbolt the retaining screws and remove the protective shield for the fuel filter.

2. Place a pan under the delivery pipe to catch the dripping fuel and slowly loosen the union bolt or flare nut to bleed off the fuel pressure.

3. Drain the remaining fuel.

4. Disconnect and plug the inlet line.

5. Unbolt and remove the fuel filter.

NOTE: When tightening the fuel line bolts to the fuel filter, use a torque wrench. The tightening torque is very important, as under or over tightening may cause fuel leakage. Insure that there is no fuel line interference and that there is sufficient clearance between it and any other parts.

6. Coat the flare nut, union nut and bolt threads with engine oil.

7. Hand tighten the inlet line to the fuel filter.

8. Install the fuel filter and then tighten the inlet bolt to 22 ft. lbs. (30 Nm).

9. Reconnect the delivery pipe using new gaskets and then tighten the union bolt to 22 ft. lbs. (30 Nm).

10. Run the engine for a few minutes and check for any fuel leaks.

11. Install the protective shield.

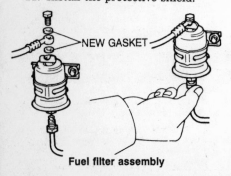

NEW GASKET

Fuel filter assembly

Mechanical Fuel Pump

The 3E, 3A-C, 4A-C and 4A-F engines use a mechanical type fuel pump. It is located on the right rear of the cylinder head.

PRESSURE TESTING

1. Remove the line which runs from the fuel pump to the carburetor.

2. Attach a pressure gauge to the outlet side of the pump.

3. Run the engine and check the pressure.

4. Check the pressure. It should be 2.6–3.5 psi.

5. If the pressure is below the specifications, check for restrictions or replace the pump.

6. Reconnect the carburetor line.

REMOVAL & INSTALLATION

1. Disconnect and plug the fuel lines to the pump.

2. Remove the nuts which hold the pump to the cylinder head.

3. Remove the pump assembly.

4. Installation is the reverse of removal. Always use a new gasket when installing a fuel pump.

TYPE I

WITH GASKET

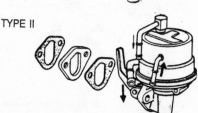

TYPE II

Typical mechanical fuel pump styles

Electric Fuel Pump

PRESSURE TESTING

——— CAUTION ———
Do not operate the fuel pump unless it is immersed in gasoline and connected to its resistor.

1. Turn the ignition switch to the **ON** position, but don't start the engine.

2. Remove the rubber cap from the fuel pump check connector and short

terminals **Fp** and **+B** with a jumper wire.

NOTE: The check connector on all engines is in a small plastic box with a flip-up lid; it is found near the strut tower or battery. The box is roughly the same size and shape for every engine and terminals Fp and +B are always in the same location.

3. Check that there is pressure in the hose to the cold start injector. On 4A-FE and 4A-GE engines, check for pressure at the regulator fuel return hose. On 2VZ-FE, 3S-FE, 5S-FE, 3S-GE 4A-GE and 3S-GTE engines, check for pressure at the fuel filter hose.

NOTE: At this time, fuel return noise from the pressure regulator should be audible.

4. If no pressure can be felt in the line, check the fuses and all other related electrical connections. If everything is all right, the fuel pump will probably require replacement.

5. Remove the jumper wire, reinstall the rubber cap and turn off the ignition switch.

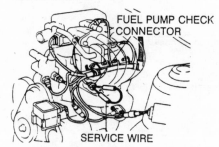

FUEL PUMP CHECK CONNECTOR

SERVICE WIRE

Shorting the fuel pump check connector—typical

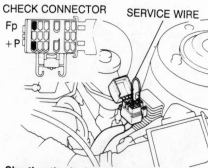

CHECK CONNECTOR   SERVICE WIRE

Fp
+P

Shorting the fuel pump check connector (3S-GE shown, check connector terminals are in same location for all engines)

REMOVAL & INSTALLATION

The fuel pump is mounted inside the fuel tank on all vehicles. On all vehicles except 1990 Celica (non-turbocharged) and 1991–92 Tercel removal of the fuel tank is necessary to remove the fuel pump. On 1990–92 Celica non-turbocharged and 1991–92 Tercel, access to the pump is gained by removing the rear seat cushion.

Except 1990–92 Celica Non-Turbocharged and 1990–92 MR2

1. Disconnect the negative battery cable.
2. Drain the fuel from the tank and then remove the fuel tank.
3. Remove the bolts and then pull the fuel pump bracket up and out of the fuel tank.
4. Remove the mounting nuts then tag and disconnect the wires at the fuel pump.
5. Pull the fuel pump out of the lower side of the bracket. Disconnect the pump from the fuel hose.
6. Remove the rubber cushion and the clip. Disconnect the fuel pump filter from the pump.
7. Installation is in the reverse order of removal procedure. Use a new fuel bracket gasket.

1990–92 Celica Non-Turbocharged and Tercel

1. Disconnect the negative battery cable.
2. Remove the rear seat cushion.
3. Remove the 5 retaining screws and floor service hole cover.
4. Disconnect all the electrical fuel pump connections at the fuel pump assembly.
5. Disconnect the fuel pipe and hose from the fuel pump bracket. Remove the fuel pump bracket assembly from the fuel tank. Remove the fuel pump from the fuel bracket.
6. Installation is the the reverse of the removal procedure. Use a new fuel bracket gasket.

1990–92 MR2

1. Disconnect the negative battery cable.
2. Drain the fuel tank into a suitable container.
3. Remove the console boxes, left lower instrument panel finish panel, ash tray and retainer.
4. Remove the center instrument panel finish panel.
5. Disconnect the fuel pump connector and fuel sender.
6. Remove the 2 screws and floor service hole cover.
7. Remove the engine under covers, front luggage under cover and fuel tank protectors.
8. Remove the parking brake intermediate lever and center floor crossmember.
9. Remove the air conditioning and radiator hoses from the body and move out of the way. If necessary, disconnect the hoses after draining the radia-

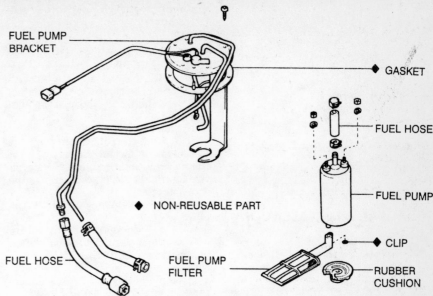

Typical fuel pump assembly—except 1990–92 Celica non-turbo

FUEL PUMP BRACKET
◆ GASKET
FUEL HOSE
FUEL PUMP
◆ NON-REUSABLE PART
◆ CLIP
RUBBER CUSHION
FUEL HOSE
FUEL PUMP FILTER

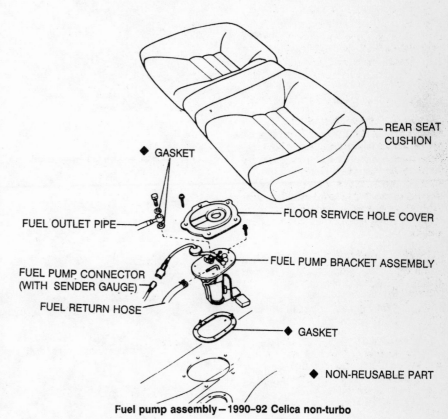

Fuel pump assembly—1990–92 Celica non-turbo

REAR SEAT CUSHION
◆ GASKET
FUEL OUTLET PIPE
FLOOR SERVICE HOLE COVER
FUEL PUMP BRACKET ASSEMBLY
FUEL PUMP CONNECTOR (WITH SENDER GAUGE)
FUEL RETURN HOSE
◆ GASKET
◆ NON-REUSABLE PART

tor and discharging the air conditioning system.
10. Remove the fuel tank heat insulators and disconnect the fuel hoses and tubes.
11. Using a suitable jack, remove the tank retainers and lower the tank.
12. Remove the pump assembly from the tank. Remove the pump from the bracket.

To install:
13. Install a new gasket, pump and retaining screws. Torque the screws to 35 inch lbs. (3.4 Nm).
14. Install the fuel tank using a suitable jack.
15. Connect all wiring and fuel hoses to the tank.
16. Torque the tank retainers to 22 ft. lbs. (29 Nm).
17. Install the remaining components, refill the tank and check for leaks.
18. Connect the battery cable.

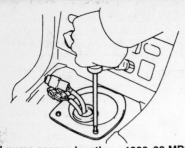

Fuel pump access location—1990–92 MR2

Carburetor

REMOVAL & INSTALLATION

NOTE: During carburetor removal, be sure to mark all hoses, lines and electrical connectors, etc., so these items may be properly reconnected during installation.

1. Disconnect the negative battery cable.
2. Remove the air cleaner housing and disconnect all air hoses from the air cleaner base.
3. Disconnect the fuel line, choke pipe, and distributor vacuum line.
4. Remove the accelerator linkage. With an automatic transaxle, also remove the throttle rod.
5. Label the vacuum hoses for ease of installation. Disconnect any remaining hoses, etc., from the carburetor.
6. Remove the 4 nuts that secure the carburetor to the manifold and lift off the carburetor and gasket.
7. Remove the carburetor heat insulator with 2 gaskets from the intake manifold.
8. Cover the open manifold with a clean rag to prevent small objects from dropping into the engine.
To install:
9. Use new gaskets, install the carburetor and torque the bolts to 12 ft. lbs. (15 Nm).
10. Reconnect all hoses and electrical connectors.
11. After the engine is started, check for fuel leaks, idle speed and float level settings.

IDLE SPEED ADJUSTMENT

The idle speed and mixture should be adjusted under the following conditions: the air cleaner must be installed, the choke fully opened, the transmission should be in N and all electrical accessories (including the electric engine cooling fan) should be turned off.

NOTE: All carbureted engines require a special type of tachometer which hooks up to the service connector wire coming out of the distributor.

1. Start the engine and allow it to reach normal operating temperature.
2. Check the float setting; the fuel level should be just about even with the spot on the sight glass. If the fuel level is too high or low, adjust the float level.
3. Stop the engine.
4. Remove the rubber cap from the IIA service connector the comes out of the distributor and connect the positive terminal of the tachometer to the connector.
5. Start the engine and check the idle speed.
6. If the idle speed is not within specifications, turn the idle speed adjusting screw until the idle speed is correct.
7. Stop the engine and disconnect the tachometer.

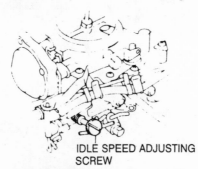

Idle speed adjusting screw—Corolla with carburetor

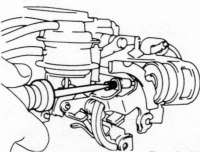

Idle speed adjusting screw—Tercel with carburetor

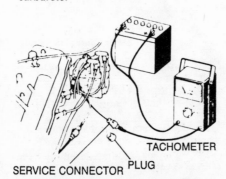

Tachometer hook-up—Corolla and Tercel with carburetor

SERVICE ADJUSTMENTS

For all carburetor service Adjustment procedures and Specifications, please refer to "Carburetor Service" in the Unit Repair Section.

Fuel Injection

IDLE SPEED ADJUSTMENT

Idle speed adjustment is performed under the following conditions:
Engine at normal operating temperature.
Air cleaner installed.
Air pipes and hoses of the air induction and EGR systems properly connected.
All vacuum lines and electrical wires connected and plugged in properly.
All electrical accessories in the OFF position.
Transaxle in the N position.

Supra, 1989–92 Cressida, Camry (2VZ-FE Engine) and 1990 Celica (3S-GTE Engine)

Idle speed is controlled by the Electronic Control Unit (ECU) and is not adjustable.

1988 Cressida

1. Connect a tachometer and timing light to the engine.
2. Start the engine and warm up to normal operating temperature.
3. Check the ignition timing. Adjust as necessary.
4. Check the idle speed.
5. If the idle speed is not within specifications, adjust by turning the idle speed adjusting screw on the throttle body.
6. Stop the engine and disconnect the tachometer.

Tercel

1. Run the engine until it reaches normal operating temperature. The cooling fan must not be running during the idle speed adjustment.
2. Connect a tachometer to the engine.

NOTE: Do not allow the tachometer or coil terminals to be grounded. This will damage the injection system.

3. On 3E-E engine, disconnect the idle up Vacuum Switching Valve (VSV) connector.
4. Run the engine at 2500 rpm for 2 minutes.
5. Adjust the idle speed by turning the idle speed adjusting screw.
6. On 3E-E engine, connect the idle

up Vacuum Switching Valve (VSV) connector.

7. Disconnect the tachometer.

Corolla

1. Run the engine until it reaches normal operating temperature. The cooling fan must not be running during the idle speed adjustment.

2. Connect a tachometer to the engine.

NOTE: Do not allow the tachometer or coil terminals to be grounded. This will damage the injection system.

3. Run the engine at 2500 rpm for 2 minutes.

4. On 1990 vehicles, short the check connector at terminals TE_1 and E_1 using a suitable jumper wire.

5. Adjust the idle speed by turning the idle speed adjusting screw.

6. On 1990–92 vehicles, remove the jumper wire from the connector terminals.

7. Disconnect the tachometer.

Camry and 1988 Celica (3S-FE Engine)

1. Run the engine until it reaches normal operating temperature. The cooling fan must not be running during the idle speed adjustment.

2. Connect a tachometer to the engine.

NOTE: Do not allow the tachometer or coil terminals to be grounded. This will damage the injection system.

3. On 1988 Camry, short the check connector at terminals T and E_1 using a suitable jumper wire. On 1989–92 Camry, short the check connector at terminals TE_1 and E_1. On Celica, short the check connector at terminals TE_1 (California) or T (except California) and E_1.

4. Run the engine at 1000–3000 rpm for 5 seconds and return the engine to idle.

5. Adjust the idle speed by turning the idle speed adjusting screw.

6. Remove the jumper wire from the connector terminals.

7. Disconnect the tachometer.

1988–89 Celica (3S-GE Engine)

1. Run the engine until it reaches normal operating temperature. The cooling fan must not be running during the idle speed adjustment.

2. Connect a tachometer to the engine.

NOTE: Do not allow the tachometer or coil terminals to be grounded. This will damage the injection system.

3. Run the engine at 2500 rpm for 2 minutes.

4. Pinch the No. 1 air intake chamber vacuum hose.

5. Adjust the idle speed by turning the idle speed adjusting screw.

6. Release the No. 1 vacuum hose.

7. Disconnect the tachometer.

1990–92 MR2 (5S-FE Engine) and 1990–92 Celica (4A-FE and 5S-FE Engine)

1. Run the engine until it reaches normal operating temperature. The cooling fan must not be running during the idle speed adjustment.

2. Connect a tachometer to the engine.

NOTE: Do not allow the tachometer or coil terminals to be grounded. This will damage the injection system.

3. On 4A-FE engine, run the engine at 2500 rpm for 2 minutes. On 5S-FE engine, run the engine at 1000–3000 rpm for 5 seconds. Allow the engine to return to idle.

4. Short the check connector at terminals TE_1 and E_1 using a suitable jumper wire.

5. Adjust the idle speed by turning the idle speed adjusting screw.

6. Remove the jumper wire from the connector terminals.

7. Disconnect the tachometer.

MR2

4A-GE ENGINE

1. Run the engine until it reaches normal operating temperature. The cooling fan must not be running during the idle speed adjustment.

2. Connect a tachometer to the engine.

NOTE: Do not allow the tachometer or coil terminals to be grounded. This will damage the injection system.

3. Run the engine at 2500 rpm for 2 minutes.

4. Adjust the idle speed by turning the idle speed adjusting screw.

5. Disconnect the tachometer.

4A-GZE ENGINE

1. Run the engine until it reaches normal operating temperature. The cooling fan must not be running during the idle speed adjustment.

2. Connect a tachometer to the engine.

NOTE: Do not allow the tachometer or coil terminals to be grounded. This will damage the injection system.

3. Short the check connector at terminals T and E_1 using a suitable jumper wire.

4. Adjust the idle speed by turning the idle speed adjusting screw.

5. Remove the jumper wire from the connector terminals.

6. Disconnect the tachometer.

3S-GTE

1. Adjust the idle speed with the engine at normal operating temperature, air cleaner installed, all accessories switched OFF and the transmission in N.

2. Connect a tachometer to the IG terminal of the check connector.

3. Using a jumper wire SST 09843-18020, jump the TE1 and E1 terminals of the check connector.

4. The idle should be 700 rpm (US with auto trans), 750 rpm (US with manual, Canada with automatic transaxle) and 850 rpm (Canada with manual transaxle).

5. If not within specifications, turn the idle adjusting screw at the throttle body to specifications.

6. Turn the ignition switch OFF and disconnect all test equipment.

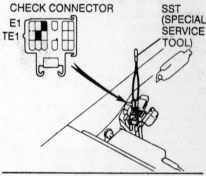

CHECK CONNECTOR
E1
TE1
SST (SPECIAL SERVICE TOOL)

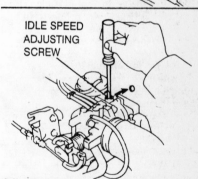

IDLE SPEED ADJUSTING SCREW

Adjusting idle speed—3S—GTE engines

IDLE MIXTURE ADJUSTMENT

On all fuel injected engines, the idle mixture is controlled by the Electronic Control Unit (ECU) and is not adjustable.

Fuel Injector

REMOVAL & INSTALLATION

Tercel

1. Relieve the fuel system pressure.

2. Disconnect the negative battery cable.

3. Disconnect and tag the PCV and vacuum hoses.

4. Remove the air intake connector.

5. Disconnect the accelerator and throttle cables.

6. Disconnect the vacuum sensing hose.

7. Remove the 2 bolts remove the dashpot and link bracket. Disconnect the spring from the dashpot and throttle linkage.

8. Remove the pulsation damper and disconnect the fuel inlet hose from it.

9. Remove the clamp and disconnect the fuel return hose.

10. Remove the cold start injector pipe.

11. Disconnect the injector harness connectors.

12. Remove the 2 bolts attaching the fuel delivery pipe to the cylinder head.

13. Pull the delivery pipe and fuel injectors from the cylinder head.

NOTE: Whe removing the delivery pipe, be careful not to drop the injectors.

14. Remove the injectors from the delivery pipe.

15. Remove the 4 spacers and insulators from the cylinder head. Remove the grommets and O-rings from the injectors. Discard these components and replace with new.

To install:

16. Install new injector grommets and O-rings. Coat the O-rings with clean fuel prior to installation. Make sure the O-ring seats properly in the injector groove. If not, the O-ring will become pinched.

17. Install new spacers and insulators into the cylinder head. Install the injectors into the delivery pipe using a moderate back and forth twisting motion.

18. Mount the injector and delivery pipe assembly onto the cylinder head.

19. Install the delivery pipe retaining bolts and torque them to 14 ft. lbs. (19 Nm). After the bolts are tight, attempt to twist each injector back and forth a small amount by hand. The injectors should rotate smoothly. If not, the injector O-ring are probably not installed properly. Replace the O-rings as required.

20. Connect the injector harness connectors.

21. Install the cold start injector pipe.

22. Connect the fuel return hose to the delivery pipe.

23. Connect the fuel inlet hose to the delivery pipe using new gaskets. Install the pulsation damper and torque to 22 ft. lbs. (29 Nm).

24. Connect the spring to the throttle linkage and dashpot. Install the dashpot and link bracket.

25. Connect the vacuum sensing hose.

26. Connect the throttle and accelerator cables.

27. Install the air intake connector.

28. Connect the PCV hoses.

29. Connect the negative battery cable.

30. Start the engine and check for fuel leaks.

Corolla and 1988–89 MR2

1. Relieve the fuel system pressure.

2. Disconnect the negative battery cable. On MR2 with 4A-GZE engine, remove the throttle body.

3. Disconnect and tag the vacuum and PCV hoses. On MR2 with 4A-GZE engine, loosen the air outlet duct and remove the throttle cable bracket.

4. Disconnect the fuel return hose from the pressure regulator. On MR2 with 4A-GZE engine, remove the fuel pressure regulator.

5. Disconnect the injector harness connectors.

6. Remove the cold start injector pipe.

7. Disconnect the fuel inlet pipe.

8. Remove the delivery pipe bolts and remove the delivery pipe and fuel injectors from the cylinder head.

NOTE: When removing the delivery pipe, be careful not to drop the injectors.

9. On Corrola and MR2 with 4A-GE engine, remove the 2 spacers and 2 insulators from the cylinder head. On MR2 with 4A-GZE engine, there are 3 spacers and 4 insulators.

10. Remove the injectors from the delivery pipe. Remove the O-rings and grommets from the injectors and discard them.

To install:

11. Install new injector grommets and O-rings. Coat the O-rings with clean fuel prior to installation. Make sure the O-ring seats properly in the injector groove. If not, the O-ring will become pinched.

12. Install new spacers and insulators into the cylinder head.

13. Install the injectors into the delivery pipe using a moderate back and forth twisting motion.

14. Mount the injector and delivery pipe assembly onto the cylinder head.

15. Install the delivery pipe retaining bolts and torque them to 11–13 ft. lbs. (15–17 Nm). After the bolts are tight, attempt to twist each injector back and forth a small amount. The injectors should rotate smoothly. If not, the injector O-ring are probably not installed properly. Replace the O-ring(s) as required.

16. Connect the fuel inlet pipe using new gaskets. Torque the union bolt to 22 ft. lbs. (29 Nm).

17. Install the cold start injector pipe.

18. Connect the injector harness connectors.

19. Connect the fuel return hose to the fuel pressure regulator.

20. Connect the vacuum and PCV hoses.

21. Connect the negative battery cable.

22. Start the engine and check for leaks.

Camry and Celica (3S-FE Engine)

1. Relieve the fuel system pressure and disconnect the negative battery cable

2. Remove the cold start injector pipe.

3. Disconnect the vacuum sensing hose from the fuel pressure regulator.

4. Disconnect the injector harness connectors.

5. Disconnect the hose from fuel return pipe.

6. Remove the fuel pressure pulsation damper.

7. Remove the 2 bolts and the delivery pipe together with the 2 injectors attached.

NOTE: When removing the delivery pipe, be careful not to drop the injectors.

8. Remove the 4 insulators and the 2 spacers from the cylinder head. Pull out the four injectors from the delivery pipe.

To install:

9. Insert 4 new insulators and 2 spacers into the injector holes in the cylinder head.

10. Install the grommet and a new O-ring to the delivery pipe end of each injector.

11. Apply a thin coat of fuel to the O-ring on each injector and then press them into the delivery pipe.

12. Install the injectors together with the delivery pipe into the cylinder head. Tighten the 2 mounting bolts to 9 ft. lbs. (13 Nm). After the bolts are tight, attempt to twist each injector back and forth a small amount. The injectors should rotate smoothly. If nct, the injector O-ring are probably not installed properly. Replace the O-ring(s) as required.

13. Install the fuel pressure pulsation damper with 2 new gaskets on the union bolt.

14. Connect the hose to the fuel return pipe.

15. Connect the injector harness connectors.

16. Connect the vacuum sensing hose to the fuel pressure regulator.

17. Install the cold start injector pipe.
18. Connect the negative battery cable.
19. Start the engine and check for leaks.

Camry (2VZ-FE Engine)

1. Relieve the fuel system pressure.
2. Disconnect the negative battery cable.
3. Drain the cooling system.
4. If equipped with automatic transaxle, disconnect the throttle cable from the bracket and throttle body.
5. Remove the air cleaner cap, air flow meter and air cleaner flexible hose.
6. Disconnect and tag all interfering vacuum hoses and electrical wiring harness connectors.
7. Remove the right engine mounting stay.
8. Disconnect the cold start injector connector.
9. Disconnect the cold start injector tube.
10. Disconnect the EGR pipe.
11. Remove the engine hanger and air intake chamber stay.
12. Remove the air intake chamber.
13. Disconnect harness connectors from the tops of the injectors.
14. Disconnect the wiring harness clamps from the left delivery pipe.
15. Disconnect the fuel hoses from the pressure regulator, fuel filter and delivery pipes.
16. Unbolt and pull the left and right delivery pipes with fuel injectors from the intake manifold.

NOTE: When removing the delivery pipes, be careful not to drop the injectors.

17. Remove the injectors from the delivery pipes.
18. Remove the 6 insulator and 4 spacers from the injector openings.
To install:
19. Insert 6 new insulators and 4 spacers into the injector openings in the intake manifold.
20. Install the grommet and a new O-ring to the delivery pipe end of each injector.
21. Apply a thin coat of fuel to the O-ring on each injector and then press them into the delivery pipe.
22. Mount the injectors together with the delivery pipes onto the intake manifold. Tighten the mounting bolts to 9 ft. lbs. (13 Nm). After the bolts are tight, attempt to twist each injector back and forth a small amount. The injectors should rotate smoothly. If not, the injector O-ring are probably not installed properly. Replace the O-ring(s) as required.
23. Connect the fuel hoses to the

pressure regulator, fuel filter and delivery pipes. Use new gaskets on union bolt connections.
24. Connect the wiring harness clamps to the left delivery pipe.
25. Connect the harness connectors to the tops of the injectors.
26. Install the air intake chamber. Torque the mounting nuts to 32 ft. lbs. (43 Nm).
27. Install the air intake chamber stay and engine hanger.
28. Connect the EGR pipe and torque the pipe union nut to 58 ft. lbs. (78 Nm).
29. Connect the cold start injector tube and plug in the connector.
30. Install the right engine mounting stay.
31. Connect all vacuum hoses and electrical wiring harness connectors.
32. Install the air cleaner cap, air flow meter and air cleaner flexible hose.
33. If equipped with automatic transaxle, connect the throttle cable to the cable bracket and throttle body.
34. Fill the cooling system and connect the negative battery cable.
35. Start the engine and check for leaks.

Celica
3S-GE ENGINE

1. Disconnect the negative battery cable.
2. Relieve the fuel system pressure.
3. Drain the cooling system.
4. Disconnect the throttle cable and the accelerator cable from the throttle linkage.
5. Disconnect the ignition coil connector and the high tension cord, then remove the suspension upper brace.
6. Disconnect the air cleaner hose.
7. Remove the ignitor.
8. Remove the throttle body.
9. Remove the No. 2 engine hanger and the No. 2 intake manifold stay.
10. Loosen the union nut of the EGR pipe.
11. Remove the cold start injector pipe.
12. Remove the EGR modulator.
13. Tag and disconnect all hoses and wires which interfere with injector removal.
14. Raise and support the vehicle safely.
15. Remove the suspension lower crossmember.
16. Disconnect the exhaust pipe.
17. Remove the No. 1 and the No. 3 intake manifold stays.
18. Disconnect the ground strap.
19. Remove the intake manifold.
20. Remove the 3 bolts and the delivery pipe with the injectors.

NOTE: When removing the delivery pipes, be careful not to drop the injectors.

21. Remove the injectors from the delivery pipes.
22. Remove the insulators and spacers from the injector openings.
To install:
23. Insert new insulators and spacers into the injector openings in the intake manifold.
24. Install the grommet and a new O-ring to the delivery pipe end of each injector.
25. Apply a thin coat of fuel to the O-ring on each injector and then press them into the delivery pipe.
26. Mount the injectors together with the delivery pipes onto the intake manifold. Tighten the mounting bolts to 9 ft. lbs. (13 Nm). After the bolts are tight, attempt to twist each injector back and forth a small amount. The injectors should rotate smoothly. If not, the injector O-ring are probably not installed properly. Replace the O-ring(s) as required.
27. Installation of the remaining components is the reverse of the removal procedure. Fill the cooling system. Start the engine and check for fuel leaks.

3S-GTE ENGINE

1. Relieve the fuel system pressure and disconnect the negative battery cable.
2. Remove the throttle body.
3. Remove the fuel pressure regulator.
4. Remove the EGR vacuum modulator
5. Disconnect the electrical connections from fuel injectors.
6. Remove the pulsation damper. Disconnect fuel inlet hose from the delivery pipe.
7. Disconnect the fuel return hose from the return pipe.
8. Remove the delivery pipe and fuel injectors and related components (insulators, spacers, O-ring and grommet).

NOTE: When removing the delivery pipes, be careful not to drop the injectors.

9. Remove the injectors from the delivery pipes.
10. Remove the insulators and spacers from the injector openings.
To install:
11. Insert new insulators and spacers into the injector openings in the intake manifold.
12. Install the grommet and a new O-ring to the delivery pipe end of each injector.
13. Apply a thin coat of fuel to the O-ring on each injector and then press them into the delivery pipe.
14. Mount the injectors together with the delivery pipes onto the intake manifold. Tighten the mounting bolts

to 9 ft. lbs. (13 Nm). After the bolts are tight, attempt to twist each injector back and forth a small amount. The injectors should rotate smoothly. If not, the injector O-ring are probably not installed properly. Replace the O-ring(s) as required.

15. Installation of the remaining components is the reverse of the removal procedure. Start the engine and check for fuel leaks.

MR2

3S-GTE ENGINE

1. Relieve the fuel system pressure.
2. Disconnect the negative battery cable.
3. Remove the throttle body.
4. Remove the left engine hood side panel.
5. Remove the air cleaner.
6. Remove the charcoal canister.
7. Remove the EGR vacuum switching valve, vacuum modulator, EGR valve and pipe.
8. Remove the cold start injector pipe and cold start injector.
9. Remove the Idle Speed Control (ISC) water bypass hoses and air hoses.
10. Disconnect the vacuum sensing hose from the vacuum sensing pipe on the injector cover.
11. Disconnect the harness connectors from the tops of the injectors.
12. Disconnect the 2 wire clamps from the mounting bolts on the No. 2 timing cover. Disconnect the 2 wire clamps from the wire brackets on the intake manifold.
13. Disconnect the fuel inlet hose from the fuel filter.
14. Disconnect the fuel return hose from the fuel pressure regulator.
15. Remove the bolt that attaches the fuel inlet hose to the water outlet.
16. Remove the 3 bolts holding the delivery pipe to the cylinder head.
17. Disconnect the fuel inlet hose from the delivery pipe.
18. Remove the delivery pipe and fuel injectors and related components (4 insulators, 3 spacers and injector O-ring and grommets).

NOTE: When removing the delivery pipes, be careful not to drop the injectors.

19. Disconnect the vacuum sensing hose from the pressure regulator and remove the cover plate from the delivery pipe. Remove the injectors from the delivery pipe using the proper tool.
To install:
20. Insert 4 new insulators and 3 spacers into the injector openings.
21. Install the grommet and a new O-ring to the delivery pipe end of each injector.
22. Apply a thin coat of fuel to the O-

ring on each injector and then press them into the delivery pipe. Make sure the injector connectors are positioned correctly.
23. Mount the injectors together with the delivery pipes. Tighten the mounting bolts to 14 ft. lbs. (19 Nm).
24. Installation of the remaining components is the reverse of the removal procedure. The injector harness connectors are color coded. The No. 1 and No. 3 injector connectors are brown and the No. 2 and No. 4 connectors are grey. Start the engine and check for fuel leaks.

Celica and MR2

5S-FE ENGINE

1. Disconnect the negative battery cable.
2. Remove the throttle body.
3. On MR2, remove the engine hood side panels, air cleaner and cruise control actuator.
4. Remove the cold start injector pipe.
5. On Celica, remove the fuel pressure regulator. On MR2, disconnect the brake booster vacuum hose from the intake manifold.
6. Disconnect necessary engine wiring and remove the left and right accelerator brackets.
7. Disconnect the electrical connectors from fuel injectors.
8. Disconnect wire retaining clamps from the No. 2 timing belt cover and and intake manifold as necessary to gain access for removal/installation of fuel injectors.
9. Disconnect the fuel return hose from the return pipe.
10. Remove the delivery pipe and fuel injectors and related components (insulators, spacers, O-ring and grommet).

NOTE: When removing the delivery pipes, be careful not to drop the injectors.

11. Remove injectors from the delivery pipe.
To install:
12. Insert new insulators and spacers into the injector openings in the intake manifold.
13. Install the grommet and a new O-ring to the delivery pipe end of each injector.
14. Apply a thin coat of fuel to the O-ring on each injector and then press them into the delivery pipe.
15. Mount the injectors together with the delivery pipes onto the intake manifold. Tighten the mounting bolts to 9 ft. lbs. (13 Nm). After the bolts are tight, attempt to twist each injector back and forth a small amount. The injectors should rotate smoothly. If not, the injector O-ring are probably not in-

stalled properly. Replace the O-ring(s) as required.

16. Installation of the remaining components is the reverse of the removal procedure. Start the engine and check for leaks.

Supra

1. Relieve the fuel system pressure.
2. Disconnect the negative battery cable.
3. Drain the cooling system.
4. Tag and disconnect all hoses and wires which interfere with injector removal.
5. Disconnect accelerator connecting rod.
6. On 7M-GE engine, remove the air intake connector. On 7M-GTE engine, remove the throttle body.
7. Remove the ISC valve and gasket.
8. Disconnect the injector connectors.
9. Disconnect the cold start injector tube from the delivery pipe.
10. Remove the pulsation damper and the 2 gaskets.
11. Remove the union bolts and 2 gaskets from the fuel return pipe support.
12. Remove the clamp bolts from the No. 1 fuel pipe and Vacuum Switching Valve (VSV).
13. Remove the union bolts and 2 gaskets from the pressure regulator.
14. Disconnect the fuel hose from the No. 2 fuel pipe.
15. Remove the clamp bolt and the return fuel pipe.
16. Loosen the locknut and remove the pressure regulator.
17. Remove the 3 bolts, and then remove the delivery pipe with the injectors.

NOTE: When removing the delivery pipe, be careful not to drop the injectors.

18. Remove the 6 insulators and the 3 spacers from the cylinder head, then pull out the injectors from the delivery pipe.
To install:
19. Before installing, apply a thin coat of gasoline to the O-ring on each injector and then press them into the delivery pipe.
20. Insert 6 new insulators into the injector hole of the cylinder head.
21. Install the black rings on the upper portion of each of the 3 spacers, then install the spacers on the delivery pipe mounting hole of the cylinder head.
22. Install the 3 spacers and bolts and torque to 13 ft. lbs. (After the bolts are tight, attempt to twist each injector back and forth a small amount. The injectors should rotate smoothly. If not, the injector O-ring are probably

not installed properly. Replace the O-ring(s) as required.

23. Fully loosen the locknut of the pressure regulator. Push the pressure regulator completely into the delivery pipe by hand, then turn the regulator counterclockwise until the outlet faces outward in the correct position. Torque the locknut to 18 ft. lbs. (24 Nm).

24. Install the No. 2 fuel pipe and clamp bolt.

25. Connect the fuel hose.

26. Install the union bolt and 2 new gaskets to the pressure regulator and torque the union bolt to 18 ft. lbs. (24 Nm).

27. Install the No. 1 fuel pipe, Vacuum Switching Valve (VSV) and clamp bolt.

28. Install the union bolt and 2 new gaskets to the support pipe and torque the union bolts to 22 ft. lbs. (30 Nm).

29. Install the pulsation damper and 2 new gaskets and torque to 29 ft. lbs. (39 Nm).

30. Connect the injector connectors.

31. Install the Idle Speed Control (ISC) valve with a new gasket and torque to 9 ft. lbs. (13 Nm).

32. Install the throttle body or the air intake connector.

33. Connect the accelerator connecting rod.

34. Connect all vacuum hoses and electrical wires.

35. Refill the cooling system and connect the negative battery cable.

36. Start the engine and check for leaks.

Cressida

5M-GE ENGINE

1. Disconnect the negative battery cable.

2. Relieve the fuel system pressure.

3. Remove the air intake chamber.

4. Remove the distributor.

5. Remove the fuel pipe.

6. Unplug the wiring connectors from the tops of the fuel injectors and remove the 2 plastic clamps that hold the wiring harness to the fuel delivery pipe.

7. Unscrew the 4 mounting bolts and remove the delivery pipe with the injectors attached. Do not remove the injector cover.

NOTE: When removing the delivery pipe, be careful not to drop the injectors.

8. Pull the injectors out of the delivery pipe.

To install:

9. Insert 6 new insulators into the injector holes on the intake manifold.

10. Install the grommet and a new O-ring to the delivery pipe end of each injector.

11. Apply a thin coat of gasoline to the O-ring on each injector and then press them into the delivery pipe.

12. Install the injectors together with the delivery pipe in the intake manifold. Tighten the mounting bolts to 13 ft. lbs. (17 Nm). After the bolts are tight, attempt to twist each injector back and forth a small amount. The injectors should rotate smoothly. If not, the injector O-ring are probably not installed properly. Replace the O-ring(s) as required.

13. Secure the injector wiring harness to the delivery pipe with the plastic clamps. Connect the harness connectors to the tops of the injectors.

14. Install the fuel pipe, distributor and air intake chamber.

15. Connect the negative battery cable.

16. Start the engine and check for leaks.

7M-GE ENGINE

1. Disconnect the negative battery cable and drain the cooling system.

2. Relieve the fuel system pressure.

3. Remove the throttle body.

4. Remove the Idle Speed Control (ISC) valve.

5. Disconnect the injector harness connectors.

6. Disconnect the cold start injector from the delivery pipe.

7. Disconnect the EGR Vacuum Switching Valve (VSV) connector.

8. Remove the union bolt and 2 gaskets from the delivery pipe and fuel filter.

9. Remove the clamp bolt and remove the No. 1 fuel pipe with the vacuum switching valve.

10. Disconnect the No. 3 PCV hose.

11. Disconnect the vacuum sensing hose.

12. Disconnect the fuel hose from the No. 2 fuel pipe.

13. Remove the union bolt and 2 gaskets from the pressure regulator.

14. Remove the clamp bolts and remove the No. 2 fuel pipe.

15. Loosen the locknut and remove the fuel pressure regulator.

16. Remove the delivery pipe attaching bolts. Remove the delivery pipe with the 6 fuel injectors.

17. Remove the 6 insulators and 3 spacers from the cylinder head.

18. Remove the injectors from the delivery pipe.

To install:

19. Install new injector grommets and O-rings. Coat the O-rings with clean fuel prior to installation. Make sure the O-ring seats properly in the injector groove. If not, the O-ring will become pinched.

20. Install new insulators into the cylinder head. Install the black rings on the upper portion of each spacer.

Then, install the spacers into the mounting holes in the head.

21. Install the injectors into the delivery pipe using a moderate back and forth twisting motion.

22. Mount the injector and delivery pipe assembly onto the cylinder head. Make sure the injector connectors are facing up.

23. Install the delivery pipe retaining bolts and torque them to 13 ft. lbs. (18 Nm). After the bolts are tight, attempt to twist each injector back and forth a small amount by hand. The injectors should rotate smoothly. If not, the injector O-ring are probably not installed properly. Replace the O-rings as required.

24. Install the fuel pipes and pressure regulator.

25. Connect the vacuum hoses and injector harness connectors. Install the Idle Speed Control (ISC) valve and throttle body.

26. Fill the cooling system to the proper level and connect the negative battery cable.

27. Start the engine and check for leaks.

DRIVE AXLE

Halfshaft

REMOVAL & INSTALLATION

Tercel

1. Raise the vehicle and support it safely.

2. Remove the cotter pin and locknut cap.

3. Have an assistant step on the brake pedal and at the same time, loosen the bearing locknut.

4. Remove the brake caliper and position it aside. Remove the brake disc.

5. Remove the cotter pin and nut from the tie rod end. Using a suitable puller, disconnect the tie rod end from the steering knuckle.

6. Matchmark the lower strut mounting bracket where it attaches to the steering knuckle, remove the mounting bolts and then disconnect the steering knuckle from the strut bracket.

7. Using a suitable puller, pull the axle hub off the outer halfshaft end.

8. Remove the stiffener plate from the left side of the transaxle assembly.

9. Using the proper tool, tap the halfshaft out of the transaxle casing.

To install:

NOTE: Be sure to cover the halfshaft input opening.

10. During installation, observe the following:

   a. Coat the oil seal in the transaxle input opening with MP grease before inserting the halfshaft.

   b. On 1988–92 vehicles, torque the bolts to 166 ft. lbs.

   c. Tighten the tie rod end nut to 36 ft. lbs. (49 Nm).

   d. Tighten the bearing locknut to 137 ft. lbs. (186 Nm).

   e. Tighten the stiffener plate bolts to 29 ft. lbs. (39 Nm).

   f. Check the front wheel alignment.

Corolla FX/FX16
1988 Corolla (4A-GE)

1. Raise and support the vehicle safely.

2. Remove the cotter pin, locknut cap and locknut from the hub.

3. Remove the engine under cover. Remove the 6 nuts attaching the halfshaft (front driveshaft) to the transaxle (differential side gear).

4. Remove the brake caliper from the steering knuckle and support it aside with wire. Remove the rotor disc.

5. Disconnect the steering knuckle from the lower arm by removing the bolt and 2 nuts, then disconnect the lower arm from the steering knuckle.

6. Using a suitable puller, pull the axle hub from the halfshaft. Be sure to cover the dust boot with a shop rag to prevent damage to the the boot.

To install:

7. During installation, observe the following:

   a. Torque the steering knuckle to 47–64 ft. lbs. (64 Nm) on 1988 FX and 105 ft. lbs. (142 Nm) on 1988 sedan and wagon.

   b. Torque the caliper bolts to 65 ft. lbs. (88 Nm).

   c. Torque the bearing nut to 137 ft. lbs. (186 Nm).

   d. Torque the halfshaft nuts to 27 ft. lbs. (36 Nm).

1988–92 Corolla (Except 1988 4A-GE)

1. Raise and safely support the vehicle.

2. Remove the cotter pin and locknut cap.

3. Have an assistant step on the brake pedal and at the same time, loosen the bearing locknut.

4. Remove the engine undercovers and then drain the gear oil or fluid.

5. Remove the cotter pin and nut from the tie rod end. Using a suitable puller, disconnect the tie rod end from the steering knuckle.

6. Remove the mounting bolts and then disconnect the steering knuckle from the lower control arm.

7. Use a rubber mallet and drive the

outer end of the shaft out of the axle hub.

8. Using the proper tools, tap or pry the halfshaft out of the transaxle casing.

To install:

NOTE: Be sure to cover the halfshaft input opening.

9. During installation, observe the following:

   a. Coat the oil seal in the transaxle input hole with MP grease before inserting the halfshaft.

   b. Tighten the steering knuckle-to-lower arm bolts to 105 ft. lbs. (142 Nm).

   c. Tighten the tie rod end nut to 36 ft. lbs. (49 Nm).

   d. Tighten the bearing locknut to 137 ft. lbs. (186 Nm).

   e. Check that there is 0.08–0.12 in. (2–3mm) axial play on each shaft.

   f. Check the front wheel alignment.

1988–89 Celica (2WD)

1. Raise and safely support the vehicle.

2. Remove the wheels.

3. Remove the cotter pin, cap and locknut from the hub.

4. Remove the engine under covers.

5. Drain the transmission fluid or the differential fluid on the GTS.

6. Remove the transaxle gravel shield on the GTS.

7. Loosen the 6 nuts attaching the inner end of the halfshaft to transaxle (all except Celica GTS).

NOTE: Wrap the exposed end of the halfshaft in an old shop cloth to prevent damage to it.

8. Remove the cotter pin from the tie end rod and then press the tie rod out of the steering knuckle. Remove the bolt and 2 nuts and disconnect the steering knuckle from the lower arm control.

9. On all but the GTS, use a 2-armed gear puller or equivalent, and press the halfshaft out of the steering knuckle.

10. On the GTS, mark a spot somewhere on the left halfshaft and measure the distance between the spot and the transaxle case. Using the proper tool, pull the halfshaft out of the transaxle.

11. On the GTS, use a 2-armed puller and press the outer end of the right halfshaft out of the steering knuckle. Use a pair of pliers to remove the snapring at the inner end and pull the halfshaft out of the center driveshaft.

12. On all but the GT-S, remove the snap-ring on the center shaft with a pair of pliers and then pull the center shaft out of the transaxle case.

To install:

13. When installing the center drive-

shaft on ST and GT vehicles, coat the transaxle oil seal with grease, insert the halfshaft through the bearing bracket and secure it with a new snap-ring.

14. Repeat Step 13 when installing the inner end of the right halfshaft on the GTS.

15. On the right halfshaft of the GTS, use a new snap-ring, coat the transaxle oil seal with grease and then press the inner end of the shaft into the differential housing. Check that the measurement made in Step 10 is the same. Check that there is 0.08–0.11 in. (2–3mm) of axial play. Check also that the halfshaft will not come out by trying to pull it back by hand.

16. Press the outer end of each halfshaft into the steering knuckle on the GTS.

17. On the ST and GT, press the outer end of the halfshafts into the steering knuckle and then finger-tighten the nuts on the inner end.

18. Connect the steering knuckle to the lower control arm and tighten the bolts to 94 ft. lbs. (127 Nm).

19. Connect the tie rod end to the steering knuckle and tighten the nut to 36 ft. lbs. (49 Nm). Install a new cotter pin.

20. Tighten the hub locknut to 137 ft. lbs. (186 Nm) while depressing the brake pedal. Install the cap and use a new cotter pin.

21. On the ST and GT, tighten the 6 nuts on the inner halfshaft ends to 27 ft. lbs. (36 Nm) while depressing the brake pedal.

22. Install the transaxle gravel shield on the GTS.

23. Fill the transaxle with gear oil or fluid.

24. Install the engine under cover.

1990 Celica (2WD)

NOTE: The hub bearing can be damaged if it is subjected to the vehicle weight such as moving the vehicle when the halfshaft is removed. If with ABS, after disconnecting halfshaft, work carefully so as not damage the sensor rotor serrations on the halfshaft.

1. Raise and safely support the vehicle. Remove the wheels.

2. Remove the cotter pin, cap and locknut (loosen locknut while depressing brake pedal) from the hub.

3. Remove the engine under covers.

4. Drain the transaxle fluid.

5. Remove the brake caliper and rotor disc.

6. Disconnect the tie rod end (remove cotter pin and nut) from the steering knuckle.

7. Disconnect steering knuckle from the lower arm.

8. Remove the halfshaft from the

steering knuckle using a suitable puller. Cover the halfshaft boot with shop cloth or equivalent to protect it from damage.

9. Remove the left side halfshaft using the proper tool.

10. Remove the the right side halfshaft. On the 5S-FE engine, remove the 2 bolts of the center bearing bracket and pull out the halfshaft with center bearing case and center halfshaft. On the 4A-FE engine, use a suitable brass punch tap out the right side halfshaft.

To install:

11. Install the left side halfshaft. Apply grease to the transaxle oil seal lip. Position the new snap-ring opening side facing downward using brass punch, tap halfshaft in until it makes contact with the pinion shaft. Install the outboard joint side of the halfshaft to the axle hub.

12. Install right side halfshaft on the 5S-FE engine using the following procedure:

a. Apply grease to the transaxle oil seal lip.

b. Insert the center halfshaft with the right side to the transaxle through the bearing bracket. When inserting the halfshaft, insert so the straight pin on the center bearing case aligns with the hole on the bearing bracket.

c. Install retaining bolts and torque to 47 ft. lbs. (64 Nm).

d. Install the outboard joint side of the halfshaft to the axle hub.

13. Install right side halfshaft on the 4A-FE engine using the following procedure:

a. Apply grease to the transaxle oil seal lip.

b. Position the new snap-ring opening side facing downward using brass punch, tap halfshaft in until it makes contact with the pinion shaft.

c. Install the outboard joint side of the halfshaft to the axle hub.

14. Check that the halfshaft will not come out by trying to pull it by hand.

15. Connect the steering knuckle to the lower control arm and tighten the bolts to 94 ft. lbs. (128 Nm).

16. Connect the tie rod end to the steering knuckle and tighten the nut to 36 ft. lbs. (49 Nm). Install a new cotter pin.

17. Install all necessary brake components. Tighten the hub locknut to 137 ft. lbs. (186 Nm) while depressing the brake pedal. Install the cap and use a new cotter pin.

18. Fill the transaxle to the proper level. Install the engine under cover. Check front wheel alignment.

Celica (AWD)
FRONT

NOTE: The hub bearing can be

damaged if it is subjected to the vehicle weight such as moving the vehicle when the halfshaft is removed. On 1990–92 vehicles with ABS, after disconnecting halfshaft work carefully so as not damage the sensor rotor serrations on the halfshaft.

1. Raise and support the vehicle safely.

2. Remove the wheels.

3. Remove the cotter pin, cap and locknut from the hub.

4. Remove the transaxle gravel shield, if with manual transmission. Remove the engine under cover and front fender apron seal.

5. Remove the cotter pin and nut from the tie rod end and then disconnect it from the steering knuckle.

6. Remove the bolt and 2 nuts and disconnect the steering knuckle from the lower control arm.

7. Loosen the 6 nuts attaching the inner end of the halfshaft to the transaxle side gear shaft.

8. Grasp the halfshaft and push the axle carrier outward until the shaft can be removed from the side gear shaft.

NOTE: Wrap the exposed end of the halfshaft in an old shop cloth to prevent damage to it.

9. Use a rubber mallet and tap the outer end of the shaft from the axle hub.

To install:

10. Press the outer end of the halfshaft into the axle hub, position the inner end and install the 6 nuts finger-tight.

11. Connect the tie rod end to the steering knuckle and tighten the nut to 36 ft. lbs. (49 Nm). Install a new cotter pin. If the cotter pin holes do not align, tighten the nut until they align. Never loosen it.

12. Connect the steering knuckle to the lower control arm and tighten to 94 ft. lbs. (127 Nm).

13. Tighten the 6 inner shaft mounting nuts to 48 ft. lbs. (65 Nm). Measure the distance between the right and left side shafts; it must be less then 27.75 in. (704.7mm).

14. With the brake pedal depressed, install the bearing locknut and tighten it to 137 ft. lbs. (186 Nm). Install the cap and a new cotter pin.

15. Install the wheels and lower the vehicle.

REAR

1. Raise and safely support the vehicle. Remove the wheels.

2. Remove the cotter pin, locknut cap and bearing nut.

3. Scribe matchmarks on the inner joint tulip and the side gear shaft

flange. Loosen and remove the 4 nuts.

4. Disconnect the inner end of the shaft by punching it upward and then pull the outer end from the axle carrier. Remove the halfshaft.

To install:

5. Position the halfshaft into the axle carrier and pull the inner end down until the matchmarks are aligned.

6. Connect the halfshaft to the side gear shaft and tighten the nuts to 51 ft. lbs. (69 Nm).

7. Install the bearing nut and tighten it to 137 ft. lbs. (186 Nm) with the brake pedal depressed. Install the cap and a new cotter pin.

8. Install the wheels and lower the vehicle.

Camry (2WD)
1988

1. Raise and support the vehicle safely.

2. Remove the wheels.

3. Remove the cotter pin, cap and locknut from the hub.

4. Remove the transaxle gravel shield, if with manual transaxle. Remove the engine under cover and front fender apron seal.

5. Loosen the 6 nuts attaching the inner end of the halfshaft to the transaxle or center shaft.

NOTE: Wrap the exposed end of the halfshaft in an old shop cloth to prevent damage to it.

6. Remove the brake caliper with the hydraulic line still attached, position it aside and suspend it with a wire. Remove the rotor.

7. Remove the 2 bolts attaching the ball joint to the steering knuckle. Pull the lower control arm down while pulling the strut outward; this will disconnect the inner end of the halfshaft from the transaxle.

8. Using a 2-armed puller, or the like, press the outer end of the halfshaft from the steering knuckle and then remove the halfshaft.

9. Drain the transaxle fluid, remove the snap-ring with pliers and pull the shaft out of the transaxle case.

To install:

10. When installing the center halfshaft, coat the transaxle oil seal with grease, insert the halfshaft through the bearing bracket and secure it with a new snap-ring.

11. Press the outer end of the halfshaft into the steering knuckle, position the inner end and install the 6 nuts finger-tight.

12. Reconnect the ball joint to the steering knuckle, if disconnected, and tighten the bolts to 94 ft. lbs. (127 Nm).

13. Install the rotor and brake cali-

per. Tighten the caliper-to-knuckle bolts to 65 ft. lbs. (88 Nm).

14. Tighten the wheel bearing locknut to 137 ft. lbs. (186 Nm) while depressing the brake pedal. Install the locknut cap and use a new cotter pin.

15. Tighten the 6 inner end nuts to 27 ft. lbs. (36 Nm) while depressing the brake pedal.

16. Install the transaxle gravel shield, if equipped.

17. Fill the transaxle to the proper level.

1989–92

1. Raise and support the vehicle safely.

2. Remove the front wheels.

3. Remove the cotter pin, cap and locknut from the hub.

4. Remove the engine under covers.

5. Drain the transmission fluid or the differential fluid on the wagon.

6. Remove the transaxle gravel shield on the wagon.

7. Loosen the 6 nuts attaching the inner end of the halfshaft to transaxle, all except wagon.

NOTE: Wrap the exposed end of the halfshaft in an old shop cloth to prevent damage to it.

8. Remove the cotter pin from the tie end rod and then press the tie rod out of the steering knuckle. Remove the bolt and 2 nuts and disconnect the steering knuckle from the lower arm control.

9. On all except 4 cylinder wagon, use a 2-armed gear puller or equivalent, and press the halfshaft out of the steering knuckle.

10. On the 4 cylinder wagon, mark a spot somewhere on the left halfshaft and measure the distance between the spot and the transaxle case. Using the proper tool, pull the halfshaft out of the transaxle.

11. On the 4 cylinder wagon, use a 2-armed puller and press the outer end of the right halfshaft out of the steering knuckle. Remove the snap-ring at the inner end and pull the halfshaft out of the center driveshaft.

12. On all except the 4 cylinder wagon, remove the snap-ring on the center shaft and pull the center shaft out of the transaxle case.

To install:

13. When installing the center driveshaft on sedan and V6 vehicles, coat the transaxle oil seal with grease, insert the halfshaft through the bearing bracket and secure it with a new snap-ring.

14. Repeat Step 13 when installing the inner end of the right halfshaft on the 4 cylinder wagon.

15. On the right halfshaft of the 4 cylinder wagon, use a new snap-ring, coat the transaxle oil seal with grease and then press the inner end of the shaft into the differential housing. Check that the measurement made in Step 10 is the same. Check that there is 0.08–0.12 in. (2–3mm) of axial play. Check also that the halfshaft will not come out by trying to pull it by hand.

16. Press the outer end of each halfshaft into the steering knuckle on the 4 cylinder wagon.

17. On all except the 4 cylinder wagon, press the outer end of the halfshafts into the steering knuckle and then finger-tighten the nuts on the inner end.

18. Connect the steering knuckle to the lower control arm and tighten the bolts to 83 ft. lbs. (113 Nm).

19. Connect the tie rod end to the steering knuckle and tighten the nut to 36 ft. lbs. (49 Nm). Install a new cotter pin.

20. Tighten the hub locknut to 137 ft. lbs. (186 Nm) while depressing the brake pedal. Install the cap and use a new cotter pin.

21. On all except the 4 cylinder wagon, tighten the 6 nuts on the inner halfshaft ends to 27 ft. lbs. (36 Nm) while depressing the brake pedal.

22. Install the transaxle gravel shield on the wagon.

23. Fill the transaxle with gear oil or fluid.

24. Install the engine under cover.

Camry (AWD)

FRONT—1988

1. Raise and safely support the vehicle.

2. Remove the wheels.

3. Remove the cotter pin, cap and locknut from the hub.

4. Remove the transaxle gravel shield. Remove the engine under cover and front fender apron seal.

5. Remove the cotter pin and nut from the tie rod end and then disconnect it from the steering knuckle.

6. Remove the bolt and 2 nuts and disconnect the steering knuckle from the lower control arm.

7. Loosen the 6 nuts attaching the inner end of the halfshaft to the transaxle side gear shaft.

8. Grasp the halfshaft and push the axle carrier outward until the shaft can be removed from the side gear shaft.

NOTE: Wrap the exposed end of the halfshaft in an old shop cloth to prevent damage to it.

9. Use a rubber mallet and tap the outer end of the shaft from the axle hub.

To install:

10. Press the outer end of the halfshaft into the axle hub, position the inner end and install the 6 nuts finger-tight.

11. Connect the tie rod end to the steering knuckle and tighten the nut to 36 ft. lbs. (49 Nm). Install a new cotter pin. If the cotter pin holes do not align, tighten the nut until they align. Never loosen it.

12. Connect the steering knuckle to the lower control arm and tighten to 94 ft. lbs. (127 Nm).

13. Tighten the 6 inner shaft mounting nuts to 48 ft. lbs. (65 Nm). Measure the distance between the right and left side shafts; it must be less then 27.75 in. (704.7mm).

14. With the brake pedal depressed, install the bearing locknut and tighten it to 137 ft. lbs. (186 Nm). Install the cap and a new cotter pin.

15. Install the wheels and lower the vehicle.

FRONT—1989–92

1. Raise and support the vehicle safely.

2. Remove the wheels.

3. Remove the cotter pin, cap and locknut from the hub.

4. Remove the engine undercovers.

5. Disconnect the tie rod end from the steering knuckle.

6. Disconnect the lower control arm at the steering knuckle and pull it down and aside.

7. Use a plastic hammer and carefully tap the outer end of the halfshaft until it frees itself from the axle hub.

8. Cover the outer boot with a rag and then remove the inner end of the halfshaft from the transaxle. Use the proper tools.

To install:

9. Coat the lip of the oil seal with grease and then carefully drive the inner end of the shaft into the transaxle until it makes contact with the pinion shaft.

NOTE: Be careful not to damage the boots when installing the halfshafts; also, position the boot snap-ring so the opening is facing downward.

10. Put the outer end of each shaft into the axle hub, being careful not to damage the boots.

11. Check that there is 0.08–0.12 in. (2–3mm) of axial play. Check also that the halfshaft will not come out by hand.

12. Connect the lower control arm to the steering arm and tighten the bolt to 83 ft. lbs. (113 Nm).

13. Connect the tie rod to the steering knuckle and tighten the nut to 36 ft. lbs. (49 Nm). Use a new cotter pin to secure it.

14. Install the axle bearing locknut and tighten it to 137 ft. lbs. (186 Nm) while stepping on the brake pedal. Install the locknut cap and then a new cotter pin.

15. Fill the transaxle with gear oil or fluid, install the undercovers and wheels. Lower the vehicle and check the front end alignment.

REAR — 1988–92

1. Raise and support the vehicle safely. Remove the wheels.

2. Remove the cotter pin, locknut cap and bearing nut.

3. Scribe matchmarks on the inner joint tulip and the side gear shaft flange. Loosen and remove the 4 nuts.

4. Disconnect the inner end of the shaft by punching it upward and then pull the outer end from the axle carrier. Remove the halfshaft.

5. Position the halfshaft into the axle carrier and pull the inner end down until the matchmarks are aligned.

6. Connect the halfshaft to the side gear shaft and tighten the nuts to 51 ft. lbs. (69 Nm).

7. Install the bearing nut and tighten it to 137 ft. lbs. (186 Nm) with the brake pedal depressed. Install the cap and a new cotter pin.

8. Install the wheels and lower the vehicle.

MR2

NOTE: On 1990–92 vehicles with ABS, after disconnecting halfshaft, work carefully so as not damage the sensor rotor serrations on the halfshaft.

1. Raise and support the vehicle safely.

2. Remove the wheels.

3. Remove the cotter pin, cap and locknut from the hub.

4. Remove the transaxle gravel shield.

5. Loosen the 6 nuts attaching the inner end of the halfshaft to transaxle.

NOTE: Wrap the exposed end of the halfshaft in an old shop cloth to prevent damage to it.

6. If equipped with automatic transaxle, remove the 2 bolts holding the ball joint to the rear axle carrier and disconnect the lower arm from the rear axle carrier.

7. Also, if equipped with automatic transaxle, remove the cotter pin and nut using the proper tool. Disconnect the suspension arm from the rear axle carrier.

8. While holding the halfshaft, knock the outer end of the wheel hub assembly. Remove the halfshaft.

To install:

9. Press the outer end of the halfshaft into the wheel hub assembly.

10. Position the inner end of the halfshaft and install the 6 nuts finger-tight.

11. Install the transaxle gravel shield.

12. Tighten the wheel bearing locknut to 137 ft. lbs. (186 Nm) while depressing the brake pedal. Install the locknut cap and use a new cotter pin.

13. Tighten the 6 inner end nuts to 27 ft. lbs. (36 Nm) while depressing the brake pedal. Torque the suspension arm nut to 36 ft. lbs. (49 Nm) and the lower arm to rear axle carrier to 83 ft. lbs. (113 Nm).

14. Fill the transaxle to the proper level.

1988 Cressida

1. Raise and support the vehicle safely.

2. Place matchmarks on the halfshaft and flanges.

3. Remove the 4 nuts retaining the halfshaft to the differential and disconnect the halfshaft from the differential.

4. Remove the 4 nuts retaining the halfshaft to the axle shaft and disconnect the halfshaft from the axle shaft. Remove the halfshaft from the under the vehicle.

5. Installation is the reverse order of the removal procedure. Be sure to align the matchmarks on the halfshaft and torque the retaining nuts to 51 ft. lbs. (69 Nm).

Supra and 1989–92 Cressida

1. Raise and support the vehicle safely. Remove the rear wheels.

2. Using a suitable jack, raise the No. 2 suspension arm until it is horizontal. Matchmarks the rear halfshaft to the side gear shaft flange.

3. Remove the 6 retaining nuts (while an assistant is depressing the brake pedal) and disconnect the rear halfshaft from the differential.

4. Remove the cotter pin and locknut cap. Loosen and remove the bearing locknut.

5. Using a suitable hammer, tap out the rear halfshaft.

6. Installation is the reverse order of the removal procedure. Tighten the bearing locknut to 203 ft. lbs. (275 Nm) and the 6 halfshaft retaining bolts to 51 ft. lbs. (69 Nm).

CV-Boot

REMOVAL & INSTALLATION

1. Mount the halfshaft in a suitable holding fixture.

2. Remove the inboard joint boot clamps.

3. Place matchmarks on the inboard joint tulip and tripod.

4. Remove the inboard joint tulip from the halfshaft.

5. Remove the tripod joint snap-ring.

6. Place matchmarks on the shaft and tripod.

7. Using a brass punch or equivalent remove the tripod joint from the halfshaft.

8. Remove inboard joint boot.

9. Remove the outboard joint boot clamps and boot.

10. Installation is the reverse of the removal procedures. Pack all CV-joints with suitable grease. Use new boot retaining clamps and snap-rings as necessary.

Driveshaft and U-Joints

REMOVAL & INSTALLATION

1. Raise and support the rear axle housing safely.

2. Matchmark the driveshaft and companion flange. Unfasten the bolts which attach the driveshaft universal joint yoke flange to the mounting flange on the differential drive pinion.

3. If equipped with 3 universal joints, perform the following:

   a. Remove the driveshaft sub-assembly from the U-joint sleeve yoke.

   b. Remove the center support bearing from its bracket.

4. Remove the driveshaft end from the transmission.

5. Plug the transmission opening to keep the transmission oil from running out.

6. Remove the driveshaft.

To install:

7. Apply multipurpose grease on the section of the U-joint sleeve which is to be inserted into the transmission.

8. Insert the driveshaft sleeve into the transmission.

NOTE: Be careful not to damage any of the seals.

9. If equipped with 3 U-joints and center bearings, perform the following:

   a. Adjust the center bearing clearance with no load placed on the drive line components; the top of the rubber center cushion should be 0.04 in. (1mm) behind the center of the elongated bolt hole.

   b. Install the center bearing assembly. Use the same number of washers on the center bearing brackets as were removed.

   c. Matchmark the arrow marks on the driveshaft and grease fittings.

10. Align the matchmarks. Secure the U-joint flange to the differential pinion flange with the mounting bolts.

NOTE: Be sure the bolts are of the same type as those removed and that they are tightened securely.

11. Remove the axle housing supports and lower the vehicle.

12. Tighten the center bearing-to-bracket bolts to 30 ft. lbs. (40 Nm) on 1988 Cressida, 27 ft. lbs. (37 Nm) on 1989–92 Cressida or 36 ft. lbs. (49 Nm) on Supra. Tighten the flange bolts to 31 ft. lbs. (42 Nm) on 1988 Cressida or 54 ft. lbs. (74 Nm) on Supra and 1989–92 Cressida.

1988–92 Camry (AWD), 1988–92 Celica (AWD)
1989–92 Corolla (AWD), 1988 Tercel Wagon (AWD)

1. Matchmark the front driveshaft flange and the front center bearing flange. Remove the 4 bolts, washers and nuts and disconnect the rear end of the front driveshaft from the front center bearing flange. Pull the shaft out of the transfer case and remove it. Plug the transfer case to prevent leakage.

2. Depress the brake pedal and loosen the cross groove set bolts ½ turn. These bolts are at the front edge of the rear driveshaft; rear edge of the rear center bearing.

3. Matchmark the rear flange of the rear driveshaft to the differential pinion flange and then disconnect them.

4. Remove the 2 mounting bolts from the front and rear center bearings and then remove the 2 center bearings, intermediate shaft and rear driveshaft as an assembly.

5. Matchmark the universal joint and the rear center bearing flange, remove the bolts and separate the rear driveshaft from the rear center bearing.

6. Pull the front and rear center bearings from the intermediate shaft.
To install:

7. Install the 2 center bearings onto the intermediate shaft ends and then temporarily install the assembly.

8. Align the matchmarks and connect the rear driveshaft to the differential. Tighten the bolts to 54 ft. lbs. (74 Nm); 27 ft. lbs. (37 Nm) on the Corolla.

9. Press the front driveshaft yoke into the transfer case, align the matchmarks at the rear of the shaft with those on the front center bearing flange and tighten the bolts to 54 ft. lbs. (74 Nm); 27 ft. lbs. (37 Nm) on the Corolla.

10. With the front edge of the rear driveshaft in position, depress the brake pedal and tighten the cross groove joint set bolts to 20 ft. lbs. (27 Nm) on all except 1989–92 Celica. On 1989–92 Celica, torque the bolts to 48 ft. lbs. (65 Nm).

11. With the vehicle in an unladen condition, adjust the distance between the rear edge of the boot cover and the rear driveshaft to 2.58–2.78 in. (65.5–70.5mm) on all except 1989–92 Celica. On 1989–92 Celica adjust the distance to 2.85–3.05 in. (72.5–77.5mm).

12. With the vehicle in an unladen condition, adjust the distance between the rear side of the center bearing housing and the rear side of the cushion to 0.45–0.53 in. (11.5–13.5mm).

13. Tighten the center bearing mounting bolts to 27 ft. lbs. (37 Nm). Make sure the center line of the bracket is at right angles to the shaft axial direction.

Rear Axle Shaft, Bearing and Seal

NOTE: These service procedures apply to rear wheel drive, four wheel drive vehicles and MR2 only. For rear axle shaft service front wheel drive vehicles (except AWD), refer to "Rear Axle Hub, Carrier and Bearing" in the Rear Suspension section.

REMOVAL & INSTALLATION

Corolla (AWD) and Tercel (AWD)

1. Raise and support the vehicle safely.

2. Drain the oil from the axle housing.

3. Remove the rear wheels.

4. Punch matchmarks on the brake drum and the axle shaft to maintain rotational balance.

5. Remove the brake drum and related components.

6. Remove the rear bearing retaining nut.

7. Remove the backing plate attachment nuts through the access holes in the rear axle shaft flange.

8. Use a slide hammer with a suitable adapter to withdraw the axle shaft from its housing.

NOTE: Use care not to damage the oil seal when removing the axle shaft.

9. Repeat the procedure for the axle shaft on the opposite side.

NOTE: Be careful not to mix the components of the 2 sides.

10. Installation is performed in the reverse order of removal. Coat the lips of the rear housing oil seal with multipurpose grease prior to installation of the rear axle shaft. Always use new nuts, as they are the self-locking type.

1988 Cressida

1. Raise and safely support the vehicle.

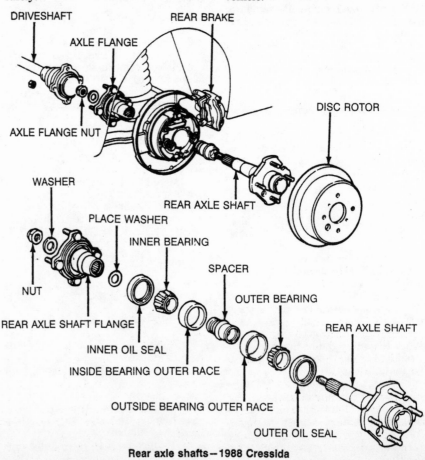

Rear axle shafts–1988 Cressida

2. Disconnect the halfshaft from the axle flange and lower the halfshaft aside.

3. Apply the parking brake completely; pulled up as far as possible.

4. Remove the axle flange nut.

NOTE: The axle flange nut is staked in place. It will be necessary to loosen the staked part of the nut with a hammer and chisel, prior to loosening the nut.

5. Using the proper tools, disconnect the axle flange from the axle shaft. Be careful not to lose the plate washer from the bearing side of the flange.

6. Remove the parking brake shoes.

7. Using the proper tools, pull out the rear axle shaft, along with the oil seal and outer bearing.

8. Clean and inspect the bearings, races, and seal. If these parts are in good condition, repack the bearings with MP grease No. 2 and proceed to Step 15 to install the axle shaft.

9. Using a hammer and chisel, increase the clearance between the axle shaft hub and the outer bearing.

10. Using a puller installed with the jaws in the gap made in Step 9, pull the outer bearing from the axle shaft and remove the oil seal.

11. Drive the outer bearing race out of the hub with a brass drift and a hammer.

NOTE: Bearing and races must be replaced in matched sets. Do not use a new bearing with an old race or vice-versa.

12. Drive the new outer bearing race into the axle shaft hub until it is completely seated.

NOTE: The inner bearing race is replaced in the same manner as Steps 11 and 12.

13. Repack and install both bearings into the hub, being careful not to mis the bearings. The bearings should be packed with No. 2 multi-purpose grease.

14. Drive the seals into place. The inner seal should be driven to a depth of 1.22 in.; the outer to 0.217 in.
To install:

15. Apply a thin coat of grease to the axle shaft flange. Install the rear axle shaft into the housing and install the flange with the plate washer.

16. Using the proper tools, draw the axle shaft into the flange.

17. Install a new axle shaft flange nut. Torque the nut to 22–36 ft. lbs. (30–49 Nm). There should be no horizontal play evident at the axle shaft.

18. Turn the axle shaft back and forth and retorque the nut to 58 ft. lbs. (78 Nm).

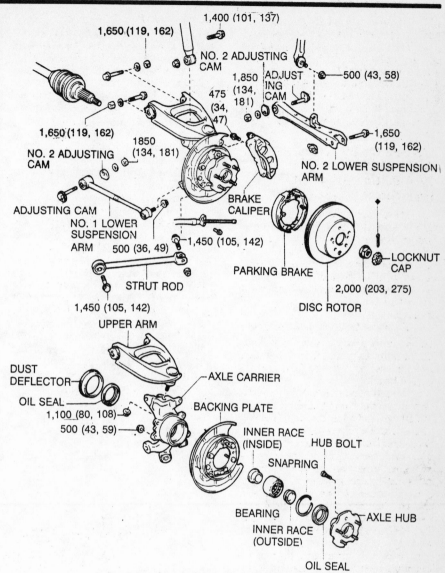

Rear axle, hub and bearing—1988–92 Cressida

19. Using a torque wrench, check the amount of torque required to turn the axle shaft. The correct rotational torque is 0.9–3.5 inch lbs.

NOTE: The shaft should be turned at a rate of 6 seconds per turn to attain a true rotational torque reading.

20. If the rotational torque is less than specified, tighten the nut 5–10 degrees at a time until the proper rotational torque is reached. Do not tighten the nut to more than 145 ft. lbs. (196 Nm).

21. If the rotational torque is greater than specified, replace the bearing spacer and repeat Steps 18–20, if necessary.

22. After the proper rotational torque is reached, restake the nut into position.

23. Install the parking brake shoes.

24. Connect the axle driveshaft to

the flange and torque the nuts to 51 ft. lbs. (69 Nm).

NOTE: If the maximum torque is exceeded while retightening the nut, replace the bearing spacer and repeat Steps 18–20. Do not back off the axle shaft nut to reduce the rotational torque.

25. Install the rear wheel and lower the vehicle.

Supra and 1989–92 Cressida

1. Raise and support the vehicle safely.

2. Remove the rear wheel and tire assembly. Remove the disc brake caliper from the rear axle carrier and suspend it with wire. Remove the rotor disc.

3. Remove the rear driveshaft. Disconnect the parking brake cable assembly.

4. Remove the bolt and nut attach-

ing the carrier to the No. 1 suspension arm. Using the proper tool, separate the No. 1 suspension arm from the axle carrier.

5. Remove the bolt and nut attaching the carrier to the No. 2 suspension arm.

6. Disconnect the strut rod from the axle carrier. Disconnect the strut assembly from the axle carrier.

7. Disconnect the upper arm from the body and remove the axle hub assembly. Remove the upper arm mounting nut and remove the upper arm from the axle carrier.

8. Separate the backing plate and axle carrier. Using a suitable puller, remove the upper arm from the axle carrier.

9. Remove the dust deflector from the axle hub. Using a suitable puller remove the inner oil seal. Remove the hole snap-ring.

10. Using a suitable press, press out the bearing outer race from the axle carrier. Be sure to always replace the bearing as an assembly.

11. Remove the bearing inner race (inside) and 2 bearings from the bearing outer race.

To install:

12. During installation, observe the following torques:

Backing plate to axle carrier nuts—43 ft. lbs. (58 Nm).

Backing plate to axle carrier bolts—19 ft. lbs. (26 Nm).

No. 1 suspension arm nut—43 ft. lbs. (59 Nm)—Supra. 36 ft. lbs. (49 Nm)—Cressida.

Upper arm mounting nut—80 ft. lbs. (108 Nm).

Strut assembly nut—101 ft. lbs. (137 Nm).

Upper arm to body bolt—121 ft. lbs. (164 Nm) for Supra or 119 ft. lbs. (162 Nm) for Cressida.

No. 2 suspension arm to axle carrier—121 ft. lbs. (164 Nm) for Supra or 119 ft. lbs. (162 Nm) for Cressida.

Strut rod to axle carrier—121 ft. lbs. (164 Nm) for Supra. 105 ft. lbs. or (142 Nm) for Cressida.

Disc brake caliper bolts—34 ft. lbs. (47 Nm).

1988–92 Camry (AWD) and Celica (AWD)

1. Raise the vehicle and support it safely.

2. Remove the rear wheel. Remove the disc brake caliper from the rear axle carrier and suspend it with wire. Remove the rotor disc.

3. Remove the rear halfshaft. Disconnect the parking brake cable assembly and remove the cable.

4. Remove the 2 axle carrier set nuts and the 2 bolts and then remove the camber adjusting cam.

5. Disconnect the strut rod at the axle carrier. Disconnect the No. 1 and No. 2 suspension arms at the axle carrier. Remove the axle carrier and hub.

6. Press the axle shaft out of the axle hub.

7. Using a 2-armed puller, remove the bearing inner race (outside) from the axle shaft. Remove the dust cover.

8. Remove the inner and outer oil seal from the axle carrier. Remove the hole snap-ring.

9. Using a suitable press, press out the bearing.

To install:

10. Please observe the following notes:

Tighten the axle carrier-to-shock bolts to 188 ft. lbs. (255 Nm).

Tighten the brake caliper mounting bolts to 34 ft. lbs. (47 Nm).

With the parking brake engaged, tighten the bearing locknut to 137 ft. lbs. (186 Nm).

With the wheels resting on the ground, tighten the strut rod bolt to 83 ft. lbs. (113 Nm); tighten the 2 suspension arms to 90 ft. lbs. (123 Nm).

Check the rear wheel alignment.

MR2

1. Raise and support the vehicle safely.

2. Remove the rear wheel and tire assembly. Remove the cotter pin, bearing locknut cap and bearing locknut.

3. Disconnect the parking brake cable. Remove the disc brake caliper from the rear axle carrier and suspend it with wire. Remove the rotor disc.

4. Disconnect the rear axle carrier from the lower arm. Remove the cotter pin and nut from the suspension arm. If equipped with ABS, remove the speed sensor from the axle carrier.

5. Using a suitable tool separate the suspension arm from the rear axle carrier.

6. Place matchmarks on the strut lower bracket and camber adjusting cam.

7. Remove the 2 axle carrier set nuts and 2 bolts with the camber adjusting cam. Remove the rear axle carrier and axle hub.

8. Remove the dust deflector from the axle hub. Using a suitable puller remove the inner oil seal. Remove the hole snap-ring.

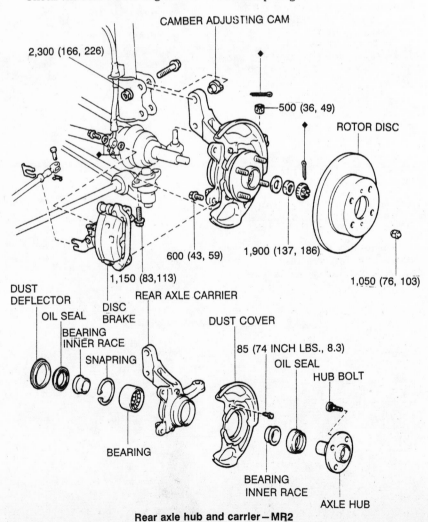

Rear axle hub and carrier—MR2

9. Remove the 3 bolts holding the disc brake dust cover to the rear axle carrier. Using a suitable puller remove the axle hub from the rear axle carrier.

10. Remove the bearing inner (inside) race. Using a suitable puller remove the bearing inner race (outside) from the rear axle hub.

11. Using a suitable puller remove the outer oil seal.

12. Remove the hub bearing by first placing the removed inner race (outside) in the bearing and using a suitable press, press out the bearing. Be sure to always replace the bearing as an assembly.

To install:

13. Installation is the reverse order of the removal procedure. During installation, observe the following torque specifications:

Two camber adjusting cam set bolts—166 ft. lbs. (226 Nm).

Suspension arm nut—36 ft. lbs. (49 Nm).

Rear axle carrier to the lower arm—59 ft. lbs. (1988-92: 83 ft. lbs.).

Brake caliper—43 ft. lbs. (59 Nm).

Bearing locknut—137 ft. lbs. (186 Nm) all except 3S-GTE engine. On 3S-GTE engine torque the rear wheel bearing locknut to 217 ft. lbs. (294 Nm).

If equipped with ABS, torque the wheel sensor bolt to 74 inch lbs. (8.3 Nm).

Front Wheel Hub, Knuckle and Bearings

REMOVAL & INSTALLATION

Front Wheel Drive Vehicles Only

1. Raise and support the vehicle safely. Remove the front wheels.

2. Remove the cotter pin from the bearing locknut cap and then remove the cap.

3. Depress the brake pedal and loosen the bearing locknut.

4. Remove the brake caliper mounting nuts, position the caliper aside with the hydraulic line still attached and suspend it with a wire.

5. Remove the brake disc.

6. Remove the cotter pin and nut from the tie rod end and then, using a tie rod end removal tool, remove the tie rod.

7. Place matchmarks on the shock absorber lower mounting bracket and the camber adjustment cam, remove the bolts and separate the steering knuckle from the strut.

8. Remove the 2 ball joint attaching nuts and disconnect the lower control arm from the steering knuckle.

9. Carefully grasp the axle hub and knuckle assembly and pull it out from the halfshaft using the proper tool.

NOTE: Cover the halfshaft boot with a shop rag to protect it from any damage.

10. Clamp the steering knuckle in a vise and remove the dust deflector. Remove the nut holding the steering knuckle to the ball joint. Press the ball joint out of the steering knuckle.

11. Remove the dust deflector from the hub.

12. Pry out the bearing inner oil seal and then remove the hole snap-ring.

13. Remove the 3 bolts attaching the steering knuckle to the disc brake dust cover.

14. Remove the axle hub from the steering knuckle using the proper tool.

15. Remove the bearing inner race (inside).

16. Remove the bearing inner race (outside).

17. Remove the oil seal from the knuckle.

18. Position an old bearing inner race (outside) on the bearing and then use a hammer and a drift to carefully knock the bearing out of the knuckle.

To install:

19. Press a new bearing into the steering knuckle.

20. Using a suitable oil seal installation tool, drive a new oil seal into the knuckle.

21. Install the disc brake dust cover onto the knuckle using liquid sealant.

22. Apply grease between the oil seal lip, oil seal and the bearing and then press the axle hub into the steering knuckle.

23. Install a new hole snap-ring into the knuckle.

24. Press a new oil seal onto the knuckle and coat the contact surface of the seal and the halfshaft with grease. Press a new dust deflector into the knuckle.

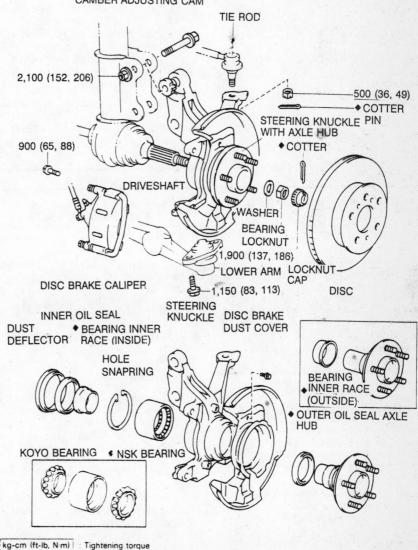

CAMBER ADJUSTING CAM

TIE ROD

2,100 (152, 206)

900 (65, 88)

DRIVESHAFT

500 (36, 49)

COTTER

STEERING KNUCKLE PIN WITH AXLE HUB

COTTER

WASHER

BEARING LOCKNUT

1,900 (137, 186)

LOWER ARM

LOCKNUT CAP

DISC

DISC BRAKE CALIPER

1,150 (83, 113)

STEERING KNUCKLE

DISC BRAKE DUST COVER

INNER OIL SEAL

DUST DEFLECTOR

BEARING INNER RACE (INSIDE)

HOLE SNAPRING

BEARING INNER RACE (OUTSIDE)

OUTER OIL SEAL AXLE HUB

KOYO BEARING

NSK BEARING

kg-cm (ft-lb, N·m) : Tightening torque

◆ : Non-reusable part

Front axle hub and steering knuckle assembly—front wheel drive models

25. Position the ball joint on the steering knuckle and tighten the nut to 14 ft. lbs. (20 Nm). Remove the nut, install a new one and tighten it to 82 ft. lbs. (111 Nm). On 1988 Camry and Corolla torque the nut to 94 ft. lbs. (127 Nm).

26. Connect the knuckle assembly to the lower strut bracket. Insert the mounting bolts from the rear and make sure the matchmarks made earlier are in alignment. Tighten the nuts as follows:

105 ft. lbs. (142 Nm) on 1988 Celica.

188 ft. lbs. (255 Nm) on 1989–92 Celica.

166 ft. lbs. (226 Nm) on 1988 Camry and 1988–92 Tercel.

224 ft. lbs. (304 Nm) on 1989–92 Camry.

194 ft. lbs. (263 Nm) on Corolla sedan and wagon.

27. Connect the tie rod end to the knuckle, tighten the nut to 36 ft. lbs. (49 Nm) and install a new cotter pin.

28. Connect the ball joint to the lower control arm and tighten the bolt to 47 ft. lbs. (64 Nm) except on the following vehicles:

1988 Camry—67 ft. lbs. (91 Nm).

1989–92 Camry—90 ft. lbs. (123 Nm).

1988–92 Celica—94 ft. lbs. (122 Nm).

1988–92 Tercel—59 ft. lbs. (80 Nm).

Corolla sedan and wagon—105 ft. lbs. (142 Nm).

29. Install the brake disc and the caliper. Tighten the caliper mounting bolts to 65 ft. lbs. (88 Nm) on all vehicles except Camry and 1988–92 Celica. On Camry torque the caliper mounting bolts to 86 ft. lbs. (117 Nm). On 1988–92 Celica, torque the caliper mounting bolts to 70 ft. lbs. (95 Nm).

30. Install the bearing locknut while having someone depress the brake pedal. Tighten it to 137 ft. lbs. (186 Nm). Install the adjusting nut cap and insert a new cotter pin.

31. Check the front end alignment.

MANUAL TRANSMISSION

For further information on transmissions/transaxles, please refer to "Chilton's Guide to Transmission Repair".

Transmission Assembly

REMOVAL & INSTALLATION

Supra

1. Disconnect the negative battery cable. Remove the center console trim panel. Remove the shift lever.

2. Raise and support the vehicle safely. Drain the transmission fluid. Remove the driveshaft.

3. Disconnect the front exhaust pipe from the tailpipe. Remove the front exhaust pipe.

4. Disconnect the speedometer cable. Disconnect the back-up light switch electrical connector. If equipped with ABS, disconnect the rear speed sensor electrical connector.

5. Remove the clutch release cylinder. Remove the starter assembly.

6. Support the engine and the transmission using the proper equipment. Remove the transmission support crossmember.

7. Remove the transmission mounting bolts. Remove the flywheel housing bolts. Carefully, move the transmission rearward, down, and out of the vehicle.

NOTE: On turbocharged vehicles, it will be necessary to remove the transmission with the clutch cover and disc. To do this pull the release fork through the left clutch housing hole and then remove the assembly.

8. Installation is the reverse of the removal procedure. Tighten the mounting bolts to 29 ft. lbs. (39 Nm).

MANUAL TRANSAXLE

For further information on transmissions/transaxles, please refer to "Chilton's Guide to Transmission Repair".

Transaxle Assembly

REMOVAL & INSTALLATION

Tercel

SEDAN (2WD)

1. Disconnect the negative battery cable. If equipped with cruise control remove the battery and cruise control actuator with mounting bracket.

2. Remove the clutch release cylinder and tube clamp. Disconnect the back-up light switch electrical connector.

3. Disconnect the transaxle shift control cables. Remove the selecting bellcrank along with the bracket from the transaxle case. Remove the upper transaxle-to-engine retaining bolts.

4. Raise the vehicle and support it safely. Remove the under covers.

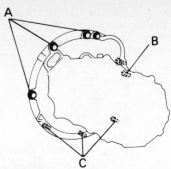

Manual transaxle mounting bolt locations on Tercel (2WD)

Drain the transaxle fluid. Disconnect the speedometer cable.

5. Disconnect both halfshafts. Remove the engine rear mounting brackets. Remove the starter assembly.

6. Support the engine and transaxle assembly using the proper equipment. Disconnect the left engine mounting.

7. Remove the remaining engine-to-transaxle retaining bolts. Carefully remove the transaxle assembly from the vehicle.

8. Installation is the reverse of the removal procedure. Tighten the transaxle-to-engine bolts to 47 ft. lbs. (64 Nm)—**A**; 34 ft. lbs. (46 Nm)—**B**; and 65 inch lbs. (7.4 Nm)—**C**. Tighten the front engine mount bracket bolts to 43 ft. lbs. (58 Nm). Tighten the rear engine mount bracket bolts to 21 ft. lbs. (28 Nm). Tighten the left engine mount bolts to 35 ft. lbs. (48 Nm) and the front and rear mount bolts to 47 ft. lbs. (64 Nm).

1988 WAGON (AWD)

1. Disconnect the negative battery cable. Remove the air cleaner assembly.

2. Remove the console. Remove the shift lever assembly. Remove the upper engine-to-transaxle retaining bolts.

3. Raise and support the vehicle safely. Drain the transaxle fluid. Remove both halfshafts.

4. On some vehicles, it will be necessary to remove the catalytic converter air inlet pipe. Remove the front exhaust pipe.

5. Disconnect the selector rod from the rear drive shift link lever. Disconnect the speedometer cable. Remove the right stiffener plate.

6. Disconnect the back-up light switch electrical connector. Disconnect the AWD and the low gear indicator switch electrical connectors.

7. Support the engine and the transaxle using the proper equipment. Remove the remaining transaxle-to-engine retaining bolts. Remove the rear crossmember assembly.

NOTE: Properly position a piece of wood between the engine

and the firewall so the assembly will not make contact with the power brake booster when it is removed.

8. Carefully remove the transaxle assembly from the vehicle.

9. Installation is the reverse of the removal procedure. Tighten the 14mm transaxle-to-engine bolts to 29 ft. lbs. (39 Nm) and the 17mm bolts to 43 ft. lbs. (59 Nm). Tighten the rear support member bolts to 70 ft. lbs. (95 Nm) and the right stiffener plate bolts to 29 ft. lbs. (39 Nm).

Corolla

2WD (EXCEPT C52 TRANSAXLE)

1. Disconnect the negative battery cable. Remove the air cleaner assembly.

2. Disconnect the back-up light switch electrical connector. Remove the speedometer cable. Disconnect the transmission control cables.

3. Raise the vehicle and support safely. Remove the water inlet from the transaxle. Remove the clutch release cylinder.

4. Remove the under cover. Remove the front and rear mounting. Remove the engine mounting center member.

5. Disconnect the halfshaft from the transaxle. Disconnect the steering knuckle from the lower control arm. Pull the steering knuckle outward and remove the left halfshaft.

6. Remove the starter. Disconnect the ground strap. Remove the No. 2 engine rear plate.

7. Support the engine and the transaxle using the proper equipment. Remove the left engine mounting.

8. Remove the engine-to-transaxle retaining bolts. Carefully remove the transaxle assembly from the vehicle.

9. Installation is the reverse of the removal procedure. Tighten the 12mm engine-to-transaxle bolts to 47 ft. lbs. (64 Nm) and the 10mm bolts to 34 ft. lbs. (46 Nm). Tighten the left engine mount bolts to 38 ft. lbs. (52 Nm). Tighten the front and rear engine mount bolts and the engine mounting center member bolts to 29 ft. lbs. (39 Nm).

2WD (C52 TRANSAXLE)

1. Disconnect the negative battery cable. Drain the radiator. Remove the air cleaner assembly. Remove the engine cooling fan assembly.

2. Disconnect the oxygen sensor electrical connector and the back-up light switch connector.

3. Remove the clutch release cylinder. It may be possible to leave the fluid lines attached to the cylinder.

4. Disconnect the water inlet from the transaxle. Disconnect the transaxle control cables. Disconnect the

speedometer cable. Disconnect the ground cable.

5. Remove the starter. Remove the engine under covers. Remove the front exhaust pipe.

6. Disconnect the front and rear engine mountings. Remove the engine mounting center member.

7. Remove the left front wheel. Loosen the 6 nuts while depressing the brake pedal. Disconnect the halfshaft from the side gear shaft. Disconnect the lower ball joint from the lower control arm. Pull the shock absorber outward. Remove the halfshaft.

8. Support the engine and the transaxle assembly using the proper equipment. Remove the No. 2 engine rear plate. Remove the left engine mounting.

9. Remove the transaxle retaining bolts. Carefully remove the transaxle assembly from the vehicle.

10. Installation is the reverse of the removal procedure. Tighten the 12mm engine-to-transaxle bolts to 47 ft. lbs. (64 Nm) and the 10mm bolts to 34 ft. lbs. (46 Nm). Tighten the left engine mount bolts to 38 ft. lbs. (52 Nm). Tighten the front and rear engine mount bolts and the engine mounting center member bolts to 29 ft. lbs. (39 Nm).

AWD

1. Remove the engine and transaxle as an assembly.

2. Remove the rear end plate.

3. Disconnect the vacuum lines and then remove the transfer case vacuum actuator.

4. Remove the right and center transfer case stiffener plates.

5. Pull the transaxle out slowly until there is approximately 2.36–3.15 in. (60–80mm) clearance between the transaxle and the engine.

6. Turn the output shaft in a clockwise direction and then remove the transaxle.

To install:

7. Install the transaxle assembly to the engine and tighten the 10mm bolts to 34 ft. lbs. (46 Nm). Tighten the 12mm bolts to 47 ft. lbs. (64 Nm).

8. Tighten the 8mm stiffener plate bolts to 14 ft. lbs. (20 Nm) and the 10mm bolts to 27 ft. lbs. (37 Nm).

9. Tighten the rear end plate mounting bolts to 17 ft. lbs. (23 Nm).

10. Install the engine/transaxle assembly.

Camry

2WD

1. Disconnect the negative battery cable. Remove the clutch release cylinder and tube clamp. Remove the clutch tube bracket.

2. Disconnect the control cables.

Disconnect the back-up light switch electrical connector. Remove the ground strap.

3. Remove the starter assembly. Remove the transaxle upper mounting bolts.

4. Raise and support the vehicle safely. Remove the under covers. Drain the transaxle fluid. Disconnect the speedometer cable.

5. Remove the suspension lower crossmember. Remove the engine mounting centermember.

6. Disconnect both driveshafts. Remove the center driveshaft. Disconnect the left steering knuckle from the lower control arm. Remove the stabilizer bar.

7. Properly support the engine and remove the left engine mount.

8. Properly support the transaxle assembly. Remove the engine-to-transaxle bolts, lower the left side of the engine and carefully ease the transaxle out of the engine compartment.

To install:

9. Install the transaxle and tighten the 12mm mounting bolts to 47 ft. lbs. (64 Nm) and the 10mm bolts to 34 ft. lbs. (46 Nm).

10. Tighten the left engine mount to 38 ft. lbs. (52 Nm). Tighten the 4 center engine mount bolts to 29 ft. lbs. (39 Nm). Tighten the front and rear engine mount bolts to 32 ft. lbs. (43 Nm).

11. Tighten the lower crossmember bolts to 153 ft. lbs. (207 Nm) – 4 outer bolts; and, 29 ft. lbs. (39 Nm) – 2 inner bolts.

AWD

1. Remove the engine/transaxle assembly.

2. Remove the transfer case stiffener plate and the exhaust pipe front brake.

3. Remove the left stiffener plate and the front engine mount.

4. Remove the left engine mount bracket and then separate the transaxle from the engine.

5. Installation is in the reverse order of removal. Tighten the 12mm transaxle-to-engine mounting bolts to 47 ft. lbs. (64 Nm) and the 10mm bolts to 34 ft. lbs. (46 Nm).

Celica

2WD

1. Disconnect the negative battery cable. On some vehicles, it may be necessary to remove the battery. Remove the air cleaner assembly.

2. Remove the clutch tube bracket. Disconnect the back-up light switch at the transaxle. Disconnect the speedometer and the engine ground strap.

3. Disconnect the transaxle control cable and position them aside.

4. Unbolt the clutch release cylin-

der. It may be possible to position it aside with the hydraulic line still attached.

5. Remove the upper transaxle retaining bolts. Raise the vehicle and support safely. Remove the engine undercover. Drain the transaxle fluid.

6. Disconnect the exhaust pipe from the manifold. Remove the lower suspension crossmember. Remove the starter assembly.

7. Properly support the engine and transaxle assembly. Remove the front and rear transaxle mounts. Remove the center engine mount.

8. Disconnect both halfshafts at the transaxle. Unbolt the steering knuckle from the suspension arm and pull it outward. Remove the left halfshaft.

9. On some vehicles, remove the No. 2 rear engine plate. With the engine properly supported remove the left engine mount.

10. Remove the engine-to-transaxle bolts, lower the left side of the engine and carefully ease the transaxle out of the engine compartment.

To install:

11. Install the transaxle and tighten the 12mm engine-to-transaxle bolts to 47 ft. lbs. (64 Nm) and the 10mm bolts to 34 ft. lbs. (46 Nm).

12. Tighten the left engine mount bolts to 38 ft. lbs. (52 Nm).

13. Tighten the center member bolts and the front and rear engine mount bolts to 29 ft. lbs. (39 Nm).

AWD

1. Remove the engine and transaxle assembly.

2. Separate the transaxle from the engine.

3. Installation is in the reverse order of removal. Tighten the 12mm engine-to-transaxle bolts to 47 ft. lbs. (64 Nm) and the 10mm bolts to 34 ft. lbs. (46 Nm). Tighten the left engine mount bolts to 38 ft. lbs. (52 Nm). Tighten the center member bolts and the front and rear engine mount bolts to 29 ft. lbs. (39 Nm).

MR2
1988–89

1. Disconnect the negative battery cable. Drain the radiator. Raise and support the vehicle safely. Drain the transaxle fluid.

2. Disconnect the back-up light switch and the speedometer cable at the transaxle. On 4A-GZE engine, remove the intercooler.

3. Loosen the mounting bolts and remove the water inlet from the transaxle.

4. Remove the engine undercover. Remove the fuel tank protector.

5. Disconnect the transaxle control cables at the transaxle and position them aside.

6. Remove the water hose clamp from the control cable bracket and then remove the No. 2 control cable bracket.

7. Remove the main control cable bracket and the clutch release cylinder. Position these components aside.

8. Disconnect the exhaust pipe from the manifold, remove the pipe bracket from the chassis and then remove the exhaust pipe assembly from the bracket.

9. Remove the transaxle protector. Disconnect the halfshaft from the side gear shaft. Remove the starter assembly.

10. Remove the No. 2 engine rear plate. Remove the front and rear engine mounts from the body.

11. Properly support the engine and remove the left engine mount.

12. Properly support the transaxle assembly. Remove the engine-to-transaxle bolts, lower the left side of the engine and carefully ease the transaxle out of the engine compartment.

13. Remove the side gear shaft from the transaxle.

14. Installation is the reverse of the removal procedure. Tighten the 12mm engine-to-transaxle bolts to 47 ft. lbs. (64 Nm) and the 10mm bolts to 34 ft. lbs. (46 Nm). On 1988–89 vehicles, tighten the left and rear engine mounts to 38 ft. lbs. (52 Nm). On 1990–92 vehicles, tighten the front engine mounting bracket thru-bolt to 71 ft. lbs. (96 Nm) and the rear engine mounting thru-bolt to 64 ft. lbs. (87 Nm).

1990–92

1. Disconnect the negative battery cable.

2. Raise the vehicle and support safely. Drain the transaxle fluid.

3. Remove the rear wheels and under covers.

4. Disconnect the halfshafts from the knuckle and stabilizer bar from the knuckle.

5. Disconnect the exhaust pipe from the manifold.

6. Remove the air cleaner.

7. Disconnect the shift cables, clutch release cylinder and speedometer cable.

8. Remove the starter motor, rear end plate and engine stiffener plate.

9. Install an engine holding fixture to support the engine and transaxle. Remove the engine and transaxle mounting brackets.

10. Place a suitable transaxle jack under the assembly, remove the bell housing bolts and remove the transaxle from the vehicle.

To install:

11. Place a suitable transaxle jack under the assembly, Install the trans-

axle bolts and torque to 47 ft. lbs. (64 Nm).

12. Install the engine and transaxle mounting brackets. Torque the engine mounts to 71 ft. lbs. (96 Nm), transaxle bolts to 47 ft. lbs. (64 Nm) and the stiffener plate bolts to 27 ft. lbs. (37 Nm).

13. Install the starter motor, rear end plate and engine stiffener plate.

14. Connect the shift cables, clutch release cylinder and speedometer cable.

15. Install the air cleaner.

16. Connect the exhaust pipe to the manifold and torque to 46 ft. lbs. (62 Nm).

17. Connect the halfshafts to the knuckle and stabilizer bar to the knuckle.

18. Install the rear wheels and under covers.

19. Lower the vehicle safely.

20. Connect the negative battery cable, refill the transaxle and check for leaks.

LINKAGE ADJUSTMENT

Manual transmission linkage adjustments are neither possible or necessary. Damaged bushings can cause shift cable problems.

CLUTCH

Clutch Assembly

REMOVAL & INSTALLATION

1. Disconnect the negative battery cable.

2. Remove the transmission or transaxle assembly from the vehicle.

NOTE: On some 1988–90 Supra's, the clutch assembly is removed along with the transmission. On the 1989–92 Corolla (AWD) and the 1988–92 Camry and Celica AWD, the engine and transaxle are removed from the vehicle as an assembly.

3. Matchmark the clutch cover to the flywheel.

4. Remove the clutch pressure plate retaining bolts small amounts in a criss-cross pattern to relieve the clutch disc spring tension.

5. Remove the clutch cover.

6. Remove the clutch disc.

7. Remove the retaining clip and withdraw the release bearing.

8. Remove the release fork and boot assembly.

To install:

9. Using a suitable clutch disc align-

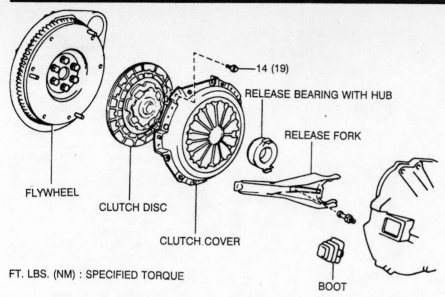

FLYWHEEL
CLUTCH DISC
CLUTCH COVER

14 (19)
RELEASE BEARING WITH HUB
RELEASE FORK
BOOT

FT. LBS. (NM) : SPECIFIED TORQUE

Tighten the clutch cover retaining bolts in a criss-cross pattern

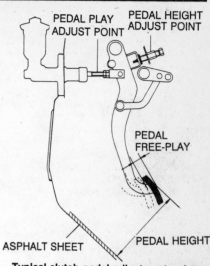

PEDAL PLAY ADJUST POINT
PEDAL HEIGHT ADJUST POINT
PEDAL FREE-PLAY
ASPHALT SHEET
PEDAL HEIGHT

Typical clutch pedal adjustment points

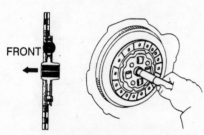

FRONT

Use a clutch pilot tool to center the clutch disc on the flywheel

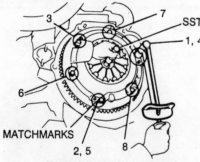

3   7   SST
1, 4
6
MATCHMARKS
2, 5   8

Typical clutch assembly

ment tool, install the clutch disc onto the flywheel.

10. Position the clutch cover onto the flywheel and align the matchmarks.

11. Install the clutch cover retaining bolts. Torque the bolts in a criss-cross pattern to 14 ft. lbs. (19 Nm).

12. Lubricate the release fork pivot and contact points, release bearing, bearing hub and input shaft spline surfaces with a suitable molybdenum disulphide lithium based or multi-purpose grease.

13. Install the boot, release fork, hub and bearing assemblies.

14. Install the transmission or transaxle.

PEDAL HEIGHT/FREE-PLAY ADJUSTMENT

Except 1988 Tercel Wagon

1. Adjust the clearance between the master cylinder piston and the pushrod to specification by loosening the pushrod locknut and rotating the pushrod while depressing the clutch pedal lightly.

2. Tighten the locknut when finished the adjustment.

3. Adjust the release cylinder free-play by loosening the release cylinder pushrod locknut and rotating the pushrod until proper specification is obtained.

4. Measure the clutch pedal free-play after performing the adjustments. If it fails to fall within specification, repeat the procedure. Pedal free-play specifications are as follows:

1988–92 Tercel – 0.2–0.59 in. (5–15mm)

1988 Corolla (RWD W/4A-LC) – 0.51–0.91 in. (13–23mm)

1988 Corolla (RWD W/4A-GE) – 0.2–0.59 in. (5–15mm)

1988 Corolla (FWD) – 0.28–0.67 in. (7–17mm)

1988–92 Corolla – 0.2–0.59 in. (5–15mm)

Celica – 0.2–0.59 in. (5–15mm)

Supra – 0.2–0.59 in. (5–15mm)

1988 Cressida – 0.2–0.59 in. (5–15mm)

MR2 – 0.2–0.59 in. (5–15mm).

1988 Tercel Wagon

1. Depress the clutch pedal several times.

2. Depress the clutch pedal until resistance is felt. Free-play should be within specification.

3. Check the clutch release sector pawl. Six notches should remain between the pawl and the end of the sector. If less than 6 notches, replace the clutch disc. If the clutch disc has been replaced, the pawl should be 3–10 notches.

4. To obtain either the used or new position, change the position of the E-ring.

Clutch Cable

REMOVAL & INSTALLATION

1988 Tercel Wagon

1. Disconnect the negative battery cable.

2. Disconnect the sector tension spring from the clutch pedal.

3. Disconnect the clutch release cable from the release fork lever.

4. Turn the release sector toward the front side and disconnect the release cable from the release sector. Remove the release cable.

5. Installation is the reverse of the removal procedure.

Clutch Master Cylinder

REMOVAL & INSTALLATION

Rear Wheel Drive Vehicles

1. Disconnect the negative battery cable. Remove the pushrod clevis pin and clip.

NOTE: On some vehicles, it will be necessary to remove the under dash panel in order to gain access to the pushrod clevis pin.

2. Disconnect the fluid line.

3. Unbolts and remove the clutch master cylinder.

4. Installation is the reverse of the removal procedure. Bleed the system.

Front Wheel Drive Vehicles

1. Disconnect the negative battery cable.

2. On the Tercel, remove the reservoir tank from the clutch master clyinder.

3. On Celica with 3S-GTE engine, remove the brace that runs across the struts. On the MR2, remove the spare tire guard and luggage compartment trim cover.

4. Remove the ABS control relay, if equipped.

5. Remove the pushrod clevis pin and clip.

NOTE: On some vehicles it will be necessary to remove the under dash panel in order to gain access to the pushrod clevis pin.

6. On the 1988–92 Corolla (4A-GE) remove the brake booster.

7. Disconnect the fluid line and plug the end of the line to prevent leakage.

8. Unbolt and remove the clutch master cylinder.

9. Installation is the reverse of the removal procedure. Bleed the system.

Clutch Slave Cylinder

REMOVAL & INSTALLATION

Except 1990–92 MR2

1. Disconnect the negative battery cable. Raise and support the vehicle safely.

2. Remove the gravel shield, if equipped.

3. Disconnect the fluid line.

4. Remove the slave cylinder retaining bolts.

5. Remove the clutch slave cylinder from the vehicle.

6. Installation is the reverse of the removal procedure. Bleed the system.

1990–92 MR2

1. Disconnect the negative battery cable.

2. Raise and support the vehicle safely.

3. Remove the engine under cover.

4. Disconnect the control cables from the transaxle.

5. Disconnect the fluid line.

6. Support the engine and transaxle.

7. From the engine side, remove the engine front mounting bracket bolts.

8. Unbolt and remove the clutch slave cylinder.

9. Installation is the reverse of the removal procedure. Bleed the system.

Hydraulic Clutch System Bleeding

1. Check and fill the clutch fluid

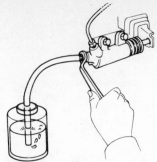

Bleeding the clutch hydraulic system

reservoir to the specified level as necessary. During the bleeding process, continue to check and replenish the reservoir to prevent the fluid level from getting lower than ½ the specified level.

2. Remove the dust cap from the bleeder screw on the clutch slave cylinder and connect a tube to the bleeder screw and insert the other end of the tube into a clean glass or metal container.

NOTE: Take precautionary measures to prevent the brake fluid from getting on any painted surfaces.

3. Pump the clutch pedal several times, hold it down and loosen the bleeder screw slowly.

4. Tighten the bleeder screw and release the clutch pedal gradually. Repeat this operation until air bubbles disappear from the brake fluid being expelled out through the bleeder screw.

5. Repeat until all evidence of air bubbles completely disappears from the fluid being pumped out of the tube.

6. When the air is completely removed tighten the bleeder screw and replace the dust cap.

7. Check and refill the master cylinder reservoir, as necessary.

8. Depress the clutch pedal several times to check the operation of the clutch and check for leaks.

AUTOMATIC TRANSMISSION

For further information on transmissions/transaxles, please refer to "Chilton's Guide to Transmission Repair".

Transmission Assembly

REMOVAL & INSTALLATION

Corolla

1. Disconnect the negative battery cable.

2. Remove the air cleaner assembly. Disconnect the transmission throttle cable. Disconnect the starter assembly electrical connections.

3. Raise the vehicle and support it safely. Drain the transmission fluid. Remove the driveshaft.

4. Remove the exhaust pipe clamp. Disconnect the exhaust pipe from the exhaust manifold.

5. Disconnect the manual shift linkage. Disconnect the oil cooler lines. Remove the starter.

6. Support the engine and transmission using the proper equipment. Remove the rear crossmember.

7. Disconnect the speedometer cable. Disconnect all necessary electrical wiring from the transmission.

8. Remove the torque converter cover. Remove the torque converter-to-engine retaining bolts.

9. Remove the bolts retaining the transmission to the engine. carefully remove the transmission from the vehicle.

To install:

10. Install the transmission into the vehicle and torque the bell housing bolts to 35 ft. lbs. (48 Nm).

11. Install the exhaust pipe, starter motor and crossmember. Torque the crossmember bolts to 40 ft. lbs. (54 Nm).

12. Install the transmission cables, linkage and driveshaft.

13. Refill the transmission with the approved fluid and check for leaks.

14. Lower the vehicle. Road test the vehicle and check operation.

Cressida

1. Disconnect the negative battery cable. Drain the radiator and remove the upper radiator hose. Remove the air cleaner assembly. Disconnect the transmission throttle cable.

2. Raise the vehicle and support it safely. Drain the transmission fluid. Remove the driveshaft along with the center bearing.

3. Remove the exhaust pipe together with the catalytic converter. Disconnect the manual shift linkage. Remove the speedometer cable.

4. Disconnect the oil cooler lines. As necessary, remove the transmission oil filler tube. As required, remove the starter assembly. Remove the speedometer cable.

5. Remove both stiffener plates and the catalytic converter cover from the transmission housing and cylinder block.

6. Support the engine and transmission properly. Remove the rear crossmember.

7. Remove the torque converter cover. Remove the torque converter-to-engine retaining bolts.

8. Remove the bolts retaining the

transmission to the engine. carefully remove the transmission from the vehicle.

To install:

9. Install the transmission into the vehicle and torque the bell housing bolts to 35 ft. lbs. (48 Nm).

10. Install the exhaust pipe, starter motor and crossmember. Torque the crossmember bolts to 47 ft. lbs. (64 Nm). Torque the torque converter bolts to 20 ft. lbs. (27 Nm).

11. Install the transmission cables, linkage and driveshaft.

12. Refill the transmission with the approved fluid and check for leaks.

13. Lower the vehicle. Road test the vehicle and check operation.

Supra

1. Disconnect the negative battery cable. Remove the air cleaner assembly. Disconnect the transmission throttle cable.

2. Raise the vehicle and support it safely. Drain the transmission fluid. Disconnect the electrical connectors for the neutral safety switch and back-up lights.

3. Remove the intermediate driveshaft along with the center bearing. Disconnect the exhaust pipe from the tail pipe.

4. Disconnect the transmission oil cooler lines. Plug the lines to prevent leakage. Disconnect the manual shift linkage and speedometer cable.

5. Remove the exhaust pipe bracket and torque converter cover.

6. Remove both stiffener brackets.

7. Support the engine and transmission using the proper equipment. Remove the rear crossmember.

8. Remove the engine under cover. Remove the torque converter-to-engine retaining bolts. Remove the starter.

9. Remove the bolts retaining the transmission to the engine. carefully remove the transmission from the vehicle.

To install:

10. Install the transmission into the vehicle and torque the bell housing bolts to 35 ft. lbs. (48 Nm).

11. Install the exhaust pipe, starter motor and crossmember. Torque the crossmember bolts to 40 ft. lbs. (54 Nm).

12. Install the transmission cables, linkage and driveshaft. Adjust the throttle cable.

13. Refill the transmission with the approved fluid and check for leaks.

14. Lower the vehicle. Road test the vehicle and check operation.

SHIFT LINKAGE ADJUSTMENT
Corolla

1. Loosen the nut on the shift linkage.

2. Push the selector lever all the way to the rear of the vehicle.

3. Return the lever 2 notches to the **N** shift position.

4. While holding the selector lever slightly toward the **R** shift position tighten the connecting rod nut.

Cressida and Supra

1. Loosen the nut on the shift linkage. Push the selector lever all the way to the rear of the vehicle.

2. Return the lever 2 notches to the **N** shift position.

3. While holding the selector lever slightly toward the **R** shift position, tighten the connecting rod nut.

THROTTLE LINKAGE ADJUSTMENT

1. Remove the air cleaner.

2. Confirm that the accelerator linkage opens the throttle fully. Adjust the linkage as necessary.

3. Peel the rubber dust boot back from the throttle cable.

4. Loosen the adjustment nuts on the throttle cable bracket (cylinder head cover) just enough to allow cable housing movement.

5. Have an assistant depress the accelerator pedal fully.

6. Adjust the cable housing so the gap between its end and the cable stop collar is 0.04 in. (1mm).

7. Tighten the adjustment nuts. Make sure the adjustment has not changed. Install the dust boot and the air cleaner.

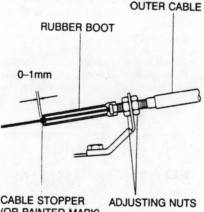

OUTER CABLE

RUBBER BOOT

0–1mm

CABLE STOPPER (OR PAINTED MARK) ADJUSTING NUTS

Throttle cable adjustment

AUTOMATIC TRANSAXLE

For further information on transmissions/transaxles, please refer to "Chilton's Guide to Transmission Repair".

Transaxle Assembly

REMOVAL & INSTALLATION

Tercel

1. Disconnect the negative battery cable. Drain the radiator and remove the upper radiator hose, as required. Remove the air cleaner assembly.

2. Raise the vehicle and support it safely. Remove both halfshafts. Drain the fluid from the transaxle and differential, if equipped.

3. Remove the torque converter cover. Remove the bolts that retain the torque converter to the crankshaft. Remove the exhaust pipe. Remove the shift lever rod.

4. Remove the speedometer cable and back-up light connector. If equipped with AWD, remove the electrical solenoid connector. Disconnect and remove all throttle linkage.

5. Remove the fluid lines from the transaxle. Remove the starter assembly, as required. On AWD vehicles, remove the rear driveshaft.

6. Support the engine and transaxle using a suitable jack. Remove the rear crossmember.

7. Remove the transaxle-to-engine retaining bolts. Separate the transaxle from the engine and carefully remove it from the vehicle.

To install:

8. Install the transaxle and tighten the transaxle-to-engine bolts to 47 ft. lbs. (64 Nm). Tighten the left engine mount bracket bolts to 32 ft. lbs. (43 Nm). Tighten the rear engine mount bracket bolts to 43 ft. lbs. (58 Nm). Tighten the torque converter mounting bolts to 13 ft. lbs. (18 Nm).

9. Install the transaxle cables, linkage and halfshafts.

10. Refill the transaxle with the approved fluid and check for leaks.

11. Lower the vehicle. Road test the vehicle and check operation.

Corolla

1. Disconnect the negative battery cable. Remove the air cleaner.

2. Disconnect the neutral start switch. Disconnect the speedometer cable.

3. Disconnect the shift control cable and throttle linkage.

4. Disconnect the oil cooler hose. Plug the end of the hose to prevent leakage.

5. Drain the radiator and remove the water inlet pipe.

6. Raise and support the vehicle safely. Drain the transaxle fluid. As required remove the exhaust front pipe.

7. Remove the engine undercover. Remove the front and rear transaxle mounts.

8. Support the engine and transaxle

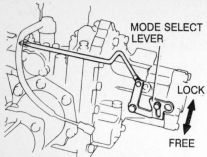

MODE SELECT LEVER

LOCK

FREE

Setting the mode selector—Corolla w/A241H

using the proper equipment. Remove the engine center support member.

9. Remove the halfshafts. Remove the starter assembly. Remove the steering knuckles, as required.

10. Remove the flywheel cover plate. Remove the torque converter bolts.

11. Remove the left engine mount. Remove the transaxle-to-engine bolts. Slowly and carefully back the transaxle away from the engine. Lower the assembly to the floor.

To install:

12. Install the transaxle and tighten the transaxle-to-engine bolts to 47 ft. lbs. (64 Nm). Tighten the left engine mount bracket bolts to 32 ft. lbs. (43 Nm). Tighten the rear engine mount bracket bolts to 43 ft. lbs. (58 Nm). Tighten the torque converter mounting bolts to 13 ft. lbs. (18 Nm).

13. When installing the A241H vehicle transaxle on AWD vehicles, be sure the mode selector lever is positioned in the **FREE** mode and attach the lock bolt.

14. Install the transaxle cables, linkage and halfshafts.

15. Refill the transaxle with the approved fluid and check for leaks.

16. Lower the vehicle. Road test the vehicle and check operation.

Camry

1. Disconnect the negative battery cable. Remove the air flow meter and the air cleaner assembly.

2. Disconnect the transaxle wire connector. Disconnect the neutral safety switch electrical connector.

3. Disconnect the transaxle ground strap. Disconnect the throttle cable from the throttle linkage.

4. Remove the transaxle case protector. Disconnect the speedometer cable and control cable.

5. Disconnect the oil cooler hoses. Remove the upper starter retaining bolts, as required remove the starter assembly. Remove the upper transaxle housing bolts. Remove the engine rear mount insulator bracket set bolt.

6. Raise and support the vehicle safely. Drain the transaxle fluid.

7. Remove the left front fender

apron seal. Disconnect both driveshafts.

8. Remove the suspension lower crossmember assembly. Remove the center driveshaft.

9. Remove the engine mounting center crossmember. Remove the stabilizer bar. Remove the left steering knuckle from the lower control arm.

10. Remove the torque converter cover. Remove the torque converter retaining bolts.

11. Properly support the engine and transaxle assembly. Remove the rear engine mounting bolts. Remove the remaining transaxle to engine retaining bolts.

12. Carefully remove the transaxle assembly from the vehicle.

To install:

13. Install the transaxle and tighten the 12mm transaxle housing bolts to 47 ft. lbs. (64 Nm); tighten the 10mm bolts to 34 ft. lbs. (46 Nm). Tighten the rear engine mount set bolts to 38 ft. lbs. (52 Nm). Tighten the torque converter mounting bolts to 20 ft. lbs. (27 Nm).

14. Install the transaxle cables, linkage and halfshafts.

15. Refill the transaxle with the approved fluid and check for leaks.

16. Lower the vehicle. Road test the vehicle and check operation.

Celica

1. Disconnect the negative battery cable. Remove the air flow meter and the air cleaner hose.

2. Disconnect the speedometer cable. Remove the starter assembly electrical connections. Disconnect the throttle cable from the throttle linkage and bracket.

3. Disconnect the ground strap. Remove the starter retaining bolts and as required remove the starter assembly.

4. Remove the upper transaxle housing retaining bolts. Remove the engine rear mount insulator bracket retaining bolt.

5. Raise and support the vehicle safely. Drain the transaxle fluid. Remove the engine under covers.

6. Remove the lower suspension crossmember. Disconnect the front and rear mounting components. Remove the engine mounting center member.

7. Remove the left halfshaft. Disconnect the right halfshaft.

8. Disconnect the exhaust pipe from the manifold. Remove the stiffener plate. Disconnect the control cable.

9. Disconnect the oil cooler hoses. Remove the torque converter cover. Remove the torque converter retaining bolts.

10. Support the engine and transaxle assembly, using the proper equipment. Remove the transaxle-to-engine

retaining bolts. Disconnect the front and rear transmission mount bolts.

11. Carefully lower the transaxle assembly to the floor.

To install:

12. Install the transaxle and tighten the 12mm engine-to-transaxle bolts to 47 ft. lbs. (64 Nm) and the 10mm bolts to 34 ft. lbs. (46 Nm). Tighten the torque converter bolts to 20 ft. lbs. (27 Nm). Tighten the left engine mount bracket bolts to 32 ft. lbs. (43 Nm). Tighten the rear engine mount bracket bolts to 43 ft. lbs. (58 Nm).

13. Install the transaxle cables, linkage and halfshafts.

14. Refill the transaxle with the approved fluid and check for leaks.

15. Lower the vehicle. Road test the vehicle and check operation.

MR2

1. Disconnect the negative battery cable. Remove the air flow meter and the air cleaner hose.

2. Remove the intercooler on the 4A-GZE.

3. Remove the water inlet set bolts. Disconnect the ground strap. Remove the transaxle mounting set bolt.

4. Disconnect the speedometer cable at the transaxle. Disconnect the throttle cable from the throttle linkage and the bracket.

5. Raise and support the vehicle safely. Drain the transaxle fluid. Remove the left tire.

6. Remove the transaxle gravel shield. Disconnect the speedometer cable at the transaxle assembly.

7. Disconnect the oil cooler lines at the transaxle. Remove the transaxle control cable clip and retainer and then disconnect the cable from the bracket. Remove the bracket.

8. Remove the starter assembly. Disconnect the exhaust pipe at the manifold. Remove the pipe.

9. Remove the stiffener plate. Remove the rear engine end plate. Remove the torque converter cover. Remove the torque converter retaining bolts.

10. Disconnect both the right and left halfshafts from their side gear shafts. Depress and hold the brake pedal while removing the halfshaft retaining nuts. Properly position the halfshaft aside.

11. Disconnect the suspension arm from the rear axle carrier, using the proper tools. Disconnect the rear axle carrier from the lower control arm.

12. Disconnect the halfshaft from the side gear shaft. Properly position the driveshaft aside.

13. Support the engine and transaxle assembly, using the proper equipment. Remove the transaxle-to-engine retaining bolts. Disconnect the front and rear transmission mount bolts.

14. Carefully lower the transaxle assembly to the floor.

To install:

15. Install the transaxle and tighten the transaxle-to-engine bolts to 47 ft. lbs. (64 Nm). Tighten the left engine mount bracket bolts to 32 ft. lbs. (43 Nm). Tighten the rear engine mount bracket bolts to 43 ft. lbs. (58 Nm). Tighten the torque converter mounting bolts to 20 ft. lbs. (27 Nm).

16. Install the transaxle cables, linkage and halfshafts.

17. Refill the transaxle with the approved fluid and check for leaks.

18. Lower the vehicle. Road test the vehicle and check operation.

SHIFT LINKAGE ADJUSTMENT

Tercel

SEDAN

1. Loosen the swivel nut on the selector lever.

2. Push the lever fully toward the right side of the vehicle.

3. Return the lever 2 notches to the **N** position.

4. Set the shift lever in the **N** position.

5. While holding the selector lever slightly toward the **R** shift position tighten the swivel nut to 48 inch lbs. (5.4 Nm).

WAGON

1. Push the selector lever all the way to the rear of the vehicle.

2. Return the lever 2 notches to the **N** shift position.

3. Set the selector lever in the **N** position.

4. While holding the selector lever slightly toward the **R** shift position tighten the connecting rod nut.

Corolla, Camry, MR2 and Celica

1. Loosen the swivel nut on the selector lever.

2. Push the lever fully toward the right side of the vehicle.

3. Return the lever 2 notches to the **N** position.

4. Set the selector lever in the **N** position.

5. While holding the selector lever slightly toward the **R** shift position tighten the swivel nut to 48 inch lbs. (5.4 Nm).

THROTTLE LINKAGE ADJUSTMENT

1. Remove the air cleaner.

2. Confirm that the accelerator linkage opens the throttle fully. Adjust the linkage as necessary.

3. Peel the rubber dust boot back from the throttle cable.

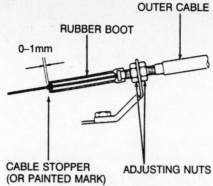

Throttle cable adjustment

4. Loosen the adjustment nuts on the throttle cable bracket (cylinder head cover) just enough to allow cable housing movement.

5. Have an assistant depress the accelerator pedal fully.

6. Adjust the cable housing so the distance between its end and the cable stop collar is 0.04 in.

7. Tighten the adjustment nuts. Make sure the adjustment has not changed. Install the dust boot and the air cleaner.

FRONT SUSPENSION

MacPherson Strut

REMOVAL & INSTALLATION

1. Remove the hubcap and loosen the lug nuts.

2. Raise and support the vehicle safely.

NOTE: Do not support the weight of the vehicle on the suspension arm; the arm will deform under its weight.

3. Unfasten the lug nuts and remove the wheel.

4. Remove the union bolt and 2 washers and disconnect the front brake line from the disc brake caliper. Remove the clip from the brake hose and pull off the brake hose from the brake hose bracket.

5. Remove the caliper and wire it aside. Matchmark on the strut lower bracket and camber adjust cam, if equipped. Remove the 2 bolts and nuts which attach the strut lower end to the steering knuckle lower arm.

6. Disconnect and remove the TEMS actuator from the top of the strut on late Cressida and Supra.

7. Unbolt the upper control arm where it attaches to the body on 1987–89 Supra.

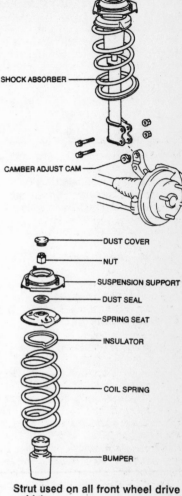

Strut used on all front wheel drive vehicles

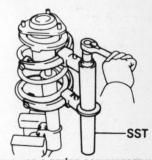

Always use a spring compressor when servicing the MacPherson strut

8. Remove the 3 nuts (4 nuts on the FX vehicles) which secure the upper strut mounting plate to the top of the wheel arch.

NOTE: Press down on the suspension lower arm, in order to remove the strut assembly. This must be done to clear the collars on the steering knuckle arm bolt holes when removing the shock/spring assembly.

To install:

9. Align the hole in the upper sus-

pension support with the shock absorber piston or end, so they fit properly.

10. Always use a new nut and nylon washer on the shock absorber piston rod end when securing it to the upper suspension support. Torque the nut to 29–40 ft. lbs. (39–54 Nm).

NOTE: Do not use an impact wrench to tighten the nut.

11. Coat the suspension support bearing with multipurpose grease prior to installation. Pack the space in the upper support with multipurpose grease, also, after installation.

12. Tighten the suspension support-to-wheel arch bolts to the following specifications:

Corolla RWD—11–16 ft. lbs. (11–22 Nm).

Corolla FWD (1988–90)—29 ft. lbs. (39 Nm).

Celica—47 ft. lbs. (64 Nm).

Camry—47 ft. lbs. (64 Nm).

Supra—26 ft. lbs. (35 Nm).

1988 Cressida—25–29 ft. lbs. (34–39 Nm).

1989–92 Cressida—32 ft. lbs. (43 Nm).

1988–90 Tercel—20–25 ft. lbs. (27–34 Nm).

1988–89 MR2—21–25 ft. lbs. (28–34 Nm).

1990–92 MR2—29 ft. lbs. (39 Nm).

13. Tighten the shock absorber-to-steering knuckle arm bolts to the following specifications:

Corolla (RWD)—50–65 ft. lbs. (68–88 Nm).

Corolla (FWD)—1988–92 sedan and wagon 194 ft. lbs. (263 Nm).

1988–92 Tercel Sedan—166 ft. lbs. (226 Nm).

1988 MR2—105 ft. lbs. (143 Nm).

1989 MR2—119 ft. lbs. (162 Nm).

1990–92 MR2—188 ft. lbs. (255 Nm).

Supra—106 ft. lbs. (144 Nm).

1988 Celica FWD—152 ft. lbs. (204 Nm).

1989–92 Celica FWD—188 ft. lbs. (255 Nm).

Cressida—80 ft. lbs. (109 Nm).

1988 Camry—166 ft. lbs. (226 Nm).

1989–92 Camry—224 ft. lbs. (304 Nm).

All others—65 ft. lbs. (88 Nm).

14. Adjust the front wheel bearing preload.

15. Bleed the brake system.

Upper Ball Joints

INSPECTION

1. Raise the front of the vehicle and

support safely. Place jackstands under the front lower control arms.

2. With the wheels straight ahead, move the lower control arm up and down and check for excessive play. The vertical movement should not exceed 0.012 in. (0.3mm) for the upper ball joint.

REMOVAL & INSTALLATION

NOTE: If equipped with both upper and lower ball joints—if both are to be removed, always remove the lower one first.

Supra

The ball joint is an integral component of the upper control arm. Ball joint replacement requires that the entire arm assembly be replaced. Please refer to Upper Control Arm.

Lower Control Arm/ Ball Joints

INSPECTION

Corolla (RWD) and Cressida

Raise the front end and position a piece of wood under the wheel and lower the vehicle until there is an ½ load on the strut. Check the front wheel play. Replace the lower ball joint if the play at the wheel rim exceeds 0.1 in. (2.54mm) vertical motion or 0.25 in. (6mm) horizontal motion. Be sure the dust covers are not torn and that they are securely glued to the ball joints.

NOTE: Do not raise the control arm on Corolla or Cressida vehicles; damage to the arm will result.

Tercel, Camry, Corolla (FWD), Celica, Supra and MR2

1. Raise the vehicle and place wooden blocks under the front wheels. The block height should be 7.09–7.87 in. (180–200 Nm).

2. Use jack stands for additional safety.

3. Make sure the front wheels are in a straight forward position.

4. Check the wheels.

5. Lower the jack until there is approximately half a load on the front springs.

6. Move the lower control arm up and down to check that there is no ball joint vertical play. Replace if any play exists.

REMOVAL & INSTALLATION

NOTE: If equipped with both upper and lower ball joints, if both ball joints are to be removed,

always remove the lower and then the upper ball joint.

Corolla (RWD)

The ball joint and control arm cannot be separated from each other. If one fails, then both must be replaced as an assembly, in the following manner:

1. Remove the stabilizer bar securing bolts.

2. Unfasten the torque strut mounting bolts.

3. Remove the control arm mounting bolt and detach the arm from the front suspension member.

4. Remove the steering knuckle arm from the control arm with a ball joint puller.

5. Inspect the suspension components, which were removed for wear or damage. Replace any parts, as required.

To install:

6. During installation, please observe the following:

a. When installing the control arm on the suspension member, tighten the bolts partially at first.

b. Complete the assembly procedure and lower the vehicle to the ground.

c. Bounce the front of the vehicle several times. Allow the suspension to settle, then tighten the lower control arm bolts to 51–65 ft. lbs. (69–88 Nm).

d. Lubricate the ball joint.

e. Check the front end alignment.

Celica

1. Raise and support the vehicle safely. Remove the wheels.

2. Disconnect the lower control arm from the steering knuckle.

3. Remove the nut and disconnect the stabilizer bar from the control arm.

4. On all but the left-side control arm, if with automatic transmissions, remove the control arm front set nut and washer. Remove the rear bracket bolts and then remove the arm.

5. On the left arm, if with automatic transmissions, remove the control arm front set nut and washer. Remove the 4 bolts and 2 nuts that attach the lower suspension crossmember to the frame and remove the crossmember. Remove the bolt and nut and lift out the lower arm with the lower arm shaft.

To install:

6. On all but the left-side control arm, if with automatic transmissions, install the lower control arm shaft washer with the tapered side toward the body. Install the lower arm with the bracket and then temporarily install the washer and nut to the lower arm shaft and bracket bolts.

7. On the left-side arm, if with automatic transmissions, position the washer on the lower arm shaft and then install them to the lower arm. Temporarily install the washer and nut to the shaft with the tapered side toward the body. Install the lower arm with the shaft to the body and temporarily install the rear brackets. Install the bolt and nut to the lower arm shaft and tighten them to 154 ft. lbs. (208 Nm). Install the crossmember to the body and tighten the 4 bolts to 154 ft. lbs. (208 Nm). Tighten the 2 nuts to 29 ft. lbs. (39 Nm).

8. Connect the lower arm to the steering knuckle and tighten the bolt and 2 nuts to 94 ft. lbs. (127 Nm).

9. Connect the stabilizer bar to the control arm and tighten the nut to 26 ft. lbs. (35 Nm).

10. Install the wheel, lower the vehicle and bounce it several times to set the suspension.

11. Tighten the front set nut to 156 ft. lbs. (212 Nm). Tighten the rear bracket bolts to 72 ft. lbs. (98 Nm).

MR2

1. Raise the vehicle and support it safely. Remove the wheel.

2. Remove the cotter pin and castle nut and then press the lower arm out of the ball joint.

3. Press the ball joint out of the steering knuckle.

4. Remove the 2 nuts and disconnect the strut bar from the control arm.

5. Remove the lower control arm-to-body bolt and remove the arm.

To install:

6. When installing the lower arm, position it in the strut bar and tighten the nuts finger-tight. Do the same thing with the arm-to-body bolt.

7. Connect the control arm to the ball joint and tighten the castle nut to 58 ft. lbs. (78 Nm). Install a new cotter pin.

8. Tighten the strut bar-to-arm bolts to 83 ft. lbs. (113 Nm).

9. Install the tires, lower the vehicle and bounce it several times to set the suspension.

10. Tighten the control arm-to-body bolt to 94 ft. lbs. (127 Nm) on 1988–89 vehicles and 87 ft. lbs. (118 Nm) on 1990–92 vehicles. Check the wheel alignment.

Cressida

1. Raise and support the vehicle safely. Remove the wheels.

2. Remove the 2 knuckle arm-to-strut bolts, pull down on the control arm and disconnect it and the knuckle arm from the strut.

3. Remove the cotter pin and nut and press the tie rod off the knuckle arm.

4. Remove the nut attaching the stabilizer bar to the control arm and disconnect the bar.

5. Remove the 2 nuts and then disconnect the strut bar from the control arm.

6. Disconnect the control arm from the crossmember and remove it and the rack boot protector as an assembly.

7. Remove the cotter pin and nut and then press the knuckle arm off the control arm.

To install:

8. Press the knuckle arm into the control arm and then install the assembly into the crossmember.

9. Connect the stabilizer bar to the control arm and tighten the nut to 13 ft. lbs. (18 Nm).

10. Connect the strut bar to the control arm and tighten the nuts to 54 ft. lbs. (73 Nm) for 1988 and 76 ft. lbs. (103 Nm) for 1989–92.

11. Connect the knuckle arm to the strut housing and tighten the bolts to 80 ft. lbs. (108 Nm).

12. Install the wheel and lower the vehicle. Bounce the vehicle several times to set the suspension and then tighten the control arm-to-body bolt to 80 ft. lbs. (108 Nm) for 1988 or 121 ft. lbs. (164 Nm) for 1989–92.

13. Check the front wheel alignment.

Supra

NOTE: This procedure is for ball joint removal only. To remove the lower control arm, please refer to the Lower Control Arm procedure.

1. Raise and support the vehicle safely.

2. Remove the wheels.

3. Remove the steering knuckle and then remove the upper control arm.

4. Remove the lower ball joint mounting nuts and the bolt. Remove the attachment plate.

5. Remove the lower ball joint.

6. Installation is in the reverse order of removal. Tighten the ball joint mounting bolt and nuts to 94 ft. lbs. (127 Nm).

Tercel

1. Raise and support the vehicle safely. Remove the wheels.

2. Remove the 2 bolts attaching the ball joint to the steering knuckle.

3. Remove the stabilizer bar nut, retainer and cushion.

4. Raise the opposite wheel until the body of the vehicle just lifts off the supports.

5. Loosen the lower control arm mounting bolt, wiggle the arm back and forth and then remove the bolt. Disconnect the lower control arm from the stabilizer bar.

NOTE: When removing the lower control arm, be careful not to lose the caster adjustment spacer.

6. Carefully mount the lower control arm in a vise and then, using a ball joint removal tool, disconnect the ball joint from the arm.

To install:

7. Tighten the ball joint-to-control arm nut to 51–65 ft. lbs. (69–88 Nm) and use a new cotter pin.

8. Tighten the steering knuckle-to-control arm bolts to 59 ft. lbs. (80 Nm).

9. Before tightening the stabilizer bar nuts, mount the wheels and lower the vehicle. Bounce the vehicle several times to settle the suspension and then tighten the stabilizer bolts to 66–90 ft. lbs. (90–122 Nm).

10. Tighten the arm-to-body bolts to 83 ft. lbs.

11. Check the front end alignment.

Corolla (FWD)

1. On all vehicles, except the left side on those with automatic transaxle, perform the following:

   a. Remove the bolt and 2 nuts attaching the ball joint to the lower arm and disconnect the lower arm from the steering knuckle.

   b. On 4A-F and 4A-FE, remove the nut holding the stabilizer bar to the lower arm and disconnect the bar from the arm.

   c. On 4A-GE, remove the lower nut on the stabilizer bar link and disconnect the link from the arm.

   d. Remove the rear bracket bolts and nut. Remove the lower arm front mounting bolt.

   e. Remove the rear bracket and the stabilizer bar bracket and lift out the lower control arm.

2. To remove the left control arm, if with automatic transaxle, perform the following:

   a. Disconnect the arm at the steering knuckle.

   b. Disconnect the stabilizer bar at the lower arm.

   c. Remove the lower arm rear brackets. Move the stabilizer bar toward the rear and remove the bracket.

   d. Remove the 6 bolts and 2 nuts and remove the suspension crossmember with the lower arm.

   e. Remove the lower arm from the crossmember.

To install:

3. To install the left control arm, if with automatic transaxle, install the lower arm on the crossmember and install the assembly to the body.

4. On all others, install the lower arm to the body, move the stabilizer bar into position and install the front mounting bolt. Install the stabilizer bar and rear brackets.

5. Connect the lower arm to the steering knuckle and tighten the bolts to 105 ft. lbs. (142 Nm).

6. On 4A-F, connect the stabilizer bar to the lower arm and tighten the nut to 13 ft. lbs. (18 Nm).

7. On 4A-GE, connect the stabilizer bar link to the lower arm and tighten the nut to 26 ft. lbs. (35 Nm).

8. Lower the vehicle and bounce it several times to stabilize the suspension. Tighten the lower arm front bolt to 174 ft. lbs. (235 Nm) for 1988 or 152 ft. lbs. (206 Nm) for 1989–92. Tighten the rear bracket bolts to 94 ft. lbs. (127 Nm) on the lower arm side, 37 ft. lbs. (50 Nm) on the stabilizer bar side and tighten the small bolt and nut to 14 ft. lbs. (19 Nm).

9. Check the front end alignment.

Camry

1. Raise and support the vehicle safely. Remove the wheels.

2. Remove the 2 bolts attaching the ball joint to the steering knuckle.

3. Remove the stabilizer bar nut, retainer and cushion.

4. Remove the nut attaching the lower arm shaft to the lower arm.

5. Remove the lower suspension crossmember (2 bolts and 4 nuts).

6. Remove the lower control arm and lower arm shaft as an assembly.

7. Grip the lower arm assembly in a vise and remove the ball joint cotter pin and retaining nut. With a ball joint removal tool, pull the ball joint out of the control arm.

To install:

8. Position the ball joint in the lower arm and tighten the nut to 67 ft. lbs. (91 Nm) for 1988 or 90 ft. lbs. (123 Nm) for 1989–92. Install a new cotter pin.

9. Install the lower arm to the stabilizer bar and then install the lower arm shaft to the body. Install the lower arm nut and retainer. Screw on a new stabilizer bar end nut and retainer.

10. Connect the ball joint to the steering knuckle and tighten the bolts to 94 ft. lbs. (127 Nm) for 1988 or 83 ft. lbs. (113 Nm) for 1989–92.

11. Install the suspension lower crossmember. Tighten the inner bolts to 32 ft. lbs. (43 Nm) and the outer ones to 153 ft. lbs. (207 Nm).

12. Install the wheels and lower the vehicle. Bounce it several times to set the suspension.

13. Tighten the stabilizer bar end nut and the lower arm shaft-to-lower arm bolt to 156 ft. lbs. (212 Nm).

Upper Control Arm

REMOVAL & INSTALLATION

Supra

1. Raise and support the vehicle safely. Remove the wheels.

2. Unclip the brake hose bracket at the steering knuckle, remove the retaining nut and press the upper arm out of the knuckle.

3. Remove the upper mounting bolt and nut and lift out the upper control arm.

4. Connect the upper arm to the body. Connect the arm to the steering knuckle.

5. Install the wheels and lower the vehicle. Bounce it several times to set the suspension and then tighten the arm-to-knuckle nut to 80 ft. lbs. (108 Nm). Tighten the arm-to-body bolt to 121 ft. lbs. (164 Nm).

Lower Control Arms

REMOVAL & INSTALLATION

Tercel and Celica

1. Raise the vehicle and support safely. Remove the front wheels.

2. Disconnect the lower arm from the steering knuckle, at the ball joint.

3. Disconnect the stabilizer bar from the lower arm.

4. Remove the lower arm by loosening the front bolt. Remove the bracket bolts, stabilizer bracket and front bolt.

5. Remove the lower arm.

To install:

6. Install the lower arm assemblies. Loosely install the lower arm bushing bolts and clamp.

7. Install the ball joint bolts and torque to 59 ft. lbs. (80 Nm) for the Tercel and 94 ft. lbs. (127 Nm) for the Celica.

8. Install the stabilizer and torque to 76 ft. lbs. (103 Nm) for the Tercel and 26 ft. lbs. (34 Nm) for the Celica.

9. Install the front wheels and lower the vehicle. Bounce up and down to stabilize the suspension. With the vehicle weight on the suspension, torque the front side to 105 ft. lbs. (142 Nm) for the Tercel and 156 ft. lbs. (212 Nm) for the Celica. Torque the rear bracket bolts to 94 ft. lbs. (126 Nm) for the Tercel and 72 ft. lbs. (96 Nm) for the Celica.

10. Align the front end.

Corolla (RWD)

1. Raise and support the vehicle safely.

2. Remove the wheel.

3. Disconnect the steering knuckle from the control arm.

4. Disconnect the tie rod, stabilizer bar and strut bar from the control arm.

5. Remove the control arm mounting bolts, and remove the arm.

6. Install in reverse of above. Tighten, but do not torque fasteners until vehicle is on the ground.

7. Lower vehicle to ground, rock it from side-to-side several times and torque control arm mounting bolts to 51–65 ft. lbs. (68–88 Nm), stabilizer bar to 16 ft. lbs. (22 Nm), strut bar to 40 ft. lbs. (54 Nm) and shock absorber to 65 ft. lbs. (88 Nm).

8. Align the front end.

Corolla (FWD)

1. Raise the vehicle and support safely. Remove the front wheels.

2. Disconnect the lower arm from the steering knuckle, at the ball joint.

3. Disconnect the stabilizer bar from the lower arm.

4. Remove the lower arm by loosening the front bolt. Remove the bracket bolts and nuts, rear and stabilizer bracket and front bolt for all FWD Corollas, except left side with an automatic transaxle.

5. Remove the lower arms with the suspension crossmember for FWD Corollas with automatic transaxle. Remove the 4 left and right lower arm rear brackets, stabilizer bar bracket and 6 crossmember bolts.

To install:

6. Install the crossmember and lower arm assemblies. Loosely install the lower arm bushing bolts.

7. Install the ball joint bolts and torque to 105 ft. lbs. (142 Nm).

8. Install the stabilizer and torque to 26 ft. lbs. (35 Nm).

9. Install the front wheels and lower the vehicle. Bounce up and down to stabilize the suspension. With the vehicle weight on the suspension, torque the lower control arm bushing bolts to 174 ft. lbs. (235 Nm) for the front bolt, 94 ft. lbs. (126 Nm) for the lower arm side, 37 ft. lbs. (50 Nm) for the stabilizer bar side and 14 ft. lbs. (19 Nm) for the small bolt and nut.

10. Align the front end.

Camry

1. Raise the vehicle and support safely. Remove the front wheels.

2. Disconnect the lower arm from the steering knuckle, at the ball joint.

3. Disconnect the stabilizer bar from the lower arm.

4. Remove the crossmember and lower arm as an assembly. Remove the lower suspension with the lower suspension arm shaft.

To install:

5. Install the crossmember and lower arm assemblies. Loosely install the lower arm bushing bolts. Torque the crossmember bolts to 112 ft. lbs. (152 Nm).

6. Install the ball joint bolts and torque to 90 ft. lbs. (123 Nm).

7. Install the stabilizer nut loosely.

8. Install the front wheels and lower the vehicle. Bounce up and down to stabilize the suspension. With the ve-

hicle weight on the suspension, torque the lower control arm bushing and stabilizer nuts to 156 ft. lbs. (212 Nm).

9. Align the front end.

Supra

1. Raise and support the vehicle safely. Remove the wheels.

2. Disconnect the stabilizer bar link from the lower control arm. Remove the locknut and press the ball joint out of the steering knuckle.

3. Disconnect the lower control arm at the strut. Matchmark the front and rear adjusting cams to the body. Remove the nuts and cams and then remove the lower arm.

4. Unbolt the ball joint from the control arm.

To install:

5. Install the ball joint to the arm and tighten the nuts to 94 ft. lbs. (127 Nm).

6. Position the lower control arm and install the adjusting cams and nuts finger-tight.

7. Connect the ball joint to the steering knuckle and tighten a conventional nut to 14 ft. lbs. (20 Nm). Install a locknut on top of the other and tighten it to 107 ft. lbs. (145 Nm).

8. Tighten the arm-to-strut bolt to 106 ft. lbs. (143 Nm). Tighten the stabilizer bar link nut to 47 ft. lbs. (64 Nm).

9. Install the wheels and lower the vehicle. Bounce the vehicle several times to set the suspension. Align the matchmarks on the adjusting cams and the body and tighten them to 177 ft. lbs. (240 Nm). Check the front alignment.

MR2

1. Raise the vehicle and support safely. Remove the front wheels.

2. Disconnect the lower arm from the steering knuckle, at the ball joint using a ball joint separator.

3. Disconnect the stabilizer and strut bar from the lower arm.

4. Remove the lower arm by loosening the front bolt.

To install:

5. Install the lower arm assembly. Loosely install the lower arm bushing bolt.

6. Install the ball joint bolts and torque to 58 ft. lbs. (78 Nm).

7. Install the stabilizer and torque to 47 ft. lbs. (64 Nm). Install the strut bar bolts and torque to 83 ft. lbs. (113 Nm).

8. Install the front wheels and lower the vehicle. Bounce up and down to stabilize the suspension. With the vehicle weight on the suspension, torque the lower control arm bushing bolts to 87 ft. lbs. (118 Nm).

9. Align the front end.

Front Wheel Bearings

NOTE: These procedures apply to rear wheel drive vehicles only. For front wheel bearing service on front wheel drive vehicles, refer to the "Drive Axle" section.

ADJUSTMENT

NOTE: The Supra uses preset hub bearings. The adjustment is obtained through the hub nut torque. The hub assembly has to be removed from the vehicle to adjust the hub nut. Make sure the bearing axial play does not exceed 0.0020 in. (0.05mm).

1. With the front hub/disc assembly installed, tighten the castellated nut to 19–23 ft. lbs. (26–31 Nm).

2. Rotate the disc back and forth, 2–3 times, to allow the bearing to seat properly.

3. Loosen the castellated nut until it is only finger-tight.

4. Tighten the nut firmly, using a box wrench.

5. Measure the bearing preload with a spring scale attached to a wheel mounting stud. Check it against the specifications.

6. Install the cotter pin.

NOTE: If the hole does not align with the nut (or cap) holes, tighten the nut slightly until it does.

7. Finish installing the brake components and the wheel.

REMOVAL & INSTALLATION

Except Supra

1. Raise the vehicle and support safely. Remove the disc/hub assembly.

2. If either the disc or the entire hub assembly is to be replaced, unbolt the hub from the disc.

NOTE: If only the bearings are to be replaced, do not separate the disc and hub.

3. Using a brass rod as a drift, tap the inner bearings cone out. Remove the oil seal and the inner bearings.

NOTE: Throw the old oil seal away.

4. Drive out the inner bearing cup.

5. Drive out the outer bearing cup. Inspect the bearings and the hub for signs of wear or damage. Replace components, as necessary.

To install:

6. Install the inner bearing cup an then the outer bearing cup, by driving them into place.

NOTE: Use care not to cock the bearing cups in the hub.

7. Pack the bearings, hub inner well and grease cap with multipurpose grease.

8. Install the inner bearing into the hub.

9. Carefully install a new oil seal with a soft drift.

10. Install the hub on the spindle. Be sure to install all of the washers and nuts which were removed.

11. Adjust the bearing preload.

12. Install the caliper assembly and the wheel.

Supra

1. Raise the vehicle and support safely.

2. Remove the front wheel, brake caliper and disc. Hange the caliper from the suspension with a piece of wire. Remove the hose bracket from the knuckle.

3. Disconnect the tie rod from the knuckle using a tie rod separator, or equivalent.

4. Disconnect the steering knuckle from the upper suspension arm using a ball joint separator, or equivalent.

5. Remove the steering knuckle from the lower ball joint using a ball joint separator.

6. Remove the bearing cap from the rear of the knuckle with a suitable prybar.

7. Remove the axle hub lock nut using a hammer and chisel to loosen the stacked part of the lock nut.

8. Press the axle hub out of the knuckle using a bearing puller or press.

9. Remove the hub bearing inner race by using a bearing puller SST 09950–20017 or equivalent.

10. Remove the 4 dust cover bolts and cover.

11. Remove the snapring and press out the hub bearing.

To install:

12. Press in the hub bearing and install the snapring.

13. Install the outer seal and disc dust cover. Torque the cover bolts to 14 ft. lbs. (19 Nm).

14. Install the axle hub using a press.

15. Install the axle lock nut and torque to 147 ft. lbs. (199 Nm). Stake the lock nut using a hammer and chisel.

16. Install the bearing cap.

17. Install the axle hub onto the vehicle. Torque the upper ball joint to 76 ft. lbs. (103 Nm), lower ball joint to 92 ft. lbs. (125 Nm) and tie rod end to 36 ft. lbs. (49 Nm).

18. Install the brake rotor, caliper and front wheel.

19. Lower the vehicle and have the front end aligned.

REAR SUSPENSION

Shock Absorbers

REMOVAL & INSTALLATION

Corolla (RWD) and 1989–92 Corolla (AWD)

1. Raise and support the vehicle safely. Support the rear axle.
2. Unfasten the upper shock absorber retaining nuts. It may be necessary to hold the shock absorber shaft with a suitable tool while removing the top retaining nut.

NOTE: **Always remove and install the shock absorbers one at a time. Do not allow the rear axle to hang in place as this may cause undue damage.**

3. Remove the lower shock retaining nut where it attaches to the rear axle housing.
4. Remove the shock absorber.
5. Inspect the shock for wear, leaks or other signs of damage.
6. Installation is in the reverse order of removal. During installation, observe the following:
    a. Tighten the upper retaining nuts to 18 ft. lbs. (25 Nm).
    b. Tighten the lower retaining nuts to 27 ft. lbs. (37 Nm).

1988 Cressida

1. Raise and support the differential housing and the suspension control arms safely.
2. Remove the brake hose clips. Disconnect the stabilizer bar end.
3. Disconnect the halfshaft at the CV-joint on the wheel side.
4. With a jackstand under the suspension control arm, unbolt the shock absorber at its lower end. Using a prybar to keep the shaft from turning, remove the nut holding the shock absorber to its upper mounting. If with TEMS, disconnect the actuator and remove it. Remove the shock.
5. Installation is in the reverse order of removal. Torque the halfshaft nuts to 44–57 ft. lbs. (60–78 Nm), the upper shock mounting nut to 14–22 ft. lbs. (19–30 Nm) and the lower shock mounting nut to 22–32 ft. lbs. (30–44 Nm).

MacPherson Strut

REMOVAL & INSTALLATION

Tercel

1. Working inside the vehicle, re-move the shock absorber cover and package tray bracket.
2. Raise the rear of the vehicle and support safely. Remove the wheel.
3. Disconnect the brake line from the wheel cylinder, if necessary. Disconnect the brake line from the flexible hose at the mounting bracket on the strut tube. Disconnect the flexible hose from the strut.
4. Loosen the nut holding the suspension support to the shock absorber; do not remove the nut.
5. Remove the bolts and nuts mounting on the strut on the axle carrier and then disconnect the strut.
6. Remove the 3 upper strut mounting nuts and carefully remove the strut assembly.

To install:

7. Install the strut assembly into the vehicle. During installation, observe the following torque specification:
    a. Tighten the upper strut retaining nuts to 23 ft. lbs. (31 Nm).
    b. Tighten the lower strut-to-axle carrier bolts to 105 ft. lbs. (143 Nm).
    c. Tighten the nut holding the suspension support to the shock absorber to 36 ft. lbs. (49 Nm).
    d. Tighten the strut-to-axle beam nut to 47 ft. lbs. (64 Nm) on 1988–92 vehicles.
8. Bleed the brakes.

Camry and Corolla (2WD)

1. On the 4-door sedan, remove the package tray and vent duct.
2. On the hatchback, remove the speaker grilles.
3. Disconnect the brake line from the wheel cylinder.
4. Remove the brake line from the brake hose.
5. Disconnect the brake hose from its bracket on the strut.
6. Remove the strut suspension support cover. Loosen, but do not remove, the nut holding the suspension support to the strut.
7. Unbolt the strut from the rear arm and or axle carrier.
8. Unbolt the strut from the body.

To install:

9. Install the strut assembly. During installation, please observe the following torque specifications:
    a. Tighten the strut-to-body bolts to 23 ft. lbs. (31 Nm) on 1988 Camry or 29 ft. lbs. (39 Nm) on 1988–89 Corolla sedan/wagon and 1989–92 Camry).
    b. Tighten the strut-to-axle carrier bolts 166 ft. lbs. on 1988–92 Camry and 105 ft. lbs. (143 Nm) on Corolla.
    c. Tighten the the suspension support-to-strut nut to 36 ft. lbs. (49 Nm).

10. Refill and bleed the brake system.

Celica (FWD)

1. Raise and support the vehicle safely. Position an hydraulic jack under the rear hub assembly; raise it just enough to support the assembly.
2. On the liftback, remove the rear speaker grilles.
3. On the coupe, remove the suspension service hole cover.
4. On the ST and GT vehicles, disconnect and plug the brake line at the backing plate. Remove the clip and E-ring and then disconnect the brake hose and tube from the strut housing.
5. On the GTS, remove the union bolts and gaskets and disconnect the brake line from the brake cylinder. Remove the clip and E-ring from the strut and then disconnect the brake hose from the strut housing.
6. Loosen, but do not remove, the nut attaching the suspension support to the strut.

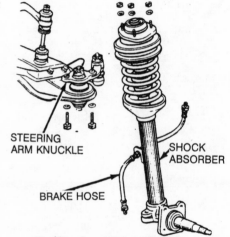

STEERING ARM KNUCKLE

SHOCK ABSORBER

BRAKE HOSE

Strut used on all rear wheel drive vehicles

7. Disconnect the stabilizer bar at the lower end of the strut housing.
8. Disconnect the strut at the axle carrier.
9. Remove the 3 strut-to-body bolts and then remove the strut.

To install:

10. Tighten the upper strut-to-body nuts to 23 ft. lbs. (31 Nm).
11. Tighten the lower strut-to-carrier bolts to 119 ft. lbs. (162 Nm).
12. Connect the stabilizer bar to the strut and tighten the bolts to 26 ft. lbs. (35 Nm).
13. Tighten the strut holding nut to 36 ft. lbs. (49 Nm). Install the dust cover onto the suspension support.
14. Reconnect the brake line and hose. Bleed the system, lower the vehicle and check the rear wheel alignment.

MR2

1. Raise and support the vehicle safely. Position a hydraulic floor jack under the rear hub assembly; raise it just enough to support the assembly.

2. Remove the union bolts and gaskets and disconnect the brake line from the brake cylinder. Remove the clip and E-ring from the strut and then disconnect the brake hose from the strut housing.

3. Matchmark the lower strut bracket and the camber adjusting cam, remove the 2 axle carrier bolts and the adjusting cam and disconnect the strut from the carrier.

4. Remove the engine hood side panel.

5. Remove the 3 upper strut-to-body nuts and then remove the strut.

To install:

6. Position the strut and tighten the upper mounting nuts to 23 ft. lbs. (31 Nm).

7. Install the engine hood side panel.

8. Connect the axle carrier to the lower strut bracket. Insert the mounting bolts from the rear and align the matchmarks made in Step 3. Tighten the nuts to 105 ft. lbs. (142 Nm) on 1988 vehicles and 166 ft. lbs. (226 Nm) on 1989–92 vehicles.

9. Connect the brake line, bleed the system and check rear wheel alignment.

Supra and 1989–92 Cressida

1. Raise and support the vehicle safely. Remove the wheels.

2. Remove the speaker grille and interior quarter panel trim, if equipped with TEMS.

3. Disconnect the strut from the axle carrier.

4. Remove the strut cap. Remove the Toyota Electronic Modulated Suspension (TEMS) actuator.

5. Remove the 3 strut mounting nuts from the body and remove the strut assembly.

6. Mount the strut assembly in a vise. Using a spring compressor, compress the coil spring.

7. Remove the strut suspension support nut. Remove the strut suspension support, remove the coil spring and bumper.

To install:

8. Mount the strut in a vise. Using a spring compressor, compress the coil spring.

9. Install the bumper to the strut, align the coil spring end with the lower seat hollow and install the coil spring.

10. Align the strut suspension support hole and piston rod and install it. Align the suspension support with the strut lower bushing.

11. Install the strut suspension sup-

port nut and torque it to 20 ft. lbs. (27 Nm). Connect the strut assembly with the 3 retaining nuts and torque them to 10 ft. lbs. (14 Nm).

12. Connect the strut assembly to the axle carrier and torque it to 101 ft. lbs. (137 Nm).

13. Install the TEMS actuator and strut cap. Install the quarter panel trim panel and speaker grille.

Coil Springs

REMOVAL & INSTALLATION

1. Loosen the rear wheel lug nuts.

2. Raise and support the rear axle housing and frame safely.

3. Remove the lug nuts and wheel.

4. If equipped, disconnect the rear stabilizer bar from the axle housing or suspension arm on 1988–89 Supra and 1988 Cressida. Remove the bolt holding the stabilizer bar bushing to the rear axle housing.

5. Unfasten the lower shock absorber end. On the Corolla RWD, Corolla FWD (AWD) and Tercel AWD, disconnect the lateral control rod from the axle.

NOTE: On Supra and 1988 Cressida with IRS suspension, remove the rear halfshafts.

6. Slowly lower the jack under the rear axle housing until the axle is at the bottom of its travel.

7. Withdraw the coil spring, complete with its insulator.

8. Inspect the coil spring and insulator for wear, cracks or weakness; replace either or both, as necessary.

9. Installation is the reverse of the removal procedure.

Lower Control Arms

REMOVAL & INSTALLATION

MR2

1. Raise and support the vehicle safely. Remove the wheels.

2. Remove the cotter pin and retaining nut from the bottom of the ball joint stem. Using a ball joint removal tool, press the ball joint out of the control arm.

3. Remove the strut rod nut and retainer from the lower control arm.

4. Remove the bolt holding the lower control arm to the body. Remove the cushion and then disconnect the lower arm from the strut rod. Remove the lower control arm.

To install:

5. Connect the lower arm to the strut rod. Install the strut rod nut, cushion and retainer.

6. Connect the lower arm to the

body and install the retaining nut fingertight.

7. Connect the lower control arm to the ball joint and tighten the retaining nut to 67 ft. lbs. (91 Nm). Install a new cotter pin.

NOTE: If the holes do not align when installing the new cotter pin, tighten the nut until they are aligned. Do not loosen the nut!

8. Install the wheel and lower the vehicle. Tighten the strut rod nut to 86 ft. lbs. (117 Nm) and the arm-to-body bolt to 94 ft. lbs. (127 Nm).

Celica, Camry and Corolla

1. Raise the vehicle and support safely. Remove the rear wheels.

2. Remove the nut from the axle carrier.

3. Place the matchmarks to the toe adjust cam and suspension member.

4. Remove the service hole cover and loosen the bolt and remove the toe adjust plate No. 2.

5. Remove the bolt with toe adjust cam and disconnect the suspension arm and remove.

To install:

6. Face the mark on the suspension arms to the rearward of the vehicle. Install the bushing with the slit side towards the rear and the small paint spot to the outside of the vehicle for the Camry.

7. Install the stamped suspension arm with the identification mark **L** for left and **R** for right. Temporarily install the suspension arms with the bolt, washer and nut. Do not tighten at this time.

8. Loosely install the bolt into the axle carrier.

9. Install the rear wheels and lower the vehicle. Bounce the suspension up and down a few times.

10. Torque suspension components with the vehicle weight loading the suspension.

11. Torque the suspension arm bolts to 83 ft. lbs. (113 Nm) and the nut to 166 ft. lbs. (226 Nm).

12. Have the wheel alignment checked.

Supra and 1989–92 Cressida

1. Raise and support the vehicle safely. Remove the wheels.

2. Remove the halfshaft.

3. Remove the nut and disconnect the No. 1 lower arm from the axle carrier. Matchmark the adjusting cam to the body, remove the cam and bolt and then lift out the No. 1 arm.

4. Remove the bolt and nut and disconnect the No. 2 lower arm from the axle carrier. Matchmark the adjusting cam to the body, remove the cam and bolt and then lift out the No. 2 arm.

To install:

5. Position the No. 2 arm and in-

stall the adjusting cam and bolt so the matchmarks are in alignment. Connect the arm to the axle carrier.

6. Position the No. 1 arm and install the adjusting cam and bolt so the matchmarks are in alignment. Connect the arm to the axle carrier. Use a new nut and tighten it to 43 ft. lbs. (59 Nm).

7. Install the halfshaft.

8. Install the wheels and lower the vehicle. Bounce it several times to set the suspension and then tighten the body-to-arm bolts and nuts to 136 ft. lbs. (184 Nm) on Supra or 134 ft. lbs. (181 Nm) on Cressida. Tighten the No. 2 arm-to-carrier bolt to 121 ft. lbs. (164 Nm) on Supra or 119 ft. lbs. (162 Nm) – Cressida. Tighten the No. 1 arm-to-carrier nut to 36 ft. lbs. (49 Nm).

9. Check the rear wheel alignment.

Upper Control Arm

REMOVAL & INSTALLATION

Supra and 1989–92 Cressida

1. Raise and support the rear of the vehicle safely. Remove the wheels.

2. Unbolt the brake caliper and suspend aside. Remove the halfshaft.

3. Disconnect the parking brake cable at the equalizer. Remove the 2 cable brackets from the body and then pull the cable through the suspension member.

4. Disconnect the 2 lower arms and the strut rod at the axle carrier. Disconnect the lower strut mount.

5. Disconnect the upper arm at the body and remove the axle hub assembly.

6. Remove the upper arm mounting nut. Remove the backing plate mounting nuts and separate the plate from the carrier. Press the upper arm out of the axle carrier.

To install:

7. Connect the upper arm to the body.

8. Connect the axle hub assembly to the arm with a new nut.

9. Connect the No. 1 lower control arm with a new nut and tighten it to 43 ft. lbs. (59 Nm) on Supra and 36 ft. lbs. (49 Nm) on Cressida. Connect the No. 2 lower arm and the strut rod.

10. Tighten the upper arm mounting nut to 80 ft. lbs. (108 Nm). Tighten the strut to 101 ft. lbs. (137 Nm).

11. Reconnect the parking brake cable and install the halfshaft. Install the brake caliper and tighten the bolts to 34 ft. lbs. (47 Nm).

12. Install the wheels and lower the vehicle. Bounce it several times to set the suspension and then tighten the upper arm-to-body bolt, the No. 2 lower arm-to-carrier and the strut rod to

121 ft. lbs. (164 Nm) on Supra and 119 ft. lbs. (162 Nm) on Cressida.

Upper and Lower Control Arms

REMOVAL & INSTALLATION

Tercel Wagon (AWD) and 1989–92 Corolla (AWD)

1. Raise and support the vehicle safely. Remove the wheels and support the rear axle safely.

2. Remove the upper control arm-to-body bolt. Remove the upper arm-to-axle bolt and lift out the upper control arm.

3. Remove the lower control arm-to-body bolt. Remove the lower arm-to-axle bolt and lift out the lower control arm.

To install:

4. Install the upper control arm with the nuts and bolts just snugged down.

5. Install the lower control arm with the nuts and bolts just snugged down.

6. Install the wheels, remove the safety stands and floor jack and then lower the vehicle.

7. Bounce the vehicle several times to stabilize the suspension and then raise the axle housing until the body is free.

8. Tighten all bolts to 83 ft. lbs. (113 Nm) on the Tercel and 72 ft. lbs. (98 Nm) on the Corolla.

Rear Wheel Bearings

NOTE: These procedures apply to front wheel drive vehicles only. For information on all rear wheel drive vehicles and all AWD vehicles, please refer to Rear Axle Shaft in the Drive Axle section.

REMOVAL & INSTALLATION

1988–89 Tercel Sedan

1. Raise and support the vehicle safely.

2. Remove the rear wheels.

3. Remove the brake drums.

4. Remove the locknut cap and cotter pin. Pry off the locknut and then remove the locknut itself.

5. Pull off the axle hub along with the outer wheel bearing and thrust washer.

6. Pry the inner bearing oil seal out of the brake drum and then remove the inner bearing.

7. Using a brass drift and hammer, drive out the bearing races.

To install:

8. Press new outer bearing races

into the axle hub and fill it and the bearing cap with grease.

9. Pack the bearing with grease.

10. Position the inner bearing into the hub and then drive in a new oil seal. Coat the seal with grease.

11. Position the axle hub/brake drum onto the axle shaft. Install the outer bearing, fill the hole with grease and position the thrust washer. Install the bearing locknut and tighten it to 22 ft. lbs. (29 Nm).

12. Spin the axle hub several times to snug down the bearing and then loosen the bearing locknut until it can be turned by hand.

NOTE: There must be absolutely no brake drag at this time.

13. Retighten the bearing locknut until there is a bearing preload of 0.9–2.2 lbs. (3.2–9.8 N) while turning the wheel.

14. Install the locknut lock, a new cotter pin and the cap. If the cotter pin hole does not align properly, align the holes by tightening the nut.

15. Lower the vehicle.

Camry (2WD), Celica (2WD) Corolla (2WD) and 1990–92 Tercel

1. Raise and support the vehicle safely.

2. Remove the rear wheel and tire assembly.

3. Remove the brake drum. On the Corolla FX, remove the disc brake caliper from the axle carrier and suspend it with a wire.

4. Disconnect and plug the brake line at the backing plate.

5. Remove the 4 axle hub-to-carrier bolts and slide off the hub and brake assembly. Remove the O-ring from the backing plate.

6. Remove the bolt and nut attaching the carrier to the strut rod.

7. Remove the bolt and nut attaching the carrier to the No. 1 suspension arm.

8. Remove the bolt and nut attaching the carrier to the No. 2 suspension arm.

9. Unbolt the carrier from the rear strut tube and remove the carrier.

10. Using a hammer and cold chisel, loosen the staked part of the hub nut and remove the nut.

11. Using a 2-armed puller or the like, press the axle shaft from the hub.

12. Remove the bearing inner race (inside).

13. Using a 2-armed puller again, pull off the bearing inner race (outside) over the bearing and then press it out of the hub.

To install:

14. Position a new bearing inner race (outside) on the bearing and then

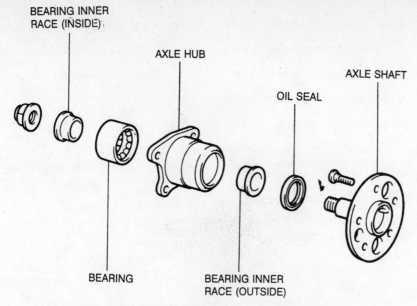

BEARING INNER RACE (INSIDE)

AXLE HUB

OIL SEAL

AXLE SHAFT

BEARING

BEARING INNER RACE (OUTSIDE)

Pressed axle hub bearing assembly

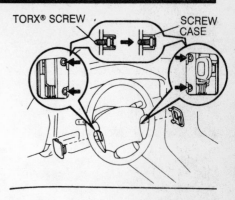

TORX® SCREW

SCREW CASE

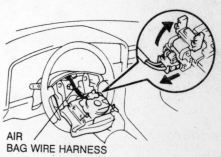

AIR BAG WIRE HARNESS

Removing the air bag

press a new oil seal into the hub. Coat the lip of the seal with grease.

15. Position a new bearing inner race (inside) on the bearing and then press the inner race with the hub onto the axle shaft.

16. Install the nut and tighten it to 90 ft. lbs. (123 Nm). Stake the nut with a brass drift.

17. Position the axle carrier on the strut tube and tighten the nuts to 119 ft. lbs. (162 Nm), 105 ft. lbs. (142 Nm) on Corolla FX and Corolla sedan and wagon; 166 ft. lbs. (226 Nm) on Camry.

18. Install the bolt and nut attaching the carrier to the No. 2 suspension arm; finger-tighten it only.

NOTE: Make sure the lip of the nut is in the hole on the arm.

19. Repeat Step 5 for the No. 1 suspension arm.

NOTE: Make sure the lip of the nut is in the hole on the arm.

20. Install the strut rod-to-carrier bolt so the lip of the nut is in the groove on the bracket.

21. Install a new O-ring onto the axle carrier. Install the axle hub and brake backing plate. Tighten the 4 bolts to 59 ft. lbs. (80 Nm).

22. Reconnect the brake line, install the brake drum and then bleed the brakes.

23. Lower the vehicle and bounce it a few times to set the rear suspension.

24. Tighten the suspension arm bolts and the strut rod bolt to 64 ft. lbs. (87 Nm). On the Camry and Celica, tighten the strut rod mounting bolts to 83 ft. lbs. (113 Nm) and the suspension arm bolts to 134 ft. lbs. (181 Nm).

STEERING

Steering Wheel

CAUTION

On vehicles equipped with an air bag, the negative battery cable must be disconnected, before working on the system. Failure to do so may result in deployment of the air bag and possible personal injury.

REMOVAL & INSTALLATION

Without Air Bag

CAUTION

Do not attempt to remove or install the steering wheel by hammering on it. Damage to the energy-absorbing steering column could result.

1. Disconnect the negative battery cable. Position the front wheels straight ahead.

2. Unfasten the horn and turn signal multi-connector(s) at the base of the steering column shroud.

3. If equipped with a 3 spoked wheel, loosen the trim pad retaining screws from the back side of the steering wheel. The 2 spoke steering wheel is removed in the same manner as the three spoke, except that the trim pad should be pried off with a small prybar. Remove the pad by lifting it toward the top of the wheel.

4. Lift the trim pad and horn button assembly from the wheel.

5. Remove the steering wheel hub retaining nut.

6. Scribe matchmarks on the hub and shaft to aid in correct installation.

7. Use a suitable puller to remove the steering wheel.

8. Installation is the reverse of removal. Tighten the wheel retaining nut to 25 ft. lbs. (34 Nm).

With Air Bag

CAUTION

Air bag equipped vehicles: Work must be started after 20 seconds or longer from the time the ignition switch is turned to the LOCK and the negative battery terminal is disconnected. If the air bag system is disconnected with the ignition switch at the ON or ACC, diagnostic codes will be recorded.

1. Disconnect the negative battery cable.

2. Place the front wheels facing straight ahead.

3. Remove the steering wheel screw covers.

4. Using a Torx® wrench T30, loosen the screws until the groove trailing the screw circumference catches on the screw case.

5. Pull the wheel pad out from the steering wheel and disconnect the air bag connector.

6. Remove the steering wheel nut. Place matchmarks on the wheel and steering shaft.

7. Using a steering wheel puller SST 09213-31021 or equivalent, remove the steering wheel.

To install:

8. Install the steering wheel, align the matchmarks and torque the nut to 25 ft. lbs. (34 Nm).

9. Connect the air bag connector and install the steering pad.

10. Torque the Torx® screws to 65 inch lbs. (7.4 Nm).

11. Install the screw covers.

12. Connect the battery cable and check operation.

Steering Column
REMOVAL & INSTALLATION
———— CAUTION ————

Air bag equipped vehicles: Work must be started after 20 seconds or longer from the time the ignition switch is turned to the LOCK and the negative battery terminal is disconnected. If the air bag system is disconnected with the ignition switch at the ON or ACC, diagnostic codes will be recorded.

1. Disconnect the negative battery cable.

2. If equipped, remove the universal joint at the steering gear and at the main shaft of the steering column assembly.

3. If equipped, disconnect upper universal joint from the intermediate shaft.

4. Remove the steering wheel.

5. Remove the instrument lower finish panels, air ducts and column covers.

6. Disconnect all electrical connections for ignition switch and combination switch. Remove the combination switch, as necessary.

7. Loosen column hole cover clamp bolt. Remove the support mounting bolt. Remove the 4 column tube mounting bolts. Pull out steering column. On some vehicles, remove the bolts from the column hole cover plate and remove 2 column bracket mounting nuts. Turn the steering column assembly clockwise and remove it from the vehicle, as necessary.

To install:

8. Place the the steering column assembly in the installed position. Tighten all necessary mounting nuts (torque evenly). Install and tighten all column cover bolts. Tighten column hole cover clamp bolt, as necessary.

9. Install combination switch. Reconnect all electrical connectors.

10. Install instrument lower finish panels, air duct and column covers.

11. Install or connect the universal joint. Insure that the retaining bolts are installed through both shaft grooves.

12. Install steering wheel and connect the negative battery cable.

Manual and Power Steering Rack
REMOVAL & INSTALLATION
Corolla (RWD)

1. Raise and support the vehicle

safely. Remove the front wheels. Remove the bolt attaching the coupling yoke (U-joint) to the steering worm.

2. Disconnect the relay rod from the pitman arm. Disconnect the cotter pin and nut holding the knuckle arm to the tie rod.

3. If equipped with power steering, remove the front exhaust pipe, disconnect and plug the hydraulic lines and wire them aside.

4. Remove the gear housing bracket set bolts and remove the steering gear housing down and to the left.

5. Install in reverse of removal. Torque the housing-to-frame bolts to 25–36 ft. lbs. (34–49 Nm); the coupling yoke bolt to 26 ft. lbs. (35 Nm) and the relay rod to 36–50 ft. lbs. (49–68 Nm).

Cressida

1. Raise and support the vehicle safely and remove the front wheels. Place matchmarks on the coupling and steering column shaft. Disconnect the solenoid connectors.

2. Disconnect the Pitman arm from the relay rod using a tie rod puller on the Pitman arm set nut. Disconnect the tie rod ends from the steering knuckles.

3. Remove the steering damper, if equipped.

4. Disconnect the steering gearbox at the coupling. Unbolt the gearbox from the chassis and remove. Remove the grommets from the gear housing.

5. Installation is in the reverse order of removal, with the exception of first aligning the matchmarks and connecting the steering shaft to the coupling before bolting the gearbox into the vehicle permanently. Tighten the steering damper bolts to 20 ft. lbs. (26 Nm). Tighten the tie rod ends to 43 ft. lbs. (59 Nm). Tighten the mounting bracket bolts to 56 ft. lbs. (76 Nm).

Corolla (FWD), Camry and Supra

1. Raise and support the vehicle safely. Remove the front wheels. Open the hood. Remove the 2 set bolts, and remove the sliding yoke from between the steering rack housing and the steering column shaft. On Supra, unbolt and remove the intermediate shaft (rack housing side first).

2. Remove the cotter pin and nut holding the knuckle arm to the tie rod end. Using a tie rod puller, disconnect the tie rod end from the knuckle arm.

3. On Corolla and Camry with power steering, remove the lower crossmember, remove the engine under cover, center engine mount member and the rear engine mount.

4. On Supra with turbocharger, remove the No. 1 air intake connector with No. 2 air hose.

5. Disconnect the power steering

lines, if equipped. Remove the steering gear housing brackets. Slide the gear housing to the right side and then to the left side to remove the housing.

To install:

6. Install the rack housing and torque to the following specifications.

7. Torque the rack housing mounting bolts to 29–39 ft. lbs. (39–53 Nm) on the Celica and Supra; 43 ft. lbs. (58 Nm) on the Corolla and Camry, and the tie rod set nuts to 37–50 ft. lbs. (50–68 Nm) on Celica and Supra; 36 ft. lbs. (49 Nm) on the the Corolla and Camry.

8. Use a new cotter pin. On Supras, install the intermediate shaft column side first, then rack side. On Corollas with power steering, tighten the rear engine mount bolts to 29 ft. lbs. (39 Nm) On Camry tighten the rear engine mounting bolts to 38 ft. lbs. (52 Nm). Tighten the center mounting member to 29 ft. lbs. (39 Nm).

9. On power steering-equipped vehicles, bleed the power steering system and check for fluid leaks.

10. Adjust toe-in on all vehicles.

Celica

1. Raise and support the vehicle safely. Remove the front wheels.

2. Remove the both engine under covers.

3. Remove the 2 bolts that connect the steering column U-joint to the rack and then disconnect the column from the rack.

4. Remove the cotter pin and nut and then using a tie rod end removal tool, disconnect the tie rod end from the steering knuckle.

5. Remove the lower suspension crossmember.

6. Remove the mounting bolts and remove the center engine mount member.

7. Disconnect the exhaust pipe from the manifold. Position it aside.

8. Tag and disconnect the 2 hydraulic lines. Position them aside and suspend on a wire.

9. Remove the rear engine mount bracket.

10. Remove the mounting bolts and brackets and lower steering rack from the vehicle.

To install:

11. Position the rack assembly, install the grommets and brackets and then tighten the 2 bolts and 2 nuts to 43 ft. lbs. (59 Nm).

12. Install the rear engine mount bracket and tighten the 2 bolts to 38 ft. lbs. (52 Nm).

13. Connect the hydraulic lines and tighten the union nuts to 29 ft. lbs. (39 Nm).

14. Connect the exhaust pipe to the manifold.

15. Install the center engine mount member and tighten the bolts to 29 ft. lbs. (39 Nm).

16. Install the lower crossmember and tighten the 5 outer bolts to 154 ft. lbs. (208 Nm). Tighten the center bolts to 29 ft. lbs.

17. Installation of the remaining components is in the reverse order of removal. Tighten the tie rod end nuts to 36 ft. lbs. (49 Nm) and use a new cotter pin. Tighten the steering column U-joint bolts to 26 ft. lbs. (35 Nm). Fill the power steering pump to the proper level, bleed the system and check the wheel alignment.

1988–89 MR2 and Tercel

1. Raise and support the vehicle safely.

2. Remove the front wheels and the engine under cover.

3. Place matchmarks on the main shaft, joint yoke and pinion shaft. Remove the intermediate shaft from the worm gear shaft.

4. If equipped with power steering, disconnect the 2 hydraulic lines. Position them aside and suspend on a wire.

5. Remove both tie rod ends.

6. Remove the lower suspension crossmember. Remove the center floor crossmember.

7. Remove the rack housing bracket mounting bolts and brackets.

NOTE: Be careful not to damage the rubber boots.

8. Remove the steering linkage.

9. Installation is the reverse of the removal procedure. Fill the power steering pump to the proper level and bleed the system. Check the wheel alignment.

1990–92 MR2

1. Disconnect the negative battery cable.

2. Position the front wheels straight ahead.

3. Raise the vehicle and support safely. Remove the front wheels and front luggage under cover.

4. Remove the dust cover. Matchmark the universal joint and control valve shaft for installation.

5. Disconnect the tie rod ends using a tie rod separator or equivalent.

6. Disconnect the power steering hoses and drain the fluid into a container.

7. Remove the housing-to-frame retaining bolts and remove the assembly.

To install:

8. Install the assembly and torque the retaining bolts to 32 ft. lbs. (43 Nm).

9. Connect the tie rods and torque to 36 ft. lbs. (49 Nm).

10. Connect the universal joint and torque to 26 ft. lbs. (35 Nm).

11. Install the dust cover and connect the power steering hoses.

12. Connect the battery cable. Start the engine and bleed the system.

13. Install the front wheels and lower the vehicle.

Power Steering Pump

REMOVAL & INSTALLATION

Tercel, Corolla, Camry, Celica, 1988–89 MR2 and 1988 Cressida

1. Raise and support the vehicle safely. Remove the fan shroud.

2. On Camry and Celica, remove the right front wheel and the engine under cover. Remove the lower suspension crossmember.

3. Unfasten the nut from the center of the pump pulley. Disconnect the vacuum hose from the air control valve, if equipped.

NOTE: Use the drive belt as a brake to keep the pulley from rotating.

4. Withdraw the drive belt. On some vehicles, it may be necessary to remove the pulley in order to remove the drive belt.

5. If equipped with an idler pulley and on the Corolla FX, push on the drive belt to hold the pulley in place and remove the pulley set nut. Loosen the idler pulley set nut and adjusting bolt. Remove the drive belt and loosen the drive pulley to remove the Woodruff key.

6. Remove the pulley and the Woodruff key from the pump shaft.

7. Detach and plug the intake and outlet hoses from the pump reservoir.

NOTE: Tie the hose ends up high so the fluid cannot flow out of them. Drain or plug the pump to prevent fluid leakage.

8. Remove the bolt from the rear mounting brace.

9. Remove the front bracket bolts and withdraw the pump.

To install:

10. Tighten the pump pulley mounting bolt to 25–39 ft. lbs. (34–53 Nm).

11. Tighten the 5 outer mounting bolts on the lower crossmember to 154 ft. lbs. (209 Nm). On Celica, tighten the center bolt to 29 ft. lbs. (39 Nm).

12. Adjust the pump drive belt tension. The belt should deflect 0.13–0.93 in. under thumb pressure applied midway between the air pump and the power steering pump.

13. Fill the reservoir with Dexron® II automatic transmission fluid. Bleed the air from the system.

Supra and 1989–92 Cressida

7M-GE ENGINE

1. Raise and support the vehicle safely. Drain the fluid from the reservoir tank.

2. Disconnect the air hose from the air control tank. Disconnect the return hose from the reservoir tank.

3. Remove the engine under cover. Disconnect and plug the pressure hose from the power steering pump.

4. Holding the power steering pump pulley, remove the pulley set nut. Remove the drive belt adjusting nut.

5. Remove the power steering pump set bolt. Remove the drive belt, pulley and Woodruff key.

6. Disconnect the oil cooler hose bracket from the power steering pump. Remove the drive belt adjust bolt and remove the power steering set bolt and power steering pump.

7. Installation is the reverse order of the removal procedure. Be sure to bleed the system upon completion of the installation procedure.

7M-GTE ENGINE

1. Raise and support the vehicle safely. Drain the fluid from the reservoir tank.

2. Remove the No. 1 and No. 2 air hoses with the No. 4 air cleaner pipe.

3. Disconnect the connector from the air flow meter. Remove the air flow meter installation bolt. Loosen the 5 clamps and disconnect the air hoses, release the 3 clips on the air cleaner case. Loosen the No. 7 air hose clamp and remove the No. 7 air cleaner hose with the air flow meter.

4. Remove the oil reservoir tank with bracket. Disconnect the 2 air hoses from the air control valve on the power steering pump.

5. Remove the adjusting strut. Remove the engine under cover.

6. Holding the power steering pump pulley, remove the pulley set nut. Remove the drive belt adjusting nut.

7. Remove the power steering pump set bolt. Remove the drive belt, pulley and Woodruff key.

8. Disconnect and plug the pressure hose from the power steering pump.

9. Remove the power steering set bolt and power steering pump.

10. Installation is the reverse order of the removal procedure. Be sure to bleed the system upon completion of the installation procedure.

1990–92 MR2

The power steering pump is not driven by a conventional drive belt. Instead the pump is driven by an electric motor. The pump and motor are combined as one unit. Removal and installation is as follows:

1. Disconnect the negative battery cable.

2. Raise and support the vehicle safely.

3. Remove the front luggage under cover.

4. Remove the pump shield and rear stay.

5. Disconnect the hydraulic lines from the pump. Plug the lines to prevent the loss of power steering fluid.

6. Disconnect the electrical wires from the top of the motor.

7. Remove the pump mounting bolts, bushings and spacers. Check the bushings for cracks and deformation. Replace as necessary.

8. Remove the pump and motor assembly.

To install:

9. Position the pump and install the mounting bolts, bushings and spacers. Torque the bolts to 19 ft. lbs. (25 Nm).

10. Connect the electrical wires to the top of the motor.

11. Connect the hydraulic lines.

12. Install the rear stay and pump shield.

13. Install the front luggage under cover.

14. Lower the vehicle and connect the negative battery cable.

15. Fill the power steering reservoir to the proper level and bleed the system.

BELT ADJUSTMENT

1. Inspect the power steering drive belt to see that it is not cracked or worn. Be sure its surfaces are free of grease or oil.

2. Push down on the belt halfway between the fan and the alternator pulleys (or crankshaft pulley) with thumb pressure. Belt deflection should be $\frac{3}{8}$–$\frac{1}{2}$ in. (10–13mm).

3. If the belt tension requires adjustment, loosen the adjusting link bolt and move the power steering pump until the proper belt tension is obtained.

4. Do not over-tighten the belt, as damage to the power steering pump bearings could result. Tighten the adjusting link bolt.

5. Drive the vehicle and re-check the belt tension. Adjust as necessary.

SYSTEM BLEEDING

1. Raise and support the vehicle safely.

2. Fill the pump reservoir with the proper fluid.

3. Rotate the steering wheel from lock-to-lock several times. Add fluid if necessary.

4. With the steering wheel turned fully to one lock, crank the starter while watching the fluid level in the reservoir.

NOTE: Do not start the engine. Operate the starter with a remote starter switch or have an assistant do it from inside the vehicle. Do not run the starter for prolonged periods.

5. Repeat Step 4 with the steering wheel turned to the opposite lock.

6. Start the engine. With the engine idling, turn the steering wheel from lock-to-lock several times.

7. Lower the front of the vehicle and repeat Step 6.

8. Center the wheel at the midpoint of its travel. Stop the engine.

9. The fluid level should not have risen more than 0.2 in. (5mm). If it does, repeat Step 7.

10. Check for fluid leakage.

Tie Rod Ends

REMOVAL & INSTALLATION

1. Scribe alignment marks on the tie rod and rack end.

2. Working at the steering knuckle arm, pull out the cotter pin and then remove the castellated nut.

3. Using a tie rod end puller, disconnect the tie rod from the steering knuckle arm.

4. Repeat the first 2 steps on the other end of the tie rod (where it attaches to the relay rod or steering rack).

To install:

5. Align the alignment marks on the tie rod and rack end.

6. Install the tie rod end.

7. Tighten the tie rod end nuts to 36 ft. lbs. (49 Nm). Install a new cotter pin.

8. Install the front wheels and lower the vehicle.

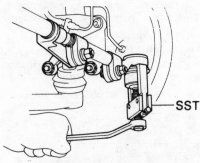

Typical tie rod end removal

BRAKES

For all brake system repair and service procedures not detailed below, please refer to "Brakes" in the Unit Repair section.

Master Cylinder and ABS Actuator

REMOVAL & INSTALLATION

1. Disconnect the negative battery cable. Label and disconnect the electrical connectors.

2. Remove the fluid in the master cylinder with a suitable syringe.

3. On MR2, remove the luggage compartment.

4. Disconnect the hydraulic lines from the master cylinder. Plug the ends of the lines to prevent loss of fluid.

5. Detach the hydraulic fluid pressure differential switch wiring connectors.

6. Loosen the master cylinder reservoir mounting nuts.

7. Unfasten the nuts and remove the master cylinder or ABS actuator assembly from the power brake unit.

To install:

8. Before tightening the master cylinder mounting nuts or bolts, screw the hydraulic line into the cylinder body a few turns.

9. Install the master cylinder or actuator. Torque the hydraulic lines to 11 ft. lbs. (15 Nm) and the mounting nuts to 25 ft. lbs. (34 Nm).

10. After installation is completed, bleed the master cylinder and the brake system.

Proportioning Valve

A proportioning valve is used to reduce the hydraulic pressure to the rear brakes because of weight transfer during high speed stops. This helps to keep the rear brakes from locking up by improving front to rear brake balance.

REMOVAL & INSTALLATION

1. Disconnect the brake lines from the valve unions.

2. Remove the valve mounting bolt, if used, and remove the valve.

NOTE: If the proportioning valve is defective, it must be replaced as an assembly; it cannot be rebuilt.

3. Installation is the reverse of removal. Bleed the brake system after it is completed.

Power Brake Booster

REMOVAL & INSTALLATION

1. Disconnect the negative battery cable. Remove the master cylinder and disconnect the vacuum hose from the brake booster.

2. Remove the instrument lower finish panel, as required.

3. On the MR2, remove the wheel guard, instrument lower finish panel and air duct.

4. Remove the brake pedal return spring.

5. Remove clip and clevis pin.

6. Remove the brake booster nuts and clevis pin.

7. Pull out the brake booster and gasket.

8. Installation is the reverse order of the removal procedure. Bleed the brake system. Torque the nuts to 25 ft. lbs. (34 Nm).

Brake Caliper

REMOVAL & INSTALLATION

1. Raise and support the vehicle safely.

2. Remove the front or rear wheels.

3. Disconnect the brake hose from the caliper. Plug the end of the hose to prevent loss of fluid.

4. Remove the bolts that attach the caliper to the torque plate.

5. Lift up and remove the caliper assembly.

6. Installation is the reverse of the removal procedure. Grease the caliper slides and bolts with Lithium grease or equivalent. Torque the caliper bolts to 20–27 ft. lbs. (27–41 Nm) on front disc brakes and 14 ft. lbs. (20 Nm) for rear disc brakes. Fill and bleed the system.

Disc Brake Pads

REMOVAL & INSTALLATION

MR2 – Rear, Cressida – Front and Rear and 1988–92 Celica (AWD) – Front

1. Raise and support the vehicle safely.

2. Remove the wheels.

3. Siphon a sufficient quantity of brake fluid from the master cylinder reservoir to prevent any brake fluid from overflowing the master cylinder when removing or installing new pads. This is necessary as the piston must be forced into the caliper bore to provide sufficient clearance when installing the pads.

4. Grasp the caliper from behind and carefully pull it to seat the piston in its bore.

5. Loosen and remove the lower caliper slide pin (mounting bolt).

6. Swivel the caliper upward and aside, exposing the brake pads. Do not disconnect the brake line.

7. Slide out the old brake pads along with any anti-squeal shims, anti-rattle springs, pad wear indicators, pad guide

plates and pad support plates. Take great care to note the position of all assorted pad hardware.

8. Check the brake disc (rotor) for thickness and run-out. Inspect the caliper and piston assembly for breaks, cracks, fluid seepage or other damage. Overhaul or replace as necessary.

To install:

9. Install the pad support plates, anti-rattle springs or guide plates into the torque plate.

10. Install the pad wear indicators onto the pad.

11. Install the anti-squeal shims on the outside of each pad and then install the pad assemblies into the torque plate.

12. Swivel the caliper back down over the pads. If it won't fit, use a C-clamp or hammer handle and carefully force the piston into its bore.

13. On the MR2 rear disc, turn the piston clockwise while pushing it in until it locks in place. Fit the protrusion on the inner pad into the groove in the piston stopper and then install the caliper. Be careful not to pinch the boot.

14. Install and tighten the lower slide pin or mounting bolt.

15. Install the parking brake cable on the Corolla with rear discs and then adjust the automatic adjuster by pulling and releasing the parking brake lever several times.

16. Install the wheel and lower the vehicle. Check the brake fluid level. Adjust the parking brake on rear disc brake vehicles.

MR2 – Front, Tercel – Front
Camry – Front and Rear
Supra – Front and Rear
1988–92 Celica (AWD) – Rear
Celica (2WD) – Front and Rear
Corolla (FWD) – Front and Rear

1. Raise and support the vehicle safely.

2. Remove the wheels.

3. Siphon a sufficient quantity of brake fluid from the master cylinder reservoir to prevent any brake fluid from overflowing the master cylinder when removing or installing new pads. This is necessary as the piston must be forced into the caliper bore to provide sufficient clearance when installing the pads.

4. Grasp the caliper from behind and carefully pull it to seat the piston in its bore.

5. Loosen and remove the 2 caliper mounting pins (bolts) and then remove the caliper assembly. Position it aside. Do not disconnect the brake line.

6. Slide out the old brake pads along with any anti-squeal shims, springs, pad wear indicators and pad support plates. Make sure to note the position of all assorted pad hardware.

To install:

7. Check the brake disc (rotor) for thickness and run-out. Inspect the caliper and piston assembly for breaks, cracks, fluid seepage or other damage. Overhaul or replace as necessary.

8. Install the pad support plates into the torque plate.

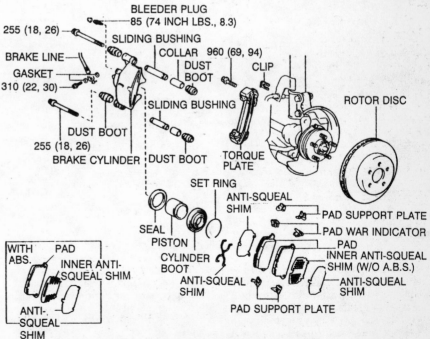

Front disc brake assembly – Celica 2WD

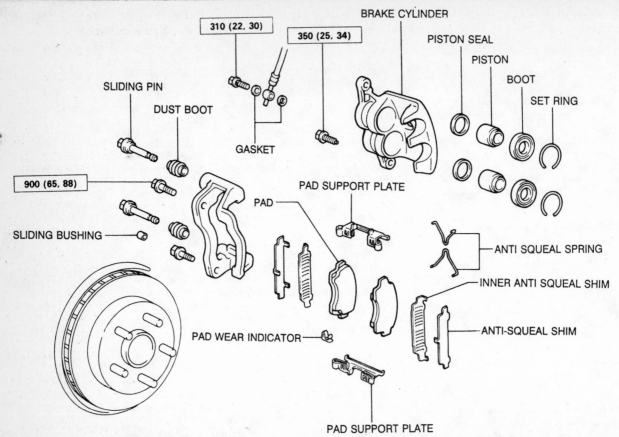

Front disc brake assembly—1990–92 MR2 with 3S-GTE engines

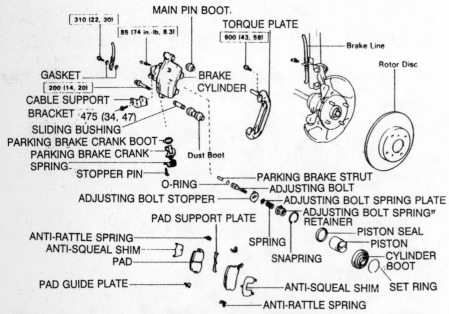

Rear disc brake assembly—MR2

Brake Rotor

REMOVAL & INSTALLATION

1. Raise and support the vehicle safely.
2. Remove the wheels.
3. Temporarily attach 2 lug nuts onto the rotor disc.
4. Unbolt the torque plate from the steering knuckle.
5. Remove the lug nuts and pull the rotor from the wheel hub.
6. Installation is the reverse of the removal procedure.

Brake Drums

REMOVAL & INSTALLATION

1. Raise and support the vehicle safely.
2. Remove the wheels.
3. Remove the brake drum. Tap the drum lightly with a rubber mallet in order to free it. If the brake drum cannot be removed easily, insert a small prybar through the hole in the backing plate and hold the automatic adjuster lever away from the adjusting bolt. Using another prybar, relieve the brake shoe tension by rotating the adjusting bolt (star wheel) in a clockwise direction. If the drum still will not come off,

9. Install the pad wear indicators onto the pads. Be sure the arrow on the indicator plate is pointing in the direction of rotation.
10. Install the anti-squeal shims on the outside of each pad and then install the pad assemblies into the torque plate.

11. Position the caliper back down over the pads. If it won't fit, use a C-clamp or hammer handle and carefully force the piston into its bore.
12. Install and tighten the caliper mounting bolts.
13. Install the wheels and lower the vehicle. Check the brake fluid level.

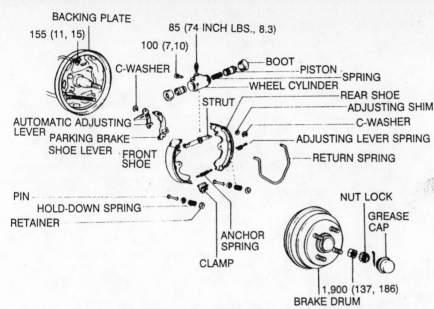

BACKING PLATE
155 (11, 15)
85 (74 INCH LBS., 8.3)
100 (7,10)
C-WASHER
BOOT — PISTON
SPRING
WHEEL CYLINDER
REAR SHOE
ADJUSTING SHIM
STRUT
AUTOMATIC ADJUSTING LEVER
C-WASHER
PARKING BRAKE SHOE LEVER
ADJUSTING LEVER SPRING
FRONT SHOE
RETURN SPRING
PIN
HOLD-DOWN SPRING
RETAINER
NUT LOCK
GREASE CAP
ANCHOR SPRING
CLAMP
1,900 (137, 186)
BRAKE DRUM

Exploded view of the rear drum brake—Tercel sedan

use a puller; but, first make sure the parking brake is released.

4. Installation is the reverse of the removal procedure.

Brake Shoes

REMOVAL & INSTALLATION

Tercel Sedan, Corolla (RWD) 1988 Cressida

1. Raise and support the vehicle safely. Remove the wheels.

2. Remove the brake drums.

NOTE: Do not depress the brake pedal once the brake drum has been removed.

3. Carefully unhook the tension spring from the leading (front) brake shoe. On the Tercel, its a return spring; also remove the clamp.

4. Press the hold-down spring retainer in and turn the pin.

5. Remove the hold-down spring, retainers and the pin. Pull out the brake shoe and unhook the anchor spring from the lower edge.

6. Remove the hold-down spring from the trailing (rear) shoe. Pull the shoe out with the adjuster strut, automatic adjuster assembly and springs attached and disconnect the parking brake cable. Remove the tension/return and anchor springs from the rear shoe.

7. Remove the adjusting strut. Unhook the adjusting lever spring from the rear shoe and then remove the automatic adjuster assembly by popping out the C-clip.

To install:

8. Inspect the shoes for signs of unusual wear or scoring.

9. Check the wheel cylinder for any sign of fluid seepage or frozen pistons.

10. Clean and inspect the brake backing plate and all other components. Check that the brake drum inner diameter is within specified limits. Lubricate the backing plate bosses and the anchor plate.

11. Mount the automatic adjuster assembly onto a new rear brake shoe. Make sure the C-clip fits properly. Connect the adjusting strut and install the spring.

12. Connect the parking brake cable to the rear shoe and then position the shoe so the lower end rides in the anchor plate and the upper end is against the boot in the wheel cylinder. Install the pin and the hold-down spring. Press the retainer down over the pin and rotate the pin so the crimped edge is held by the retainer. Install the anchor spring between the front and rear shoes and then stretch the spring enough so the front shoe will fit as the rear did in Step 10. Install the hold-down spring, pin and retainer. Stretch the tension/return spring between the 2 shoes and connect it so it rides freely. Don't forget the return spring clamp on the Tercel.

13. Check that the automatic adjuster is operating properly; the adjusting bolt should turn when the parking brake lever, in the brake assembly, not in the vehicle!, is moved. Adjust the strut as short as possible and then install the brake drum. Set and release the parking brake several times.

14. Install the wheel and lower the vehicle. Check the level of brake fluid in the master cylinder.

Camry, Celica, Tercel Wagon and Corolla (FWD)

1. Raise and support the vehicle safely. Remove the wheels.

2. Remove the brake drums.

NOTE: Do not depress the brake pedal once the brake drum has been removed.

3. Carefully unhook the return spring from the leading (front) brake shoe. Grasp the hold-down spring pin with pliers and turn it until its in line with the slot in the hold-down spring. Remove the hold-down spring and the pin. Pull out the brake shoe and unhook the anchor spring from the lower edge.

4. Remove the hold-down spring from the trailing (rear) shoe. Pull the shoe out with the adjuster strut, automatic adjuster assembly and springs attached and disconnect the parking brake cable. Unhook the return spring and then remove the adjusting strut. Remove the anchor spring.

5. Remove the adjusting strut. Unhook the adjusting lever spring from the rear shoe and then remove the automatic adjuster assembly by popping out the C-clip.

To install:

6. Inspect the shoes for signs of unusual wear or scoring.

7. Check the wheel cylinder for any sign of fluid seepage or frozen pistons.

8. Clean and inspect the brake backing plate and all other components. Check that the brake drum inner diameter is within specified limits. Lubricate the backing plate bosses and the anchor plate.

9. Mount the automatic adjuster assembly onto a new rear brake shoe. Make sure the C-clip fits properly. Connect the adjusting strut/return spring and then install the adjusting spring.

10. Connect the parking brake cable to the rear shoe and then position the shoe so the lower end rides in the anchor plate and the upper end is against the boot in the wheel cylinder. Install the pin and the hold-down spring. Rotate the pin so the crimped edge is held by the retainer.

11. Install the anchor spring between the front and rear shoes and then stretch the spring enough so the front shoe will fit as the rear did in Step 10. Install the hold-down spring and pin. Connect the return spring/adjusting strut between the 2 shoes and connect it so it rides freely.

12. Check that the automatic adjuster is operating properly; the adjusting bolt should turn when the parking brake lever (in the brake assembly, not in the vehicle!) is moved. Adjust the strut as short as possible and then in-

stall the brake drum. Set and release the parking brake several times.

13. Install the wheel and lower the vehicle. Check the level of brake fluid in the master cylinder.

Wheel Cylinder

REMOVAL & INSTALLATION

1. Plug the master cylinder inlet to prevent hydraulic fluid from leaking.
2. Remove the brake drums and shoes.
3. Working from behind the backing plate, disconnect the hydraulic line from the wheel cylinder.
4. Unfasten the screws retaining the wheel cylinder and withdraw the cylinder.

To install:

5. Installation is performed in the reverse order of removal. However, once the hydraulic line has been disconnected from the wheel cylinder, the union seat must be replaced. To replace the seat, proceed in the following manner:

a. Use a screw extractor with a diameter of 0.1 in. or equivalent, having reverse threads, to remove the union seat from the wheel cylinder.

b. Drive in the new union seat with a $5/16$ in. bar, used as a drift.

6. Bleed the brake system after completing wheel cylinder, brake shoe and drum installation.

Parking Brake Cable

ADJUSTMENT

1. Slowly pull the parking brake lever upward, without depressing the button on the end of it, while counting the number of notches required until the parking brake is applied.

NOTE: Two "clicks" are equal to 1 notch.

2. Check the number of notches against the following specifications.
　　Tercel (2WD)—5–8 clicks.
　　1988–92 Tercel (2WD)—7–9 clicks.
　　1988 Tercel (AWD)—6–8 clicks.
　　1988–90 Celica—4–7 clicks.
　　1988–90 Corolla FWD—4–7 clicks (FX16 and vehicles w/rear disc—5–8).
　　MR2—5–8 clicks.
　　Cressida—5–8 clicks.
　　Camry—5–8 clicks.
　　Supra—5–8 clicks.

3. If the brake system requires adjustment, loosen the cable adjusting nut cap which is located at the rear of the parking brake lever.

NOTE: On some vehicles, the adjustment and lock nuts are lo- cated under the vehicle, beneath the lever assembly.

4. Take up the slack in the parking brake cable by rotating the adjusting nut with another open end wrench.

a. If the number of notches is less than specified, turn the nut counterclockwise.

b. If the number of notches is more than specified, turn the nut clockwise.

5. Tighten the adjusting cap, using care not to disturb the setting of the adjusting nut.

6. Check the rotation of the rear wheels to be sure the brakes are not dragging.

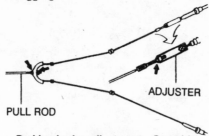

Parking brake adjustment—Cressida

Brake System Bleeding

Master Cylinder

1. Check the fluid level in the master cylinder. Add fluid as necessary. Never allow the master cylinder to run dry.
2. Disconnect the brake tubes from the master cylinder.
3. Slowly depress the brake pedal and hold it.
4. Close off the outlet opening on the master cylinder with finger pressure and release the brake pedal.
5. Repeat Steps 3 and 4 several times to bleed all the air from the master cylinder.

Brake Lines

1. Bleed the caliper or wheel cylinder with the longest hydraulic line. Connect a vinyl tube the to the bleeder screw on the brake cylinder and submerge the other end of the tube in a transparent container half filled with clean brake fluid.
2. Pump the brake pedal several times and loosen the bleeder screw with the pedal held down.
3. When brake fluid stops coming out of the tube, tighten the bleeder screw and release the brake pedal.
4. Repeat Steps 2 and 3 until no air bubbles can be seen in the container.
5. Repeat the procedure for each wheel.
6. Check the level in the master cylinder. Add fluid as necessary.

Anti-Lock Brake System

PRECAUTIONS

● When welding with an electric welding unit, unplug the electric control unit.
● During paint jobs, the electronic control unit may be exposed to a maximum of 203°F (95°C) for up to 2 hours or 185°F (85°C) if more time is needed.
● When removing the rear axle centerpiece, make sure the correct toothed wheel make with the correct ratio for the wheel speed sensor is installed. If a wheel with the wrong number of teeth is installed, this fault will not show up when checking the system with the ABS tester. The stopping distance, however, will be increased during controlled braking.
● If work was done to non-ABS brake components, a simple operational test will be sufficient. This means that after driving about 5 mph, the warning light on the instrument panel should go out if the ABS system is intact.
● If ABS components have been replaced, the entire system should be checked using the appropriate tester in combination with brake test bench or an adaptor in combination with a multimeter.

RELIEVING ANTI–LOCK BRAKE SYSTEM PRESSURE

Pump the brake pedal at least 20 times with the ignition key in the OFF position. Place a shop rag around the hydraulic line fitting and wear safety glasses when disconnecting the hydraulic lines.

Anti-Lock Brake Actuator

REMOVAL & INSTALLATION

1. Disconnect the negative battery cable.
2. Remove the brake fluid from the master cylinder with a syringe.
3. Remove the plastic cover from the actuator.
4. Disconnect the hydraulic lines from the actuator. Plug the lines to prevent loss of fluid.
5. Disconnect the electrical connectors from the actuator.
6. Remove the actuator from the bracket.
7. Installation is the reverse of the removal procedure. Fill and bleed the system.
8. Connect the battery cable, start

the engine and check brake system operation before driving the vehicle.

CHASSIS ELECTRICAL

─ CAUTION ─

On vehicles equipped with an air bag, the negative battery cable must be disconnected, before working on the system. Failure to do so may result in deployment of the air bag and possible personal injury.

Air Bag

DISARMING

Work must be started after about 20 seconds or longer from the time the ignition switch is turned to the **LOCK** position and the battery cable is disconnected from the battery.

Heater Blower Motor

NOTE: On most vehicles, the air conditioner assembly is integral with the heater assembly (including the blower motor) and therefore the blower motor removal may differ from the procedures detailed below. In some case it may be necessary to remove the air conditioning-heater housing and assembly to remove the blower motor. Due to the lack of information available at the time of this publication, a general blower motor removal and installation procedure is outlined. The removal steps can be altered, as required.

REMOVAL & INSTALLATION

Celica and Supra

1. Disconnect the negative battery cable.
2. Working from under the instrument panel, unfasten the defroster hoses from the heater box.
3. Unplug the multi-connector.
4. Loosen the mounting screws and withdraw the blower assembly.
5. Installation is the reverse of the removal procedure.
6. Check the blower for proper operation at all speeds.

Cressida

1. Disconnect the negative battery cable.
2. Remove the instrument panel undercover and cowl side trim panel.

3. Remove the air duct and the glove box.
4. Disconnect the heater control cable from the blower motor and remove the blower duct.
5. Disconnect the heater relay from the heater relay electrical connector.
6. Remove the retaining screws from the blower motor assembly. Remove the assembly from the vehicle.
7. Remove the blower motor from the blower motor assembly.
8. Installation is the reverse order of the removal procedure.
9. Check the blower for proper operation at all speeds.

Corolla and Tercel

1. Disconnect the negative battery cable.
2. Remove the under tray, if equipped.
3. Remove the blower duct and air duct. Before removing the air duct, remember to remove the 2 attaching clamps.
4. Remove the glove box and the heater control cable.
5. Disconnect the electrical connector on the blower motor.
6. Remove the blower motor retaining bolts and remove the blower motor.
7. Installation is the reverse of the removal procedure.
8. Check the blower for proper operation at all speeds.

Camry and MR2

1. Disconnect the negative battery cable.
2. Remove the 3 screws attaching the retainer.
3. Remove the glove box. Remove the duct between the blower motor assembly and the heater assembly.
4. Disconnect the blower motor wire connector at the blower motor case.
5. Disconnect the air source selector control cable at the blower motor assembly.
6. Loosen the nuts and bolts attaching the blower motor to the blower case, remove the blower motor from the vehicle.
7. Installation is the reverse of the removal procedure.
8. Check the blower for proper operation at all speeds.

Front Windshield Wiper Motor

REMOVAL & INSTALLATION

Tercel and Corolla

1. Disconnect the negative battery terminal.

2. Remove the wiper arm.
3. Insert a small prybar between the linkage and the motor. Pry up to separate the linkage from the motor.
4. Disconnect the electrical connector from the motor.
5. Remove the mounting bolts and remove the motor.
6. Installation is the reverse of the removal procedure.

Celica, Supra, Camry and Cressida

1. Remove the access hole cover.
2. Separate the wiper and motor by prying gently with a small prybar.
3. Remove the left and right cowl ventilators.
4. Remove the wiper arms and the linkage mounting nuts. Push the linkage pivot ports into the ventilators.
5. Loosen the wiper link connectors at their ends and with the linkage from the cowl ventilator.
6. Start the wiper motor and turn the ignition key to the **OFF** when the crank is a position best suited for removal of the motor.

NOTE: The wiper motor is difficult to remove when it is in the parked position. If the motor is turned off at the wiper switch, it will automatically return to this position.

7. Unplug the wiper motor connector.
8. Loosen the motor mounting bolts and withdraw the motor.
To install:
9. Be sure to install the wiper motor with it in the park position by connecting the multi-connector and operating the wiper control switch.
10. Assemble the crank, connect the wiring and check operation.

MR2

1. With the wiper arms in the up position and the wiper switch on **LOW**, turn the ignition switch to the **OFF** position.
2. Disconnect the negative battery cable. Disconnect the wiper motor electrical connector, then remove the light retractor relay from the wiper bracket.
3. Remove the wiper motor set bolts. Manually lower the wiper arms, then connect the wiper link hook to the dash panel service hole.
4. Disconnect the wiper motor link. Remove the wiper motor attaching bolts then remove the wiper motor.
5. Installation is the reverse order of the removal procedure.

Rear Wiper Motor

REMOVAL & INSTALLATION

1. Disconnect the negative battery cable.

2. Remove the wiper arm and rear door trim cover.

3. Disconnect the wiper motor wire connector.

4. Remove the wiper motor bracket attaching bolts and the wiper motor along with the bracket.

5. Installation is the reverse of the removal procedure.

Instrument Cluster

REMOVAL & INSTALLATION

Cressida

1. Disconnect the negative battery cable.

2. Remove the cluster finish panel.

3. Loosen the instrument cluster retaining screws and tilt the panel forward.

4. Detach the speedometer cable and wiring connectors.

5. Remove the entire cluster assembly.

6. Remove the instruments from the panel as required.

7. Installation is the reverse of the removal proecedure.

Tercel and Corolla

1. Disconnect the negative battery cable.

2. Remove the steering column cover.

NOTE: Be careful not to damage the collapsible steering column mechanism.

3. On Tercel, remove the heater control knob and the center instrument cluster finish panel.

4. Remove the switches and hole cover from the cluster hood.

5. Remove the cluster hood.

6. Remove the cluster attaching screws and pull the unit forward.

7. Disconnect the speedometer and any other electrical connections that are necessary.

8. Remove the instruments from the panel as required.

9. Installation is the reverse of the removal procedure.

Camry, Celica and Supra

1. Disconnect the negative battery cable.

2. Remove the fuse box cover from under the left side of the instrument panel.

3. Remove the heater control knobs.

4. Carefully pry off the heater control panel.

5. Remove the cluster hood.

6. Unscrew the cluster finish panel retaining screws and pull out the bottom of the panel.

7. Unplug the electrical connectors and disconnect the speedometer cable.

8. Remove the instrument cluster.

9. Remove the instruments from the panel as required.

10. Installation is the reverse of the removal procedure.

MR2

1. Disconnect the negative battery cable.

2. Remove the steering column covers.

3. Pull the rheostat knob from the cluster finish panel and remove the nut from the rheostat. Remove the cluster finish panel and disconnect the rheostat multi-connector.

4. Remove the cluster hood.

5. Remove the cluster attaching screws and pull the unit forward.

6. Disconnect the speedometer and any other electrical connections that are necessary.

7. Remove the instruments from the panel as required.

8. Installation is the reverse of the removal procedure.

Radio

REMOVAL & INSTALLATION

Celica, Supra, Camry and Cressida

1. Disconnect the negative battery cable.

2. Remove the ash tray and cigarette lighter, if required.

3. Remove the radio finish panel.

4. Detach the antenna lead from the jack on the radio case.

5. Remove the cowl air intake duct.

6. Detach the power and speaker leads. Label the leads for assembly.

7. Remove the radio support nuts and bolts.

8. Remove the radio from beneath the dashboard.

9. Installation is the reverse of the removal procedure.

Corolla and Tercel

1. Disconnect the negative battery cable.

2. Remove the 2 screws from the top of the dashboard center trim panel.

3. Lift the center panel out far enough to gain access to the cigarette lighter wiring and disconnect the wiring. Remove the trim panel.

4. Unfasten the screws which secure the radio to the instrument panel braces.

5. Lift out the radio and disconnect the leads from it. Remove the radio.

6. Installation is the reverse of the removal procedure.

MR2

1. Disconnect the negative battery cable.

2. Remove the ash tray and cigarette lighter.

3. Remove the radio finish panel.

4. Remove the radio bolts and pull the radio out far enough to disconnect all necessary electrical wiring.

5. Remove the radio.

6. Installation is the reverse of the removal procedure.

Concealed Headlights
—— CAUTION ——

Before attempting to manually operate the concealed (retractable) headlights, first pull the fuse or disconnect the negative battery cable. Otherwise the headlights and motor shaft may suddenly move and catch hand and fingers. When opening and closing retractable headlights, make sure nobody is near them, otherwise personal injury may result.

MANUAL OPERATION

NOTE: If the headlights are frozen and inoperative, carefully melt the ice before attempting the manual operation procedure. Operation of a frozen headlight will drain the battery and may cause damage to the motor and operating linkages.

1. Switch the headlight and retractable headlight switches to the **OFF** position.

2. Pull the retractable headlight fuse or disconnect the negative battery cable.

3. Remove the rubber cap from the manual operation knob.

4. Manually turn the knob clockwise until the headlights are in the desired position (open or closed).

5. Install the rubber cap.

6. Install the fuse.

7. Make sure the lights work properly.

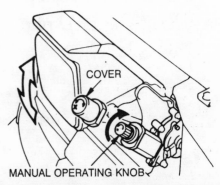

Manual operation of concealed headlights

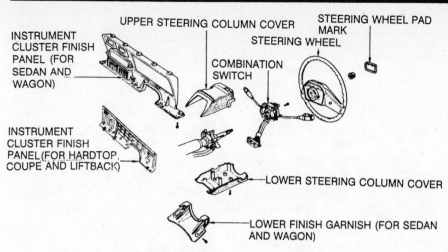

INSTRUMENT CLUSTER FINISH PANEL (FOR SEDAN AND WAGON)

UPPER STEERING COLUMN COVER

STEERING WHEEL PAD MARK

STEERING WHEEL

COMBINATION SWITCH

INSTRUMENT CLUSTER FINISH PANEL (FOR HARDTOP, COUPE AND LIFTBACK)

LOWER STEERING COLUMN COVER

LOWER FINISH GARNISH (FOR SEDAN AND WAGON)

Typical combination switch mounting

Combination Switch

REMOVAL & INSTALLATION

1. Disconnect the negative battery cable.
2. Remove the steering column garnish.
3. Remove the upper and lower steering column covers.
4. Remove the steering wheel.
5. Trace the switch wiring harness to the multi-connector. Push in the lock levers and pull apart the connectors.
6. If equipped with electronic modulated suspension (TEMS), remove the steering sensor. On air bag equipped vehicles, disconnect the cable connectors, remove the spiral cable housing attaching screws and slide the cable assembly from the front of the combination switch.
7. Unscrew the mounting screws and slide the combination switch from the steering column.
8. Installation is the reverse of the removal procedure. Check all switch functions for proper operation.

Ignition Lock/Switch

REMOVAL & INSTALLATION

1. Disconnect the negative battery cable.
2. Unfasten the ignition switch connector under the instrument panel.
3. Remove the screws which secure the upper and lower halves of the steering column cover.
4. Turn the lock cylinder to the **ACC** position with the ignition key.
5. Push the lock cylinder stop in with a small, round object (cotter pin, punch, etc.).

NOTE: On some vehicles, it may

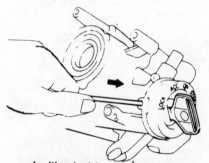

Ignition lock/switch removal

be necessary to remove the steering wheel and combination switch first.

6. Withdraw the lock cylinder from the lock housing while depressing the stop tab.
7. To remove the ignition switch, unfasten its securing screws and withdraw the switch from the lock housing.
To install:
8. Align the locking cam with the hole in the ignition switch and insert the switch into the lock housing.
9. Secure the switch with its screw(s).
10. Make sure both the lock cylinder and column lock are in the **ACC** position. Slide the cylinder into the lock housing until the stop tab engages the hole in the lock.
11. Install the steering column covers.
12. Connect the ignition switch connector.
13. Connect the negative battery cable.

Stoplight Switch

ADJUSTMENT

1. Remove the instrument lower finish panel and the air duct if required to gain access to the stoplight switch.

2. Disconnect the stoplight switch connector.
3. Loosen the switch locknut.
4. Turn the stoplight switch until the end of the switch lightly contacts the pedal stopper.
5. Hold the switch and tighten the locknut.
6. Connect the switch connector.
7. Depress the brake pedal and verify that the brake lights illuminate.
8. Install the air duct and the lower finish panel, if removed.

REMOVAL & INSTALLATION

1. Disconnect the negative battery cable.
2. Remove the instrument lower finish panel and the air duct if required to gain access to the stoplight switch.
3. Disconnect the stoplight switch connector.
4. Remove the switch mounting nut, then slide the switch from the mounting bracket on the pedal.
To install:
5. Install the switch into the mounting bracket and adjust.
6. Connect the switch connector.
7. Depress the brake pedal and verify that the brake lights illuminate.
8. Install the air duct and the lower finish panel, if removed.

Clutch Switch

ADJUSTMENT

1. Attempt to start the engine when the clutch pedal is released. The engine should not start.
2. Depress the clutch pedal fully and attempt to start the engine. The engine should start.
3. If the engine does not start, depress the clutch pedal fully. With the clutch pedal depressed, loosen the switch locknut.
4. Use the adjusting nut to turn the switch until the tip of the switch contacts the clutch pedal stop.
5. Tighten the locknut and attempt to start the engine. Re-adjust as necessary.

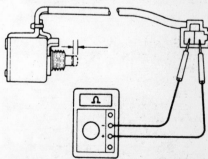

Checking clutch start switch continuity

6. If the switch cannot be adjusted, check the switch continuity with a suitable ohmmeter. There should be continuity between the switch terminals when the switch is on (tip pushed in) and no continuity when the switch is off (tip released). If the continuity is not as specified, replace the switch.

REMOVAL & INSTALLATION

1. Disconnect the negative battery cable.
2. Disconnect the switch connector.
3. Remove the switch adjusting nut.
4. Withdraw the switch from the mounting bracket.
5. Installation is the reverse of the removal procedure. Adjust the switch.

Neutral Safety Switch

The shift lever is adjusted properly if the engine will not start in any position other than N or P.

ADJUSTMENT

1. Loosen the neutral start switch bolt. Position the selector in the N position.
2. Align the switch shaft groove with the neutral base line which is located on the switch.
3. Tighten the bolt to 48 inch lbs. (5.4 Nm). on all vehicles except Tercel wagon. On the Tercel wagon, tighten to 9 ft. lbs. (13 Nm).

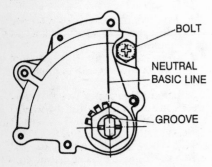

Neutral safely switch adjustment—Cressida and Supra

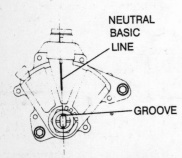

Neutral safety switch adjustment—Tercel, MR2, Camry and Celica

REMOVAL & INSTALLATION

1. Disconnect the negative battery cable.
2. Unplug the switch wiring connectors.
3. Disconnect the transmission control cable from the manual shift lever.
4. Remove the manual shift lever.
5. Pry the C-washer from the manual shaft nut. Discard the washer and replace with new.
6. Remove the manual shaft nut and washer.
7. Remove the manual shaft lever packing (1990–92 Celica with A241E transaxle).
8. Remove the retaining bolts and withdraw the switch from the transaxle case.
9. Installation is the reverse of the removal procedure. Adjust the switch.

Circuit Breakers

Camry

The automatic shoulder belt and defogger circuit breakers are located in the junction block No. 1. The heater circuit breaker is located in the relay block No. 4.

Celica

The power and door lock circuit breakers are located in the relay block No. 1. The defogger circuit breaker is located in the junction block No. 1. The heater circuit breaker is located in the relay block No. 4.

Corolla

The heater circuit breaker is located in the relay block No. 4. The defogger and power circuit breakers are located in the junction block No. 1, left kick panel.

Cressida and Supra

The heater circuit breaker is located in the relay block No. 4. The defogger and power circuit breakers are located in the junction block No. 1, left kick panel. The automatic shoulder belt and door lock circuit breakers are located in the relay block No. 3, left kick panel.

MR2

The door lock circuit breaker is located in the relay block No. 1, left kick panel, upper section.

Tercel

The defogger circuit breaker is located in the junction block No. 1, left kick panel.

Fusible Links

Camry

The condenser fan and radiator fan 30 amp fusible link is located in the relay block No. 2. The ignition, headlight and alternator fusible link is located in the fusible link box near the battery. The auto seat belt, starter relay and voltage regulator 0.5 amp fusible link is located near the battery, green.

1988–89 Celica

The condenser fan link is located in the relay block No. 5. The starter and alternator links are located in the fusible link box near the battery.

1990-92 Celica

The condenser fan link is located in the relay block No. 5. The radiator fan link is located in the junction block No. 2. The ignition, starter, alternator and Anti-lock brake system links are located in the fusible link box near the battery.

Corolla

The condenser fan link is located in the relay block No. 5. The radiator fan link is located in the junction block No. 2. The A/C 7.5 amp fuse is located in the bottom of the relay block No. 4.

Cressida

The condenser fan, ignition switch, Anti-lock brake and alternator link is located in the junction block No. 2, near the battery.

1988–89 MR2

The heater, ignition switch and headlight links are located in the relay block No. 2, left engine compartment. The A/C and heater fuses are located in the relay block No. 4, right kick panel.

1990–92 MR2

The ignition switch, heater and alternator links are located in the relay block No. 5, right side luggage compartment. The Anti-lock Brake System (ABS) link is located in the relay block No. 5, right side luggage compartment.

Supra

The ECU-battery 15 amp Fuse is located in the fuse block. The ignition switch, condenser fan, Anti-lock Brake System (ABS) and alternator link is located in the junction block No. 2, left fender.

1988–90 Tercel

The condenser fan and radiator fan links are located in the relay block No. 2, left inner fender. The A/C 10 amp fuse is located in the relay block No. 4.

1991–92 Tercel

The defogger and heater links are located in the relay block No. 6, left kick panel. The ignition switch and alternator links are located in the relay block No. 2.

Relays, Sensors, Modules and Computer Location

Camry

- **A/C Fan Relays**—are located in the relay block No. 2.
- **Air Flow Meter**—is located on the air cleaner assembly.
- **Air Intake Temperature Sensor**—is located at the air cleaner assembly.
- **Anti-lock Brake System Control Unit**—is located on the right side rear compartment.
- **Blower Control Relay**—is located in the relay block No. 4, on the right side kick panel.
- **Coolant Temperature Sensor**—is located in the thermostat housing, to the right of the start injector time switch.
- **Diagnostic Check Connector**—is located near the master cylinder.
- **EFI Main Relay**—is located in the relay block No. 2, a round relay with 4 wires.
- **EFI Water Temperature Sensor**—is located at the intake manifold.
- **Electronic Controlled Transmission (ECT) Control Unit (4WD)**—is located behind the console.
- **Electronic Controlled Transmission (ECT) Control Unit (FWD)**—is located behind the instrument panel, to the right.
- **Engine Electronic Control Unit (ECU)**—is located behind the radio.
- **Fuel Pump Relay**—refer to the circuit opening relay.
- **Hazard Flasher**—is located within the turn signal flasher.
- **Knock Sensor**—is threaded into the cylinder block, under the intake manifold.
- **Start Injector Time Switch**—is located in the thermostat housing, to the left of the coolant temperature sensor.
- **Throttle Position Sensor**—is located on the throttle body.
- **Turn Signal Flasher**—is located in the relay block No. 1, left kick panel, upper section, 3 wires.

1988–89 Celica

- **A/C Dual and High Pressure Switches**—is located at the right inner fender.
- **A/C Clutch Relay**—is located in the relay block No. 5.

- **A/C Fan Relays**—are located in the relay block No. 5.
- **Air Flow Meter**—is located on the air cleaner assembly.
- **Air Intake Temperature Sensor**—is located at the air cleaner assembly.
- **Anti-lock Brake System Computer (FWD)**—is located on top of the ECU.
- **Blower Relay**—is located behind the relay block No. 4.
- **Cold Start Injector**—is located at the air intake plenum.
- **Diagnostic Check Connector**—is located rearward of the left strut tower.
- **EFI Resistor**—is located rearward of the EGR valve.
- **EFI Water Temperature Sensor**—is located at the intake manifold.
- **Electronic Controlled Transmission (ECT) Control Unit**—is located under console, in front of shifter.
- **Engine Electronic Control Unit (ECU)**—is located under the radio, behind the console.
- **Fuel Pump Control Relay**—is located in engine compartment, center of firewall.
- **Hazard Flasher**—is located within the turn signal flasher.
- **Heater and Horn Relays**—are located in the relay block No. 4.
- **Ignitor**—is located in the center of the firewall.
- **Knock Sensor**—is threaded into the cylinder block, under the intake manifold.
- **Oil Pressure Switch**—is located at the transmission side of the engine.
- **Start Injector Time Switch**—is located in the thermostat housing, to the left of the coolant temperature sensor.
- **Throttle Position Sensor**—is located on the throttle body.
- **Turn Signal Flasher**—is located in the relay block No. 1, upper position.
- **Vehicle Speed Sensor**—is located inside the speedometer head.
- **Water Temperature Sender**—is located on the left side of the engine cylinder head.

1990–92 Celica

- **Air Flow Meter and Air Intake Temperature Sensor**—is located on the air cleaner assembly.
- **Anti-lock Brake System Computer**—is located in the rear seat area, right side.
- **Blower Control Relay**—is located at the blower housing.
- **Cold Start Injector**—is located at the air intake plenum.
- **Coolant Temperature Sensor**—is located in the thermostat housing.

- **Diagnostic Check Connector**—is located rearward of the left strut tower.
- **EFI Water Temperature Sensor**—is located at the left side of the cylinder head, above the transmission.
- **Engine Control Unit (ECU)**—is located behind the console.
- **Fuel Pump Relay**—refer to the circuit opening relay.
- **Ignitor**—is located at the firewall.
- **Knock Sensor**—is threaded into the cylinder block.
- **Oil Pressure Switch**—is located at the driver's side of the engine.
- **Radiator Fan Relay**—is located in the junction block No. 2.
- **Start Injector Time Switch**—is located in the thermostat housing.
- **Throttle Position Sensor**—is located on the throttle body.
- **Water Temperature Sender**—is located at the driver's side of the engine.

Corolla

- **Air Flow Meter and Air Intake Temperature Sensor**—is located on the air cleaner assembly.
- **Cold Start Injector**—is located at the air intake plenum.
- **Coolant Temperature Sensor**—is located in the thermostat housing.
- **Diagnostic Check Connector**—is located rearward of the left strut tower.
- **EFI Water Temperature Sensor**—is located at the left side of the cylinder head.
- **Engine Electronic Control Unit (ECU)**—is located behind the radio.
- **Oil Pressure Switch**—is located at the front of the engine, near the distributor.
- **Start Injector Time Switch**—is located in the thermostat housing.
- **Throttle Position Sensor**—is located on the throttle body.
- **Turn Signal Flasher**—is located to the right of the junction block No. 1, under the instrument panel.
- **Hazard Flasher**—is located within the turn signal flasher.
- **Water Temperature Sender (4A-GE engine)**—is located at the right side of the intake manifold.

Cressida

- **Air Flow Meter and Air Intake Temperature Sensor**—is located on the air cleaner assembly.
- **Anti-lock Brake System (ABS) Control Unit**—is located behind the trunk trim panel, on the right side.
- **Diagnostic Check Connector**—is located rearward of the left strut tower.

- **EFI Water Temperature Sensor** – is located at the right side of the cylinder head.
- **Engine and Transmission Control Unit** – is located behind the glove compartment.
- **Fuel Pump Relay** – is located in the relay block No. 4, right kick panel.
- **Hazard Flasher** – is located within the turn signal flasher.
- **Oil Pressure Switch** – is located at the right side of the engine.
- **Starter Relay (US)** – is located in the relay block No. 3.
- **Throttle Position Sensor** – is located on the throttle body.
- **Toyota Diagnostic Communication Link (TDCL)** – is located the left of the steering column.
- **Turn Signal Flasher** – is located in the relay block No. 3, left kick panel.
- **Water Temperature Sender** – is located at the right side of the engine.

1988–89 MR2

- **Air Flow Meter and Air Intake Temperature Sensor** – is located on the air cleaner assembly.
- **Blower Resistor** – is located behind the instrument panel, in the center.
- **Coolant Temperature Sensor** – is located in the thermostat housing.
- **Diagnostic Check Connector** – is located at the firewall.
- **EFI Water Temperature Sensor** – is located at the cylinder head, left side.
- **Engine Electronic Control Unit (ECU)** – is located on center of the firewall.
- **Hazard Flasher** – is located within the turn signal flasher.
- **Heater Relay** – is located in the relay block No. 4,
- **Oil Pressure Switch** – is located at the rear of the engine, on the right side.
- **Throttle Position Sensor** – is located on the throttle body.
- **Turn Signal Flasher** – is located behind the instrument panel, on the left side.
- **Water Temperature Sender** – is located at the right side of the intake manifold.

1990–92 MR2

- **Anti-lock Brake System (ABS) Control Unit** – is located under the instrument panel.
- **Blower Resistor** – is located behind the instrument panel.
- **Coolant Temperature Sensor** – is located in the thermostat housing.
- **Cooling Fan Relay** – is located in the relay block No. 2.
- **Diagnostic Check Connector** – is located at the rear firewall, on the left side.
- **EFI Water Temperature Sensor** – is located at the cylinder head.
- **Engine Electronic Control Unit (ECU)** – is located on the firewall.

- **Oil Pressure Switch** – is located at the rear of the engine, on the left side.
- **Turn Signal Flasher** – is located in the relay block No. 1, left kick panel, at the top.

Supra

- **Air Flow Meter** – is located on the air cleaner assembly.
- **Anti-lock Brake System (ABS) Control Unit** – is located behind the glove compartment.
- **Blower Resistor** – is located near the relay block.
- **Diagnostic Check Connector** – is located at the left strut tower.
- **EFI Water Temperature Sensor** – is located under the upper radiator hose.
- **Electronic Fuel Injection (EFI) Relay** – is located in the junction block No. 2.
- **Engine Electronic Control Unit (ECU)** – is located behind the glove compartment.
- **Fuel Pump Relay** – is located at the right strut tower.
- **Oil Pressure Switch** – is located at the right of the engine, near the throttle body.
- **Start Injector Time Switch** – is located in the thermostat housing.
- **Turn Signal and Hazard Flasher** – is located in the relay block No. 5, left kick panel.
- **Water Temperature Sender** – is located at the right side of the intake manifold.

SERIAL NUMBER IDENTIFICATION

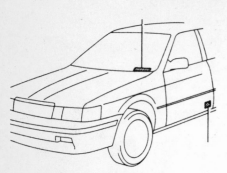

Vehicle identification plate

Vehicle Identification Plate

The vehicle identification plate is lo-

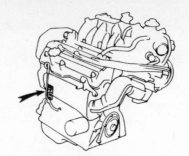

Engine serial number — ES250

cated at the top of the left instrument panel.

Engine Number

The engine identification on the side of

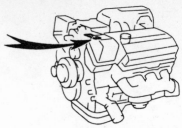

Engine serial number — LS400

the engine block on the ES250 and the top of the engine block on the LS400.

Vehicle Identification Number

The vehicle identification number is also located on the left front door.

SPECIFICATIONS

ENGINE IDENTIFICATION

Year	Model	Engine Displacement cu. in. (cc/liter)	Engine Series Identification	No. of Cylinders	Engine Type
1990	ES250	153 (2508/2.5)	2V2-FE	6	DOHC
	LS400	242.1 (3969/4.0)	1UZ-FE	8	DOHC
1991–92	ES250	153 (2508/2.5)	2V2-FE	6	DOHC
	LS400	242.1 (3969/4.0)	1UZ-FE	8	DOHC

GENERAL ENGINE SPECIFICATIONS

Year	Model	Engine Displacement cu. in. (cc)	Fuel System Type	Net Horsepower @ rpm	Net Torque @ rpm (ft. lbs.)	Bore × Stroke (in.)	Compression Ratio	Oil Pressure @ rpm
1990	ES250	153 (2508)	EFI	156 @ 5600	160 @ 4400	3.44 × 2.74	9.0:1	43–78 @ 3000
	LS400	242.1 (3969)	EFI	250 @ 5600	260 @ 4400	3.44 × 3.25	10.0:1	36–71 @ 3000
1991–92	ES250	153 (2508)	EFI	156 @ 5600	160 @ 4400	3.44 × 2.74	9.0:1	43–78 @ 3000
	LS400	242.1 (3969)	EFI	250 @ 5600	260 @ 4400	3.44 × 3.25	10.0:1	36–71 @ 3000

GASOLINE ENGINE TUNE-UP SPECIFICATIONS

Year	Model	Engine Displacement cu. in. (cc)	Spark Plugs Type	Spark Plugs Gap (in.)	Ignition Timing (deg.) MT	Ignition Timing (deg.) AT	Compression Pressure (psi)	Fuel Pump (psi)	Idle Speed (rpm) MT	Idle Speed (rpm) AT	Valve Clearance In.	Valve Clearance Ex.
1990	ES250	153 (2508)	①	0.043	10B	10B	142	38–44	600–700	650–750	③	③
	LS400	242.1 (3969)	②	0.043	8–12B	8–12B	142	38–44	600–700	600–700	④	④

GASOLINE ENGINE TUNE-UP SPECIFICATIONS

Year	Model	Engine Displacement cu. in. (cc)	Spark Plugs Type	Gap (in.)	Ignition Timing (deg.) MT	AT	Compression Pressure (psi)	Fuel Pump (psi)	Idle Speed (rpm) MT	AT	Valve Clearance In.	Ex.
1991	ES250	153 (2508)	①	0.043	10B	10B	142	38–44	600–700	650–750	③	③
	LS400	242.1 (3969)	②	0.043	8–12B	8–12B	142	38–44	600–700	600–700	④	④
1992		SEE UNDERHOOD SPECIFICATIONS STICKER										

① BCPR6EP11
② BKR6EP11
③ Intake—0.005 in. cold
   Exhaust—0.015 in. cold
④ Intake—0.006–0.010 in. cold
   Exhaust—0.010–0.014 in. cold

FIRING ORDERS

NOTE: To avoid confusion, always replace spark plug wires one at a time.

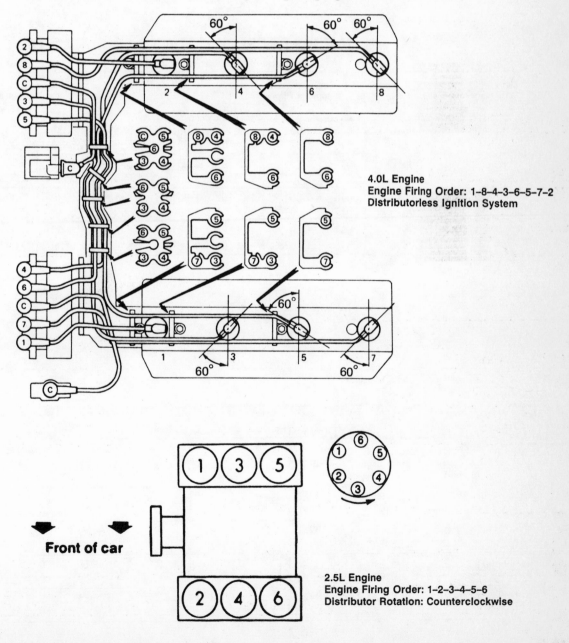

4.0L Engine
Engine Firing Order: 1–8–4–3–6–5–7–2
Distributorless Ignition System

Front of car

2.5L Engine
Engine Firing Order: 1–2–3–4–5–6
Distributor Rotation: Counterclockwise

CAPACITIES

Year	Model	Engine Displacement cu. in. (cc)	Engine Crankcase (qts.) with Filter	Engine Crankcase (qts.) without Filter	Transmission (pts.) 4-Spd	Transmission (pts.) 5-Spd	Transmission (pts.) Auto.	Drive Axle (pts.)	Fuel Tank (gal.)	Cooling System (qts.)
1990	ES250	153 (2508)	4.1	3.9	—	4.4	2.6	1.1	15.9	①
	LS400	242.1 (3969)	5.3	5	—	—	2.1	1.3	22.5	11.2
1991-92	ES250	153 (2508)	4.1	3.9	—	4.4	2.6	1.1	15.9	①
	LS400	242.1 (3969)	5.3	5	—	—	2.1	1.3	22.5	11.2

① MT—10.0 qts.
  AT—9.9 qts.

CAMSHAFT SPECIFICATIONS

All measurements given in inches.

Year	Engine Displacement cu. in. (cc)	Journal Diameter 1	Journal Diameter 2	Journal Diameter 3	Journal Diameter 4	Journal Diameter 5	Lobe Lift In.	Lobe Lift Ex.	Bearing Clearance	Camshaft End Play
1990	153 (2508)	1.0610–1.0616	1.0610–1.0616	1.0610–1.0616	1.0610–1.0616	1.0610–1.0616	1.5555–1.5594	1.5339–1.5378	0.0014–0.0028	0.0012–0.0031
	242.1 (3969) ①	1.0612–1.0618	1.0612–1.0618	1.0612–1.0618	1.0612–1.0618	1.0612–1.0618	1.6421–1.6461	1.6500–1.6539	0.0012–0.0026	0.0016–0.0035
1991-92	153 (2508)	1.0610–1.0616	1.0610–1.0616	1.0610–1.0616	1.0610–1.0616	1.0610–1.0616	1.5555–1.5594	1.5339–1.5378	0.0014–0.0028	0.0012–0.0031
	242.1 (3969) ①	1.0612–1.0618	1.0612–1.0618	1.0612–1.0618	1.0612–1.0618	1.0612–1.0618	1.6421–1.6461	1.6500–1.6539	0.0012–0.0026	0.0016–0.0035

① The exhaust camshaft thrust portion of journal diameter is 0.9433–0.9439 in.

CRANKSHAFT AND CONNECTING ROD SPECIFICATIONS

All measurements are given in inches.

Year	Engine Displacement cu. in. (cc)	Crankshaft Main Brg. Journal Dia.	Crankshaft Main Brg. Oil Clearance	Crankshaft Shaft End-play	Crankshaft Thrust on No.	Connecting Rod Journal Diameter	Connecting Rod Oil Clearance	Connecting Rod Side Clearance
1990①	153 (2508)	2.5191–2.5197	0.0011–0.0022	0.0008–0.0087	3	1.8892–1.8898	0.0011–0.0026	0.0059–0.0130
	242.1 (3969)	2.6373–2.6378	0.0010–0.0018	0.0008–0.0087	3	2.0465–2.0472	0.0011–0.0021	0.0063–0.0114
1991-92	153 (2508)	2.5191–2.5197	0.0011–0.0022	0.0008–0.0087	3	1.8892–1.8898	0.0011–0.0026	0.0059–0.0130
	242.1 (3969)	2.6373–2.6378	0.0010–0.0018	0.0008–0.0087	3	2.0465–2.0472	0.0011–0.0021	0.0063–0.0114

VALVE SPECIFICATIONS

Year	Engine Displacement cu. in. (cc)	Seat Angle (deg.)	Face Angle (deg.)	Spring Test Pressure (lbs.)	Spring Installed Height (in.)	Stem-to-Guide Clearance (in.)		Stem Diameter (in.)	
						Intake	Exhaust	Intake	Exhaust
1990	153 (2508)	NA	44.5	41.0–47.2	1.677	0.0010–0.0024	0.0012–0.0026	0.2350–0.2356	0.2348–0.2354
	242.1 (3969)	NA	44.5	41.9–46.3	1.717	0.0010–0.0024	0.0012–0.0026	0.2350–0.2356	0.2348–0.2354
1991–92	153 (2508)	NA	44.5	41.0–47.2	1.677	0.0010–0.0024	0.0012–0.0026	0.2350–0.2356	0.2348–0.2354
	242.1 (3969)	NA	44.5	41.9–46.3	1.717	0.0010–0.0024	0.0012–0.0026	0.2350–0.2356	0.2348–0.2354

NA—Not Available

PISTON AND RING SPECIFICATIONS

All measurements are given in inches.

Year	Engine Displacement cu. in. (cc)	Piston Clearance	Ring Gap			Ring Side Clearance		
			Top Compression	Bottom Compression	Oil Control	Top Compression	Bottom Compression	Oil Control
1990	153 (2508)	0.0018–0.0026	0.0118–0.0213	0.0138–0.0244	0.0079–0.0224	0.0004–0.0031	0.0012–0.0028	—
	242.1 (3969)	0.0008–0.0016	0.0098–0.0177	0.0138–0.0236	0.0059–0.0197	0.0008–0.0024	0.0006–0.0022	—
1991–92	153 (2508)	0.0018–0.0026	0.0118–0.0213	0.0138–0.0244	0.0079–0.0224	0.0004–0.0031	0.0012–0.0028	—
	242.1 (3969)	0.0008–0.0016	0.0098–0.0177	0.0138–0.0236	0.0059–0.0197	0.0008–0.0024	0.0006–0.0022	—

TORQUE SPECIFICATIONS

All readings in ft. lbs.

Year	Engine Displacement cu. in. (cc)	Cylinder Head Bolts	Main Bearing Bolts	Rod Bearing Bolts	Crankshaft Pulley Bolts	Flywheel Bolts	Manifold		Spark Plugs
							Intake	Exhaust	
1990	153 (2508)	①	②	③	181	61	13	29	13
	242.1 (3969)	④	⑤	③	181	72	13	29	13
1991–92	153 (2508)	①	②	③	181	61	13	29	13
	242.1 (3969)	④	⑤	③	181	72	13	29	13

① Tighten in 3 steps:
   1—tighten to 25 ft. lbs.
   2—turn 90 degrees
   3—turn 90 degrees
② Tighten in 2 steps:
   1—tighten to 45 ft. lbs.
   2—turn 90 degrees
③ Tighten in 2 steps:
   1—tighten to 18 ft. lbs.
   2—turn 90 degrees
④ Tighten in 2 steps:
   1—tighten to 29 ft. lbs.
   2—turn 90 degrees
⑤ Tighten in 2 steps:
   1—tighten to 20 ft. lbs.
   2—turn 90 degrees

BRAKE SPECIFICATIONS

All measurements in inches unless noted.

Year	Model	Lug Nut Torque (ft. lbs.)	Master Cylinder Bore	Brake Disc Minimum Thickness	Brake Disc Maximum Runout	Standard Brake Drum Diameter	Minimum Lining Thickness Front	Minimum Lining Thickness Rear
1990	ES250	76	NA	0.945②	0.0028③	NA	0.039	0.039
	LS400	76	NA	0.906①	0.0020	NA	0.039	0.039
1991–92	ES250	76	NA	0.945②	0.0028③	NA	0.039	0.039
	LS400	76	NA	0.906①	0.0020	NA	0.039	0.039

NA—Not available
① Rear—0.591 in.
② Rear—0.354 in.
③ Rear—0.0059 in.

WHEEL ALIGNMENT

Year	Model	Caster Range (deg.)	Caster Preferred Setting (deg.)	Camber Range (deg.)	Camber Preferred Setting (deg.)	Toe-in (in.)	Steering Axis Inclination (deg.)
1990	ES250 Front	1¹¹/₁₂P–2⁵/₁₂P	1²/₃P	¼N–1¼P	½P	¹/₁₆N–¾	22
	Rear	—	—	1⁵/₁₂N–¼P	²/₃N	¹³/₃₂–¹⁵/₁₆	—
	LS400 Front	8½P–10P②	9¼P	²/₃N–1⁵/₆P①	¹/₁₂P	0–¹³/₁₆	32²/₃③
	Rear	—	—	¾N–¾P	0④	0–¹³/₁₆⑤	—
1991–92	ES250 Front	1¹¹/₁₂P–2⁵/₁₂P	1²/₃P	¼N–1¼P	½P	¹/₁₆N–¾	22
	Rear	—	—	1⁵/₁₂N–¼P	²/₃N	¹³/₃₂–¹⁵/₁₆	—
	LS400 Front	8½P–10P②	9¼P	²/₃N–1⁵/₆P①	¹/₁₂P	0–¹³/₁₆	32²/₃③
	Rear	—	—	¾N–¾P	0④	0–¹³/₁₆⑤	—

① W/air suspension ⁵/₆N–²/₃P—Range ¹/₁₂N—Preferred
② W/air suspension 9¼P–10⁷/₁₂P—Range 9⁵/₆P—Preferred
③ W/air suspension 32½ degree
④ W/air suspension 1½N–0—Range ¾N—Preferred
⑤ W/air suspension ¹¹/₃₂–⅞ in.

ENGINE MECHANICAL

NOTE: Disconnecting the negative battery cable on some vehicles may interfere with the functions of the on board computer systems and may require the computer to undergo a relearning process, once the negative battery cable is reconnected.

Engine Assembly

REMOVAL & INSTALLATION

ES250

1. Disconnect the negative battery cable. Disconnect the positive battery cable.

2. Remove the hood assembly. Remove the battery from the vehicle and disconnect the ground cable.
3. Remove the engine undercovers and drain the cooling system.
4. Raise and safely support the vehicle. Drain the engine oil in a suitable container. Lower the vehicle.
5. Remove the suspension upper brace.
6. Disconnect the igniter connector, noise filter connector and the high tension electrical cord.
7. Remove the ignition coil, igniter and bracket assembly.
8. Remove the radiator, alternator, alternator belt and the adjusting bar.
9. Remove the radiator reservoir tank and disconnect the accelerator cable from the throttle body.
10. Disconnect the throttle cable from the throttle body, on vehicles equipped with automatic transaxle.
11. Remove the cruise control actuator.

12. Disconnect the air flow meter connector, ISC valve air hose and the vacuum pipe air hose.
13. Disconnect the air cleaner hose, air cleaner cap, hoses and the air flow meter. Remove the mounting bolts and the air cleaner assembly.
14. Disconnect the following:
   a. Check connector
   b. Ground straps from the left side fender apron
   c. Connectors from the relay box
   d. Engine compartment wire connector.
15. Disconnect the following hoses:
   a. Brake booster vacuum hose to the air intake chamber.
   b. Air conditioning control valve vacuum hose.
   c. Charcoal canister vacuum hose.
16. Disconnect the ground strap from the transaxle.
17. Disconnect the heater hoses and

the fuel line hoses. Use a suitable container to catch any excess fuel.

18. Remove the starter, if equipped with manual transaxle.

19. Remove the clutch release cylinder and tube clamp, do not disconnect the tube, if equipped with manual transaxle.

20. Disconnect the speedometer cable and the transaxle control cable(s).

21. Remove the engine undercover and glove compartment box.

22. Disconnect the following connectors:

a. 3 engine and ECT Electronic Control Unit (ECU) connectors.

b. Circuit opening relay connector.

c. Cowl wire connector.

d. Instrument panel wire connector.

23. Remove the engine wire from the cowl panel.

24. Raise and safely support the vehicle. Remove the bolts and the suspension lower crossmember.

25. Disconnect the front exhaust pipe.

26. Disconnect the air conditioning wire connectors and remove the mounting bolts. Position and support the compressor aside.

27. Remove the halfshafts.

28. Remove the power steering pump and support. Do not disconnect the hoses.

29. Remove the mounting bolts and the engine cross member. Support the engine with a suitable jacking device.

30. Remove the front, center and rear engine mounting insulators and brackets.

31. Remove the power steering reservoir tank, without disconnecting the hoses. Remove the ground strap.

32. Remove the mounting brackets on both sides of the engine assembly. Lower the vehicle, keeping the engine supported.

33. Attach a suitable engine hoist to the engine hangers. Disconnect the clamps of the power steering oil cooler lines.

34. Remove the mounting insulators on both sides of the engine. Lift the engine and transaxle out of the vehicle.

35. Remove the starter, if equipped with automatic transaxle.

36. Separate the engine from the transaxle.

To install:

37. Assemble the engine to the transaxle. Install the starter, if equipped with an automatic transaxle. Tighten the mounting bolts to 29 ft. lbs. (40 Nm).

38. Lower the engine and the transaxle into the vehicle with a suitable engine hoist. Tilt the transaxle downward to clear the left side mounting bracket.

39. Align the right side and the left side mounting with the body bracket. Attach the right side mounting insulator to the mounting bracket and install the bolts.

40. Install the left side mounting bracket to the transaxle case and tighten the mounting bolts to 38 ft. lbs. (52 Nm).

41. Attach the left side mounting insulator to the mounting bracket. Tighten the mounting bolts to 38 ft. lbs. (52 Nm) and the through bolt to 64 ft. lbs. (87 Nm).

42. Tighten the nuts and bolts of the right side mounting insulator. Tighten the bolts to 47 ft. lbs. (64 Nm). Tighten the bracket nuts to 38 ft. lbs. (52 Nm) and the body nuts to 65 ft. lbs. (88 Nm).

43. Remove the engine hoist from the engine and connect the power steering cooler line clamp.

44. Connect the right side engine mounting brackets. Tighten the bolts to 38–48 ft. lbs. (52-65 Nm) and the nut 38 ft. lbs. (52 Nm).

45. Connect the left side engine mounting bracket. Tighten the bolt to 14 ft. lbs. (19 Nm) and the nuts to 38 ft. lbs. (52 Nm).

46. Install the automatic transaxle mounting bracket, if equipped. Tighten the 12mm nut to 15 ft. lbs. (20 Nm) and the 14mm nut to 38 ft. lbs. (52 Nm).

47. Install the power steering reservoir tank and connect the ground strap.

48. Install the front engine mounting bracket and insulator. Tighten the bolts to 57 ft. lbs. (77 Nm).

49. Install the center mounting bracket and insulator. Tighten the mounting bolts to 38 ft. lbs. (52 Nm).

50. Install the rear engine mounting bracket and insulator. Tighten the mounting bolts to 57 ft. lbs. (77 Nm).

51. Install the engine mounting cross member and tighten the bolts to 29 ft. lbs. (40 Nm).

52. Install and tighten the bolts holding the insulators to the cross member. Tighten the bolts to 54 ft. lbs. (73 Nm). Tighten the mounting insulator through bolts to 64 ft. lbs. (87 Nm).

53. Install the power steering pump and the halfshafts.

54. Install the air conditioning compressor and tighten the mounting bolts. Connect the electrical connection.

55. Raise and safely support the vehicle. Install the front exhaust pipe . Tighten the manifold nuts to 46 ft. lbs. (62 Nm) and the converter nuts 32 ft. lbs. (43 Nm).

56. Install the suspension cross member and tighten the bolts and nuts to 153 ft. lbs. (207 Nm). Lower the vehicle.

57. Push in the engine wire through the cowl panel.

58. Connect the following connectors:

a. 3 engine and ECT Electronic Control Unit (ECU) connectors

b. Circuit opening relay connector

c. Cowl wire connector

d. Instrument panel wire connector

59. Install the glove compartment box and the engine undercover.

60. Connect the transaxle control cables and the speedometer cable.

61. Install the clutch release cylinder and tube clamp, if equipped with manual transaxle.

62. Install the starter, if equipped with manual transaxle.

63. Connect the heater hoses and fuel line hoses. Connect the ground strap to the transaxle case.

64. Connect the following hoses:

a. Brake booster vacuum hose to the air intake chamber

b. Air conditioning control valve vacuum hose

c. Charcoal canister vacuum hose

65. Connect the following:

a. Check connector

b. Ground straps from the left side fender apron

c. Connectors from the relay box

d. Engine compartment wire connector

66. Install the air cleaner case and tighten the mounting bolt. Connect the air cleaner hose, air cleaner cap, hoses and the air flow meter.

67. Install the cruise control actuator.

68. Install the throttle control cable and adjust, if equipped with automatic transaxle.

69. Install the accelerator cable and adjust. Replace the radiator reservoir tank.

70. Install the alternator belt adjusting bar, the alternator and belt.

71. Install the radiator assembly.

72. Replace the ignition coil, igniter and bracket assembly. Connect the igniter connector, noise filter connector and the high tension electrical cord.

73. Install the suspension upper brace and tighten the mounting bolts to 47 ft. lbs.

74. Install the battery and refill the cooling system. Fill the engine crankcase to the proper oil level.

75. Install the engine undercovers and the hood assembly.

76. Connect the battery cables. Start the engine and check for leaks. Check the ignition timing. Adjust, if necessary. Recheck the cooling system and the oil level.

LS400

1. Disconnect the negative battery

cable and the positive battery cable. Remove the hood assembly.

2. Remove the dust covers and the air duct above the radiator assembly. Drain the cooling system.

3. Remove the battery from the vehicle. Raise and safely support the vehicle.

4. Remove the engine undercover and drain the engine oil, in a suitable container. Lower the vehicle.

5. Disconnect the radiator upper hose from the water inlet. Loosen the nuts holding the fluid coupling to the fan bracket.

6. Loosen the drive belt tension by turning the belt tensioner counterclockwise. Remove the drive belt.

7. Remove the radiator assembly.

8. Disconnect the air flow meter connector, the mounting bolts and the the air cleaner hose. Remove the air cleaner, the air flow meter and hose assembly.

9. Remove the igniter cover and disconnect the igniter connectors.

10. Remove the bolts, nut and the throttle body cover. Disconnect the accelerator and cruise control actuator cables.

11. Disconnect the air hose from the ISC valve and the power steering air control valve.

12. Disconnect the air connector pipe from the throttle body and remove the air connector pipe. Remove the bolt and connector pipe bracket.

13. Disconnect the air hose from the air intake chamber. Remove the power steering pump mounting bolts and nut. Position the pump aside.

14. Disconnect the coolant level sensor connector and remove the radiator reservoir tank. Remove the mounting bolts and reservoir tank bracket.

15. Disconnect the following hoses:
    a. Heater and bypass hoses
    b. Fuel hoses (plug the open end and catch the fuel in a suitable container)
    c. Vacuum hose from the brake booster on the air intake chamber
    d. Air conditioning control valve vacuum hoses
    e. EVAP and BVSV vacuum hoses

16. Remove the relay box cover. Disconnect the connector and ground cables from the engine compartment relay box. Remove the ground straps from under the fender aprons.

17. Remove the cruise control actuator cover.

18. Remove the instrument panel undercover and the lower the trim panel the ECU for the engine and transmission.

19. Disconnect the glove box door, the glove box light and remove the glove box assembly.

20. Disconnect the ABS ECU and the heater air duct.

21. Disconnect the following connectors:
    a. The 3 engine and Electronic Controlled Transmission (ECT) ECU connectors
    b. Circuit opening relay connector
    c. Cowl wire connector
    d. Instrument panel wire connector

22. Remove the mounting bolts and pull out the engine wire from the cowl panel.

23. Raise and safely support the vehicle.

24. Remove the mounting bolts and disconnect the power steering oil cooler pipe from the oil pan.

25. Remove the mounting bolts and the steering damper.

26. Disconnect the grommet from the floor and the sub-oxygen sensor from the exhaust pipe. Disconnect the 2 sub-oxygen sensors.

27. Remove the sub-oxygen sensor covers and the exhaust pipe. Remove the exhaust pipe support brackets.

28. Remove the catalytic converters and the exhaust pipe heat insulator.

29. Remove the center floor crossmember braces. Remove the driveshaft.

30. Disconnect the shift control rod from the shift lever. Lower the vehicle.

31. Attach a suitable engine hoist to the engine hangers and support the engine.

32. Remove the nuts holding the engine mounting insulators to the front suspension crossmember.

33. Remove the rear engine mounting member. Disconnect the ground strap.

34. Lift out the engine with the transmission attached. Place the engine assembly on a suitable holding fixture. Separate the engine from the transmission.

To install:

35. Connect the engine to the transmission. Attach a suitable engine hoist to the engine hangers.

36. Lower the engine assembly into the vehicle. Insert the stud bolts of the front engine mounting brackets into the stud bolt holes of the front suspension crossmember.

37. Install the rear engine mounting member and tighten the bolts to 19 ft. lbs. (26 Nm) and the nuts to 10 ft. lbs. (14 Nm). Install the ground strap.

38. Remove the engine hoist. Raise and safely support the vehicle.

39. Install the nuts holding the engine mounting brackets to the front suspension crossmember. Tighten the nuts to 43 ft. lbs. (58 Nm).

40. Connect the transmission control rod to the shift lever. Install the propeller shaft.

41. Install the center floor crossmember braces. Tighten the bolts to 9 ft. lbs. (11 Nm).

42. Install the exhaust pipe heat insulator. Replace the catalytic converters and tighten the bolts to 46 ft. lbs. (65 Nm).

43. Install the front exhaust pipe and the sub-oxygen sensor covers. Tighten the bolts to 32 ft. lbs. (44 Nm). Install the sub-oxygen sensors to the exhaust pipe and tighten to 33 ft. lbs. (45 Nm).

44. Install the steering damper and tighten the mounting bolts to 20 ft. lbs. (27 Nm). Connect the engine wire to the wire bracket on the front suspension crossmember.

45. Install the power steering oil cooler pipe to the engine oil pan. Lower the vehicle.

46. Push in the engine wire through the cowl panel and install the wire retainer.

47. Connect the following connectors:
    a. 3 engine and ECT ECU connectors
    b. Circuit opening relay connector
    c. Cowl wire connector
    d. Instrument panel wire connector

48. Install the heater duct and the glove compartment.

49. Install the right side lower instrument panel pad and the engine and ECT electronic control units. Replace the right side instrument panel undercover.

50. Install the cruise control actuator. Connect the connectors and ground cables to the relay box.

51. Install the upper cover to the relay box. Connect the 2 ground straps to the underside of the fender aprons.

52. Connect the following hose:
    a. The heater bypass water hoses
    b. Fuel hoses
    c. Vacuum hose to the brake booster on the air intake chamber
    d. Vacuum hose to the EVAP BVSV

53. Install the air conditioning compressor. Tighten the bolts to 36 ft. lbs. (49 Nm) and the nut to 22 ft. lbs. (30 Nm). Connect the electrical connectors.

54. Install the radiator reservoir tank bracket, the reservoir and connect the coolant level sensor connector.

55. Install the power steering pump and tighten the mounting bolts to 29 ft. lbs. (40 Nm) and the nuts to 32 ft. lbs. (44 Nm). Connect the air hose to the air intake manifold.

56. Connect the air connector pipe to the throttle body. Connect the air hose

to the ISC valve and the power steering control valve.

57. Connect the accelerator cable and the cruise control actuator cable to the throttle body. Install the throttle body cover.

58. Connect the igniter connectors and install the igniter cover.

59. Connect the air cleaner hose to the intake air connector pipe.

60. Install the air cleaner, the air flow meter and hose assembly. Connect the air flow meter connector.

61. Install the radiator.

62. Temporarily install the fan pulley, the fan and the fluid coupling assembly. Install the drive belt by turning the belt tensioner counterclockwise.

63. Tighten the bolts holding the fluid coupling to the fan bracket to 16 ft. lbs. (22 Nm).

64. Install the battery. Replace the air ducts and dust covers.

65. Refill the cooling system and the crankcase to the proper levels.

66. Install the engine undercover and hood assembly.

67. Connect the battery cables. Start the engine and check for leaks. Check the timing.

68. Recheck the fluid levels.

Engine Mounts

REMOVAL & INSTALLATION

ES250

FRONT

1. Raise and safely support the vehicle.

2. Support the engine with a suitable jacking device.

3. Remove the nut and the through bolt.

4. Remove the insulator. Remove the mounting bolts and the bracket, if necessary.

5. The installation is the reverse of the removal procedure. Tighten the bolts to 57 ft. lbs. (77 Nm).

CENTER

1. Raise and safely support the vehicle.

2. Support the engine with a suitable jacking device.

3. Remove the nut and the through bolt.

4. Remove the insulator. Remove the mounting bolts and the bracket, if necessary.

5. The installation is the reverse of the removal procedure. Tighten the bolts to 38 ft. lbs. (52 Nm).

REAR

1. Raise and safely support the vehicle.

2. Support the engine with a suitable jacking device.

3. Remove the nut and the through bolt.

4. Remove the insulator. Remove the mounting bolts and the bracket, if necessary.

5. The installation is the reverse of the removal procedure. Tighten the bolts to 57 ft. lbs. (77 Nm). Remove the jacking device and lower the vehicle.

LS400

1. Raise and safely support the vehicle.

2. Support the engine with a suitable jacking device.

3. Remove the mounting nuts and the mounting insulator. Remove the mounting bracket, if necessary.

4. The installation is the reverse of the removal procedure. Tighten the bolts to 19 ft. lbs. (26 Nm) and the nuts to 10 ft. lbs. (14 Nm). Remove the jacking device and lower the vehicle.

Cylinder Head

REMOVAL & INSTALLATION

ES250

1. Disconnect the negative battery cable. Drain the cooling system.

2. Remove the suspension upper brace. Disconnect the throttle cable from the throttle body, if equipped with an automatic transaxle.

3. Disconnect the accelerator cable from the throttle body. Remove the cruise control actuator and the vacuum pump.

4. Disconnect the ISC hose, the vacuum pipe hose and the air cleaner hose.

5. Raise and safely support the vehicle. Remove the right side engine undercover.

6. Remove the suspension lower crossmember and the front exhaust pipe. Lower the vehicle.

7. Remove the alternator, the ISC valve and the throttle body.

8. Remove the EGR pipe, the EGR valve and the vacuum modulator.

9. Disconnect the (4) BVSV vacuum hoses, the fuel pressure VSV hose and the air conditioning control valve vacuum hose.

10. Remove the distributor and the exhaust crossover pipe.

11. Disconnect the cold start injector connector and remove the cold start injector tube.

12. Disconnect the following hoses:
    a. PCV hose
    b. Vacuum sensing hose
    c. Fuel pressure VSV hose
    d. Air conditioning control valve vacuum hose

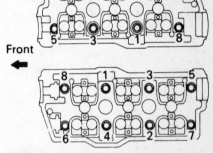

12 Pointed Head

Cylinder head tightening sequence— ES250

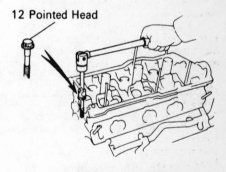

12 Pointed Head

Front

Cylinder head bolt removal sequence— ES250

e. EGR gas temperature sensor connector—California only

13. Remove the mounting bolts and the brackets from the intake chamber. Remove the air intake chamber.

14. Remove the delivery pipes and the injectors.

15. Disconnect the water tempera-

ture sensor connector and the upper radiator hose. Remove the water outlet.

16. Disconnect the following connectors and hose:

    a. Cold start injector time switch connector

    b. Water temperature sensor connector

    c. Heater water bypass hose

17. Remove the water bypass outlet. Remove the crossover pipe insulator from the water bypass outlet.

18. Remove the cylinder head rear plate and the idler pulley bracket.

19. Remove the mounting bolts and nuts. Lift off the intake manifold.

20. Disconnect the main oxygen sensor connector. Remove outside heat insulator.

21. Remove the mounting nuts and the right side exhaust manifold. Remove the inside heat insulator.

22. Remove the heat insulator, the mounting nuts and the left side exhaust manifold.

23. Remove the spark plugs.

24. Remove the timing belt, camshaft pulleys and the No. 2 idler pulley.

25. Remove the No. 3 timing cover.

26. Remove the mounting nuts and the cylinder head covers. Remove the spark plug tube gaskets.

27. Remove the exhaust camshaft of the right side cylinder head by:

    a. Align the timing marks, 2 pointed marks, of the camshaft drive and the drive gear by turning the camshaft with a wrench.

    b. Secure the exhaust camshaft sub-gear to the drive gear with a service bolt.

    c. Uniformly, loosen and remove the bearing cap bolts, in sequence.

    d. Remove the bearing caps and the camshaft.

28. Uniformly, loosen and remove the 10 bearing cap bolts on the right side cylinder head in the proper sequence. Remove the 5 bearing caps and remove the intake camshaft.

29. Remove the exhaust camshaft of the left side cylinder head by performing the following:

    a. Align the timing marks, 1 pointed mark, of the camshaft drive and the drive gear by turning the camshaft with a wrench.

    b. Secure the exhaust camshaft sub-gear to the drive gear with a service bolt.

    c. Uniformly, loosen and remove the 8 bearing cap bolts in the proper sequence.

    d. Remove the 4 bearing caps and the exhaust camshaft.

30. Uniformly, loosen and remove the bearing cap bolts on the right side cylinder head, in sequence. Remove

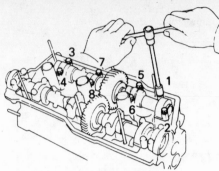

Right hand exhaust camshaft bolt removal sequence—ES250

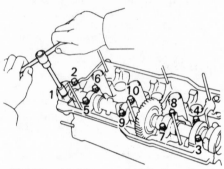

Right hand intake camshaft bolt removal sequence—ES250

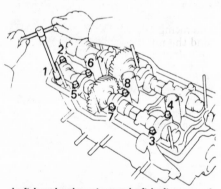

Left hand exhaust camshaft bolt removal sequence—ES250

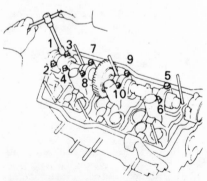

Left hand intake camshaft bolt removal sequence—ES250

the bearing caps and remove the intake camshaft.

31. Remove the recessed head bolts with the proper tool.

32. Uniformly, loosen and remove the cylinder head bolts in the proper sequence. Lift the cylinder head from the dowels on the cylinder block and place on wooden blocks.

NOTE: Be careful not to damage the contact surfaces of the cylinder head and cylinder block.

To install:

33. Place the cylinder head gasket in position on the cylinder block.

34. Install the cylinder head onto the cylinder block, aligning the dowels.

35. Install the head bolts and tighten, in sequences, using 3 steps:

    a. Apply a light coat of engine oil on the threads of the bolts and install. Uniformly, tighten the cylinder head bolts in several passes, in sequence, to 25 ft. lbs. (34 Nm).

    b. Mark the front of the cylinder head bolt with paint. Retighten the cylinder head bolts 90 degrees in the proper sequence.

    c. Retighten the cylinder head bolts by an additional 90 degrees.

36. Apply a light coat of engine oil on the recessed head bolts. Install the head bolts and tighten to 13 ft. lbs. (18 Nm).

37. Install the left side engine hanger and tighten to 27 ft. lbs. (37 Nm).

NOTE: Since the thrust clearance of the camshaft is small, the camshaft must be held level while it is being installed. If the camshaft is not level, the portion of the cylinder head receiving the shaft thrust may crack or be damaged, causing the camshaft to seize or break.

38. Install the intake camshaft of the right side cylinder head by:

    a. Apply a suitable multi-purpose grease to the thrust portion of the camshaft.

    b. Apply seal packing to the No. 1 bearing cap. Install the bearing caps.

    c. Apply a light coat of oil on the threads of the bearing cap bolts.

    d. Install and uniformly tighten the bearing cap bolts to 12 ft. lbs. (16 Nm).

39. Install the exhaust camshaft of the right side cylinder head by:

    a. Apply a suitable multi-purpose grease to the thrust portion of the camshaft.

    b. Align the timing marks, 2 pointed marks, of the camshaft and the drive gears.

    c. Place the camshaft on the cylinder head and install the bearing caps.

    d. Apply a light coat of oil on the threads of the bearing cap bolts.

e. Install and uniformly tighten the bearing cap bolts to 12 ft. lbs. (16 Nm). Remove the service bolt.

40. Install the intake camshaft on the left side cylinder head by:

a. Apply a suitable multi-purpose grease to the thrust portion of the camshaft.

b. Place the intake camshaft at a 90 degree angle of the timing mark on the cylinder head, 1 pointed mark.

c. Apply seal packing to the No. 1 bearing cap. Install the bearing caps.

d. Apply a light coat of oil on the threads of the bearing cap bolts.

e. Install and uniformly tighten the bearing cap bolts to 12 ft. lbs. (16 Nm).

41. Install the exhaust camshaft on the left side cylinder head by:

a. Apply MP grease to the thrust portion of the camshaft.

b. Align the timing marks, 1 pointed mark, of the camshaft and the drive gears.

c. Place the camshaft on the cylinder head and install the bearing caps.

d. Apply a light coat of oil on the threads of the bearing cap bolts.

e. Install and uniformly tighten the bearing cap bolts to 12 ft. lbs. (16 Nm). Remove the service bolt.

42. Turn the camshaft and position the cam lobe upward. Check and adjust the valve clearance.

43. Apply a suitable multi-purpose grease to the the new camshaft oil seals and install with the proper tool.

44. Install the spark plug tube gaskets. Install the proper seal packing to the cylinder heads.

45. Install the cylinder head cover gasket and install with the seal washers and nuts. Tighten to 52 inch lbs.

46. Install the No. 3 timing belt cover and tighten the bolts 65 inch lbs.

47. Install the No. 2 idler pulley, the camshaft timing pulleys and the timing belt.

48. Install the spark plugs and tighten to 13 ft. lbs. (18 Nm).

49. Install the right side heat insulator and the right side exhaust manifold with a new gasket. Tighten the nuts to 29 ft. lbs. (40 Nm).

50. Install the outside heat insulator and connect the oxygen sensor connector.

51. Install the left side exhaust manifold with a new gasket and tighten the nuts to 29 ft. lbs. (40 Nm). Install the heat insulator.

52. Install the intake manifold with new gaskets. Tighten the nuts and bolts to 13 ft. lbs. (18 Nm).

53. Install the No. 2 idler pulley bracket and tighten to 13 ft. lbs. (18

Nm). Replace the rear cylinder head plate.

54. Install the crossover pipe heat insulator. Replace the water bypass outlet and tighten the mounting bolts to 14 ft. lbs. (19 Nm).

55. Connect the heater water bypass hose, cold start injector time switch connector and the water temperature switch.

56. Connect the upper radiator hose, the water temperature sensor connector and install the water outlet with a new gasket. Tighten the bolts to 73 inch lbs.

57. Install the fuel injectors and the delivery pipes.

58. Install the air intake chamber with new gaskets and tighten the mounting bolts to 32 ft. lbs. (44 Nm).

59. Install the intake chamber brackets.

60. Connect the following hoses:

a. PCV hose

b. Vacuum sensing hose

c. EGR gas temperature sensor connector (California only)

d. Air conditioning control valve air hose

61. Install the cold start injector tube and connect the cold start injector connector.

62. Install the exhaust crossover pipe with new gaskets and tighten the mounting bolts to 25 ft. lbs. (34 Nm) and the nuts to 29 ft. lbs. (40 Nm).

63. Install the distributor.

64. Install the vacuum pipe and connect the BVSV vacuum hoses and the air conditioning control valve vacuum hose.

65. Install the EGR valve and vacuum modulator. Tighten the mounting bolts to 13 ft. lbs. (18 Nm).Install the vacuum pipe hoses.

66. Install the EGR pipe with a new gasket and tighten the mounting bolts to 13 ft. lbs. (18 Nm) and the nut to 58 ft. lbs. (79 Nm).

67. Install the throttle body and ISC valve.

68. Raise and support the vehicle. Install the front exhaust pipe and the right side engine undercover. Lower the vehicle.

69. Install the air cleaner hose and cruise control actuator.

70. Install the accelerator cable and connect the throttle cable, if equipped with an automatic transaxle.

71. Install the suspension upper brace and fill the cooling system.

72. Connect the negative battery cable. Run the engine and check for leaks.

73. Adjust the timing. Recheck the fluid levels.

LS400

1. Disconnect the negative battery cable. Drain the cooling system.

2. Remove the camshaft timing pulleys.

3. Disconnect the accelerator cable, the throttle control cable, if equipped with automatic transmission and the cruise control actuator cable.

4. Remove the high tension cord cover and the right side ignition coil.

5. Remove the water inlet housing mounting bolts and disconnect the water bypass hose from the ISC valve.

6. Remove the water inlet and inlet housing assemblies. Remove the O-ring from the water inlet housing.

7. Remove the EGR pipe.

8. Disconnect the following:

a. VSV connector

b. Vacuum pipe hose

c. EGR water bypass pipe

d. Fuel pressure VSV

9. Disconnect the EGR vacuum hoses and remove the EGR VSV.

10. Disconnect the following hoses:

a. Water bypass pipe hose from the ISC valve.

b. Water bypass joint hose.

c. Vacuum pipe hoses.

11. Disconnect the EGR gas temperature sensor (California only). Remove the EGR valve adapter.

12. Disconnect the following:

a. Fuel pressure regulator vacuum hose.

b. Air intake chamber vacuum hose.

c. Vacuum hose from the EVAP BVSV.

13. Remove the mounting bolts, hoses and the vacuum pipe.

14. Remove the ISC valve.

15. Remove the throttle body sensor connectors and the water bypass pipe from the rear water bypass joint.

16. Remove the mounting bolts/nuts and disconnect the PCV valve. Remove the throttle body and gasket.

17. Disconnect the accelerator cable bracket and the brake booster vacuum union and hose.

18. Disconnect the cold start injector connector and the cold start injector tube from the right side delivery pipe.

19. Disconnect the check connector from the intake chamber and remove the mounting nuts and bolts.

20. Remove the air intake chamber and the cold start injector, tube and wire assembly.

21. Disconnect the engine wire from the intake manifold and from the right side cylinder head. Disconnect the heater hoses.

22. Remove the delivery pipes and the fuel injectors. Remove the mounting bolts and nuts. Lift up the intake manifold.

23. Remove the front and rear water bypass joint.

24. Raise and safely support the vehicle. Remove the front exhaust pipe

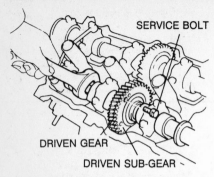

Securing the camshaft gears – LS400

and the main catalytic converters. Lower the vehicle.

25. Disconnect the right side oxygen sensor. Remove the mounting bolts and nuts and remove the right side exhaust manifold.

26. Remove the oil dipstick and guide. Disconnect the left side oxygen sensor.

27. Remove the mounting bolts and nuts and remove the left side exhaust manifold.

28. Remove the 2 engine hangers and the wire brackets from the right side cylinder head.

29. Remove the bolts, washers and the cylinder head cover. Remove the semi-circular plugs, if necessary.

30. Remove the exhaust camshaft of the right side cylinder head by:

a. Position the service bolt hole of the drive sub-gear to the upright position. Secure the camshaft sub-gear to drive gear with a service bolt.

b. Set the timing mark, 1 pointed mark, of the camshaft drive gear at approximately 10 degrees, by turning the camshaft with the proper tool.

c. Alternately, loosen and remove the bearing cap bolts holding the intake camshaft side of the oil feed pipe to the cylinder head.

d. Uniformly, loosen and remove the bearing cap bolts in sequence.

e. Remove the oil feed pipe and the bearing caps. Remove the camshaft.

31. Remove the intake camshaft from the right side cylinder head by:

a. Set the timing mark, 1 pointed mark, of the camshaft drive gear at approximately 45 degrees, by turning the camshaft with the proper tool.

c. Uniformly, loosen and remove the bearing cap bolts in the proper sequence.

d. Remove the bearing caps, oil seal and the intake camshaft.

32. Remove the exhaust camshaft of the left side cylinder head by:

a. Position the service bolt hole of the drive sub-gear to the upright po-

sition. Secure the camshaft sub-gear to drive gear with a service bolt.

NOTE: When removing the camshaft, make sure the torsional spring force of the sub-gear has been eliminated.

b. Set the timing mark, 2 pointed marks, of the camshaft drive gear at approximately 15 degrees, by turning the camshaft with the proper tool.

c. Alternately, loosen and remove the bearing cap bolts holding the intake camshaft side of the oil feed pipe to the cylinder head.

d. Uniformly, loosen and remove the bearing cap bolts in the proper sequence.

e. Remove the oil feed pipe and the bearing caps. Remove the camshaft.

33. Remove the intake camshaft from the left side cylinder head by:

a. Set the timing mark, 1 pointed mark, of the camshaft drive gear at approximately 60 degrees, by turning the camshaft with the proper tool.

c. Uniformly, loosen and remove the bearing cap bolts in the proper sequence.

d. Remove the bearing caps, oil seal and the intake camshaft.

34. Disconnect the ground straps and clamp of the engine wire from the rear of the cylinder heads.

35. Uniformly, loosen the head bolts of 1 side of the cylinder head then on the other side. Remove the head bolts and washers.

36. Lift out the cylinder head from the dowels on the cylinder block and place on a suitable holding fixture. Remove the gasket and clean the mounting surface.

To install:

37. Place new cylinder gasket into position on the cylinder block. Install the cylinder head.

38. The cylinder head bolts are tightened in 2 steps:

a. Apply a light coat of engine oil on the threads of the bolts. Temporarily, install the washers and bolts. Uniformly, tighten the head bolts one 1 side of the cylinder head in the proper sequence then the other side. Tighten the bolts to 29 ft. lbs. (40 Nm).

b. Mark the front the cylinder head bolt with paint. Retighten the cylinder head bolts in the proper se-

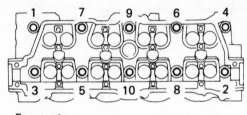

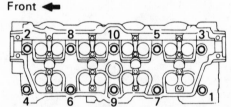

Cylinder head bolt removal sequence – LS400

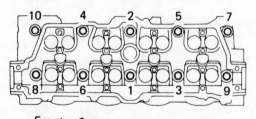

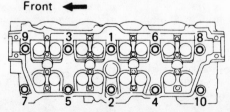

Cylinder head tightening sequence – LS400

quence 90 degrees. Check that the painted mark is at a 90 degree angle to the front.

39. Connect the engine wire to the cylinder head(s), tighten the clamps.

40. Install the circular plugs on the cylinder head with the cup side facing forward.

41. Remove any old packing and apply new seal packing to the bearing caps.

42. Install the bearing cap on the right side cylinder head, marked I1, in position with the arrow mark facing the rear. Install the bearing cap on the left side cylinder head, marked I6, in position with the arrow mark facing the front.

43. Apply a light coat of oil on the threads of the cap bolts. Install the nearing cap bolts with new washers and tighten to 12 ft. lbs. (16 Nm).

44. Install the right side cylinder head intake camshaft by:

a. Apply MP grease to the thrust portion of the camshaft.

b. Place the intake camshaft at a 45 degree angle of the timing mark, 1 pointed mark, on the cylinder head.

c. Remove any old packing and apply new seal packing to the bearing cap marked I6 and install the front bearing cap, marked I6 with the arrow facing rearward.

d. Align the arrows at the front and rear of the cylinder head with the bearing cap.

e. Install the remaining bearing caps in the proper sequence with the arrow mark facing rearward. Install the oil feed pipe and the mounting bolts.

f. Uniformly, tighten the bearing cap bolts in the proper sequence to 12 ft. lbs. (16 Nm).

45. Install the right side cylinder head exhaust camshaft by:

a. Set the timing mark, 1 pointed mark, of the camshaft drive gear at a 10 degree angle by turning the intake camshaft with the proper tool.

b. Apply MP grease to the thrust portion of the camshaft.

c. Align the timing marks, 1 pointed mark, of the camshaft drive and driven gears.

d. Place the exhaust camshaft ion the cylinder head. Install the rear bearing cap with the arrow mark facing rearward.

e. Align the arrow marks at the front and rear of the cylinder head with the mark on the bearing cap. Apply a light coat of oil on the threads of the bearing cap bolts.

f. Uniformly, tighten the bearing cap bolts in the proper sequence to 12 ft. lbs.

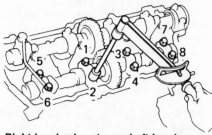

Right hand intake camshaft bearing cap bolt tightening sequence—LS400

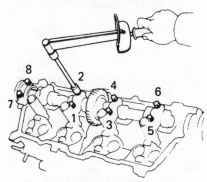

Right hand exhaust camshaft bearing cap bolt tightening sequence—LS400

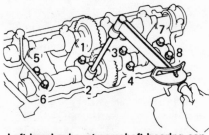

Left hand intake camshaft bearing cap bolt tightening sequence—LS400

Left hand exhaust camshaft bearing cap bolt tightening sequence—LS400

g. Bring the service bolt installed upward by turning the camshaft with the proper tool. Remove the service bolt.

46. Install the left side cylinder head intake camshaft by:

a. Apply MP grease to the thrust portion of the camshaft.

b. Place the intake camshaft at a

60 degree angle of the timing mark (1 pointed mark) on the cylinder head.

c. Remove any old packing and apply new seal packing to the bearing cap marked I6 and install the front bearing cap, marked I1 with the arrow facing rearward.

d. Align the arrows at the front and rear of the cylinder head with the bearing cap. Apply a light coat of oil on the threads of the bearing cap bolts.

e. Install the remaining bearing caps in the proper sequence with the arrow mark facing rearward. Install the oil feed pipe and the mounting bolts.

f. Uniformly, tighten the bearing cap bolts in the proper sequence to 12 ft. lbs. (16 Nm).

47. Install the left side cylinder head exhaust camshaft by:

a. Set the timing mark, 2 dot marks, of the camshaft drive gear at a 15 degree angle by turning the intake camshaft with the proper tool.

b. Apply MP grease to the thrust portion of the camshaft.

c. Align the timing marks, 2 dot marks, of the camshaft drive and driven gears.

d. Place the exhaust camshaft ion the cylinder head. Install the rear bearing cap with the arrow mark facing rearward.

e. Align the arrow marks at the front and rear of the cylinder head with the mark on the bearing cap. Apply a light coat of oil on the threads of the bearing cap bolts.

f. Uniformly, tighten the bearing cap bolts in the proper sequence to 12 ft. lbs. (16 Nm).

g. Bring the service bolt installed upward by turning the camshaft with the proper tool. Remove the service bolt.

48. Install the camshaft oil seals with the proper tool. Install the semi-circular plugs with the proper seal packing.

49. Install the cylinder head covers with the proper seal packing and gasket. Tighten the mounting bolts to 52 inch lbs.

50. Install the engine wire bracket and hangers. Tighten the hanger bolts to 27 ft. lbs. (37 Nm).

51. Install the right side exhaust manifold with a new gasket and tighten the mounting bolts to 29 ft. lbs. (40 Nm). Connect the right side oxygen sensor connector.

52. Install the right side exhaust manifold with a new gasket and tighten the mounting bolts to 29 ft. lbs. (40 Nm). Connect the right side oxygen sensor connector.

53. Install the left side exhaust manifold with a new gasket and tighten the

mounting bolts to 29 ft. lbs. (40 Nm). Connect the left side oxygen sensor connector.

54. Install the oil dipstick and guide. Raise and safely support the vehicle.

55. Install the catalytic converters and front exhaust pipe. Lower the vehicle.

56. Install the front and rear water bypass joints. Tighten the mounting bolts to 13 ft. lbs. (18 Nm).

57. Install the intake manifold, using new gaskets. Tighten the mounting nuts and bolts to 13 ft. lbs. (18 Nm).

58. Install the delivery pipes and fuel injectors. Install the fuel return pipe with new gaskets. Tighten the union bolt to 26 ft. lbs. (35 Nm).

59. Connect the fuel hoses and the injector connectors. Connect the engine wire to the delivery pipes.

60. Connect the connectors on the left side delivery pipe, the water temperature sensor connector, cold start injector time switch connector and the water temperature sender gauge connector.

61. Connect the heater hoses and engine wire bracket. Install the engine wire to the bracket.

62. Install the cold start injector, tube and wire assembly. Tighten the mounting bolts to 69 inch lbs.

63. Install the air intake chamber with new gaskets and tighten the mounting bolts to 13 ft. lbs. (18 Nm).

64. Connect the cold start injector tube to the right side delivery pipe and tighten the union bolt to 11 ft. lbs. (15 Nm).

65. Connect the cold start injector connector. Install the accelerator cable bracket.

66. Install the brake booster union and connect the vacuum hose. Tighten the union bolt to 22 ft. lbs. (30 Nm).

67. Connect the water bypass hose to the throttle body and the PCV hose to the cylinder head cover.

68. Install the throttle body, using a new gasket. Tighten the mounting bolts to 13 ft. lbs. (18 Nm).

69. Install the water bypass pipe and connect the sensor connectors. Install the ISC valve and tighten the mounting bolts to 13 ft. lbs. (18 Nm). Connect the water bypass hose.

70. Install the vacuum pipe and the following hoses:
   a. Fuel pressure regulator vacuum hose.
   b. Vacuum hose to the upper port of the EVAP BVSV.
   c. Air intake chamber vacuum hose.
   d. Throttle body vacuum hoses.

71. Install the EGR valve adapter with a new gasket. Tighten the mounting bolts to 13 ft. lbs. (18 Nm).

72. Connect the EGR gas temperature sensor connector (California only).

73. Install the EGR valve and vacuum modulator. Connect the water bypass hoses and the vacuum hoses.

74. Install the EGR and fuel pressure VSV and connect the hoses and connectors. Replace the EGR pipe and tighten the mounting bolts to 13 ft. lbs. (18 Nm).

75. Install the timing belt rear plates and tighten the bolts to 69 inch lbs. Install the water inlet and inlet hosing and tighten the bolts to 13 ft. lbs. (18 Nm).

76. Install the right side ignition coil and the high tension cord cover.

77. Connect and adjust the accelerator cable, the automatic transmission throttle cable and the cruise control actuator cable. Install the camshaft timing pulley.

78. Fill the cooling system and connect the negative battery cable. Start the engine and check for leaks.

79. Recheck all the fluid levels and check the ignition timing.

Valve Lifters

REMOVAL & INSTALLATION

1. Disconnect the negative battery cable.

2. Remove the cylinder head from the cylinder block.

3. Remove the lifters and the adjusting shims, using the proper tool.

4. Install the valve lifters and shims.

5. Check that the valve lifter rotates smoothly.

6. Install the cylinder head.

7. Connect the negative battery cable.

Valve Lash

ADJUSTMENT

ES250

NOTE: Adjust the valve clearance when the engine is cold.

1. Disconnect the negative battery cable.

2. Remove the air intake chamber. Remove the cylinder head covers.

3. Turn the crankshaft pulley and align it's groove with the timing mark 0 of the No. 1 timing cover.

4. Check that the valve lifters on the No. 1 intake and exhaust are tight.

5. Measure the clearance between the valve lifter and the camshaft. Record the measurements on valves No. 1, 2, 3 and 6.

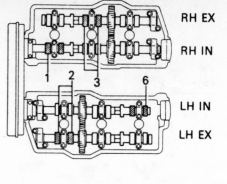

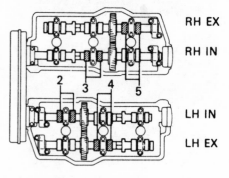

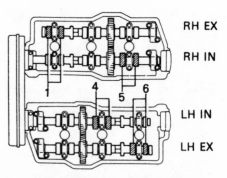

Valve adjusting sequence—ES250

a. The intake valve clearance cold is 0.005-0.009 in. (0.13-0.23mm).

b. The exhaust valve clearance cold is 0.011-0.015 in. (0.27-0.37mm).

6. Turn the crankshaft $2/3$ of a revolution and check the clearance on valves No. 2, 3, 4 and 5 and record.

7. Turn the crankshaft another $2/3$ of a revolution and check valves; 1, 4, 5 and 6 and record.

8. Remove the adjusting shim and turn the crankshaft to position the cam lobe of the camshaft on the adjusting valve upward. Press down the valve lifter with the proper tool and place the proper tool between the camshaft and the valve lifter. Remove the tool.

9. Remove the adjusting shim with the proper tool.

10. Install the specified valve shim on the valve lifter with the proper tool.

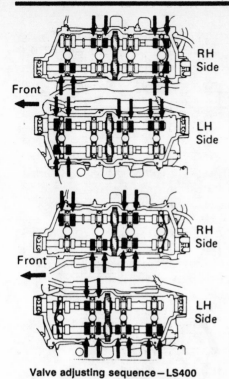

Valve adjusting sequence—LS400

11. Recheck the valve clearance.

12. Install the cylinder head covers and intake chamber.

13. Connect the negative battery cable.

LS400

1. Disconnect the negative battery cable.

2. Remove the No. 3 timing belt covers.

3. Disconnect the spark plug wires and remove the cylinder head covers.

4. Turn the crankshaft pulley and align it's groove with the timing mark 0 of the No. 1 timing cover. Check that the timing marks of the camshaft timing pulleys and timing belt rear plates are aligned. If not, turn the crankshaft 1 revolution (360 degrees) and align the mark.

5. Measure the clearance between the valve lifter and the camshaft on the valves in step 1 and record.

   a. The intake valve clearance cold is 0.006-0.010 in. (0.15-0.25mm).

   b. The exhaust valve clearance cold is 0.010-0.014 in. (0.25-0.35mm).

6. Turn the crankshaft 1 full revolution (360 degrees) and align the mark.

7. Measure the clearance between the valve lifter and the camshaft on the valves in step 2 and record.

8. Remove the adjusting shim and turn the crankshaft to position the cam lobe of the camshaft on the adjusting valve upward. Press down the valve lifter with the proper tool and

place the proper tool between the camshaft and the valve lifter. Remove the tool.

9. Remove the adjusting shim with the proper tool.

10. Install the specified valve shim on the valve lifter with the proper tool.

11. Recheck the valve clearance. Install the cylinder head covers.

12. Connect the spark plug wires and install the No. 3 timing belt covers.

13. Connect the negative battery cable.

Intake Manifold

REMOVAL & INSTALLATION

ES250

1. Disconnect the negative battery cable. Drain the cooling system.

2. Remove the suspension upper brace. Disconnect the throttle cable from the throttle body, if equipped with an automatic transaxle.

3. Disconnect the accelerator cable from the throttle body. Remove the cruise control actuator and the vacuum pump.

4. Disconnect the ISC hose, the vacuum pipe hose and the air cleaner hose.

5. Raise and safely support the vehicle. Remove the right side engine undercover.

6. Remove the suspension lower crossmember and the front exhaust pipe. Lower the vehicle.

7. Remove the alternator, the ISC valve and the throttle body.

8. Remove the EGR pipe, the EGR valve and the vacuum modulator.

9. Disconnect the (4) BVSV vacuum hoses, the fuel pressure VSV hose and the air conditioning control valve vacuum hose.

10. Remove the distributor and the exhaust crossover pipe.

11. Disconnect the cold start injector connector and remove the cold start injector tube.

12. Disconnect the following hoses:

   a. PCV hose

   b. Vacuum sensing hose

   c. Fuel pressure VSV hose

   d. Air conditioning control valve vacuum hose

   e. EGR gas temperature sensor connector—California only

13. Remove the mounting bolts and the brackets from the intake chamber. Remove the air intake chamber.

14. Remove the delivery pipes and the injectors.

15. Disconnect the water temperature sensor connector and the upper radiator hose. Remove the water outlet.

16. Disconnect the following connectors and hose:

   a. Cold start injector time switch connector

   b. Water temperature sensor connector

   c. Heater water bypass hose

17. Remove the water bypass outlet. Remove the crossover pipe insulator from the water bypass outlet.

18. Remove the cylinder head rear plate and the idler pulley bracket.

19. Remove the mounting bolts and nuts. Lift off the intake manifold.

To install:

20. Install the intake manifold with new gaskets. Tighten the nuts and bolts to 13 ft. lbs. (18 Nm).

21. Install the No. 2 idler pulley bracket and tighten to 13 ft. lbs. (18 Nm). Replace the rear cylinder head plate.

22. Install the crossover pipe heat insulator. Replace the water bypass outlet and tighten the mounting bolts to 14 ft. lbs. (19 Nm).

23. Connect the heater water bypass hose, cold start injector time switch connector and the water temperature switch.

24. Connect the upper radiator hose, the water temperature sensor connector and install the water outlet with a new gasket. Tighten the bolts to 73 inch lbs.

25. Install the fuel injectors and the delivery pipes.

26. Install the air intake chamber with new gaskets and tighten the mounting bolts to 32 ft. lbs. (44 Nm).

27. Install the intake chamber brackets.

28. Connect the following hoses:

   a. PCV hose

   b. Vacuum sensing hose

   c. EGR gas temperature sensor connector (California only)

   d. Air conditioning control valve air hose

29. Install the cold start injector tube and connect the cold start injector connector.

30. Install the exhaust crossover pipe with new gaskets and tighten the mounting bolts to 25 ft. lbs. (34 Nm) and the nuts to 29 ft. lbs. (40 Nm).

31. Install the distributor.

32. Install the vacuum pipe and connect the BVSV vacuum hoses and the air conditioning control valve vacuum hose.

33. Install the EGR valve and vacuum modulator. Tighten the mounting bolts to 13 ft. lbs. (18 Nm). Install the vacuum pipe hoses.

34. Install the EGR pipe with a new gasket and tighten the mounting bolts to 13 ft. lbs. (18 Nm) and the nut to 58 ft. lbs. (79 Nm).

35. Install the throttle body and ISC valve.

Installed Shim thickness (mm) chart — Exhaust valve adjusting shim selection

Installed Shim thickness (mm) column headers, left to right:

2.500 · 2.525 · 2.550 · 2.575 · 2.600 · 2.620 · 2.640 · 2.650 · 2.660 · 2.680 · 2.700 · 2.720 · 2.740 · 2.750 · 2.760 · 2.780 · 2.800 · 2.820 · 2.840 · 2.850 · 2.860 · 2.880 · 2.900 · 2.920 · 2.940 · 2.950 · 2.960 · 2.980 · 3.000 · 3.020 · 3.040 · 3.050 · 3.060 · 3.080 · 3.100 · 3.120 · 3.140 · 3.150 · 3.160 · 3.180 · 3.200 · 3.225 · 3.250 · 3.275 · 3.300

For each measured clearance range, the required shim numbers are read across the installed-shim-thickness columns (values shown in reading order, left to right):

Measured clearance (mm)	Shim numbers (across installed-thickness columns, left → right)
0.000 – 0.025	02 02 02 02 02 02 04 04 04 06 08 08 08 10 12 12 12 12 14 14 16 16 16 16 18 18 20 20 22
0.026 – 0.050	02 02 02 02 02 02 04 04 04 06 06 06 08 08 08 10 10 10 12 12 12 14 14 14 16 16 16 18 18 18 20 20 22 22
0.051 – 0.075	02 02 02 02 02 02 04 04 06 06 06 06 08 10 10 10 10 12 12 14 14 14 14 16 16 16 18 18 18 18 20 20 22 22 24
0.076 – 0.100	02 02 02 02 02 04 04 04 06 06 06 08 08 10 10 10 12 12 12 14 14 16 16 16 16 18 18 18 20 20 20 22 22 24 24
0.101 – 0.125	02 02 02 02 02 04 04 04 04 06 06 08 08 08 10 10 12 12 12 12 14 14 16 16 16 16 18 18 20 20 20 20 22 22 24 24 26
0.126 – 0.150	02 02 02 02 02 02 04 04 04 06 06 06 08 08 08 10 10 10 12 12 12 14 14 14 16 16 16 18 18 18 18 20 20 20 22 22 22 24 24 26 26
0.151 – 0.175	02 02 02 02 02 02 04 04 06 06 06 06 08 08 10 10 10 10 12 12 14 14 14 14 16 16 18 18 18 18 20 20 22 22 22 22 24 24 26 26 28
0.176 – 0.200	02 02 02 02 02 04 04 04 06 06 06 08 08 08 10 10 12 12 12 14 14 14 16 16 16 18 18 18 20 20 20 22 22 24 24 24 24 26 26 28 28
0.201 – 0.225	02 02 02 02 04 04 04 04 06 08 08 08 10 10 12 12 12 12 14 14 16 16 16 18 18 18 20 20 20 22 22 24 24 24 24 26 26 28 28 30
0.226 – 0.250	02 02 02 02 04 04 04 06 06 06 08 08 10 10 10 12 12 12 14 14 14 16 16 16 18 18 18 20 20 20 22 22 22 24 24 26 26 26 28 28 30 30
0.251 – 0.269	02 02 02 04 04 06 06 06 08 08 10 10 10 10 12 12 14 14 14 14 16 16 16 18 18 18 18 18 20 20 22 22 22 22 24 24 26 26 26 28 28 28 30 30 32
0.270 – 0.370	
0.371 – 0.375	04 06 06 08 08 08 10 10 10 12 12 12 14 14 14 16 16 16 18 18 18 20 20 20 22 22 22 24 24 24 26 26 26 28 28 28 30 30 30 32 32 34 34 34
0.376 – 0.400	04 06 06 08 08 10 10 10 12 12 12 14 14 14 16 16 16 18 18 18 20 20 20 22 22 22 24 24 24 26 26 26 28 28 28 30 30 30 32 32 32 34 34 34
0.401 – 0.425	06 06 08 08 10 10 12 12 12 12 14 14 16 16 16 16 18 18 20 20 20 20 22 22 24 24 24 24 26 26 28 28 28 28 30 30 32 32 32 32 34 34 34
0.426 – 0.450	06 08 08 10 10 12 12 12 14 14 14 16 16 16 18 18 18 20 20 20 22 22 22 24 24 24 26 26 26 28 28 28 30 30 30 32 32 32 34 34 34 34
0.451 – 0.475	08 08 10 10 12 12 14 14 14 14 16 16 18 18 18 18 20 20 22 22 22 22 24 24 26 26 26 26 28 28 30 30 30 30 32 32 34 34 34 34 34
0.476 – 0.500	08 10 10 12 12 14 14 14 16 16 16 18 18 18 20 20 20 22 22 22 24 24 24 26 26 26 28 28 28 30 30 30 32 32 32 34 34 34 34 34
0.501 – 0.525	10 10 12 12 14 14 16 16 16 16 18 18 20 20 20 20 22 22 24 24 24 24 26 26 28 28 28 28 30 30 32 32 32 32 34 34 34 34 34
0.526 – 0.550	10 12 12 14 14 16 16 16 18 18 18 20 20 20 22 22 22 24 24 24 26 26 26 28 28 28 30 30 30 32 32 32 34 34 34 34 34
0.551 – 0.575	12 12 14 14 16 16 18 18 18 18 20 20 22 22 22 22 24 24 26 26 26 26 28 28 30 30 30 30 32 32 34 34 34 34 34
0.576 – 0.600	12 14 14 16 16 18 18 18 20 20 20 22 22 22 24 24 24 26 26 26 28 28 28 30 30 30 32 32 32 34 34 34 34 34
0.601 – 0.625	14 14 16 16 18 18 20 20 20 20 22 22 24 24 24 26 26 28 28 28 28 30 30 32 32 32 32 34 34 34 34 34
0.626 – 0.650	14 16 16 18 18 20 20 20 22 22 22 24 24 26 26 26 28 28 28 30 30 30 32 32 32 34 34 34 34 34
0.651 – 0.675	16 16 18 18 20 20 22 22 22 22 24 24 26 26 26 28 28 30 30 30 30 32 32 34 34 34 34 34
0.676 – 0.700	16 18 18 20 20 22 22 22 24 24 24 26 26 28 28 28 30 30 30 32 32 32 34 34 34 34 34
0.701 – 0.725	18 18 20 20 22 22 24 24 24 24 26 26 28 28 28 30 30 32 32 32 32 34 34 34 34 34
0.726 – 0.750	18 20 20 22 22 24 24 26 26 26 28 28 30 30 30 32 32 32 34 34 34 34 34
0.751 – 0.775	20 20 22 22 24 24 26 26 26 28 28 30 30 30 30 32 32 34 34 34 34 34 34
0.776 – 0.800	20 22 22 24 24 26 26 28 28 28 30 30 30 32 32 32 34 34 34 34 34
0.801 – 0.825	22 22 24 24 26 26 28 28 28 28 30 30 32 32 32 32 34 34 34 34 34
0.826 – 0.850	22 24 24 26 26 28 28 30 30 30 32 32 32 34 34 34 34 34
0.851 – 0.875	24 24 26 26 28 28 30 30 30 30 32 32 34 34 34 34 34
0.876 – 0.900	24 26 26 28 28 30 30 30 32 32 32 34 34 34 34 34
0.901 – 0.925	26 26 28 28 30 30 32 32 32 32 34 34 34 34 34
0.926 – 0.950	26 28 28 30 30 32 32 32 34 34 34 34 34
0.951 – 0.975	28 28 30 30 32 32 34 34 34 34 34
0.976 – 1.000	28 30 30 32 32 34 34 34 34 34
1.001 – 1.025	30 30 32 32 34 34 34 34 34
1.026 – 1.050	30 32 32 34 34 34 34
1.051 – 1.075	32 32 34 34 34
1.076 – 1.100	32 34 34 34
1.101 – 1.125	34 34 34
1.126 – 1.150	34 34
1.151 – 1.170	34

New shim thicknesses — mm (in.)

Shim No.	Thickness	Shim No.	Thickness
02	2.500 (0.0984)	20	2.950 (0.1161)
04	2.550 (0.1004)	22	3.000 (0.1181)
06	2.600 (0.1024)	24	3.050 (0.1201)
08	2.650 (0.1043)	26	3.100 (0.1220)
10	2.700 (0.1063)	28	3.150 (0.1240)
12	2.750 (0.1083)	30	3.200 (0.1260)
14	2.800 (0.1102)	32	3.250 (0.1280)
16	2.850 (0.1122)	34	3.300 (0.1299)
18	2.900 (0.1142)		

Exhaust valve adjusting shim selection using chart — LS400

Installed shim thickness (mm)

Measured clearance (mm)	2.500	2.525	2.550	2.575	2.600	2.620	2.640	2.650	2.660	2.680	2.700	2.720	2.740	2.750	2.760	2.780	2.800	2.820	2.840	2.850	2.860	2.880	2.900	2.920	2.940	2.950	2.960	2.980	3.000	3.020	3.040	3.050	3.060	3.080	3.100	3.120	3.140	3.150	3.160	3.180	3.200	3.225	3.250	3.275	3.300
0.000 – 0.025						02	02	02	02	02	04	04	04	06	06	08	08	08	10	10	10	12	12	12	14	14	14	16	16	16	18	18	18	20	20	20	22	22	22	24	24	26	26		28
0.026 – 0.050							02	02	02	02	02	04	04	06	06	06	08	08	10	10	10	10	12	12	14	14	14	14	16	16	18	18	18	18	20	20	22	22	22	24	24	26	26	28	28
0.051 – 0.075					02	02	02	02	02	04	04	04	06	06	06	08	08	10	10	10	12	12	12	14	14	14	16	16	16	18	18	18	20	20	20	22	22	22	24	24	24	26	28	28	30
0.076 – 0.100				02	02	02	04	04	04	04	06	06	08	08	08	08	10	10	12	12	12	12	14	14	16	16	16	16	18	18	20	20	20	20	22	22	24	24	24	26	26	28	28	30	30
0.101 – 0.125		02	02	02	04	04	04	06	06	06	08	08	08	10	10	10	12	12	12	14	14	14	16	16	16	18	18	18	20	20	20	22	22	22	24	24	24	26	26	28	28	30	30		32
0.126 – 0.129		02	02	04	04	04	06	06	06	08	08	08	10	10	10	12	12	12	14	14	14	16	16	16	18	18	18	20	20	20	22	22	22	24	24	24	26	26	28	28	28	30	30		32
0.130 – 0.230																																													
0.231 – 0.250	04	06	06	08	08	10	10	10	10	12	12	14	14	14	14	16	16	18	18	18	20	20	22	22	22	22	24	24	26	26	26	28	28	30	30	30	30	32	32	34	34	34			
0.251 – 0.275	06	06	08	08	10	10	10	12	12	12	14	14	14	16	16	16	18	18	20	20	20	22	22	24	24	24	26	26	28	28	28	30	30	30	32	32	32	34	34	34	34				
0.276 – 0.300	06	08	08	10	10	12	12	12	14	14	16	16	16	18	18	20	20	20	22	22	22	24	24	24	26	26	28	28	28	30	30	30	32	32	32	34	34	34	34						
0.301 – 0.325	08	08	10	10	12	12	12	14	14	14	16	16	18	18	18	20	20	20	22	22	22	24	24	26	26	26	28	28	30	30	30	30	32	32	34	34	34	34	34						
0.326 – 0.350	08	10	10	12	12	14	14	14	14	16	16	18	18	18	20	20	22	22	22	24	24	24	26	26	28	28	28	30	30	30	32	32	32	34	34	34	34								
0.351 – 0.375	10	10	12	12	14	14	14	16	16	16	18	18	18	20	20	20	22	22	22	24	24	26	26	28	28	28	30	30	30	32	32	32	34	34	34	34									
0.376 – 0.400	10	12	12	14	14	16	16	16	16	18	18	20	20	20	22	22	24	24	24	26	26	28	28	28	30	30	32	32	32	32	34	34	34	34											
0.401 – 0.425	12	12	14	14	16	16	16	18	18	18	20	20	20	22	22	22	24	24	26	26	26	28	28	30	30	30	32	32	34	34	34	34													
0.426 – 0.450	12	14	14	16	16	18	18	18	18	20	20	22	22	22	24	24	26	26	26	28	28	30	30	30	30	32	32	34	34	34	34														
0.451 – 0.475	14	14	16	16	18	18	18	20	20	20	22	22	22	24	24	24	26	26	28	28	28	30	30	32	32	32	34	34	LS34	34															
0.476 – 0.500	14	16	16	18	18	20	20	20	20	22	22	24	24	24	24	26	26	28	28	28	28	30	30	32	32	32	32	34	34	34	34														
0.501 – 0.525	16	16	18	18	20	20	20	22	22	22	24	24	24	26	26	26	28	28	28	30	30	30	32	32	32	34	34	34	34																
0.526 – 0.550	16	18	18	20	20	22	22	22	22	24	24	26	26	26	28	28	30	30	30	30	32	32	34	34	34	34	34	34																	
0.551 – 0.575	18	18	20	20	22	22	22	24	24	24	26	26	26	28	28	28	30	30	32	32	32	34	34	34	34	34																			
0.576 – 0.600	18	20	20	22	22	24	24	24	24	26	26	28	28	28	30	30	32	32	32	34	34	34	34																						
0.601 – 0.625	20	20	22	22	24	24	24	26	26	26	28	28	28	30	30	30	32	32	32	34	34	34	34	34																					
0.626 – 0.650	20	22	22	24	24	26	26	26	28	28	30	30	30	30	32	32	34	34	34	34	34	34																							
0.651 – 0.675	22	22	24	24	26	26	26	28	28	30	30	30	32	32	32	34	34	34	34	34																									
0.676 – 0.700	22	24	24	26	26	28	28	28	28	30	30	32	32	32	32	34	34	34	34																										
0.701 – 0.725	24	24	26	26	28	28	28	30	30	30	32	32	32	34	34	34	34																												
0.726 – 0.750	24	26	26	28	28	30	30	30	30	32	32	34	34	34	34	34																													
0.751 – 0.775	26	26	28	28	30	30	30	32	32	32	34	34	34	34																															
0.776 – 0.800	26	28	28	30	30	32	32	32	32	34	34	34	34	34																															
0.801 – 0.825	28	28	30	30	32	32	32	34	34	34	34																																		
0.826 – 0.850	28	30	30	32	32	32	34	34	34	34																																			
0.851 – 0.875	30	30	32	32	34	34	34	34	34																																				
0.876 – 0.900	30	32	32	34	34	34	34	34																																					
0.901 – 0.925	32	32	34	34	34	34																																							
0.926 – 0.950	32	34	34	34	34																																								
0.951 – 0.975	34	34	34	34																																									
0.976 – 1.000	34	34	34																																										
1.001 – 1.025	34	34																																											
1.026 – 1.030	34																																												

New shim thicknesses   mm (in.)

Shim No.	Thickness	Shim No.	Thickness
02	2.500 (0.0984)	20	2.950 (0.1161)
04	2.550 (0.1004)	22	3.000 (0.1181)
06	2.600 (0.1024)	24	3.050 (0.1201)
08	2.650 (0.1043)	26	3.100 (0.1220)
10	2.700 (0.1063)	28	3.150 (0.1240)
12	2.750 (0.1083)	30	3.200 (0.1260)
14	2.800 (0.1102)	32	3.250 (0.1280)
16	2.850 (0.1122)	34	3.300 (0.1299)
18	2.900 (0.1142)		

Intake valve adjusting shim selection using chart – LS400

New shim thickness   mm (in.)

Shim No.	Thickness	Shim No.	Thickness
01	2.50 (0.0984)	38	2.95 (0.1161)
63	2.55 (0.1004)	43	3.00 (0.1181)
06	2.60 (0.1024)	48	3.05 (0.1201)
66	2.65 (0.1043)	51	3.10 (0.1220)
13	2.70 (0.1063)	77	3.15 (0.1240)
18	2.75 (0.1083)	56	3.20 (0.1260)
23	2.80 (0.1102)	80	3.25 (0.1280)
28	2.85 (0.1122)	61	3.30 (0.1299)
33	2.90 (0.1142)		

Exhaust valve adjusting shim selection using chart — ES250

New shim thickness mm (in.)

Shim No.	Thickness	Shim No.	Thickness
01	2.50 (0.0984)	38	2.95 (0.1161)
63	2.55 (0.1004)	43	3.00 (0.1181)
06	2.60 (0.1024)	48	3.05 (0.1201)
66	2.65 (0.1043)	51	3.10 (0.1220)
13	2.70 (0.1063)	77	3.15 (0.1240)
18	2.75 (0.1083)	56	3.20 (0.1260)
23	2.80 (0.1102)	80	3.25 (0.1280)
28	2.85 (0.1122)	61	3.30 (0.1299)
33	2.90 (0.1142)		

Installed shim thickness mm (in.) across top of chart: 2.50 (0.0984), 2.52 (0.0992), 2.54 (0.1000), 2.55 (0.1004), 2.56 (0.1008), 2.58 (0.1016), 2.60 (0.1024), 2.62 (0.1031), 2.64 (0.1039), 2.65 (0.1043), 2.66 (0.1047), 2.67 (0.1051), 2.68 (0.1055), 2.69 (0.1059), 2.70 (0.1063), 2.71 (0.1067), 2.72 (0.1071), 2.73 (0.1075), 2.74 (0.1079), 2.75 (0.1083), 2.76 (0.1087), 2.77 (0.1091), 2.78 (0.1094), 2.79 (0.1098), 2.80 (0.1102), 2.81 (0.1106), 2.82 (0.1110), 2.83 (0.1114), 2.84 (0.1118), 2.85 (0.1122), 2.86 (0.1126), 2.87 (0.1130), 2.88 (0.1134), 2.89 (0.1138), 2.90 (0.1142), 2.91 (0.1146), 2.92 (0.1150), 2.93 (0.1154), 2.94 (0.1157), 2.95 (0.1161), 2.96 (0.1165), 2.97 (0.1169), 2.98 (0.1173), 2.99 (0.1177), 3.00 (0.1181), 3.01 (0.1185), 3.02 (0.1189), 3.03 (0.1193), 3.04 (0.1197), 3.05 (0.1201), 3.06 (0.1205), 3.08 (0.1213), 3.10 (0.1220), 3.12 (0.1228), 3.14 (0.1236), 3.15 (0.1240), 3.16 (0.1244), 3.18 (0.1252), 3.20 (0.1260), 3.22 (0.1268), 3.24 (0.1276), 3.25 (0.1280), 3.26 (0.1283), 3.28 (0.1291), 3.30 (0.1299)

Measured clearance mm (in.):

Measured clearance mm (in.)
0.000 – 0.020 (0.0000 – 0.0008)
0.021 – 0.040 (0.0008 – 0.0016)
0.041 – 0.060 (0.0016 – 0.0024)
0.061 – 0.080 (0.0024 – 0.0031)
0.081 – 0.100 (0.0032 – 0.0039)
0.101 – 0.120 (0.0040 – 0.0047)
0.121 – 0.140 (0.0048 – 0.0055)
0.141 – 0.149 (0.0056 – 0.0059)
0.150 – 0.250 (0.0059 – 0.0098)
0.251 – 0.260 (0.0099 – 0.0102)
0.261 – 0.281 (0.0103 – 0.0110)
0.281 – 0.300 (0.0111 – 0.0118)
0.301 – 0.320 (0.0119 – 0.0126)
0.321 – 0.340 (0.0126 – 0.0134)
0.341 – 0.360 (0.0134 – 0.0142)
0.361 – 0.380 (0.0142 – 0.0150)
0.381 – 0.400 (0.0150 – 0.0157)
0.401 – 0.420 (0.0158 – 0.0165)
0.421 – 0.440 (0.0166 – 0.0173)
0.441 – 0.460 (0.0174 – 0.0181)
0.461 – 0.480 (0.0181 – 0.0189)
0.481 – 0.500 (0.0189 – 0.0197)
0.501 – 0.520 (0.0197 – 0.0205)
0.521 – 0.540 (0.0205 – 0.0213)
0.541 – 0.560 (0.0213 – 0.0220)
0.561 – 0.580 (0.0221 – 0.0228)
0.581 – 0.600 (0.0229 – 0.0236)
0.601 – 0.620 (0.0237 – 0.0244)
0.621 – 0.640 (0.0244 – 0.0252)
0.641 – 0.660 (0.0252 – 0.0260)
0.661 – 0.680 (0.0260 – 0.0268)
0.681 – 0.700 (0.0268 – 0.0276)
0.701 – 0.720 (0.0276 – 0.0283)
0.721 – 0.740 (0.0284 – 0.0291)
0.741 – 0.760 (0.0292 – 0.0299)
0.761 – 0.780 (0.0300 – 0.0307)
0.781 – 0.800 (0.0307 – 0.0315)
0.801 – 0.820 (0.0315 – 0.0323)
0.821 – 0.840 (0.0323 – 0.0331)
0.841 – 0.860 (0.0331 – 0.0339)
0.861 – 0.880 (0.0339 – 0.0346)
0.881 – 0.900 (0.0346 – 0.0354)
0.901 – 0.920 (0.0355 – 0.0362)
0.921 – 0.940 (0.0363 – 0.0370)
0.941 – 0.960 (0.0370 – 0.0378)
0.961 – 0.980 (0.0378 – 0.0386)
0.981 – 1.000 (0.0386 – 0.0394)
1.001 – 1.020 (0.0394 – 0.0402)
1.021 – 1.040 (0.0402 – 0.0409)
1.041 – 1.050 (0.0410 – 0.0413)

The body of the chart is a large matrix of two-digit shim numbers (01, 63, 06, 66, 13, 18, 23, 28, 33, 38, 43, 48, 51, 77, 56, 80, 61) cross-referencing the measured clearance (rows) against the installed shim thickness (columns) to determine the required replacement shim number.

Intake valve adjusting shim selection using chart — ES250

36. Raise and support the vehicle. Install the front exhaust pipe and the right side engine undercover. Lower the vehicle.

37. Install the air cleaner hose and cruise control actuator.

38. Install the accelerator cable and connect the throttle cable, if equipped with automatic transaxle.

39. Install the suspension upper brace and fill the cooling system.

40. Connect the negative battery cable. Run the engine and check for leaks.

41. Adjust the timing. Recheck the fluid levels.

LS400

1. Disconnect the negative battery cable. Drain the cooling system.

2. Remove the camshaft timing pulleys.

3. Disconnect the accelerator cable, the throttle control cable, if equipped with automatic transmission and the cruise control actuator cable.

4. Remove the high tension cord cover and the right side ignition coil.

5. Remove the water inlet housing mounting bolts and disconnect the water bypass hose from the ISC valve.

6. Remove the water inlet and inlet housing assemblies. Remove the O-ring from the water inlet housing.

7. Remove the EGR pipe.

8. Disconnect the following:
   a. VSV connector
   b. Vacuum pipe hose
   c. EGR water bypass pipe
   d. Fuel pressure VSV

9. Disconnect the EGR vacuum hoses and remove the EGR VSV.

10. Disconnect the following hoses:
   a. Water bypass pipe hose from the ISC valve.
   b. Water bypass joint hose.
   c. Vacuum pipe hoses.

11. Disconnect the EGR gas temperature sensor, California only. Remove the EGR valve adapter.

12. Disconnect the following:
   a. Fuel pressure regulator vacuum hose.
   b. Air intake chamber vacuum hose.
   c. Vacuum hose from the EVAP BVSV.

13. Remove the mounting bolts, hoses and the vacuum pipe.

14. Remove the ISC valve.

15. Remove the throttle body sensor connectors and the water bypass pipe from the rear water bypass joint.

16. Remove the mounting bolts/nuts and disconnect the PCV valve. Remove the throttle body and gasket.

17. Disconnect the accelerator cable bracket and the brake booster vacuum union and hose.

18. Disconnect the cold start injector connector and the cold start injector

tube from the right side delivery pipe.

19. Disconnect the check connector from the intake chamber and remove the mounting nuts and bolts.

20. Remove the air intake chamber and the cold start injector, tube and wire assembly.

21. Disconnect the engine wire from the intake manifold and from the right side cylinder head. Disconnect the heater hoses.

22. Remove the delivery pipes and the fuel injectors. Remove the mounting bolts and nuts. Lift up the intake manifold.

To install:

23. Install the intake manifold, using new gaskets. Tighten the mounting nuts and bolts to 13 ft. lbs. (18 Nm).

24. Install the delivery pipes and fuel injectors. Install the fuel return pipe with new gaskets. Tighten the union bolt to 26 ft. lbs. (35 Nm).

25. Connect the fuel hoses and the injector connectors. Connect the engine wire to the delivery pipes.

26. Connect the connectors on the left side delivery pipe, the water temperature sensor connector, cold start injector time switch connector and the water temperature sender gauge connector.

27. Connect the heater hoses and engine wire bracket. Install the engine wire to the bracket.

28. Install the cold start injector, tube and wire assembly. Tighten the mounting bolts to 69 inch lbs.

29. Install the air intake chamber with new gaskets and tighten the mounting bolts to 13 ft. lbs. (18 Nm).

30. Connect the cold start injector tube to the right side delivery pipe and tighten the union bolt to 11 ft. lbs. (15 Nm).

31. Connect the cold start injector connector. Install the accelerator cable bracket.

32. Install the brake booster union and connect the vacuum hose. Tighten the union bolt to 22 ft. lbs. (30 Nm).

33. Connect the water bypass hose to the throttle body and the PCV hose to the cylinder head cover.

34. Install the throttle body, using a new gasket. Tighten the mounting bolts to 13 ft. lbs. (18 Nm).

35. Install the water bypass pipe and connect the sensor connectors. Install the ISC valve and tighten the mounting bolts to 13 ft. lbs. (18 Nm). Connect the water bypass hose.

36. Install the vacuum pipe and the following hoses:
   a. Fuel pressure regulator vacuum hose.
   b. Vacuum hose to the upper port of the EVAP BVSV.
   c. Air intake chamber vacuum hose.
   d. Throttle body vacuum hoses.

37. Install the EGR valve adapter with a new gasket. Tighten the mounting bolts to 13 ft. lbs. (18 Nm).

38. Connect the EGR gas temperature sensor connector (California only).

39. Install the EGR valve and vacuum modulator. Connect the water bypass hoses and the vacuum hoses.

40. Install the EGR and fuel pressure VSV and connect the hoses and connectors. Replace the EGR pipe and tighten the mounting bolts to 13 ft. lbs. (18 Nm).

41. Install the timing belt rear plates and tighten the bolts to 69 inch lbs. Install the water inlet and inlet hosing and tighten the bolts to 13 ft. lbs. (18 Nm).

42. Install the right side ignition coil and the high tension cord cover.

43. Connect and adjust the accelerator cable, the automatic transmission throttle cable and the cruise control actuator cable. Install the camshaft timing pulley.

44. Fill the cooling system and connect the negative battery cable. Start the engine and check for leaks.

45. Recheck all the fluid levels and check the ignition timing.

Exhaust Manifold

REMOVAL & INSTALLATION

ES250

1. Disconnect the negative battery cable. Drain the cooling system.

2. Remove the suspension upper brace. Disconnect the throttle cable from the throttle body, if equipped with an automatic transaxle.

3. Disconnect the accelerator cable from the throttle body. Remove the cruise control actuator and the vacuum pump.

4. Disconnect the ISC hose, the vacuum pipe hose and the air cleaner hose.

5. Raise and safely support the vehicle. Remove the right side engine undercover.

6. Remove the suspension lower crossmember and the front exhaust pipe. Lower the vehicle.

7. Remove the alternator, the ISC valve and the throttle body.

8. Remove the EGR pipe, the EGR valve and the vacuum modulator.

9. Disconnect the (4) BVSV vacuum hoses, the fuel pressure VSV hose and the air conditioning control valve vacuum hose.

10. Remove the distributor and the exhaust crossover pipe.

11. Disconnect the cold start injector connector and remove the cold start injector tube.

12. Disconnect the following hoses:

a. PCV hose
b. Vacuum sensing hose
c. Fuel pressure VSV hose
d. Air conditioning control valve vacuum hose
e. EGR gas temperature sensor connector—California only

13. Remove the mounting bolts and the brackets from the intake chamber. Remove the air intake chamber.

14. Remove the delivery pipes and the injectors.

15. Disconnect the water temperature sensor connector and the upper radiator hose. Remove the water outlet.

16. Disconnect the following connectors and hose:
a. Cold start injector time switch connector
b. Water temperature sensor connector
c. Heater water bypass hose

17. Remove the water bypass outlet. Remove the crossover pipe insulator from the water bypass outlet.

18. Remove the cylinder head rear plate and the idler pulley bracket.

19. Remove the mounting bolts and nuts. Lift off the intake manifold.

20. Disconnect the main oxygen sensor connector. Remove outside heat insulator.

21. Remove the mounting nuts and the right side exhaust manifold. Remove the inside heat insulator. It may be necessary to remove the cylinder on the right side to provide ample clearance.

22. Remove the heat insulator, the mounting nuts and the left side exhaust manifold.

To install:

23. Install the right side heat insulator and the right side exhaust manifold with a new gasket. Tighten the nuts to 29 ft. lbs. (40 Nm).

24. Install the outside heat insulator and connect the oxygen sensor connector.

25. Install the left side exhaust manifold with a new gasket and tighten the nuts to 29 ft. lbs. (40 Nm). Install the heat insulator.

26. Install the intake manifold with new gaskets. Tighten the nuts and bolts to 13 ft. lbs. (18 Nm).

27. Install the No. 2 idler pulley bracket and tighten to 13 ft. lbs. (18 Nm). Replace the rear cylinder head plate.

28. Install the crossover pipe heat insulator. Replace the water bypass outlet and tighten the mounting bolts to 14 ft. lbs. (19 Nm).

29. Connect the heater water bypass hose, cold start injector time switch connector and the water temperature switch.

30. Connect the upper radiator hose, the water temperature sensor connec-

tor and install the water outlet with a new gasket. Tighten the bolts to 73 inch lbs.

31. Install the fuel injectors and the delivery pipes.

32. Install the air intake chamber with new gaskets and tighten the mounting bolts to 32 ft. lbs. (44 Nm).

33. Install the intake chamber brackets.

34. Connect the following hoses:
a. PCV hose
b. Vacuum sensing hose
c. EGR gas temperature sensor connector (California only)
d. Air conditioning control valve air hose

35. Install the cold start injector tube and connect the cold start injector connector.

36. Install the exhaust crossover pipe with new gaskets and tighten the mounting bolts to 25 ft. lbs. (34 Nm) and the nuts to 29 ft. lbs. (40 Nm).

37. Install the distributor.

38. Install the vacuum pipe and connect the BVSV vacuum hoses and the air conditioning control valve vacuum hose.

39. Install the EGR valve and vacuum modulator. Tighten the mounting bolts to 13 ft. lbs. (18 Nm). Install the vacuum pipe hoses.

40. Install the EGR pipe with a new gasket and tighten the mounting bolts to 13 ft. lbs. (18 Nm) and the nut to 58 ft. lbs. (79 Nm).

41. Install the throttle body and ISC valve.

42. Raise and support the vehicle. Install the front exhaust pipe and the right side engine undercover. Lower the vehicle.

43. Install the air cleaner hose and cruise control actuator.

44. Install the accelerator cable and connect the throttle cable, if equipped with an automatic transaxle.

45. Install the suspension upper brace and fill the cooling system.

46. Connect the negative battery cable. Run the engine and check for leaks.

47. Adjust the timing. Recheck the fluid levels.

LS400

1. Disconnect the negative battery cable. Drain the cooling system.

2. Remove the camshaft timing pulleys.

3. Disconnect the accelerator cable, the throttle control cable, if equipped with automatic transaxle and the cruise control actuator cable.

4. Remove the high tension cord cover and the right side ignition coil.

5. Remove the water inlet housing mounting bolts and disconnect the water bypass hose from the ISC valve.

6. Remove the water inlet and inlet housing assemblies. Remove the O-ring from the water inlet housing.

7. Remove the EGR pipe.

8. Disconnect the following:
a. VSV connector
b. Vacuum pipe hose
c. EGR water bypass pipe
d. Fuel pressure VSV

9. Disconnect the EGR vacuum hoses and remove the EGR VSV.

10. Disconnect the following hoses:
a. Water bypass pipe hose from the ISC valve.
b. Water bypass joint hose.
c. Vacuum pipe hoses.

11. Disconnect the EGR gas temperature sensor, California only. Remove the EGR valve adapter.

12. Disconnect the following:
a. Fuel pressure regulator vacuum hose.
b. Air intake chamber vacuum hose.
c. Vacuum hose from the EVAP BVSV.

13. Remove the mounting bolts, hoses and the vacuum pipe.

14. Remove the ISC valve.

15. Remove the throttle body sensor connectors and the water bypass pipe from the rear water bypass joint.

16. Remove the mounting bolts/nuts and disconnect the PCV valve. Remove the throttle body and gasket.

17. Disconnect the accelerator cable bracket and the brake booster vacuum union and hose.

18. Disconnect the cold start injector connector and the cold start injector tube from the right side delivery pipe.

19. Disconnect the check connector from the intake chamber and remove the mounting nuts and bolts.

20. Remove the air intake chamber and the cold start injector, tube and wire assembly.

21. Disconnect the engine wire from the intake manifold and from the right side cylinder head. Disconnect the heater hoses.

22. Remove the delivery pipes and the fuel injectors. Remove the mounting bolts and nuts. Lift up the intake manifold.

23. Remove the front and rear water bypass joint.

24. Raise and safely support the vehicle. Remove the front exhaust pipe and the main catalytic converters. Lower the vehicle.

25. Disconnect the right side oxygen sensor. Remove the mounting bolts and nuts and remove the right side exhaust manifold.

26. Remove the oil dipstick and guide. Disconnect the left side oxygen sensor.

27. Remove the mounting bolts and nuts and remove the left side exhaust manifold.

To install:

28. Install the right side exhaust manifold with a new gasket and tighten the mounting bolts to 29 ft. lbs. (40 Nm). Connect the right side oxygen sensor connector.

29. Install the left side exhaust manifold with a new gasket and tighten the mounting bolts to 29 ft. lbs. (40 Nm). Connect the left side oxygen sensor connector.

30. Install the oil dipstick and guide. Raise and safely support the vehicle.

31. Install the catalytic converters and front exhaust pipe. Lower the vehicle.

32. Install the front and rear water bypass joints. Tighten the mounting bolts to 13 ft. lbs. (18 Nm).

33. Install the intake manifold, using new gaskets. Tighten the mounting nuts and bolts to 13 ft. lbs. (18 Nm).

34. Install the delivery pipes and fuel injectors. Install the fuel return pipe with new gaskets. Tighten the union bolt to 26 ft. lbs. (35 Nm).

35. Connect the fuel hoses and the injector connectors. Connect the engine wire to the delivery pipes.

36. Connect the connectors on the left side delivery pipe, the water temperature sensor connector, cold start injector time switch connector and the water temperature sender gauge connector.

37. Connect the heater hoses and engine wire bracket. Install the engine wire to the bracket.

38. Install the cold start injector, tube and wire assembly. Tighten the mounting bolts to 69 inch lbs.

39. Install the air intake chamber with new gaskets and tighten the mounting bolts to 13 ft. lbs. (18 Nm).

40. Connect the cold start injector tube to the right side delivery pipe and tighten the union bolt to 11 ft. lbs. (15 Nm).

41. Connect the cold start injector connector. Install the accelerator cable bracket.

42. Install the brake booster union and connect the vacuum hose. Tighten the union bolt to 22 ft. lbs. (30 Nm).

43. Connect the water bypass hose to the throttle body and the PCV hose to the cylinder head cover.

44. Install the throttle body, using a new gasket. Tighten the mounting bolts to 13 ft. lbs. (18 Nm).

45. Install the water bypass pipe and connect the sensor connectors. Install the ISC valve and tighten the mounting bolts to 13 ft. lbs. (18 Nm). Connect the water bypass hose.

46. Install the vacuum pipe and the following hoses:
   a. Fuel pressure regulator vacuum hose.
   b. Vacuum hose to the upper port of the EVAP BVSV.
   c. Air intake chamber vacuum hose.
   d. Throttle body vacuum hoses.

47. Install the EGR valve adapter with a new gasket. Tighten the mounting bolts to 13 ft. lbs. (18 Nm).

48. Connect the EGR gas temperature sensor connector (California only).

49. Install the EGR valve and vacuum modulator. Connect the water bypass hoses and the vacuum hoses.

50. Install the EGR and fuel pressure VSV and connect the hoses and connectors. Replace the EGR pipe and tighten the mounting bolts to 13 ft. lbs. (18 Nm).

51. Install the timing belt rear plates and tighten the bolts to 69 inch lbs. Install the water inlet and inlet hosing and tighten the bolts to 13 ft. lbs. (18 Nm).

52. Install the right side ignition coil and the high tension cord cover.

53. Connect and adjust the accelerator cable, the automatic transmission throttle cable and the cruise control actuator cable. Install the camshaft timing pulley.

54. Fill the cooling system and connect the negative battery cable. Start the engine and check for leaks.

55. Recheck all the fluid levels and check the ignition timing.

Timing Belt Front Cover

REMOVAL & INSTALLATION

ES250

1. Disconnect the negative battery cable. Remove the suspension upper brace.

2. Remove the cruise control actuator.

3. Remove the power steering oil reservoir tank. Do not disconnect the hoses.

4. Raise and safely support the vehicle. Remove the right front tire and wheel assembly. Lower the vehicle.

5. Remove the alternator and power steering belts. Remove the right side fender apron seal.

6. Remove the right side engine support brackets. Raise the engine enough to remove the weight from the engine mounting on the right side, using a suitable jacking device.

7. Disconnect the power steering oil cooler lines and remove the right side engine mount.

8. Remove the air intake chamber and remove the spark plugs.

9. Remove the No. 2 timing belt cover and the right side engine mount bracket.

10. Turn the crankshaft pulley and align it's groove with the timing mark 0 of the No. 1 timing cover. Check that the timing marks of the camshaft timing pulleys and the No. 3 timing belt cover are aligned. If not, turn the crankshaft 1 full revolution (360 degrees).

11. Remove the timing belt tensioner and dust boot. Using the proper tool, loosen the tension between the left side and right side timing pulleys by slightly turning the right side camshaft timing pulley clockwise.

12. Remove the timing belt from the camshaft pulleys.

13. Remove the bolt, timing pulley and lock pin with the proper tool. Remove the 2 timing pulleys.

14. Remove the bolt and the No. 2 idler pulley.

15. Remove the crankshaft pulley bolt and the pulley, using the proper tool.

16. Remove the No. 1 timing cover.

To install:

17. Install the No. 1 timing belt cover and gasket.

18. Install the crankshaft pulley by aligning the pulley set key with the key groove of the pulley. Install the bolt with the proper tool and tighten to 181 ft. lbs. (245 Nm).

19. Install the No. 2 idler pulley and tighten the bolt to 29 ft. lbs. (40 Nm).

20. Install the left side camshaft timing pulley with the flange side facing outward. Align the knock pin hole of the camshaft with the pin groove of the timing pulley. Tighten the pulley bolt to 80 ft. lbs. (108 Nm).

21. Set the No. 1 cylinder to TDC compression by:
   a. Turn the crankshaft pulley and align it's groove with the 0 timing mark and the No. 1 timing cover.
   b. Turn the right side camshaft and align the knock pin hole of the camshaft with the timing mark of the No. 3 timing belt cover.
   c. Turn the left side camshaft and align the timing marks of the camshaft pulley with the timing mark of the No. 3 timing belt cover.

22. Install the timing belt to the left side camshaft timing pulley by:
   a. Check that the installation mark on the timing belt matches the end of the No. 1 timing cover. If not aligned, shift the meshing of the timing belt and the crankshaft pulley until they align.
   b. Using the proper tool, slightly, turn the left side camshaft timing pulley clockwise. Align the installation mark on the timing belt with the timing mark of the camshaft pulley and hang the timing belt on the left side camshaft pulley.
   c. Using the proper tool, align the timing marks of the left side cam-

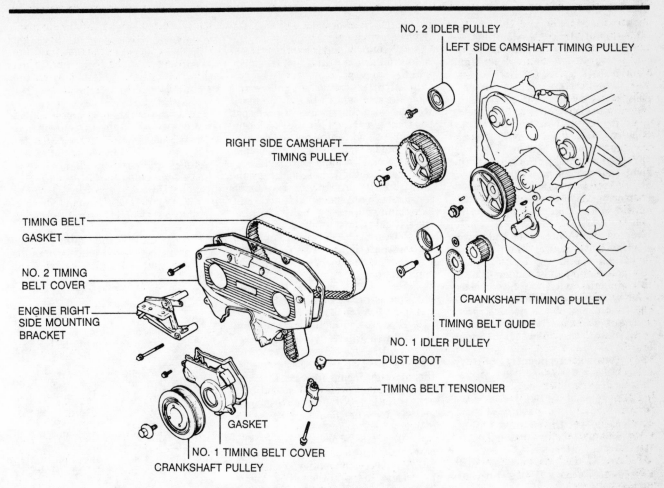

NO. 2 IDLER PULLEY

LEFT SIDE CAMSHAFT TIMING PULLEY

RIGHT SIDE CAMSHAFT TIMING PULLEY

TIMING BELT

GASKET

NO. 2 TIMING BELT COVER

ENGINE RIGHT SIDE MOUNTING BRACKET

CRANKSHAFT TIMING PULLEY

TIMING BELT GUIDE

NO. 1 IDLER PULLEY

DUST BOOT

TIMING BELT TENSIONER

GASKET

NO. 1 TIMING BELT COVER

CRANKSHAFT PULLEY

Exploded view of the timing belt assembly — ES250

Aligning the timing belt — ES250

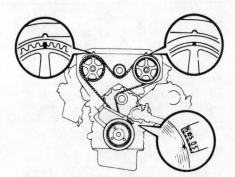

Aligning the timing belt and timing belt marks — ES250

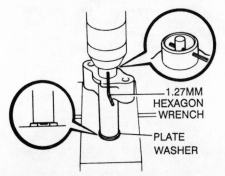

1.27MM HEXAGON WRENCH

PLATE WASHER

Setting the timing belt tensioner — ES250

shaft pulley and the No. 3 timing belt cover.

d. Check that the timing belt has tension between the crankshaft timing and the left side camshaft timing pulleys.

23. Install the timing belt to the right side camshaft timing pulley by:

a. Align the timing mark on the timing belt with the timing mark of the right side camshaft timing pulley.

b. Hang the timing belt on the right side camshaft timing pulley with the flange side facing inward.

c. Align the timing marks of the right side camshaft timing pulley and the No. 3 timing belt cover.

24. Slide the right side camshaft timing pulley on the camshaft. Align the knock pin hole of the camshaft with the knock pin groove of the pulley and install the knock pin. Tighten the bolt to 55 ft. lbs. (74 Nm).

25. The timing belt tensioner must be set prior to installation. The tensioner can be set by:

a. Place a plate washer between the tensioner and a block. Using a suitable press, press in the pushrod using 220–2205 lbs. of pressure.

b. Align the holes of the pushrod and housing, pass the proper tool through the holes to keep the setting position of the pushrod.

c. Release the press and install the dust boot to the tensioner.

26. Install the tensioner and tighten the bolts to 20 ft. lbs. (27 Nm). Remove the tool from the tensioner.

27. Turn the crankshaft pulley 2 revolutions from TDC to TDC. Always turn the crankshaft clockwise. Check that each pulley aligns with the timing marks.

28. Install the right side engine

mounting bracket and tighten the bolts to 30 ft. lbs.

29. Install the No. 2 timing belt cover and gasket. Install the spark plugs and tighten to 13 ft. lbs. (18 Nm).

30. Install the air intake chamber.

31. Install the right side engine mount and tighten the nut-to-bracket to 38 ft. lbs. (52 Nm) and the nut-to-body to 65 ft. lbs. (88 Nm). Do not tighten the mounting bolt.

32. Lower the engine. Install the right side engine mounting brackets and tighten the bolts. Fasten the power steering cooler pipes and tighten the engine mount bolt to 47 ft. lbs. (64 Nm).

33. Install the alternator and power steering belts. Replace the right side fender apron seal.

34. Raise and safely support the vehicle. Install the RF tire and wheel assembly. Lower the vehicle.

35. Install the power steering reservoir tank and cruise control actuator.

36. Install the suspension upper brace.

37. Connect the negative battery cable.

LS400

1. Disconnect the negative battery cable and the positive battery cable. Remove the battery.

2. Remove the air duct and dust covers. Remove the engine undercover.

3. Drain the cooling system. Remove the drive belt, fan, fluid coupling and fan pulley.

4. Remove the radiator, air cleaner and throttle body cover. Remove the air intake connector pipe.

5. Remove the air conditioning compressor and power steering pump. Do not disconnect the hoses.

6. Remove the upper high tension cord cover and the right side engine wire cover.

7. Disconnect the PCV hose and remove the left side engine wire cover. Remove the right side No. 3 timing cover.

8. Disconnect and tag the vacuum hoses and remove the left side engine wire cover. Disconnect the spark plug wires.

9. Remove the bolt, cover plate and idler pulley.

10. Disconnect the crank position sensor connector and remove the right side No. 2 timing belt cover.

11. Disconnect and remove the ignition coil. Disconnect the hoses and wires from the water bypass pipe. Remove the water bypass pipe.

12. Disconnect the crank position sensor connector and remove the No. 2 timing belt cover.

13. Remove the distributor caps and rotors. Disconnect and remove both distributor housings.

14. Disconnect and remove the alternator. Remove the drive belt tensioner and the spark plugs.

15. Turn the crankshaft pulley and align it's groove with the timing mark 0 of the No. 1 timing cover. Check that the timing marks of the camshaft timing pulleys and timing belt rear plates are aligned. If not, turn the crankshaft 1 full revolution (360 degrees).

16. Remove the timing belt tensioner. Using the proper tool, loosen the tension between the left side and right side timing pulleys by slightly turning the left side camshaft clockwise.

17. Disconnect the timing belt from the camshaft timing pulleys. Using the proper tool, remove the bolt and the timing pulleys.

18. Remove the bolt and the crankshaft pulley with the proper tool. Remove the fan bracket.

19. Remove the mounting bolts and the No. 1 timing belt cover.

To install:

20. Install the timing belt guide with the cup side facing forward. Replace the timing belt cover spacer.

21. Install the No. 1 timing belt cover and tighten the mounting bolts. Install the fan bracket.

22. Align the pulley set key on the crankshaft with the key groove of the pulley. Install the pulley, using the proper tool to tap in the pulley. Tighten the pulley bolt to 181 ft. lbs. (245 Nm).

23 Align the knock pin on the right side camshaft with the knock pin of the timing pulley. Slide on the timing pulley with the right side mark facing forward. Tighten the bolt to 80 ft. lbs. (108 Nm).

24. Align the knock pin on the left side camshaft with the knock pin of the timing pulley. Slide on the timing pulley with the left side mark facing forward. Tighten the bolt to 80 ft. lbs. (108 Nm).

25. Turn the crankshaft pulley and align it's groove with the 0 timing mark on the No. 1 timing belt cover. Using the proper tool, turn the crankshaft timing pulley and align the timing marks of the camshaft timing pulley and the timing belt rear plate.

26. Install the timing belt to the left side camshaft timing pulley by:

a. Using the proper tool, slightly turn the left side timing pulley clockwise. Align the installation mark of the timing belt with the timing mark of the camshaft timing pulley and hang the timing belt on the left side camshaft pulley.

b. Using the proper tool, align the timing marks of the left side camshaft pulley and the timing belt rear plate.

c. Check that the timing belt has tension between crankshaft timing pulley and the left side camshaft pulley.

27. Install the timing belt to the right side camshaft timing pulley by:

a. Using the proper tool, slightly turn the right side timing pulley clockwise. Align the installation mark of the timing belt with the timing mark of the camshaft timing pulley and hang the timing belt on the right side camshaft pulley.

b. Using the proper tool, align the timing marks of the right side camshaft pulley and the timing belt rear plate.

c. Check that the timing belt has tension between crankshaft timing pulley and the right side camshaft pulley.

28. The timing belt tensioner must be set prior to installation. The tensioner can be set by:

a. Place a plate washer between the tensioner and a block. Using a suitable press, press in the pushrod using 220–2205 lbs. of pressure.

b. Align the holes of the pushrod and housing, pass the proper tool through the holes to keep the setting position of the pushrod.

c. Release the press and install the dust boot to the tensioner.

29. Install the tensioner and tighten the bolts to 20 ft. lbs. (27 Nm). Remove the tool from the tensioner.

30. Turn the crankshaft pulley 2 revolutions from TDC to TDC. Always turn the crankshaft clockwise. Check that each pulley aligns with the timing marks.

31. Install the spark plugs and tighten to 13 ft. lbs. (18 Nm). Install the drive belt tensioner and tighten the bolt to 12 ft. lbs. (16 Nm).

32. Install the alternator and engine wire bracket. Tighten the nut and bolt to 26 ft. lbs. (35 Nm). Connect the electrical connections at the alternator.

33. Install both distributor housings and tighten the mounting bolts to 13 ft. lbs. (18 Nm). Replace the distributor rotors and caps.

34. Install the right side No. 2 timing belt cover and tighten the 10mm bolts to 69 inch lbs. and the 12mm bolts to 12 ft. lbs. (16 Nm). Connect the crank position sensor connector.

35. Install the left side No. 2 timing belt cover and connect the crank position sensor connector.

36. Install the water bypass pipe and connect the hoses and connectors.

37. Replace the left side ignition coil and connect the coil connector. Install the idler pulley and cover plate. Tighten the bolt to 27 ft. lbs. (37 Nm).

38. Install and secure the ignition wires. Install the right side No. 3 timing belt cover.

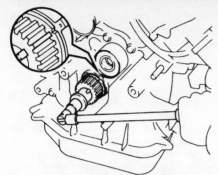

Aligning the timing mark of the crankshaft timing pulley and the oil pump body — LS400

Align the installation mark on the timing belt with the drilled mark of the crankshaft timing pulley (if the belt is marked) — LS400

Aligning the timing marks of the camshaft timing pulley and timing belt rear plate — LS400

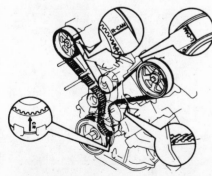

Checking the timing belt installation marks on a reinstalled used timing belt — LS400

39. Install the left side No. 3 timing belt cover and connect the vacuum hose and connectors. Install the right side engine wire cover.

40. Install the left side engine wire cover and connect the vacuum hoses.

41. Install the upper high tension cord covers. Fit the front side claw groove of the upper cover to claw of the lower cover.

42. Install the power steering pump and the air conditioning compressor.

43. Install the throttle body cover and the air cleaner.

44. Install the radiator, fan pulley, fan coupling, fan and drive belt.

45. Install the engine undercover and replace the battery.

46. Install the air ducts and dust covers. Connect the battery cables.

47. Refill the cooling system. Check the ignition timing.

OIL SEAL REPLACEMENT

The front cover oil seal replacement can and should be done when the front timing belt cover has been removed.

1. With the timing cover containing the oil seal removed, use a suitable seal removal tool and remove the front oil seal from the cover.

2. To install, apply a light coat of oil to the crankshaft and the lip of the new oil seal.

3. Using a suitable seal driver tool, drive in the new oil seal into the cover until the end of the seal sits squarely with the cover.

4. Reinstall the cover on the vehicle.

Timing Belt and Tensioner

REMOVAL & INSTALLATION

ES250

1. Disconnect the negative battery cable. Remove the suspension upper brace.

2. Remove the cruise control actuator.

3. Remove the power steering oil reservoir tank. Do not disconnect the hoses.

4. Raise and safely support the vehicle. Remove the right front tire and wheel assembly. Lower the vehicle.

5. Remove the alternator and power steering belts. Remove the right side fender apron seal.

6. Remove the right side engine support brackets. Raise the engine enough to remove the weight from the engine mounting on the right side, using a suitable jacking device.

7. Disconnect the power steering oil cooler lines and remove the right side engine mount.

8. Remove the air intake chamber and remove the spark plugs.

9. Remove the No. 2 timing belt cover and the right side engine mount bracket.

10. Turn the crankshaft pulley and align it's groove with the timing mark **0** of the No. 1 timing cover. Check that the timing marks of the camshaft timing pulleys and the No. 3 timing belt cover are aligned. If not, turn the crankshaft 1 full revolution (360 degrees).

11. Remove the timing belt tensioner and dust boot. Using the proper tool, loosen the tension between the left side and right side timing pulleys by slightly turning the right side camshaft timing pulley clockwise.

12. Remove the timing belt from the camshaft pulleys.

13. Remove the bolt, timing pulley and lock pin with the proper tool. Remove the 2 timing pulleys.

14. Remove the bolt and the No. 2 idler pulley.

15. Remove the crankshaft pulley bolt and the pulley, using the proper tool.

16. Remove the No. 1 timing cover.

17. Remove the timing belt guide and lift off the timing belt.

To install:

18. Align the installation mark on the timing belt with the drilled mark of the crankshaft timing pulley. Install the timing belt on the crankshaft timing pulley, No. 1 idler pulley and the No. 2 idler pulley.

19. Install the timing belt guide with the cup side facing outward.

20. Install the No. 1 timing belt cover and gasket.

21. Install the crankshaft pulley by aligning the pulley set key with the key groove of the pulley. Install the bolt with the proper tool and tighten to 181 ft. lbs. (245 Nm).

22. Install the No. 2 idler pulley and tighten the bolt to 29 ft. lbs. (40 Nm).

23. Install the left side camshaft timing pulley with the flange side facing outward. Align the knock pin hole of the camshaft with the pin groove of the timing pulley. Tighten the pulley bolt to 80 ft. lbs. (108 Nm).

24. Set the No. 1 cylinder to TDC compression by:

　a. Turn the crankshaft pulley and align it's groove with the 0 timing mark and the No. 1 timing cover.

　b. Turn the right side camshaft and align the knock pin hole of the camshaft with the timing mark of the No. 3 timing belt cover.

　c. Turn the left side camshaft and align the timing marks of the camshaft pulley with the timing mark of the No. 3 timing belt cover.

25. Install the timing belt to the left side camshaft timing pulley by:

a. Check that the installation mark on the timing belt matches the end of the No. 1 timing cover. If not aligned, shift the meshing of the timing belt and the crankshaft pulley until they align.

b. Using the proper tool, slightly, turn the left side camshaft timing pulley clockwise. Align the installation mark on the timing belt with the timing mark of the camshaft pulley and hang the timing belt on the left side camshaft pulley.

c. Using the proper tool, align the timing marks of the left side camshaft pulley and the No. 3 timing belt cover.

d. Check that the timing belt has tension between the crankshaft timing and the left side camshaft timing pulleys.

26. Install the timing belt to the right side camshaft timing pulley by:

a. Align the timing mark on the timing belt with the timing mark of the right side camshaft timing pulley.

b. Hang the timing belt on the right side camshaft timing pulley with the flange side facing inward.

c. Align the timing marks of the right side camshaft timing pulley and the No. 3 timing belt cover.

27. Slide the right side camshaft timing pulley on the camshaft. Align the knock pin hole of the camshaft with the knock pin groove of the pulley and install the knock pin. Tighten the bolt to 55 ft. lbs. (74 Nm).

28. The timing belt tensioner must be set prior to installation. The tensioner can be set by:

a. Place a plate washer between the tensioner and a block. Using a suitable press, press in the pushrod using 220–2205 lbs. of pressure.

b. Align the holes of the pushrod and housing, pass the proper tool through the holes to keep the setting position of the pushrod.

c. Release the press and install the dust boot to the tensioner.

29. Install the tensioner and tighten the bolts to 20 ft. lbs. (27 Nm). Remove the tool from the tensioner.

30. Turn the crankshaft pulley 2 revolutions from TDC to TDC. Always turn the crankshaft clockwise. Check that each pulley aligns with the timing marks.

31. Install the right side engine mounting bracket and tighten the bolts to 30 ft. lbs. (41 Nm).

32. Install the No. 2 timing belt cover and gasket. Install the spark plugs and tighten to 13 ft. lbs. (18 Nm).

33. Install the air intake chamber.

34. Install the right side engine mount and tighten the nut-to-bracket to 38 ft. lbs. (52 Nm) and the nut-to-

body to 65 ft. lbs. (88 Nm). Do not tighten the mounting bolt.

35. Lower the engine. Install the right side engine mounting brackets and tighten the bolts. Fasten the power steering cooler pipes and tighten the engine mount bolt to 47 ft. lbs. (64 Nm).

36. Install the alternator and power steering belts. Replace the right side fender apron seal.

37. Raise and safely support the vehicle. Install the tire and wheel assembly. Lower the vehicle.

38. Install the power steering reservoir tank and cruise control actuator.

39. Install the suspension upper brace.

40. Connect the negative battery cable.

LS400

1. Disconnect the negative battery cable and the positive battery cable. Remove the battery.

2. Remove the air duct and dust covers. Remove the engine undercover.

3. Drain the cooling system. Remove the drive belt, fan, fluid coupling and fan pulley.

4. Remove the radiator, air cleaner and throttle body cover. Remove the air intake connector pipe.

5. Remove the air conditioning compressor and power steering pump. Do not disconnect the hoses.

6. Remove the upper high tension cord cover and the right side engine wire cover.

7. Disconnect the PCV hose and remove the left side engine wire cover. Remove the right side No. 3 timing cover.

8. Disconnect and tag the vacuum hoses and remove the left side engine wire cover. Disconnect the spark plug wires.

9. Remove the bolt, cover plate and idler pulley.

10. Disconnect the crank position sensor connector and remove the right side No. 2 timing belt cover.

11. Disconnect and remove the ignition coil. Disconnect the hoses and wires from the water bypass pipe. Remove the water bypass pipe.

12. Disconnect the crank position sensor connector and remove the No. 2 timing belt cover.

13. Remove the distributor caps and rotors. Disconnect and remove both distributor housings.

14. Disconnect and remove the alternator. Remove the drive belt tensioner and the spark plugs.

15. Turn the crankshaft pulley and align it's groove with the timing mark 0 of the No. 1 timing cover. Check that the timing marks of the camshaft tim-

ing pulleys and timing belt rear plates are aligned. If not, turn the crankshaft 1 full revolution (360 degrees).

16. Remove the timing belt tensioner. Using the proper tool, loosen the tension between the left side and right side timing pulleys by slightly turning the left side camshaft clockwise.

17. Disconnect the timing belt from the camshaft timing pulleys. Using the proper tool, remove the bolt and the timing pulleys.

18. Remove the bolt and the crankshaft pulley with the proper tool. Remove the fan bracket.

19. Remove the mounting bolts and the No. 1 timing belt cover.

20. Remove the timing belt guide and lift off the timing belt.

To install:

21. Align the installation mark on the timing belt with the drilled mark of the crankshaft timing pulley. Install the timing belt on the crankshaft timing pulley, No. 1 idler pulley and the No. 2 idler pulley.

22. Install the timing belt guide with the cup side facing forward. Replace the timing belt cover spacer.

23. Install the No. 1 timing belt cover and tighten the mounting bolts. Install the fan bracket.

24. Align the pulley set key on the crankshaft with the key groove of the pulley. Install the pulley, using the proper tool to tap in the pulley. Tighten the pulley bolt to 181 ft. lbs. (245 Nm).

25. Align the knock pin on the right side camshaft with the knock pin of the timing pulley. Slide on the timing pulley with the right side mark facing forward. Tighten the bolt to 80 ft. lbs. (108 Nm).

26. Align the knock pin on the left side camshaft with the knock pin of the timing pulley. Slide on the timing pulley with the left side mark facing forward. Tighten the bolt to 80 ft. lbs. (108 Nm).

27. Turn the crankshaft pulley and align it's groove with the 0 timing mark on the No. 1 timing belt cover. Using the proper tool, turn the crankshaft timing pulley and align the timing marks of the camshaft timing pulley and the timing belt rear plate.

28. Install the timing belt to the left side camshaft timing pulley by:

a. Using the proper tool, slightly turn the left side timing pulley clockwise. Align the installation mark of the timing belt with the timing mark of the camshaft timing pulley and hang the timing belt on the left side camshaft pulley.

b. Using the proper tool, align the timing marks of the left side camshaft pulley and the timing belt rear plate.

c. Check that the timing belt has tension between crankshaft timing pulley and the left side camshaft pulley.

29. Install the timing belt to the right side camshaft timing pulley by:

a. Using the proper tool, slightly turn the right side timing pulley clockwise. Align the installation mark of the timing belt with the timing mark of the camshaft timing pulley and hang the timing belt on the right side camshaft pulley.

b. Using the proper tool, align the timing marks of the right side camshaft pulley and the timing belt rear plate.

c. Check that the timing belt has tension between crankshaft timing pulley and the right side camshaft pulley.

30. The timing belt tensioner must be set prior to installation. The tensioner can be set by:

a. Place a plate washer between the tensioner and a block. Using a suitable press, press in the pushrod using 220–2205 lbs. of pressure.

b. Align the holes of the pushrod and housing, pass the proper tool through the holes to keep the setting position of the pushrod.

c. Release the press and install the dust boot to the tensioner.

31. Install the tensioner and tighten the bolts to 20 ft. lbs. (27 Nm). Remove the tool from the tensioner.

32. Turn the crankshaft pulley 2 revolutions from TDC to TDC. Always turn the crankshaft clockwise. Check that each pulley aligns with the timing marks.

33. Install the spark plugs and tighten to 13 ft. lbs. (18 Nm). Install the drive belt tensioner and tighten the bolt to 12 ft. lbs. (16 Nm).

34. Install the alternator and engine wire bracket. Tighten the nut and bolt to 26 ft. lbs. (35 Nm). Connect the electrical connections at the alternator.

35. Install both distributor housings and tighten the mounting bolts to 13 ft. lbs. (18 Nm). Replace the distributor rotors and caps.

36. Install the right side No. 2 timing belt cover and tighten the 10mm bolts to 69 inch lbs. and the 12mm bolts to 12 ft. lbs. (16 Nm). Connect the crank position sensor connector.

37. Install the left side No. 2 timing belt cover and connect the crank position sensor connector.

38. Install the water bypass pipe and connect the hoses and connectors.

39. Replace the left side ignition coil and connect the coil connector. Install the idler pulley and cover plate. Tighten the bolt to 27 ft. lbs. (37 Nm).

40. Install and secure the ignition wires. Install the right side No. 3 timing belt cover.

41. Install the left side No. 3 timing belt cover and connect the vacuum hose and connectors. Install the right side engine wire cover.

42. Install the left side engine wire cover and connect the vacuum hoses.

43. Install the upper high tension cord covers. Fit the front side claw groove of the upper cover to claw of the lower cover.

44. Install the power steering pump and the air conditioning compressor.

45. Install the throttle body cover and the air cleaner.

46. Install the radiator, fan pulley, fan coupling, fan and drive belt.

47. Install the engine undercover and replace the battery.

48. Install the air ducts and dust covers. Connect the battery cables.

49. Refill the cooling system. Check the ignition timing.

Timing Sprockets

REMOVAL & INSTALLATION

ES250

1. Disconnect the negative battery cable.

2. Remove the timing belt assembly.

3. Remove the idler pulley with the proper tool.

4. Remove the crankshaft pulley.

NOTE: If the pulley cannot be removed by hand, carefully pry off the pulley with a suitable pry bar or puller.

To install:

5. Align the crankshaft timing pulley set key with the groove on the timing pulley. Slide on the crankshaft pulley with the flange side facing inward.

6. Install the No. 1 idler pulley, using the proper adhesive on the threads of the mounting bolt end. Install the bolt and washer with the proper tool. Tighten the bolt to 25 ft. lbs. (34 Nm). Check that the pulley bracket moves smoothly.

7. Align the installation mark on the timing belt with the drilled mark on the crankshaft pulley.

8. Install the timing belt assembly.

9. Connect the negative battery cable.

LS400

1. Disconnect the negative battery cable.

2. Remove the timing belt.

3. Remove the pulley bolt and the No. 2 idler pulley. Using the proper tool, remove the bolt and No. 1 idler pulley.

4. Remove the crankshaft timing pulley with the proper tool.

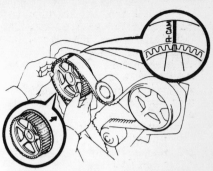

Installing the right hand camshaft pulley – ES250

To install:

5. When installing the right hand camshaft timing pulley, use the following procedure:

a. Align the knock pin on the camshaft with the knock pin groove of the timing pulley.

b. Slide the timing pulley, facing the **RH** mark forward.

c. Holding the camshaft gear still and torque the pulley bolt to 80 ft. lbs. (108 Nm).

6. When installing the right hand camshaft timing pulley, use the following procedure:

a. Align the knock pin on the camshaft with the knock pin groove of the timing pulley.

b. Slide the timing pulley, facing the **LH** mark forward.

c. Holding the camshaft gear still

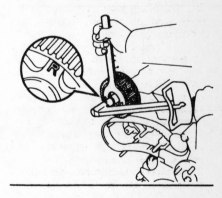

Installing the right hand and left hand camshaft pulley's – LS400

and torque the pulley bolt to 80 ft. lbs. (108 Nm).

7. Install the No. 1 idler pulley, using a suitable adhesive on the threads of the pulley bolt end. Tighten the pulley bolt to 25 ft. lbs. (34 Nm).

8. Turn the crankshaft and align the timing marks of the crankshaft timing pulley and the oil pump body.

9. Align the installation mark on the timing belt with the drilled mark of the crankshaft timing pulley (if the belt is so marked).

10. Install the timing belt assembly and connect the negative battery cable

Camshaft

REMOVAL & INSTALLATION

ES250

1. Disconnect the negative battery cable.

NOTE: Since the thrust clearance of the camshaft is small, the camshaft must be held level while it is being removed. If the camshaft is not kept level, the portion of the camshaft head receiving the shaft thrust may crack or be damaged, causing the camshaft to seize or break.

2. Remove the suspension upper brace. Disconnect the throttle cable from the throttle body, if equipped with an automatic transaxle.

3. Disconnect the accelerator cable from the throttle body. Remove the cruise control actuator and the vacuum pump.

4. Disconnect the ISC hose, the vacuum pipe hose and the air cleaner hose.

5. Raise and safely support the vehicle. Remove the right side engine undercover.

6. Remove the suspension lower crossmember and the front exhaust pipe. Lower the vehicle.

7. Remove the alternator, the ISC valve and the throttle body.

8. Remove the EGR pipe, the EGR valve and the vacuum modulator.

9. Disconnect the (4) BVSV vacuum hoses, the fuel pressure VSV hose and the air conditioning control valve vacuum hose.

10. Remove the distributor and the exhaust crossover pipe.

11. Disconnect the cold start injector connector and remove the cold start injector tube.

12. Disconnect the following hoses:
   a. PCV hose
   b. Vacuum sensing hose
   c. Fuel pressure VSV hose
   d. Air conditioning control valve vacuum hose

   e. EGR gas temperature sensor connector – California only

13. Remove the mounting bolts and the brackets from the intake chamber. Remove the air intake chamber.

14. Remove the delivery pipes and the injectors.

15. Disconnect the water temperature sensor connector and the upper radiator hose. Remove the water outlet.

16. Disconnect the following connectors and hose:
   a. Cold start injector time switch connector
   b. Water temperature sensor connector
   c. Heater water bypass hose

17. Remove the water bypass outlet. Remove the crossover pipe insulator from the water bypass outlet.

18. Remove the cylinder head rear plate and the idler pulley bracket.

19. Remove the mounting bolts and nuts. Lift off the intake manifold.

20. Disconnect the main oxygen sensor connector. Remove outside heat insulator.

21. Remove the mounting nuts and the right side exhaust manifold. Remove the inside heat insulator.

22. Remove the heat insulator, the mounting nuts and the left side exhaust manifold.

23. Remove the spark plugs.

24. Remove the timing belt, camshaft pulleys and the No. 2 idler pulley.

25. Remove the No. 3 timing cover.

26. Remove the mounting nuts and the cylinder head covers. Remove the spark plug tube gaskets.

27. Remove the exhaust camshaft of the right side cylinder head by:
   a. Align the timing marks, 2 pointed marks, of the camshaft drive and the drive gear by turning the camshaft with a wrench.
   b. Secure the exhaust camshaft sub-gear to the drive gear with a service bolt.
   c. Uniformly, loosen and remove the 8 bearing cap bolts in the proper sequence.
   d. Remove the 4 bearing caps and the camshaft.

28. Uniformly, loosen and remove the 10 bearing cap bolts on the right side cylinder head, in sequence. Remove the 5 bearing caps and remove the intake camshaft.

29. Remove the exhaust camshaft of the left side cylinder head by performing the following:
   a. Align the timing marks, 1 pointed mark, of the camshaft drive and the drive gear by turning the camshaft with a wrench.
   b. Secure the exhaust camshaft sub-gear to the drive gear with a service bolt.

   c. Uniformly, loosen and remove the 8 bearing cap bolts in the proper sequence.
   d. Remove the 4 bearing caps and the exhaust camshaft.

30. Uniformly, loosen and remove the 10 bearing cap bolts on the right side cylinder head, in sequence. Remove the 5 bearing caps and remove the intake camshaft.

To install:

31. Install the intake camshaft of the right side cylinder head by:
   a. Apply a suitable multi-purpose grease to the thrust portion of the camshaft.
   b. Apply seal packing to the No. 1 bearing cap. Install the bearing caps.
   c. Apply a light coat of oil on the threads of the bearing cap bolts.
   d. Install and uniformly tighten the bearing cap bolts to 12 ft. lbs. (16 Nm).

32. Install the exhaust camshaft of the right side cylinder head by:
   a. Apply a suitable multi-purpose grease to the thrust portion of the camshaft.
   b. Align the timing marks (2 pointed marks) of the camshaft and the drive gears.
   c. Place the camshaft on the cylinder head and install the bearing caps.
   d. Apply a light coat of oil on the threads of the bearing cap bolts.
   e. Install and uniformly tighten the bearing cap bolts to 12 ft. lbs. (16 Nm). Remove the service bolt.

33. Install the intake camshaft on the left side cylinder head by:
   a. Apply MP grease to the thrust portion of the camshaft.
   b. Place the intake camshaft at a 90 degree angle of the timing mark on the cylinder head, 1 pointed mark.
   c. Apply seal packing to the No. 1 bearing cap. Install the bearing caps.
   d. Apply a light coat of oil on the threads of the bearing cap bolts.
   e. Install and uniformly tighten the bearing cap bolts to 12 ft. lbs. (16 Nm).

34. Install the exhaust camshaft on the left side cylinder head by:
   a. Apply a suitable multi-purpose grease to the thrust portion of the camshaft.
   b. Align the timing marks, 1 pointed mark, of the camshaft and the drive gears.
   c. Place the camshaft on the cylinder head and install the bearing caps.
   d. Apply a light coat of oil on the threads of the bearing cap bolts.
   e. Install and uniformly tighten

the bearing cap bolts to 12 ft. lbs. (16 Nm). Remove the service bolt.

35. Turn the camshaft and position the cam lobe upward. Check and adjust the valve clearance.

36. Apply a suitable multi-purpose grease to the the new camshaft oil seals and install with the proper tool.

37. Install the spark plug tube gaskets. Install the proper seal packing to the cylinder heads.

38. Install the cylinder head cover gasket and install with the seal washers and nuts. Tighten to 52 inch lbs.

39. Install the No. 3 timing belt cover and tighten the bolts 65 inch lbs.

40. Install the No. 2 idler pulley, the camshaft timing pulleys and the timing belt.

41. Install the spark plugs and tighten to 13 ft. lbs. (18 Nm).

42. Install the right side heat insulator and the right side exhaust manifold with a new gasket. Tighten the nuts to 29 ft. lbs. (40 Nm).

43. Install the outside heat insulator and connect the oxygen sensor connector.

44. Install the left side exhaust manifold with a new gasket and tighten the nuts to 29 ft. lbs. (40 Nm). Install the heat insulator.

45. Install the intake manifold with new gaskets. Tighten the nuts and bolts to 13 ft. lbs. (18 Nm).

46. Install the No. 2 idler pulley bracket and tighten to 13 ft. lbs. (18 Nm). Replace the rear cylinder head plate.

47. Install the crossover pipe heat insulator. Replace the water bypass outlet and tighten the mounting bolts to 14 ft. lbs. (19 Nm).

48. Connect the heater water bypass hose, cold start injector time switch connector and the water temperature switch.

49. Connect the upper radiator hose, the water temperature sensor connector and install the water outlet with a new gasket. Tighten the bolts to 73 inch lbs.

50. Install the fuel injectors and the delivery pipes.

51. Install the air intake chamber with new gaskets and tighten the mounting bolts to 32 ft. lbs. (44 Nm).

52. Install the intake chamber brackets.

53. Connect the following hoses:
    a. PCV hose
    b. Vacuum sensing hose
    c. EGR gas temperature sensor connector, California only
    d. Air conditioning control valve air hose

54. Install the cold start injector tube and connect the cold start injector connector.

55. Install the exhaust crossover pipe with new gaskets and tighten the

mounting bolts to 25 ft. lbs. (34 Nm) and the nuts to 29 ft. lbs. (40 Nm).

56. Install the distributor.

57. Install the vacuum pipe and connect the BVSV vacuum hoses and the air conditioning control valve vacuum hose.

58. Install the EGR valve and vacuum modulator. Tighten the mounting bolts to 13 ft. lbs. (18 Nm). Install the vacuum pipe hoses.

59. Install the EGR pipe with a new gasket and tighten the mounting bolts to 13 ft. lbs. (18 Nm) and the nut to 58 ft. lbs. (77 Nm).

60. Install the throttle body and ISC valve.

61. Raise and support the vehicle. Install the front exhaust pipe and the right side engine undercover. Lower the vehicle.

62. Install the air cleaner hose and cruise control actuator.

63. Install the accelerator cable and connect the throttle cable, if equipped with an automatic transaxle.

64. Install the suspension upper brace and fill the cooling system.

65. Connect the negative battery cable. Run the engine and check for leaks.

66. Adjust the timing. Recheck the fluid levels.

LS400

1. Disconnect the negative battery cable. Drain the cooling system.

2. Remove the camshaft timing pulleys.

3. Disconnect the accelerator cable, the throttle control cable, if equipped with automatic transmission and the cruise control actuator cable.

4. Remove the high tension cord cover and the right side ignition coil.

5. Remove the water inlet housing mounting bolts and disconnect the water bypass hose from the ISC valve.

6. Remove the water inlet and inlet housing assemblies. Remove the O-ring from the water inlet housing.

7. Remove the EGR pipe.

8. Disconnect the following:
    a. VSV connector
    b. Vacuum pipe hose
    c. EGR water bypass pipe
    d. Fuel pressure VSV

9. Disconnect the EGR vacuum hoses and remove the EGR VSV.

10. Disconnect the following hoses:
    a. Water bypass pipe hose from the ISC valve.
    b. Water bypass joint hose.
    c. Vacuum pipe hoses.

11. Disconnect the EGR gas temperature sensor, California only. Remove the EGR valve adapter.

12. Disconnect the following:
    a. Fuel pressure regulator vacuum hose.

    b. Air intake chamber vacuum hose.
    c. Vacuum hose from the EVAP BVSV.

13. Remove the mounting bolts, hoses and the vacuum pipe.

14. Remove the ISC valve.

15. Remove the throttle body sensor connectors and the water bypass pipe from the rear water bypass joint.

16. Remove the mounting bolts/nuts and disconnect the PCV valve. Remove the throttle body and gasket.

17. Disconnect the accelerator cable bracket and the brake booster vacuum union and hose.

18. Disconnect the cold start injector connector and the cold start injector tube from the right side delivery pipe.

19. Disconnect the check connector from the intake chamber and remove the mounting nuts and bolts.

20. Remove the air intake chamber and the cold start injector, tube and wire assembly.

21. Disconnect the engine wire from the intake manifold and from the right side cylinder head. Disconnect the heater hoses.

22. Remove the delivery pipes and the fuel injectors. Remove the mounting bolts and nuts. Lift up the intake manifold.

23. Remove the front and rear water bypass joint.

24. Raise and safely support the vehicle. Remove the front exhaust pipe and the main catalytic converters. Lower the vehicle.

25. Disconnect the right side oxygen sensor. Remove the mounting bolts and nuts and remove the right side exhaust manifold.

26. Remove the oil dipstick and guide. Disconnect the left side oxygen sensor.

27. Remove the mounting bolts and nuts and remove the left side exhaust manifold.

28. Remove the 2 engine hangers and the wire brackets from the right side cylinder head.

29. Remove the bolts, washers and the cylinder head cover. Remove the semi-circular plugs, if necessary.

30. Remove the exhaust camshaft of the right side cylinder head by:

    a. Position the service bolt hole of the drive sub-gear to the upright position. Secure the camshaft sub-gear to drive gear with a service bolt.

    b. Set the timing mark, 1 pointed mark, of the camshaft drive gear at approximately 10 degrees, by turning the camshaft with the proper tool.

    c. Alternately, loosen and remove the bearing cap bolts holding the intake camshaft side of the oil feed pipe to the cylinder head.

d. Uniformly, loosen and remove the bearing cap bolts, in sequence.

e. Remove the oil feed pipe and the bearing caps. Remove the camshaft.

31. Remove the intake camshaft from the right side cylinder head by:

a. Set the timing mark, 1 pointed mark, of the camshaft drive gear at approximately 45 degrees, by turning the camshaft with the proper tool.

c. Uniformly, loosen and remove the bearing cap bolts in the proper sequence.

d. Remove the bearing caps, oil seal and the intake camshaft.

32. Remove the exhaust camshaft of the left side cylinder head by:

a. Position the service bolt hole of the drive sub-gear to the upright position. Secure the camshaft sub-gear to drive gear with a service bolt.

NOTE: When removing the camshaft, make sure the torsional spring force of the sub-gear has been eliminated.

b. Set the timing mark, 2 pointed marks, of the camshaft drive gear at approximately 15 degrees, by turning the camshaft with the proper tool.

c. Alternately, loosen and remove the bearing cap bolts holding the intake camshaft side of the oil feed pipe to the cylinder head.

d. Uniformly, loosen and remove the bearing cap bolts in the proper sequence.

e. Remove the oil feed pipe and the bearing caps. Remove the camshaft.

33. Remove the intake camshaft from the left side cylinder head by:

a. Set the timing mark, 1 pointed mark, of the camshaft drive gear at approximately 60 degrees, by turning the camshaft with the proper tool.

c. Uniformly, loosen and remove the bearing cap bolts, in sequence.

d. Remove the bearing caps, oil seal and the intake camshaft.

To install:

34. Remove any old packing and apply new seal packing to the bearing caps.

35. Install the bearing cap on the right side cylinder head, marked I1, in position with the arrow mark facing the rear. Install the bearing cap on the left side cylinder head, marked I6, in position with the arrow mark facing the front.

36. Apply a light coat of oil on the threads of the cap bolts. Install the nearing cap bolts with new washers and tighten to 12 ft. lbs. (16 Nm).

37. Install the right side cylinder head intake camshaft by:

a. Apply MP grease to the thrust portion of the camshaft.

b. Place the intake camshaft at a 45 degree angle of the timing mark (1 pointed mark) on the cylinder head.

c. Remove any old packing and apply new seal packing to the bearing cap marked I6 and install the front bearing cap, marked I6 with the arrow facing rearward.

d. Align the arrows at the front and rear of the cylinder head with the bearing cap.

e. Install the remaining bearing caps in the proper sequence with the arrow mark facing rearward. Install the oil feed pipe and the mounting bolts.

f. Uniformly, tighten the bearing cap bolts in the proper sequence to 12 ft. lbs. (16 Nm).

38. Install the right side cylinder head exhaust camshaft by:

a. Set the timing mark, 1 pointed mark, of the camshaft drive gear at a 10 degree angle by turning the intake camshaft with the proper tool.

b. Apply MP grease to the thrust portion of the camshaft.

c. Align the timing marks, 1 pointed mark, of the camshaft drive and driven gears.

d. Place the exhaust camshaft in the cylinder head. Install the rear bearing cap with the arrow mark facing rearward.

e. Align the arrow marks at the front and rear of the cylinder head with the mark on the bearing cap. Apply a light coat of oil on the threads of the bearing cap bolts.

f. Uniformly, tighten the bearing cap bolts in the proper sequence to 12 ft. lbs. (16 Nm).

g. Bring the service bolt installed upward by turning the camshaft with the proper tool. Remove the service bolt.

39. Install the left side cylinder head intake camshaft by:

a. Apply MP grease to the thrust portion of the camshaft.

b. Place the intake camshaft at a 60 degree angle of the timing mark, 1 pointed mark, on the cylinder head.

c. Remove any old packing and apply new seal packing to the bearing cap marked I6 and install the front bearing cap, marked I1 with the arrow facing rearward.

d. Align the arrows at the front and rear of the cylinder head with the bearing cap. Apply a light coat of oil on the threads of the bearing cap bolts.

e. Install the remaining bearing caps in the proper sequence with the arrow mark facing rearward. Install the oil feed pipe and the mounting bolts.

f. Uniformly, tighten the bearing cap bolts in the proper sequence to 12 ft. lbs. (16 Nm).

40. Install the left side cylinder head exhaust camshaft by:

a. Set the timing mark, 2 dot marks, of the camshaft drive gear at a 15 degree angle by turning the intake camshaft with the proper tool.

b. Apply MP grease to the thrust portion of the camshaft.

c. Align the timing marks, 2 dot marks, of the camshaft drive and driven gears.

d. Place the exhaust camshaft ion the cylinder head. Install the rear bearing cap with the arrow mark facing rearward.

e. Align the arrow marks at the front and rear of the cylinder head with the mark on the bearing cap. Apply a light coat of oil on the threads of the bearing cap bolts.

f. Uniformly, tighten the bearing cap bolts in the proper sequence to 12 ft. lbs. (16 Nm).

g. Bring the service bolt installed upward by turning the camshaft with the proper tool. Remove the service bolt.

41. Install the camshaft oil seals with the proper tool. Install the semi-circular plugs with the proper seal packing.

42. Install the cylinder head covers with the proper seal packing and gasket. Tighten the mounting bolts to 52 inch lbs.

43. Install the engine wire bracket and hangers. Tighten the hanger bolts to 27 ft. lbs. (37 Nm).

44. Install the right side exhaust manifold with a new gasket and tighten the mounting bolts to 29 ft. lbs. (40 Nm). Connect the right side oxygen sensor connector.

45. Install the right side exhaust manifold with a new gasket and tighten the mounting bolts to 29 ft. lbs. (40 Nm). Connect the right side oxygen sensor connector.

46. Install the left side exhaust manifold with a new gasket and tighten the mounting bolts to 29 ft. lbs. (40 Nm). Connect the left side oxygen sensor connector.

47. Install the oil dipstick and guide. Raise and safely support the vehicle.

48. Install the catalytic converters and front exhaust pipe. Lower the vehicle.

49. Install the front and rear water bypass joints. Tighten the mounting bolts to 13 ft. lbs. (18 Nm).

50. Install the intake manifold, using new gaskets. Tighten the mounting nuts and bolts to 13 ft. lbs. (18 Nm).

51. Install the delivery pipes and fuel injectors. Install the fuel return pipe

with new gaskets. Tighten the union bolt to 26 ft. lbs. (36 Nm).

52. Connect the fuel hoses and the injector connectors. Connect the engine wire to the delivery pipes.

53. Connect the connectors on the left side delivery pipe, the water temperature sensor connector, cold start injector time switch connector and the water temperature sender gauge connector.

54. Connect the heater hoses and engine wire bracket. Install the engine wire to the bracket.

55. Install the cold start injector, tube and wire assembly. Tighten the mounting bolts to 69 inch lbs.

56. Install the air intake chamber with new gaskets and tighten the mounting bolts to 13 ft. lbs. (18 Nm).

57. Connect the cold start injector tube to the right side delivery pipe and tighten the union bolt to 11 ft. lbs. (15 Nm).

58. Connect the cold start injector connector. Install the accelerator cable bracket.

59. Install the brake booster union and connect the vacuum hose. Tighten the union bolt to 22 ft. lbs. (30 Nm).

60. Connect the water bypass hose to the throttle body and the PCV hose to the cylinder head cover.

61. Install the throttle body, using a new gasket. Tighten the mounting bolts to 13 ft. lbs. (18 Nm).

62. Install the water bypass pipe and connect the sensor connectors. Install the ISC valve and tighten the mounting bolts to 13 ft. lbs. (18 Nm). Connect the water bypass hose.

63. Install the vacuum pipe and the following hoses:
   a. Fuel pressure regulator vacuum hose.
   b. Vacuum hose to the upper port of the EVAP BVSV.
   c. Air intake chamber vacuum hose.
   d. Throttle body vacuum hoses.

64. Install the EGR valve adapter with a new gasket. Tighten the mounting bolts to 13 ft. lbs. (18 Nm).

65. Connect the EGR gas temperature sensor connector (California only).

66. Install the EGR valve and vacuum modulator. Connect the water bypass hoses and the vacuum hose.

67. Install the EGR and fuel pressure VSV and connect the hoses and connectors. Replace the EGR pipe and tighten the mounting bolts to 13 ft. lbs. (18 Nm).

68. Install the timing belt rear plates and tighten the bolts to 69 inch lbs. Install the water inlet and inlet housing and tighten the bolts to 13 ft. lbs. (18 Nm).

69. Install the right side ignition coil and the high tension cord cover.

70. Connect and adjust the accelerator cable, the automatic transmission throttle cable and the cruise control actuator cable. Install the camshaft timing pulley.

71. Fill the cooling system and connect the negative battery cable. Start the engine and check for leaks.

72. Recheck all the fluid levels and check the ignition timing.

Piston and Connecting Rod

Positioning

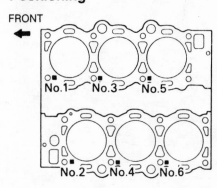

Piston Installation location – ES250

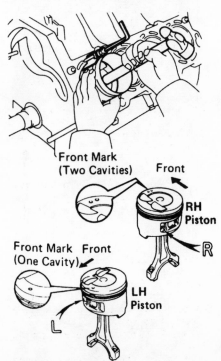

Piston installation location – LS400

ENGINE LUBRICATION

Oil Pan

REMOVAL & INSTALLATION

ES250

1. Disconnect the negative battery cable. Remove the hood assembly.
2. Raise and safely support the vehicle. Remove the engine undercovers.
3. Drain the engine oil, using a suitable container. Remove the suspension lower crossmember.
4. Disconnect the front exhaust pipe and remove the center engine support mount. Remove the front engine mount and bracket.
5. Remove the stiffener plate. Remove the oil dipstick.
6. Remove the mounting bolts and the oil pan assembly. Remove any old packing or sealer from the mounting surfaces.
7. The installation is the reverse of the removal procedure. Tighten the oil pan mounting bolts to 52 inch lbs. Tighten the stiffener plate mounting bolts to 27 ft. lbs. (37 Nm).

LS400

1. Disconnect the negative battery cable.
2. Raise and safely support the vehicle. Remove the engine undercover.
3. Drain the engine oil, using a suitable container.
4. Remove the mounting bolts and drop down the oil pan. Remove any old packing or sealer from the mounting surfaces.
5. The installation is the reverse of the removal procedure. Tighten the oil pan mounting bolts to 69 inch lbs.

Oil Pump

REMOVAL & INSTALLATION

ES250

1. Disconnect the negative battery cable. Remove the oil pan.
2. Remove the oil strainer and gasket.
2. Raise the engine using a suitable chain hoist. Remove the timing belt and pulleys.
3. Remove the alternator and the air conditioning compressor and bracket. Do not disconnect the refrigerant lines.
4. Remove the oil pump from the engine.

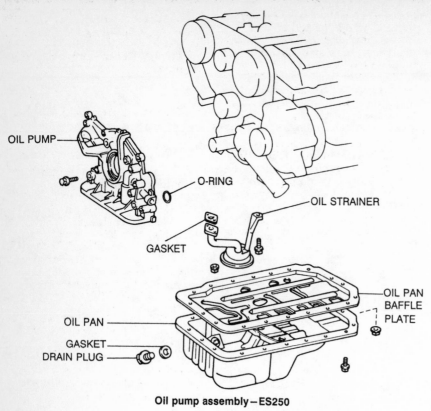

OIL PUMP

O-RING

OIL STRAINER

GASKET

OIL PAN BAFFLE PLATE

OIL PAN

GASKET
DRAIN PLUG

Oil pump assembly—ES250

5. Installation is the reverse of the removal procedure.

LS400

1. Disconnect the negative battery cable. Raise and safely support the vehicle.

2. Remove the engine undercover.

3. Drain the oil, using a suitable container. Remove the oil dipstick.

4. Remove the mounting bolts and pull down the oil pan. Remove the baffle plate and the oil strainer.

5. Remove the timing belt, No. 1 and No. 2 idler and crankshaft pulleys. Remove the bolt and pickup sensor.

6. Remove the stud bolts and mounting bolts. Remove the oil pump with the proper tool to pry away from the cylinder block. Clean and remove any packing from the mounting surfaces.

To install:

7. Install a new O-ring and align the oil drive rotor groove with the pump body mark.

8. Install the pump to the crankshaft with the spine teeth of the drive gear engaged with the large teeth of the crankshaft. Tighten the mounting bolts; 12mm bolts to 12 ft. lbs. (16

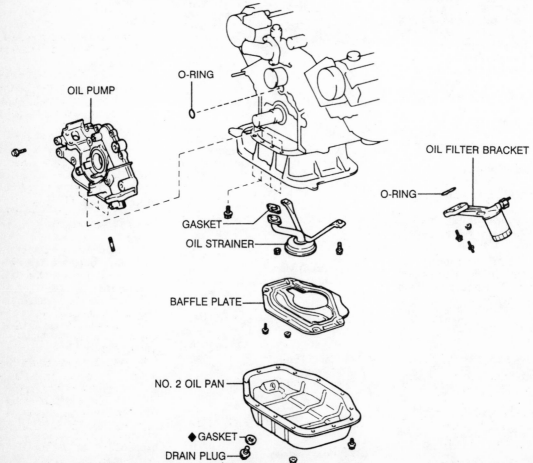

OIL PUMP

O-RING

OIL FILTER BRACKET

O-RING

GASKET

OIL STRAINER

BAFFLE PLATE

NO. 2 OIL PAN

◆ GASKET

DRAIN PLUG

Exploded view of the oil pan—LS400

Nm) and 14mm bolts to 22 ft. lbs. (30 Nm).

9. Install the pickup sensor with the bolt and tighten to 56 ft. lbs. (76 Nm). Replace the stud bolt.

10. Install the crankshaft timing pulley, No. 1 and No. 2 idler pulley and the timing belt. Install the oil strainer and tighten to 69 inch lbs.

11. Install the baffle plate and the oil pan. Tighten the mounting bolts to 69 inch lbs.

12. Install the engine undercover. Lower the vehicle.

13. Install the dipstick and refill the crankcase.

CHECKING

1. Measure the body clearance between the driven motor and the pump body with the proper tool. The standard body clearance is 0.0039–0.0069 in. The maximum clearance is 0.0118 in. If the body clearance is greater then maximum replace the rotors as a set.

2. Measure the rotor tip clearance between the drive and the driven gears with the proper tool. The standard clearance is 0.0043–0.0118 in. The maximum clearance is 0.0138 in.

3. Measure the side clearance between the rotors, using the proper tool. The standard side clearance is 0.0012–0.0035 in. The maximum side clearance is 0.0059 in. If the side clearance is greater than maximum, replace the rotors as set. If necessary, replace the oil pump assembly.

Rear Main Bearing Oil Seal

REMOVAL & INSTALLATION

1. Disconnect the negative battery cable. Raise and safely support the vehicle.

2. Remove the transmission.

3. Remove the clutch cover assembly and flywheel, if equipped with manual transaxle. Remove the drive plate, if equipped with an automatic transmission or transaxle.

4. Remove the oil seal retaining plate, complete with the oil seal.

5. Using a suitable tool pry the old seal from the retaining plate. Be careful not to damage the plate. The seal is solid type seal.

6. Install the new seal, carefully, by using a block of wood to drift it into place. Do not damage the seal as a leak will result.

7. Lubricate the lips of the seal with multipurpose grease. Installation is the reverse of removal.

ENGINE COOLING

NOTE: On the LS400, a design change has been made to the cylinder head coolant core plug and gasket. The newly designed plug has eliminated the sealant on the flat surface of the plug. There was also some molybdenum disulfide coating added to the plug gasket. Torque the new coolant core plug to 58 ft. lbs. (79 Nm). Always replace the screw plugs and gaskets as a set and only with the new style parts.

Radiator

REMOVAL & INSTALLATION

ES250

1. Disconnect both battery cables and remove the battery. Drain the cooling system.

2. Remove the ignition coil, igniter and bracket assembly.

3. Disconnect the radiator reservoir hose and the radiator hoses.

4. Remove the engine undercover and disconnect the cooling fan connectors.

5. Disconnect and plug the oil cooler lines, if equipped with an automatic transaxle.

6. Remove the mounting bolts, the supports and the radiator with the cooling fans attached.

7. Disconnect the cooling fans from the radiator.

8. The installation is the reverse of the removal procedure.

LS400

1. Disconnect the negative battery cable. Drain the cooling system.

2. Remove the air intake duct and disconnect and plug the automatic transmission lines.

3. Disconnect the cooling fan motor connector.

4. Disconnect the water hose from the coolant reservoir and remove both radiator hoses.

5. Remove the radiator supports and the radiator.

6. Remove both fan shrouds.

7. The installation is the reverse of the removal procedure.

Electric Cooling Fan

TESTING

1. Disconnect the fan motor connector.

2. Connect a suitable jumper wire between the battery and the fan motor connector.

3. If the fan does not run, replace the motor.

REMOVAL & INSTALLATION

ES250

1. Disconnect both battery cables and remove the battery. Drain the cooling system.

2. Remove the ignition coil, igniter and bracket assembly.

3. Disconnect the radiator reservoir hose and the radiator hoses.

4. Remove the engine undercover and disconnect the cooling fan connectors.

5. Disconnect and plug the oil cooler lines, if equipped with an automatic transaxle.

6. Remove the mounting bolts, the supports and the radiator with the cooling fans attached.

7. Disconnect the cooling fans from the radiator.

8. The installation is the reverse of the removal procedure.

LS400

1. Disconnect the negative battery cable. Remove the air cleaner cover on the right side.

2. Remove the clearance light and disconnect the electrical connector.

3. Remove the mounting bolts and nut. Remove the headlight together with the fog light.

4. Remove the parking light and disconnect the wire connector. Remove the engine undercover.

5. Disconnect the tube from the wind guide. Remove the mounting screws and lift off the guide.

6. Disconnect the mounting bolts. Remove the bumper and bumper retainer.

7. Remove the bumper reinforcement and both horn assemblies.

8. Disconnect the fan motor connectors.

9. Remove the mounting bolts and disconnect the wire from the mounting brackets. Remove the cooling fan(s).

To install:

10. Install the cooling fan(s) and tighten the mounting bolts. Connect the wire to the mounting brackets.

11. Connect the fan motor connector. Install the horn assemblies.

12. Replace the bumper reinforcement and replace the bumper and retainer.

13. The remainder of the installation is the reverse of the removal procedure.

Heater Core

REMOVAL & INSTALLATION

ES250

1. Disconnect the negative battery cable. Drain the cooling system.

2. Remove the console, if equipped, by removing the shift knob (manual), wiring connector and console attaching screws.

3. Remove the carpeting from the tunnel.

4. If necessary, remove the cigarette lighter and ash tray.

5. Remove the package tray, if access to the heater core is difficult.

6. Remove the bottom cover/intake assembly screws and withdraw the assembly.

7. Remove the cover from the water valve.

8. Remove the water valve.

9. Remove the hose clamps and remove the hoses from the core.

10. Remove the heater core.

11. Installation is the reverse of the removal procedure. Fill the cooling system to the proper level. Operate the heater and check for leaks.

LS400

1. Disconnect the negative battery cable. Drain the cooling system.

2. Properly discharge the air conditioning system.

3. Disconnect the mounting nuts and hoses. Remove the heater valve.

4. Remove the cooling and blower unit.

5. Disconnect the inlet and outlet water hoses. Remove the mounting nuts and the insulator retainer.

6. Remove the instrument cluster and the radio assembly with the air conditioning control attached.

7. Remove the undercover.

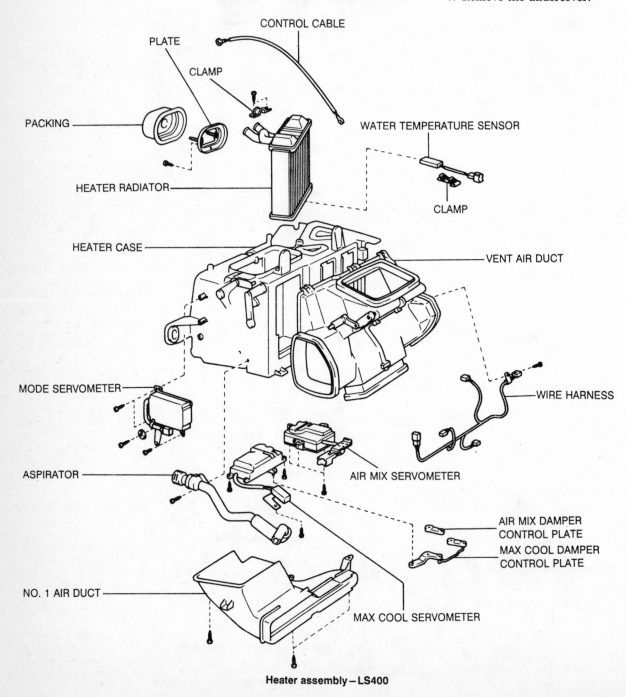

Heater assembly—LS400

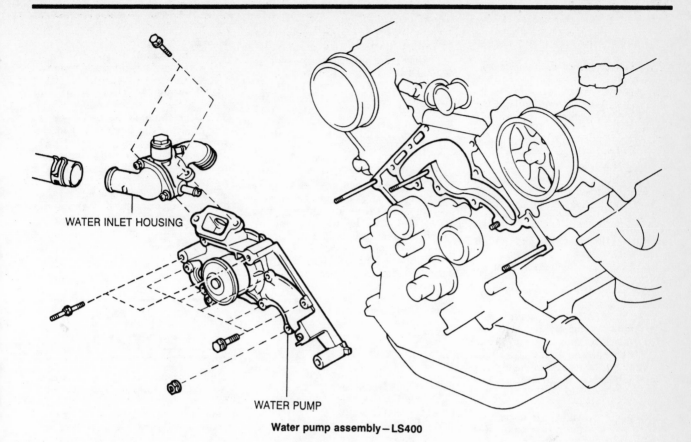

WATER INLET HOUSING

WATER PUMP

Water pump assembly—LS400

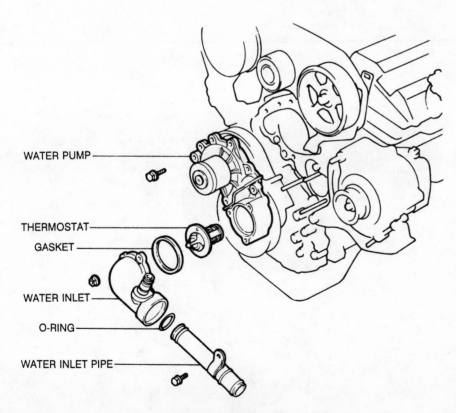

WATER PUMP

THERMOSTAT

GASKET

WATER INLET

O-RING

WATER INLET PIPE

Water pump assembly—ES250

8. Remove the glove box by performing the following:

a. Remove the glove box compartment panel and disconnect the left side check arm from the door.

b. Remove the retaining clips. Insert the proper tool between the upper side of the compartment and the safety pad, pry out the compartment to remove.

c. Disconnect the connector from the glove box compartment.

9. Disconnect the mounting bolts and remove the right side lower pad. remove the connectors from the pad.

10. Remove the glove box door. Disconnect the connectors and remove the ABS Electronic Control Unit (ECU).

11. Remove the air ducts.

12. On the driver's side, loosen the lock bolt and disconnect the junction block. Disconnect the retaining clips at the floor. Remove the combination switch.

13. On the passenger side, disconnect the connectors and the bond cable. Disconnect the retaining clips at the floor carpet.

14. Remove the bolts and nut from the safety pad and lift out the pad.

15. Remove the heater ducts.

16. Remove the mounting screws

and lift out the heater unit. Disconnect the connector and remove the servo motor.

17. Remove the mounting screws and remove the aspirator.

18. Remove the packing, screw, plate and the retaining screws. Pull out the heater core assembly.

To install:

19. Install the heater core. Replace the retaining screws, plate, screw and install the packing.

20. Install the aspirator and the servo motor. Connect the connector.

21. Install the heater unit and tighten the retaining screws. Replace the heater ducts.

22. Install the safety pad and tighten the mounting bolts and nut.

23. On the passenger side, connect the connectors and the bond cable. Connect the retaining clips at the floor carpet.

24. On the driver's side, connect the junction block and tighten the lock bolt. Connect the retaining clips at the floor. Replace the combination switch.

25. Replace the glove box door. Connect the connectors and replace the ABS Electronic Control Unit (ECU).

26. Install the air ducts.

27. Replace the right side lower pad and tighten the mounting bolts. Replace the connectors to the pad.

28. Replace the glove box by performing the following:

   a. Connect the connector to the glove box compartment.

   b. Install the glove box and replace the retaining clips.

   c. Replace the glove box compartment panel and connect the left side check arm from the door.

29. Install the undercover.

30. Replace the instrument cluster and the radio assembly with the air conditioning control attached.

31. Connect the inlet and outlet water hoses. Replace the mounting nuts and the insulator retainer.

32. Replace the cooling and blower unit.

33. Install the heater valve. Connect the mounting nuts and hoses.

34. Properly, evacuate and recharge the air conditioning system.

35. Refill the cooling system and connect the negative battery cable. Run the engine and check for leaks.

Water Pump

REMOVAL & INSTALLATION

NOTE: Work must be started after approximately 20 seconds or longer from the time the ignition switch is turned to the LOCK position and the negative battery cable is disconnected from the battery.

1. Disconnect the negative battery cable. Drain the cooling system.

2. Disconnect the radiator inlet hose from the inlet pipe. Remove the timing belt from the water pump pulley.

3. Remove the right side ignition coil on the LS400.

4. Disconnect the water inlet pipe on the ES250.

5. Remove the water inlet housing and thermostat, if necessary.

6. Remove the mounting bolts and studs. Lift out the water pump by carefully prying between the pump and the cylinder head.

7. Remove all the old packing and clean the mounting surfaces.

To install:

8. Install new seal packing to the water pump groove and a new O-ring to the water bypass pipe.

9. Install the water pump and tighten the mounting bolts to 13–14 ft. lbs. (18-19 Nm).

10. Install the water inlet housing and thermostat, if necessary.

11. Install the water inlet pipe with a new O-ring on the ES250. Tighten the mounting bolt to 14 ft. lbs. (19 Nm).

12. Replace the ignition coil on the LS400.

13. Install the timing belt and connect the inlet hose to the inlet pipe.

14. Refill the cooling system and connect the negative battery cable.

15. Run the engine and check for leaks.

Thermostat

REMOVAL & INSTALLATION

NOTE: Work must be started after approximately 20 seconds or longer from the time the ignition switch is turned to the LOCK position and the negative battery cable is disconnected from the battery.

1. Disconnect the negative battery cable. Drain the cooling system.

2. Remove the water inlet pipe and disconnect the water temperature sensor on the ES250.

3. Remove the water inlet from the inlet housing.

4. Remove the thermostat and gasket.

5. Clean the mounting surfaces. Be sure to align the jiggle valve of the thermostat with the stud bolt (pendicular) and insert the thermostat into the water inlet housing with a new gasket.

6. The installation is the reverse of the removal procedure. Tighten the water inlet bolts to 13–14 ft. lbs. (18-19 Nm).

ENGINE ELECTRICAL

NOTE: Disconnecting the negative battery cable on some vehicles may interfere with the functions of the on board computer systems and may require the computer to undergo a relearning process, once the negative battery cable is reconnected.

Distributor

REMOVAL

ES250

1. Disconnect the negative battery cable. Remove the upper bracket.

2. Remove the air cleaner top. Disconnect the air flow meter and the air cleaner hose.

3. Remove the wires from the distributor cap and disconnect the distributor wire connector.

4. Remove the hold-down bolts and pull out the distributor. Remove the O-ring.

5. Remove the distributor.

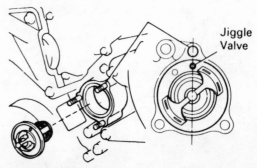

Jiggle Valve

Installing the thermostat with the jiggle valve correctly align

LS400
RIGHT

1. Disconnect the negative battery cable. Remove the air duct assembly.

2. Disconnect the air flow meter connector and air hose.

3. Remove the air flow meter assembly and the throttle body cover.

4. Disconnect the ISC and power steering idle-up air hose. Remove the No. 1 air hose.

5. Remove the high tension cable upper cover and the right side engine wire cover.

6. Disconnect the mounting bolts and remove the No. 3 timing belt cover.

7. Disconnect the sensor connector. Remove the mounting bolts, the wire cover and take off the No. 2 timing belt cover.

8. Disconnect the electrical wires from the distributor cap.

9. Remove the distributor cap and the rotor.

10. Remove the mounting bolts and lift out the distributor housing.

LEFT

1. Disconnect the negative battery cable. Drain the cooling system and remove the engine wire cover.

2. Remove the No. 2 junction block cover and the No. 3 timing belt cover.

3. Disconnect the water inlet housing hose and the reservoir tank hose.

4. Remove the mounting bolts and the water pipe from the No. 2 timing belt cover.

5. Disconnect the sensor connector, the connector boot and remove the No. 2 timing belt cover.

6. Disconnect the electrical connections from the cap and the housing. Remove the distributor cap and the rotor.

7. Remove the mounting screws and lift out the distributor housing.

INSTALLATION

Timing Not Disturbed

ES250

1. Install a new O-ring in the housing. Lubricate the O-ring with engine oil.

2. Align the cutout marks of the coupling and the housing.

3. Insert the distributor, aligning the line of the distributor housing with the cutout of the distributor attachment bearing cap. Tighten the left side hold-down bolts.

4. Install the rotor and the distributor cap. Connect the cables to the distributor cap. Align the spline of the distributor cap with the spline groove of the holder.

5. Connect the electrical connec-

tions to the distributor. Replace the air cleaner cap, air flow meter and the air cleaner hose.

6. Install the upper bracket and connect the negative battery cable.

7. Adjust the timing.

LS400 – RIGHT

1. Install the distributor housing. Replace the rotor and the distributor cap.

2. Connect the ignition cables to the distributor cap.

3. Connect the sensor connector. Install the No. 2 timing belt cover and boots.

4. Install the No. 3 timing belt cover and engine wire cover. Tighten the mounting bolts.

5. Install the electrical connection upper cover and the upper throttle body cover.

6. Replace the No. 1 air hose, the PS idle-up air hose and the ISC air hose.

7. Install the air flow meter assembly and connect the air hose and the meter connector.

8. Connect the negative battery cable. Adjust the timing.

LS400 – LEFT

1. Install the distributor housing and tighten the mounting bolts.

2. Install the rotor and replace the distributor cap.

3. Connect the ignition cables to the distributor cap.

4. Connect the sensor connector. Install the No. 2 timing belt cover and boots.

5. Connect the water inlet housing hose and the reservoir tank hose.

4. Replace the water pipe to the No. 2 timing belt cover.

6. Install the No. 3 timing belt cover and replace the No. 2 junction block cover.

7. Install the engine wire cover and tighten the mounting bolts.

8. Refill the cooling system and connect the negative battery cable. Adjust the timing.

Timing Disturbed

1. Turn the crankshaft pulley and the groove on the pulley with the timing mark **0** of the No. 1 timing belt cover.

2. Check that the timing marks of the camshaft timing pulleys and the No. 3 timing belt cover are aligned. If not, turn the crankshaft 1 revolution (360 degrees).

3. Position the slit of the intake camshaft (right side cylinder head) in the proper position.

4. Install the distributor following the proper procedure.

Ignition Timing

ADJUSTMENT

1. Allow the engine to reach normal operating temperature.

2. Connect a tachometer to terminal IG (-) of the check connector.

NOTE: Never allow the tachometer test probe to touch ground as it could result in damage to the igniter and or ignition coil. As some tachometers are not compatible with this ignition system, it is recommended to confirm the compatibility of the unit before use.

3. Check the idle speed, 650–750 rpm on the ES250 or 600–700 rpm on the LS400.

4. Connect the proper jumper wire to terminals TE1 and E1 of the check connector.

5. Connect the timing light to No. 1 cylinder on the ES250 and No. 6 cylinder on the LS400.

6. Start the engine and check the timing with the transmission in **N** position.

7. The ignition timing should be 10 degrees BTDC at idle on the ES250 or 8–12 degrees BTDC at idle on the LS400.

8. If not within specifications, loosen the hold-down bolts and adjust the timing by turning the distributor.

9. Tighten the hold-down bolts and recheck the timing and idle speed, adjust as necessary. Remove the jumper wires and test equipment.

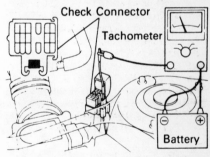

Connecting the tachometer into the check connector—ES250

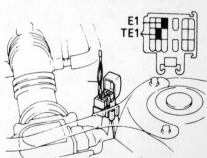

Installing the jumper wire into the TE1 and E1 terminals of the check connector—ES250

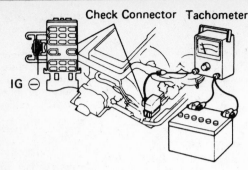

Connecting the tachometer into the check connector – LS400

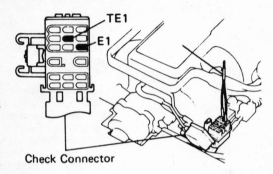

Installing the jumper wire into the TE1 and E1 terminals of the check connector – LS400

Alternator

PRECAUTIONS

Several precautions must be observed with alternator equipped vehicles to avoid damage to the unit.

• If the battery is removed for any reason, make sure it is reconnected with the correct polarity. Reversing the battery connections may result in damage to the one-way rectifiers.

• When utilizing a booster battery as a starting aid, always connect the positive to positive terminals and the negative terminal from the booster battery to a good engine ground on the vehicle being started.

• Never use a fast charger as a booster to start vehicles.

• Disconnect the battery cables when charging the battery with a fast charger.

• Never attempt to polarize the alternator.

• Do not use test lamps of more than 12 volts when checking diode continuity.

• Do not short across or ground any of the alternator terminals.

• The polarity of the battery, alternator and regulator must be matched and considered before making any electrical connections within the system.

• Never separate the alternator on an open circuit. Make sure all connections within the circuit are clean and tight.

• Disconnect the battery ground terminal when performing any service on electrical components.

• Disconnect the battery if arc welding is to be done on the vehicle.

BELT TENSION ADJUSTMENT

A belt tensioner is used to maintain the proper amount of pressure on the drive belt on the LS400. On the ES250:

1. Disconnect the negative battery cable.
2. Loosen the adjusting lock bolt and the pivot bolt.
3. Move the alternator to adjust the belt tension to 5 lbs. on a new belt or 20 lbs. on a used belt.
4. Tighten the adjusting and pivot bolts.
5. Connect the negative battery cable.

REMOVAL & INSTALLATION

ES250

1. Disconnect the negative battery cable.

NOTE: Work must be started after approximately 20 seconds or longer from the time the ignition switch is turned to the LOCK position and the negative battery cable is disconnected.

2. Remove the mounting bolts and disconnect the electrical connections at the alternator.
3. Disconnect the wiring harness from the mounting clip.
4. Remove the drive belt.
5. Remove the adjusting lock bolt, the pivot bolt and remove the alternator.

To install:

6. Mount the alternator on the alternator bracket with the pivot bolt and the adjusting bolt. Do not tighten.
7. Install the drive belt and adjust the belt tension.
8. Connect the electrical connections at the alternator and the wiring harness to the mounting clip.
9. Replace the mounting bolts and tighten to 64 ft. lbs. (87 Nm).
10. Connect the negative battery cable.

LS400

1. Disconnect the negative battery cable. Turn the ignition switch to the **LOCK** position.
2. Remove the drive belt and the engine undercover.
3. Disconnect the electrical connections at the alternator.
4. Remove the through bolt and nut. Remove the alternator.
5. The installation is the reverse of the removal procedure.

NOTE: On the LS400, there has been a problem on occasion with the alternator connector not being fully inserted into its socket. Should this condition exist, the vehicle may experience diminished alternator output which in turn will effect the battery charging system. When installing the connector to the alternator socket, be sure to reconnect the connector to the socket until a click is heard. Fasten the rubber boot and check to see that the rubber boot is fastened properly.

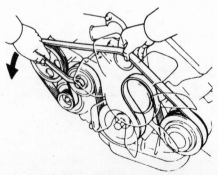

Installing the drive belt – LS400

Starter

REMOVAL & INSTALLATION

ES250

1. Disconnect both battery cables. Remove the battery and battery tray.

NOTE: Work must be started after approximately 20 seconds or longer from the time the ignition switch is turned to the LOCK position and the negative battery cable is disconnected.

2. Disconnect the noise filter connector, the igniter connector, the coil wire and the ground strap.
3. Disconnect the harness clamp and remove the igniter bracket.
4. Disconnect the electrical connections at the starter.
5. Remove the mounting bolts and the starter motor.

To install:

6. Install the starter and tighten the mounting bolts to 29 ft. lbs. (40 Nm).
7. Connect the electrical connections at the starter.
8. Install the igniter bracket and connect the harness clamp.
9. Connect noise filter connector, igniter connector, the coil wire and the ground strap.
10. Install the battery tray and the battery.
11. Connect the battery cables.

LS400

1. Disconnect the negative battery cable.

NOTE: Work must be started after approximately 20 seconds or longer from the time the ignition switch is turned to the LOCK position and the negative battery cable is disconnected.

2. Remove the intake chamber and the intake manifold.
3. Disconnect the electrical connections at the starter.
4. Remove the mounting bolts and the starter motor.

To install:

5. Install the starter and tighten the mounting bolts to 29 ft. lbs. (40 Nm).
6. Connect the electrical connections at the starter.
7. Install the intake manifold and the intake manifold.
8. Connect the negative battery cable.

EMISSION CONTROLS

Please refer to "Emission Controls" in the Unit Repair section for system maintenance procedures. Due to the complex nature of modern electronic engine control systems, comprehensive diagnosis and testing procedures fall outside the confines of this repair manual. For complete information on diagnosis, testing and repair procedures concerning all modern engine and emission control systems, please refer to "Chilton's Guide to Fuel Injection and Electronic Engine Controls".

FUEL SYSTEM

Fuel System Service Precautions

Safety is the most important factor when performing not only fuel system maintenance but any type of maintenance. Failure to conduct maintenance and repairs in a safe manner may result in serious personal injury or death. Maintenance and testing of the vehicle's fuel system components can be accomplished safely and effectively by adhering to the following rules and guidelines.

• To avoid the possibility of fire and personal injury, always disconnect the negative battery cable unless the repair or test procedure requires that battery voltage be applied.

• Always relieve the fuel system pressure prior to disconnecting any fuel system component (injector, fuel rail, pressure regulator, etc.), fitting or fuel line connection. Exercise extreme caution whenever relieving fuel system pressure to avoid exposing skin, face and eyes to fuel spray. Please be advised that fuel under pressure may penetrate the skin or any part of the body that it contacts.

• Always place a shop towel or cloth around the fitting or connection prior to loosening to absorb any excess fuel due to spillage. Ensure that all fuel spillage (should it occur) is quickly removed from engine surfaces. Ensure that all fuel soaked cloths or towels are deposited into a suitable waste container.

• Always keep a dry chemical (Class B) fire extinguisher near the work area.

• Do not allow fuel spray or fuel vapors to come into contact with a spark or open flame.

• Always use a backup wrench when loosening and tightening fuel line connection fittings. This will prevent unnecessary stress and torsion to fuel line piping. Always follow the proper torque specifications.

• Always replace worn fuel fitting O-rings with new. Do not substitute fuel hose or equivalent where fuel pipe is installed.

RELIEVING FUEL SYSTEM PRESSURE

1. Be sure the engine is cold.
2. Relieve the fuel pressure by slowly loosening the connection at the pressure regulator.
3. Be sure to place a rag under the pressure regulator to prevent the fuel from spilling on the engine.
4. Tighten the connection.

Fuel Tank

REMOVAL & INSTALLATION

NOTE: Before removing fuel system parts, clean them with a spray-type engine cleaner. Follow the instructions on the cleaner. Do not soak fuel system parts in liquid cleaning solvent.

—— CAUTION ——

The fuel injection system is under pressure. Release pressure slowly and contain spillage. Observe no smoking/no open flame precautions. Have a Class B-C (dry powder) fire extinguisher within arm's reach at all times.

ES250

1. Relieve the fuel system pressure. Remove the filler cap.
2. Using a siphon or pump, drain the fuel from the tank and store it in a proper metal container with a tight cap.
3. Disconnect the fuel pump and sending unit wiring at the connector.
4. Raise the vehicle and safely support it.
5. Loosen the clamp and remove the filler neck and overflow pipe from the tank.
6. Remove the supply hose from the tank. Wrap a rag around the fitting to collect escaping fuel. Disconnect the breather hose from the tank, again using a rag to control spillage.
7. Cover or plug the end of each dis-

connected line to keep dirt out and fuel in.

8. Support the fuel tank with a floor jack or transmission jack. Use a broad piece of wood to distribute the load. Be careful not to deform the bottom of the tank.

9. Remove the fuel tank protectors.

10. Remove the fuel tank support strap bolts.

11. Swing the straps away from the tank and lower the jack. Balance the tank. The tank is bulky and may have some fuel left in it. If its balance changes suddenly, the tank may fall.

12. Remove the fuel filler pipe extension, the breather pipe assembly and the sending unit assembly. Keep these items in a clean, protected area away from the car.

13. While the tank is out and disassembled, inspect it for any signs of rust, leakage or metal damage. If any problem is found, replace the tank. Clean the inside of the tank with water and a light detergent and rinse the tank thoroughly several times.

14. Inspect all of the lines, hoses and fittings for any sign of corrosion, wear or damage to the surfaces. Check the pump outlet hose and the filter for restrictions.

15. When reassembling, always replace the sealing gaskets with new ones. Also replace any rubber parts showing any sign of deterioration.

To install:

16. Connect the breather pipe assembly and the filler pipe extension.

NOTE: Tighten the breather pipe screw to 17 inch lbs. (2.0 Nm)

and all other attaching screws to 35 inch lbs. (3.9 Nm).

17. Place the fuel tank on the jack and elevate it into place within the car. Attach the straps and install the strap bolts, tightening them to 29 ft. lbs. (40 Nm).

18. Install the fuel tank protectors.

19. Connect the breather hose to the tank pipe, the return hose to the tank pipe and the supply hose to its tank pipe. tighten the supply hose fitting to 21 ft. lbs. (28 Nm).

20. Connect the filler neck and overflow pipe to the tank. Make sure the clamps are properly seated and secure.

21. Lower the vehicle to the ground.

22. Connect the pump and sending unit electrical connectors to the harness.

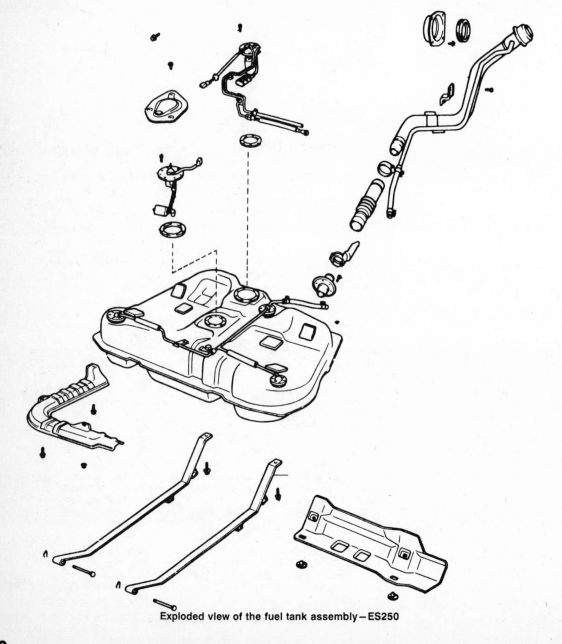

Exploded view of the fuel tank assembly—ES250

23. Using a funnel, pour the fuel that was drained from its container into the fuel filler.

24. Install the fuel filler cap.

25. Start the engine and check carefully for any sign of leakage around the tank and lines.

LS400

1. Relieve the fuel system pressure. Remove the filler cap.

2. Using a siphon or pump, drain the fuel from the tank and store it in a proper metal container with a tight cap.

3. Disconnect the fuel pump and sending unit wiring at the connectors.

4. Raise the vehicle and safely support it.

5. Loosen the clamp and remove the filler neck and overflow pipe from the tank.

6. Remove the supply hose from the tank. Wrap a rag around the fitting to collect escaping fuel. Disconnect the breather hose from the tank, again using a rag to control spillage.

7. Cover or plug the end of each disconnected line to keep dirt out and fuel in.

8. Support the fuel tank with a floor jack or transmission jack. Use a broad piece of wood to distribute the load. Be careful not to deform the bottom of the tank.

9. Remove the No. 1 and No. 2 fuel tank cover cases.

10. Remove the fuel tank retaining nuts and bolts.

11. Lower the jack and balance the tank with by hand. The tank is bulky and may have some fuel left in it. If its balance changes suddenly, the tank may fall.

12. Remove the fuel filler pipe extension, the breather pipe assembly and the sending unit assembly. Keep these items in a clean, protected area away from the car.

13. While the tank is out and disassembled, inspect it for any signs of rust, leakage or metal damage. If any problem is found, replace the tank. Clean the inside of the tank with water and a light detergent and rinse the tank thoroughly several times.

14. Inspect all of the lines, hoses and fittings for any sign of corrosion, wear or damage to the surfaces. Check the pump outlet hose and the filter for restrictions.

15. When reassembling, always replace the sealing gaskets and O-rings with new ones. Also replace any rubber parts showing any sign of deterioration.

To install:

16. Connect the breather pipe assembly and the filler pipe extension.

NOTE: Tighten the breather pipe screw to 26 in. lbs. (2.9 Nm) and all other attaching screws to 35 in. lbs. (3.9 Nm).

17. Place the fuel tank on the jack and elevate it into place within the car. Attach the retaining nuts and bolts, tightening them to 18 ft. lbs. (25 Nm).

18. Install the No. 1 and No. 2 fuel tank cover cases.

19. Connect the breather hose to the tank pipe, the return hose to the tank pipe and the supply hose to its tank pipe. Torque the inlet tank pipe screws to 26 in. lbs. (2.9 Nm). Install the fuel main tank tube and torque the bolt to

22 ft. lbs. (29 Nm). Install the No. 2 fuel return tank tube and torque the bolt to 18 ft. lbs. (25 Nm).

20. Connect the filler neck and overflow pipe to the tank. Make sure the clamps are properly seated and secure.

21. Lower the vehicle to the ground.

22. Connect the pump and sending unit electrical connectors to the harness.

23. Using a funnel, pour the fuel that was drained from its container into the fuel filler.

24. Install the fuel filler cap.

25. Start the engine and check carefully for any sign of leakage around the tank and lines.

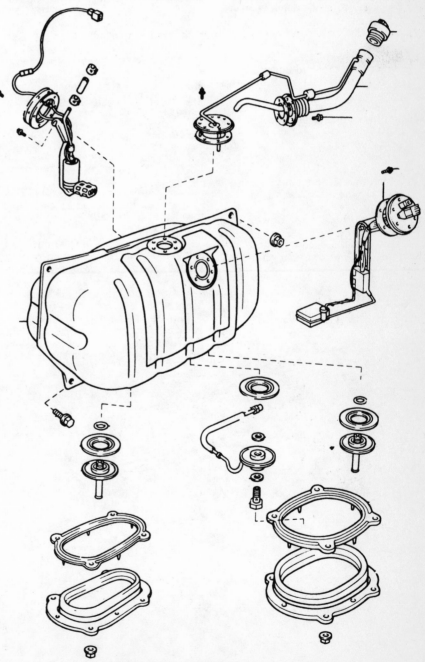

Exploded view of the fuel tank assembly—LS400

Fuel Filter

REMOVAL & INSTALLATION

ES250

The fuel filter is located under the hood, on the driver's side, by the fender well.

1. Disconnect the negative battery cable.
2. Disconnect and plug the fuel lines to the filter. Place a rag under the filter to catch any fuel that may spill.
3. Disconnect the mounting bolt(s) and remove the fuel filter.
4. Install a new filter and tighten the line connections.
5. Connect the negative battery cable. Start the engine and check for leaks.

LS400

The fuel filter is located under the vehicle on the driver's side before the rear axle.

1. Disconnect the negative battery cable. Raise and safely support the vehicle.
2. Disconnect and plug the fuel lines to the filter. Place a rag under the filter to catch any fuel that may spill.
3. Disconnect the mounting bolt(s) and remove the fuel filter.
4. Install a new filter and tighten the line connections. Lower the vehicle.
5. Connect the negative battery cable. Start the engine and check for leaks.

Electric Fuel Pump

PRESSURE TESTING

ES250

1. Check that the battery voltage is above 12 volts. Disconnect the negative battery cable.
2. Disconnect the cold start injector.
3. Place a suitable container or shop towel under the cold start injector connector.
4. Remove the union bolt and gaskets. Disconnect the cold start injector tube from the left side delivery pipe.
5. Install the proper pressure gauge to the left side delivery pipe and tighten the washers and bolt to 13 ft. lbs. (18 Nm).
6. Reconnect the negative battery cable and turn the ignition switch to the **ON** position.
7. Measure the fuel pressure; 38–44 psi.
8. Remove the pressure gauge and tighten the union bolt with the gaskets to 13 ft. lbs.
9. Connect the cold start injector tube to the left side delivery pipe. Install the cold start injector.
10. Start the engine and check for leaks.

LS400

1. Check that the battery voltage is above 12 volts. Disconnect the negative battery cable.
2. Disconnect the VSV for the EGR.
3. Place a suitable container or shop towel under the rear end of the left side delivery pipe.
4. Slowly, loosen the bolt on the left side of the rear fuel pipe and remove the bolt and gaskets from the delivery pipe.
5. Drain the fuel in the left side delivery pipe.
6. Install the proper pressure gauge and connect with the gaskets and bolt.
7. Reconnect the negative battery cable. Connect terminals FP and +B of the check connector with the proper tool.
8. Turn the ignition switch to the **ON** position.
9. Measure the fuel pressure; 38–44 psi.
10. Remove the pressure gauge. Install the bolt and gasket to the delivery pipe.
11. Connect the VSV for the EGR. Start the engine and check for leaks.

REMOVAL & INSTALLATION

ES250

1. Disconnect the negative battery cable. Raise and safely support the vehicle.
2. Drain the fuel from the fuel tank. Disconnect and clamp the fuel lines.
3. Remove both fuel tank protector shields. Support the tank with a suitable jacking device.
4. Remove the mounting bolts and take down both tank band straps. Lower the tank and disconnect the electrical connections.
5. Remove the fuel tank.
6. Remove the fuel pump bracket. Disconnect the fuel line and electrical connector.
7. Remove the mounting nuts and the pump assembly.
To install:
8. Install the fuel pump to the mounting bracket. Connect the mounting nuts, fuel line and electrical connectors.
9. Raise the fuel tank and connect the electrical connections.
10. Install the tank band straps and tighten the strap bolts to 29 ft. lbs.
11. Install both fuel tank protector shields. Connect the fuel lines.

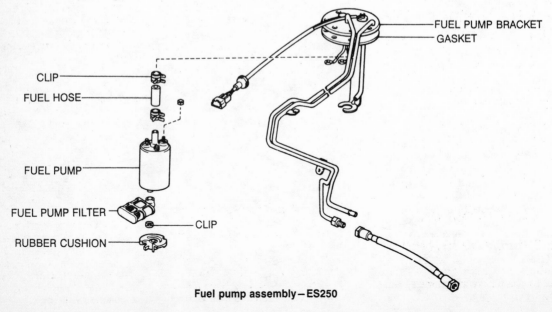

CLIP
FUEL HOSE
FUEL PUMP
FUEL PUMP FILTER
RUBBER CUSHION
CLIP
FUEL PUMP BRACKET
GASKET

Fuel pump assembly—ES250

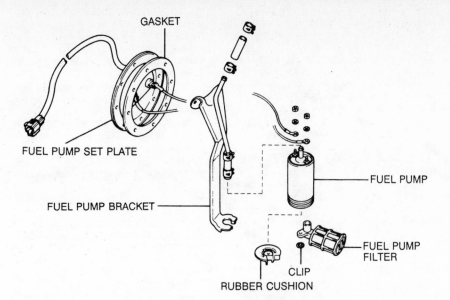

GASKET

FUEL PUMP SET PLATE

FUEL PUMP BRACKET

FUEL PUMP

FUEL PUMP FILTER

CLIP

RUBBER CUSHION

Fuel pump assembly—LS400

12. Remove the jacking device. Lower the vehicle.

13. Connect the negative battery cable. Fill the fuel tank.

LS400

1. Disconnect the negative battery cable. Drain the fuel from the fuel tank.

2. Remove the front trim panel from inside the trunk.

3. Remove the seat back assembly by performing the following:

a. Disconnect the hook and remove the clips, using the proper tool.

b. Disconnect the wire harness clamp and connectors.

c. Disconnect the side hooks and remove the mounting bolts and seat back assembly.

4. Remove the partition panel plug. Disconnect the fuel pump connector.

5. Remove the mounting bolts and the fuel pump set plate. Move the clips aside and disconnect and plug the fuel hose.

6. Remove the mounting bolts and the fuel pump assembly.

7. The installation is the reverse of the removal procedure. Tighten the fuel pump mounting bolts to 48 inch lbs. and the fuel pump set plate bolts to 26 inch lbs.

8. Connect the negative battery cable.

Fuel Injection

IDLE SPEED ADJUSTMENT

The idle speed adjustment is controlled by the Electronic Control Unit (ECU).

IDLE MIXTURE ADJUSTMENT

The idle mixture adjustment is controlled by the Electronic Control Unit (ECU).

Fuel Injector

REMOVAL & INSTALLATION

ES250

1. Disconnect the negative battery cable. Drain the cooling system.

2. Disconnect the throttle cable from the throttle body and bracket on vehicles equipped with automatic transaxle.

3. Disconnect the accelerator and bracket from the throttle body and air intake chamber.

4. Remove the air cleaner cap, the air flow meter and the air cleaner hose.

5. .Disconnect and tag the following hoses:

a. PCV hoses

b. Vacuum sensing hose

c. Water bypass hose

d. Fuel pressure VSV hose

e. Emission control vacuum hoses

f. ISC connector

g. Throttle position sensor connector

h. EGR gas temperature sensor for California only.

6. Remove the right side mounting bracket.

7. Disconnect the cold start injector and the cold start injector tube.

8. Disconnect and tag the following:

a. Brake booster vacuum hose

b. Power steering vacuum and air hoses

c. Cruise control vacuum hose

d. Ground strap connector

e. Wire harness and clamp

9. Disconnect the EGR pipe. Disconnect the engine hanger and the air intake chamber stay from the air intake chamber.

10. Remove the air intake chamber. Disconnect and tag the cold start injector connector, water temperature sensor connector and the 6 injector connectors.

11. Disconnect the wire harness from the left side delivery pipe.

12. Disconnect the fuel inlet and return hoses. Remove the fuel line.

13. Remove the 2 bolts and the left side delivery pipe with the 3 injectors. Remove the 3 bolts and the right side delivery pipe with the fuel pipe and the injectors attached.

14. Pull out the 6 injectors from the delivery pipe. Remove the 6 insulators and the 4 spacers from the intake manifold.

To install:

15. Install a new grommet and O-ring to the injector. Apply a light coat of gasoline to the new O-ring.

16. Install the injectors to the delivery pipes, while turning to left and right.

17. Place the insulators and the spacers in the proper position on the intake manifold.

18. Install the 3 injectors together with the right side delivery pipe and the fuel pipe in the proper position on the intake manifold.

19. Install the 3 injectors together with the left side delivery pipe in the proper position on the intake manifold.

NOTE: Make sure the injectors rotate smoothly. If the injectors do not rotate smoothly, the probable cause is the incorrect installation of the O-rings. Replace the O-rings.

20. Position the injector connector upward and install the mounting bolts. Tighten to 9 ft. lbs. (12 Nm)

21. Install the No. 2 fuel pipe with new gaskets. Tighten the mounting bolts to 24 ft. lbs. (33 Nm).

22. Connect the fuel inlet and return hoses. Connect the wire harness clamps to the left side delivery pipe.

23. Connect the 6 injector connectors, the cold start injector connector and the water temperature connector.

24. Install the air intake chamber, using a new gasket. Tighten the mounting bolts to 32 ft. lbs. (44 Nm).

25. Connect the EGR pipe and tighten to 58 ft. lbs. (79 Nm). Connect the wire harness clamp.

26. Connect the air intake chamber stay and tighten the bolts to 27 ft. lbs. (37 Nm).

27. Install the engine hanger and tighten the mounting bolts to 27 ft. lbs. (37 Nm).

28. Connect the following:
   a. Brake booster vacuum hose
   b. Power steering vacuum and air hoses
   c. Cruise control vacuum hose
   d. Ground strap connector
   e. Wire harness and clamp

29. Connect the cold start injector and the cold start injector tube.

30. Replace the right side mounting stay and tighten the bolts to 38 ft. lbs. (52 Nm).

31. Connect the following hoses:
   a. PCV hoses
   b. Vacuum sensing hose
   c. Water bypass hose
   d. Fuel pressure VSV hose
   e. Emission control vacuum hoses
   f. ISC connector
   g. Throttle position sensor connector
   h. EGR gas temperature sensor for California only.

32. Replace the air cleaner cap, the air flow meter and the air cleaner hose.

33. Connect the accelerator and bracket to the throttle body and air intake chamber.

34. Connect the throttle cable and bracket, if equipped with automatic transaxle.

35. Refill the cooling system. Connect the negative battery cable.

LS400

RIGHT

1. Disconnect the negative battery cable. Remove the air cleaner with the air flow meter.

2. Remove the throttle body by performing the following:
   a. Disconnect the vacuum hoses from the throttle body and the pressure regulator.
   b. Remove the upper electrical wire cover and the bypass pipe from the ISC valve.
   c. Disconnect the throttle valve motor connector, the sub-throttle position sensor connector and the main throttle position sensor, if equipped.
   d. Remove the nuts and bolts. Separate the throttle body from the intake chamber.
   e. Remove the PCV hose from the cylinder head cover and the bypass pipe from the throttle body.
   f. Remove the throttle body.

3. Remove the left side and right side engine wire covers.

4. Remove the right side and left side No. 3 timing belt covers.

5. Remove the right side ignition coil and the lower electrical wire cover.

6. Disconnect the wire harness from the delivery pipe.

7. Disconnect the water temperature sensor connector, water temperature sender gauge connector, start injector time switch connector and 4 injector connectors.

8. Disconnect the fuel return pipe, cold start injector tube and the front fuel pipe.

9. Disconnect the rear fuel pipe, leave on the vehicle.

10. Disconnect the vacuum hoses from the pressure regulator.

11 Remove the delivery pipe with the injectors attached. Remove the insulators from the intake manifold.

12. Remove the injectors from the delivery pipe.

To install:

13. Install a new grommet and O-ring to the injector. Apply a light coat of gasoline to the new O-ring.

14. Install the injectors to the delivery pipes, while turning to left and right. Make sure the injectors rotate smoothly.

15. Place the insulators and the spacers in the proper position on the intake manifold.

16. Install the injectors together with the delivery pipe in the proper position on the intake manifold.

NOTE: Make sure the injectors rotate smoothly. If the injectors do not rotate smoothly, the probable cause is the incorrect installation of the O-rings. Replace the O-rings.

17. Temporarily, install the rear fuel pipe and the front fuel pipe. Tighten the delivery pipe first, then the front and rear fuel pipes. Tighten the delivery pipe to 13 ft. lbs. (18 Nm) and the front and rear fuel pipes to 22 ft. lbs. (30 Nm).

18. Connect the cold start injector tube, tighten to 11 ft. lbs. (15 Nm). Install the fuel return pipe and tighten the union bolt to 22 ft. lbs. (30 Nm).

19. Connect the fuel injector connectors, the start injector time switch connector, the water temperature sensor connector and the water temperature sender gauge connector.

20. Install the wire harness to the delivery pipe and tighten the mounting bolts to 74 inch lbs.

21. Install the lower electrical wire cover and the right side ignition coil.

22. Install the right side and left side No. 3 timing covers.

23. Install the throttle body by performing the following:

   a. Install the bypass hoses and the PCV hose. Install a new gasket to the air intake chamber.
   b. Connect the PCV hose to the cylinder head cover and the bypass hose to the throttle body.
   c. Install the throttle body to the air intake chamber. Tighten the nuts and bolts to 13 ft. lbs. (18 Nm).
   d. Connect the main throttle position sensor connector to the throttle body. Connect the sub-throttle position sensor and the throttle valve motor connector, if equipped.
   e. Install the bypass hose to the ISC valve and the upper electrical wire cover.
   f. Connect the vacuum hoses to the throttle body and the pressure regulator.

24. Install the intake air connector pipe and tighten to 45 inch lbs.

25. Install the air cleaner and the air flow meter. Refill the engine coolant.

26. Connect the negative battery cable. Run the engine and check for leaks.

LEFT

1. Disconnect the negative battery cable. Drain the cooling system.

2. Disconnect the PCV hose. Remove the upper electrical wire cover and the left side wire cover.

3. Disconnect the following hoses:
   a. Vacuum hoses from the air pipe
   b. Vacuum hose from the BVSV
   c. Vacuum hose from the VSV for the EGR
   d. EGR vacuum modulator hose

4. Remove the EGR vacuum modulator with the mounting bracket. Disconnect the check connector from the bracket.

5. Disconnect the connectors and remove the mounting bolts from the 2 VSV's. Remove the VSV's.

6. Remove the water bypass pipes.

7. Disconnect the engine wire from the delivery pipe.

8. Disconnect the following connectors:
   a. Distributor connector
   b. Engine speed sensor connector
   c. EGR gas temperature sensor connector, California only
   d. Injector connectors

9. Remove the bolt, pulsation damper and disconnect the fuel inlet hose from the delivery pipe.

10. Disconnect the front fuel pipe and disconnect the rear fuel pipe.

11. Remove the delivery pipe with the injectors attached. Remove the insulators from the intake manifold.

12. Remove the injectors from the delivery pipe.

To install:

13. Install a new grommet and O-ring to the injector. Apply a light coat of gasoline to the new O-ring.

14. Install the injectors to the delivery pipes, while turning to left and right. Make sure the injectors rotate smoothly.

15. Place the insulators and the spacers in the proper position on the intake manifold.

16. Install the injectors together with the delivery pipe in the proper position on the intake manifold.

NOTE: Make sure the injectors rotate smoothly. If the injectors do not rotate smoothly, the probable cause is the incorrect installation of the O-rings. Replace the O-rings.

17. Temporarily, install the rear fuel pipe and the front fuel pipe. Tighten the delivery pipe first, then the front and rear fuel pipes. Tighten the delivery pipe to 13 ft. lbs. (18 Nm) and the front and rear fuel pipes to 22 ft. lbs. (30 Nm).

18. Connect the cold start injector tube, tighten to 11 ft. lbs. (15 Nm) Install the fuel return pipe and tighten the union bolt to 22 ft. lbs. (30 Nm).

19. Connect the fuel inlet hose to the delivery pipe and replace the pulsation damper and bolt. Tighten the damper to 22 ft. lbs. (30 Nm) and the bolt to 69 inch lbs.

20. Connect the following connectors:
    a. Distributor connector
    b. Engine speed sensor connector
    c. EGR gas temperature sensor connector, California only
    d. Injector connectors

21. Connect the engine wire to the delivery pipe and tighten the mounting bolts to 74 inch lbs.

22. Install the water bypass pipes.

23. Install the VSV's and tighten the bolts to 13 ft. lbs. (18 Nm). Connect the connectors.

24. Connect the check connector to the bracket. Replace the EGR vacuum modulator with the mounting bracket.

25. Connect the following hoses:
    a. Vacuum hoses from the air pipe
    b. Vacuum hose from the BVSV
    c. Vacuum hose from the VSV for the EGR
    d. EGR vacuum modulator hose

26. Connect the PCV hose. Replace the upper electrical wire cover and the left side wire cover.

27. Refill the cooling system. Connect the negative battery cable.

28. Run the engine and check for leaks.

DRIVE AXLE

Halfshaft

REMOVAL & INSTALLATION

ES250

FRONT

1. Raise and safely support the vehicle. Remove the front tire and wheel assembly. Remove the cotter pin and locknut cap. Loosen the bearing locknut while an assistant applies the brake.

2. Disconnect the steering knuckle from the lower ball joint with the proper tool.

3. Disconnect the tie rod end from the steering knuckle with the proper tool.

4. Place matchmarks on the halfshaft and center halfshaft. Using the proper tool, loosen the mounting bolts.

NOTE: Do not remove the bolts. Finger tighten them so the halfshaft does not fall.

5. Disconnect the halfshaft from the axle hub. Remove the left side halfshaft and the joint cover gasket.

6. Remove the bearing lock bolt and the snapring with the proper tool. Pull out the left side halfshaft with the center halfshaft.

NOTE: If the halfshaft cannot be pulled out, tap out the driveshaft with the proper tool.

7. Push the side gear shaft to the differential, in order to replace. Measure and note the distance between the transaxle case and the side gear shaft. Remove the side gear shaft with the proper tool.

8. If necessary, replace the side gear gear shaft oil seal.

To install:

9. Install the left side gear shaft, using a suitable tool to tap in the driveshaft until it makes contact with the pinion shaft. Ensure a new snapring is positioned securely in the groove of the side gear shaft.

10. Check that the side gear will not come out by hand. Push the side gear shaft to the differential and measure the distance between side gear shaft and the transaxle case. Make sure the distance is the same measurement taken before removing the side gear shaft.

11. Pack the side gear shaft with a suitable grease.

12. Align the matchmarks on the side gear shaft and the halfshaft. Install the left side halfshaft and finger tighten the bolts.

13. Install the right side halfshaft with the center halfshaft to the transaxle through the bearing bracket. Install the snapring.

14. Install the outboard joint side of the halfshaft to the axle hub. Temporarily, connect the steering knuckle to the lower ball joint.

15. Connect the tie rod end to the steering knuckle and tighten the bolt to 36 ft. lbs. (49 Nm). Tighten the lower ball joint mounting bolts to 83 ft. lbs. (113 Nm).

16. Tighten the hexagon bolts to 48 ft. lbs. (65 Nm).

17. Replace the front tire and wheel assembly. Lower the vehicle.

LS400

REAR

1. Raise and safely support the vehicle. Remove the rear tire and wheel assembly.

2. Remove the tail pipe O-rings and suspend the tail pipe, using a piece of wire.

3. Disconnect the height control sensor, if equipped.

4. Place matchmarks on the driveshaft and the side gear shaft. Remove the hexagon bolts and washers with the proper tool.

5. Hold the inboard joint side of the halfshaft so the outboard joint side does not bend too much. Tap the end of the halfshaft with the proper tool and disengage the axle hub.

6. Remove the halfshaft.

To install:

7. Insert the outboard joint side of the halfshaft and align the matchmarks on the side gear shaft and the halfshaft.

8. Install the hexagon bolts and tighten to 61 ft. lbs.

9. Connect the height control sensor, if equipped. Replace the O-rings supporting the tail pipe.

10. Install the bearing locknut and tighten to 253 ft. lbs. Replace the rear tire and wheel assembly.

11. Lower the vehicle.

CV-Boot

REMOVAL & INSTALLATION

ES250

FRONT

1. Raise and safely support the vehicle. Remove the front tire and wheel assembly.

2. Disconnect the center halfshaft from the right side halfshaft, using the proper tool. Remove the inboard joint clamp.

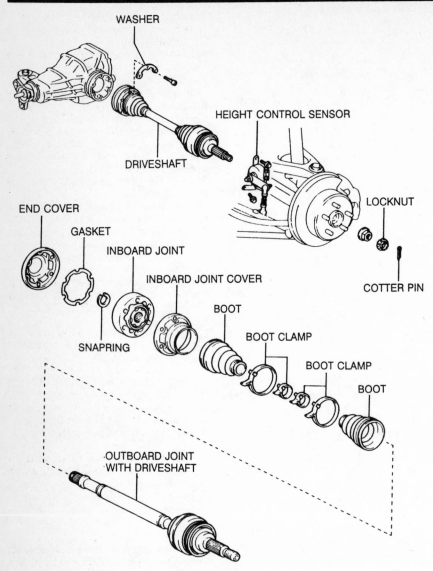

WASHER

HEIGHT CONTROL SENSOR

DRIVESHAFT

END COVER

GASKET

INBOARD JOINT

INBOARD JOINT COVER

LOCKNUT

BOOT

SNAPRING

BOOT CLAMP

BOOT CLAMP

COTTER PIN

BOOT

OUTBOARD JOINT
WITH DRIVESHAFT

Rear axle shaft assembly—LS400

3. Place matchmarks on the inboard joint and the halfshaft. Remove the snapring with the proper tool.

4. Remove the inboard joint from the halfshaft, using the proper tool to press out the joint.

5. Remove the inboard joint from the inboard joint cove. When lifting the inboard joint, hold on to the inner race and outer race.

6. Remove the inboard joint boot and the outboard joint boot.

7. Reverse the removal procedure for installation. Pack the boot with a suitable grease prior to installation.

LS400
REAR

1. Raise and safely support the vehicle. Remove the rear tire and wheel assembly.

2. Remove the tail pipe O-rings and

suspend the tail pipe, using a piece of wire.

3. Disconnect the height control sensor, if equipped.

4. Remove the halfshaft.

5. Secure the halfshaft in a suitable holding fixture. Tap out the end cover with the proper tools.

6. Remove the boot clamps from the inboard and outboard joint boots.

7. Place matchmarks on the inboard joint and halfshaft. Remove the snapring , using the proper pliers.

8. Press out the inboard joint from the halfshaft with the proper tools. Secure the inboard joint in a suitable holding fixture.

9. Tap out the inboard joint cover. Remove both the inboard and outboard joint boots.

10. The installation is the reverse of the removal procedure. Pack the boots

and the end cover with a suitable grease prior to installation.

Driveshaft and U-Joints

REMOVAL & INSTALLATION

LS400

1. Raise and safely support the vehicle.

2. Remove the front exhaust pipe and the heat insulator.

3. Remove the front and rear center floor crossmember braces.

4. Using the proper tool, loosen the adjusting nut on the driveshaft. Place matchmarks on the transmission flange and the flexible coupling.

5. Remove the bolts inserted from the transmission side.

NOTE: The bolts inserted from the driveshaft side should not be removed.

6. Place matchmarks on the differential flange and the flexible coupling.

7. Remove the bolts inserted from the differential side. Separate the flexible couplings from the transmission and the differential.

8. Remove the center support bearing set bolts and the adjusting washers, if equipped.

NOTE: When removing the set bolts, support the center support bearing so the transmission and intermediate shaft and the driveshaft and differential remain in straight line.

9. Push the rear driveshaft forward to compress the driveshaft and pull out the driveshaft from the centering pin of the differential.

10. Remove the driveshaft by pulling out toward the rear of the vehicle.

To install:

11. Apply a suitable grease to the flexible coupling centering bushings.

12. Insert the propeller shaft from the rear of the vehicle and connect the transmission and differential.

13. Temporarily, install the center support bearing set bolts with the adjusting washers, if equipped.

14. Align the matchmarks and connect the propeller shaft to the transmission. Insert the bolts from the transmission side and tighten to 58 ft. lbs. (79 Nm).

15. Align the matchmarks and install the driveshaft to the differential. Insert the bolts from the differential side and tighten to 58 ft. lbs. (79 Nm).

16. Tighten the center bearing support bolts to 27 ft. lbs. (37 Nm). Tighten the adjusting nut with the proper tool.

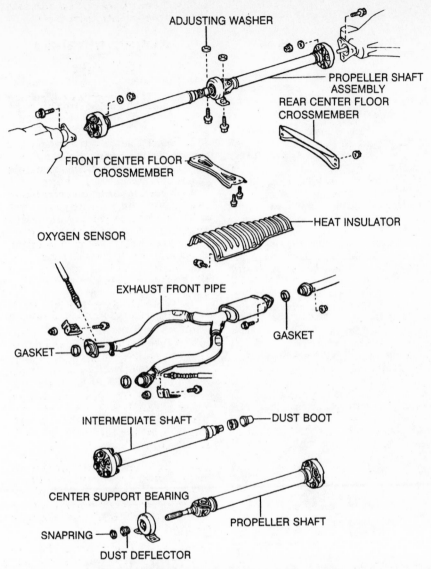

ADJUSTING WASHER

PROPELLER SHAFT ASSEMBLY

REAR CENTER FLOOR CROSSMEMBER

FRONT CENTER FLOOR CROSSMEMBER

OXYGEN SENSOR

HEAT INSULATOR

EXHAUST FRONT PIPE

GASKET

GASKET

INTERMEDIATE SHAFT

DUST BOOT

CENTER SUPPORT BEARING

PROPELLER SHAFT

SNAPRING

DUST DEFLECTOR

Exploded view of the driveshaft assembly—LS400

17. Install the front and rear cross-member braces and tighten to 9 ft. lbs. (12 Nm).

18. Install the heat insulator and the front exhaust pipe.

19. Lower the vehicle.

Front Wheel Hub, Knuckle and Bearings

REMOVAL & INSTALLATION

ES250

1. Raise and safely support the vehicle. Remove the front tire and wheel assembly.

2. Remove the brake caliper and support, using a piece of wire. Place match marks on the rotor disc and the axle hub. Remove the rotor.

3. Disconnect the ABS speed sensor.

4. Disconnect the lower ball joint from the steering knuckle with the proper tool. Disconnect the tie rod end from the steering knuckle.

5. Disconnect the steering knuckle from the shock absorber.

6. Remove the steering knuckle with the axle hub from the halfshaft.

7. Remove the dust protector and pry out the inner oil seal with the proper tool.

8. Remove the snapring from the steering knuckle and separate the dust cover from the steering knuckle.

9. Push out the axle hub and remove the inner race (inside) from the bearing, using the proper tool.

10. Remove the sensor control rotor from the axle hub. Remove the bearing inner race (outside) from the axle hub.

11. Pry out the oil seal with the proper tool and tap out the bearing.

To install:

12. Install the sensor control rotor to the axle hub and press in a new bearing into the steering knuckle.

13. Install the outer oil seal and the disc brake dust cover.

14. Apply a suitable multi-purpose grease between the oil seal lip, the oil seal and the bearing. Install the axle hub into the steering knuckle, using the proper tool.

15. Install the snapring into the steering knuckle.

16. Install the oil seal into the steering knuckle with a suitable seal installer tool. Apply a suitable multi-purpose grease to the oil seal lip.

17. Tap the dust deflector into the steering knuckle

18. Install the steering knuckle with the axle hub to the halfshaft. Tighten the lower mounting nuts to 224 ft. lbs. (304 Nm).

19. Connect the tie rod end to the steering knuckle and tighten the bolts to 36 ft. lbs. (49 Nm).

20. Tighten the lower ball joint mounting bolts to 83 ft. lbs. (113 Nm). Connect the ABS speed sensor.

21. Align the matchmarks and install the brake rotor. Replace the brake caliper and tighten the mounting bolts to 79 ft. lbs. (107 Nm).

22. Tighten the bearing locknut to 137 ft. lbs. (186 Nm). Replace the tire and wheel assembly.

23. Lower the vehicle.

LS400

1. If equipped with air suspension, move the height control switch in the trunk area to the **OFF** position.

2. Raise and safely support the vehicle. Remove the front tire and wheel assembly.

3. Disconnect the brake caliper from the steering knuckle and support with a piece of wire.

4. Place matchmarks on the disc brake rotor and the axle hub. Remove the brake rotor.

5. Remove the speed sensor from the steering knuckle. Loosen the axle shaft nut.

6. Remove the steering knuckle from the lower ball joint, using the proper tool. Remove the steering knuckle from the upper ball joint.

7. Remove the steering knuckle with the axle hub.

8. Remove the nut and the speed sensor rotor. Using the proper tool, remove the axle hub from the steering knuckle.

9. Remove the outside inner race from the axle and the oil seal from the steering knuckle, using the proper tools.

10. Remove the snapring and remove the bearing from the steering knuckle.
To install:

11. Using the proper tool, install the bearing to the steering knuckle. Replace the snapring.

12. Install the inner race (outside) and press in a new oil seal until it is flush with the end surface of the steering knuckle.

13. Install the brake dust cover to the steering knuckle and using the proper tools, press the axle hub to the steering knuckle.

14. Install the speed sensor. Install the steering knuckle to the lower ball joint and temporarily, tighten the bolts.

15. Install the steering knuckle to the upper arm and tighten the nut to 48 ft. lbs. (65 Nm).

16. Align the matchmarks and install disc rotor to the axle hub. Install the brake caliper and tighten the mounting bolts top 87 ft. lbs. (118 Nm).

17. Tighten the axle shaft nut to 147 ft. lbs. (199 Nm). Install the speed sensor to the steering knuckle.

18. Replace the front tire and wheel assembly. Lower the vehicle.

19. If equipped with air suspension, turn the height control switch to the **ON** position.

Rear Axle Shaft Bearing and Seal

REMOVAL & INSTALLATION

LS400

1. If equipped with air suspension, move the height control switch in the trunk area to the **OFF** position.

2. Raise and safely support the vehicle. Remove the rear tire and wheel assembly.

3. Disconnect the brake caliper from the rear axle carrier and support with a piece of wire.

4. Place matchmarks on the disc brake rotor and the axle hub. Remove the brake rotor.

5. Remove the speed sensor. Remove the strut rod and lower suspension rods.

6. Remove the nut on the lower side of the shock absorber. Do not remove the bolt.

7. Remove the upper arm set bolts and the bolt on the lower side of the shock absorber. Remove the axle with the arm.

8. Remove the upper arm and the dust deflector, using the proper tools. Remove the inner oil seal.

9. Remove the axle hub and the backing plate. Remove the inner race (outside) from the axle hub.

10. Pry out the outer oil seal. Re-move the snapring and the bearing, using the proper tools.
To install:

11. Install the bearing to the axle carrier.

NOTE: If the inner races come loose from the bearing outer race, be sure to install them on the same side as before.

12. Install the snapring. Replace the backing plate to the axle carrier and tighten the mounting bolts 43 ft. lbs. (58 Nm).

13. Install the inner race (outside) and a new oil seal.

14. Install the inner race (inside) and press in the axle hub with the proper tools.

15. Install the inner oil seal. Align the holes for the speed sensor in the dust deflector and axle carrier. Install the dust deflector.

16. Install the upper arm to the axle carrier. Tighten the nut and bolt to 80 ft. lbs. (108 Nm).

17. Replace the nut on the lower side of the shock absorber.

18. Install the speed sensor. Replace the strut rod and lower suspension rods.

19. Install the brake rotor. Connect the brake caliper to the rear axle carrier.

20. Replace the rear tire and wheel assembly. Lower the vehicle.

MANUAL TRANSAXLE

For further information on transmissions/transaxles, please refer to "Chilton's Guide to Transmission Repair".

Transaxle Assembly

REMOVAL & INSTALLATION

ES250

1. Disconnect the negative battery cable. Remove the clutch release cylinder and tube clamp. Remove the clutch tube bracket.

2. Disconnect the control cables. Disconnect the back-up light switch electrical connector. Remove the ground strap.

3. Remove the starter assembly. Re-move the transaxle upper mounting bolts.

4. Raise and support the vehicle

safely. Remove the undercovers. Drain the transaxle fluid. Disconnect the speedometer cable.

5. Remove the suspension lower crossmember. Remove the engine mounting center member.

6. Disconnect both halfshafts. Remove the center halfshaft. Disconnect the left steering knuckle from the lower control arm. Remove the stabilizer bar.

7. Properly support the engine and remove the left engine mount.

8. Properly support the transaxle assembly. Remove the engine-to-transaxle bolts, lower the left side of the engine and carefully ease the transaxle out of the engine compartment.

9. Installation is the reverse of the removal procedure. Please note the following:

    a. Tighten the 12mm mounting bolts to 47 ft. lbs. (64 Nm) and the 10mm bolts to 34 ft. lbs. (46 Nm).

    b. Tighten the left engine mount to 38 ft. lbs. (52 Nm).

    c. Tighten the 4 center engine mount bolts to 29 ft. lbs. (39 Nm).

    d. Tighten the front and rear engine mount bolts to 32 ft. lbs. (43 Nm).

    e. Tighten the lower crossmember bolts to 153 ft. lbs. (207 Nm) – 4 outer bolts; and, 29 ft. lbs. (39 Nm) – 2 inner bolts.

CLUTCH

Clutch Assembly

REMOVAL & INSTALLATION

ES250

1. Disconnect the negative battery cable. Remove the transaxle assembly from the vehicle.

2. Place matchmarks on the flywheel and clutch cover. Remove the clutch pressure plate retaining bolts. Remove the pressure plate assembly.

3. Remove the clutch disc.

4. Installation is the reverse of the removal procedure.

5. Tighten the pressure plate mounting bolts to 14 ft. lbs. (19 Nm).

FREE-PLAY ADJUSTMENT

ES250

1. Adjust the clearance between the master cylinder piston and the pushrod to specification by loosening the pushrod locknut and rotating the

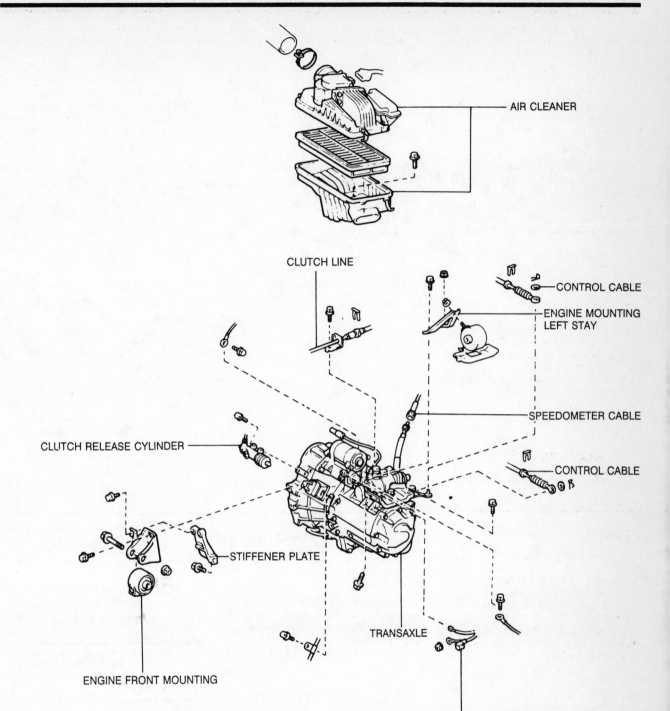

CLUTCH LINE

AIR CLEANER

CONTROL CABLE

ENGINE MOUNTING LEFT STAY

SPEEDOMETER CABLE

CONTROL CABLE

CLUTCH RELEASE CYLINDER

STIFFENER PLATE

TRANSAXLE

ENGINE FRONT MOUNTING

CONNECTOR (FOR STARTER)

Manual transaxle assembly—ES250

pushrod while depressing the clutch pedal lightly.

2. Tighten the locknut when finished with the adjustment.

3. Adjust the release cylinder free-play by loosening the release cylinder pushrod locknut and rotating the pushrod until proper specification is obtained.

4. Measure the clutch pedal free-play after performing the adjust-

ments. If it fails to fall within specification, repeat the procedure.

Clutch Master Cylinder

REMOVAL & INSTALLATION

ES250

1. Disconnect the negative battery cable.

2. Remove the ABS control relay on vehicles so equipped.

3. Remove the pushrod clevis pin and clip.

NOTE: On some vehicles it will be necessary to remove the under dash panel in order to gain access to the pushrod clevis pin.

5. Disconnect the fluid line. Remove the clutch master cylinder retaining

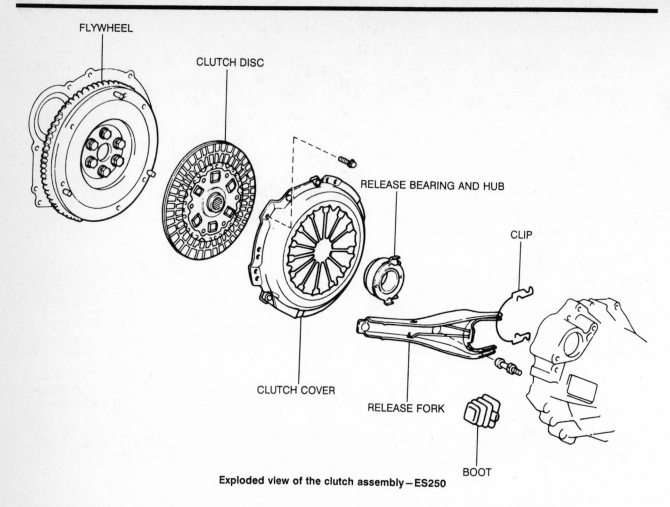

FLYWHEEL

CLUTCH DISC

RELEASE BEARING AND HUB

CLIP

CLUTCH COVER

RELEASE FORK

BOOT

Exploded view of the clutch assembly—ES250

bolts. Remove the component from the vehicle.

6. Installation is the reverse of the removal procedure. Bleed the system as required.

Clutch Slave Cylinder

REMOVAL & INSTALLATION

ES250

1. Disconnect the negative battery cable. Raise and support the vehicle safely.
2. Remove the gravel shield, if equipped. Disconnect the fluid line from the assembly.
3. Remove the slave cylinder retaining bolts. Remove the clutch slave cylinder from the vehicle.
4. Installation is the reverse of the removal procedure. Bleed the system as required.

Hydraulic Clutch System Bleeding

ES250

1. Check and fill the clutch fluid

reservoir to the specified level as necessary. During the bleeding process, continue to check and replenish the reservoir to prevent the fluid level from getting lower than ½ the specified level.

2. Remove the dust cap from the bleeder screw on the clutch slave cylinder and connect a tube to the bleeder screw and insert the other end of the tube into a clean glass or metal container.

NOTE: Take precautionary measures to prevent the brake fluid from getting on any painted surfaces.

3. Pump the clutch pedal several times, hold it down and loosen the bleeder screw slowly.
4. Tighten the bleeder screw and release the clutch pedal gradually. Repeat this operation until air bubbles disappear from the brake fluid being expelled out through the bleeder screw.
5. Repeat until all evidence of air bubbles completely disappears from the fluid being pumped out of the tube.
6. When the air is completely re-

moved, tighten the bleeder screw and replace the dust cap.

7. Check and refill the master cylinder reservoir as necessary.
8. Depress the clutch pedal several times to check the operation of the clutch and check for leaks.

AUTOMATIC TRANSMISSION

For further information on transmissions/transaxles, please refer to "Chilton's Guide to Transmission Repair".

Transmission Assembly

REMOVAL & INSTALLATION

LS400

1. Disconnect the negative battery cable. Remove the air cleaner assem-

bly. Disconnect the transmission throttle cable.

2. Raise and support the vehicle safely. Drain the transmission fluid. Remove the driveshaft along with the center bearing.

3. Remove the exhaust pipe together with the catalytic converter. Disconnect the manual shift linkage. Remove the speedometer cable.

4. Disconnect the oil cooler lines. As necessary, remove the transmission oil filler tube. As required, remove the starter assembly. Remove the speedometer cable.

5. Remove both stiffener plates and the catalytic converter cover from the transmission housing and cylinder block.

6. Support the engine and transmission using the proper jacking device. Remove the rear crossmember.

7. Remove the torque converter cover. Remove the torque converter-to-engine retaining bolts.

8. Remove the bolts retaining the transmission to the engine. Carefully remove the transmission from the vehicle.

9. Installation is the reverse of the removal procedure. Tighten the transmission housing bolts to 47 ft. lbs. (64 Nm). Tighten the torque converter bolts to 20 ft. lbs. (27 Nm).

SHIFT LINKAGE ADJUSTMENT

1. Loosen the nut on the shift linkage. Push the selector lever all the way to the rear of the vehicle.

2. Return the lever 2 notches to the **N** shift position.

3. While holding the selector lever slightly toward the **R** shift position, tighten the connecting rod nut.

THROTTLE CABLE ADJUSTMENT

1. Remove the air cleaner.

2. Confirm that the accelerator linkage opens the throttle fully. Adjust the linkage as necessary.

3. Peel the rubber dust boot back from the throttle cable.

4. Loosen the adjustment nuts on the throttle cable bracket (cylinder head cover) just enough to allow cable housing movement.

5. Have an assistant depress the accelerator pedal fully.

6. Adjust the cable housing so the distance between its end and the cable stop collar is 0.04 in.

7. Tighten the adjustment nuts. Make sure the adjustment hasn't changed. Install the dust boot and the air cleaner.

NOTE: When reinstalling the transmission assembly after repair or replacement, it is critical that the correct torque converter mounting bolts are used. The use of a torque converter set bolt longer than specified will deform the front cover of the torque converter. This deformation can internally damage the torque converter lock-up disc. Material from the lock-up clutch may contaminate the transmission and valve body requiring additional repairs. This damage will not be immediately detected by a technician.

AUTOMATIC TRANSAXLE

For further information on transmissions/transaxles, please refer to "Chilton's Guide to Transmission Repair".

Transaxle Assembly

REMOVAL & INSTALLATION

ES250

1. Disconnect the negative battery cable. Remove the air flow meter and the air cleaner assembly.

2. Disconnect the transaxle wire connector. Disconnect the neutral safety switch electrical connector.

3. Disconnect the transaxle ground strap. Disconnect the throttle cable from the throttle linkage.

4. Remove the transaxle case protector. Disconnect the speedometer cable. Disconnect the control cable.

5. Disconnect the oil cooler hoses. Remove the upper starter retaining bolts, as required remove the starter assembly. Remove the upper transaxle housing bolts. Remove the engine rear mount insulator bracket set bolt.

6. Raise and support the vehicle safely. Drain the transaxle fluid.

7. Remove the left front fender apron seal. Disconnect both halfshafts.

8. Remove the suspension lower crossmember assembly. Remove the center halfshaft.

9. Remove the engine mounting center crossmember. Remove the stabilizer bar. Remove the left steering knuckle from the lower control arm.

10. Remove the torque converter cover. Remove the torque converter retaining bolts.

11. Properly support the engine and transaxle assembly. Remove the rear engine mounting bolts. Remove the remaining transaxle-to-engine retaining bolts.

12. Carefully remove the transaxle assembly from the vehicle.

13. Installation is the reverse of the removal procedure. Tighten the 12mm transaxle housing bolts to 47 ft. lbs. (64 Nm); tighten the 10mm bolts to 34 ft. lbs. (46 Nm). Tighten the rear engine mount set bolts to 38 ft. lbs. (52 Nm). Tighten the torque converter mounting bolts to 20 ft. lbs. (27 Nm).

SHIFT CABLE ADJUSTMENT

ES250

1. Loosen the swivel nut on the selector lever.

2. Push the lever fully toward the right side of the vehicle.

3. Return the lever 2 notches to the **N** position.

4. Set the shift lever in the **N** position.

5. While holding the selector lever slightly toward the **R** shift position tighten the swivel nut to 48 inch lbs. (5.4 Nm).

THROTTLE CABLE ADJUSTMENT

ES250

1. Remove the air cleaner.

2. Confirm that the accelerator linkage opens the throttle fully. Adjust the linkage as necessary.

3. Peel the rubber dust boot back from the throttle cable.

4. Loosen the adjustment nuts on

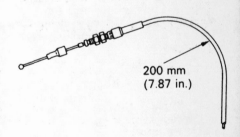

200 mm
(7.87 in.)

0.8 – 1.5 mm
(0.031 – 0.059 in.)

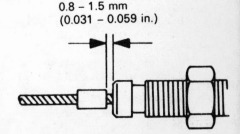

Throttle cable adjustment–ES250

the throttle cable bracket (cylinder head cover) just enough to allow cable housing movement.

5. Depress the accelerator pedal fully.

6. Adjust the cable housing so the distance between its end and the cable stop collar is 0.04 in.

7. Tighten the adjustment nuts. Make sure the adjustment hasn't changed. Install the dust boot and the air cleaner.

FRONT SUSPENSION

Pneumatic Cylinder

REMOVAL & INSTALLATION

LS400

1. Move the height control switch, located in the trunk area to the **OFF** position.

2. Raise and safely support the vehicle. Remove the steering knuckle from the upper ball joint with the proper tool.

3. Support the steering knuckle using a piece of wire. Disconnect speed sensor wire connector.

4. Remove the height control sensor link from the shock absorber lower bracket.

5. Disconnect the shock absorber from the lower mounting bracket. Remove the grommet and disconnect the air tube from the shock absorber.

6. Remove the mounting bolts and the actuator cover. Remove the mounting bolts and the actuator.

7. Remove the 3 upper mounting nuts and remove the shock absorber from the vehicle.

8. The installation is the reverse of the removal procedure. Tighten the upper mounting nuts to 27 ft. lbs. (39 Nm), actuator cover mounting nuts to 27 ft. lbs. (39 Nm) and the lower shock mount nut to 106 ft. lbs. (144 Nm).

MacPherson Strut

REMOVAL & INSTALLATION

ES250

1. Raise and safely support the vehicle. Remove the tire and wheel assembly.

2. Disconnect the ABS speed sensor connector. Disconnect the brake hose from the brake caliper.

3. Disconnect the steering knuckle

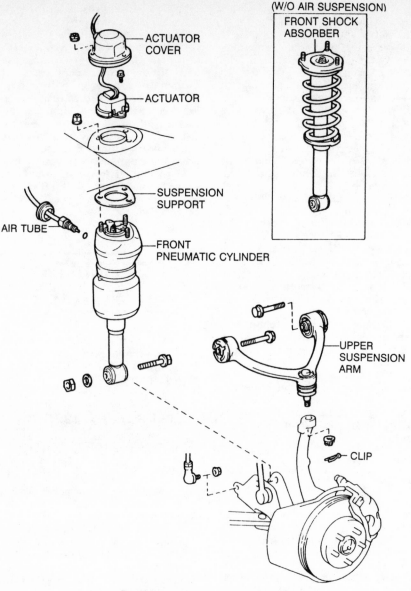

Front suspension assembly—LS400

and the strut assembly from the lower mount.

4. Remove the upper mounting nuts from the top suspension mount and remove the strut assembly.

5. The installation is the reverse of the removal procedure. Tighten the upper strut mount nuts to 47 ft. lbs. (64 Nm) and the lower strut mount nuts to 224 ft. lbs. (304 Nm).

NOTE: On the ES250, the front strut bar cushion retainers have been changed to a new resin formed casting cover to improve the front suspension noise during suspension compression and rebound.

LS400

1. Raise and safely support the vehicle. Remove the tire and wheel assembly.

2. Remove the steering knuckle from the upper ball joint with the proper tool. Support the steering knuckle using a piece of wire.

3. Disconnect the strut assembly from the lower strut bracket. Remove the plug from the upper strut mount.

4. Loosen the nut on the middle of the strut mount support. Do not remove it.

5. Remove the other 3 mounting nuts and remove the strut assembly with the coil spring from the vehicle.

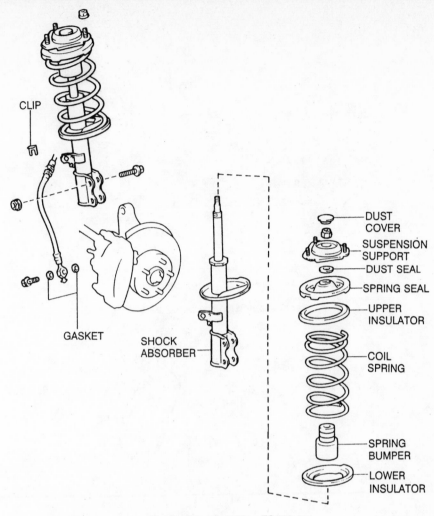

CLIP

GASKET

SHOCK ABSORBER

DUST COVER

SUSPENSION SUPPORT

DUST SEAL

SPRING SEAL

UPPER INSULATOR

COIL SPRING

SPRING BUMPER

LOWER INSULATOR

Front strut assembly—ES250

6. The installation is the reverse of the removal procedure. Tighten the upper strut mount nuts to 27 ft. lbs. (37 Nm) and the lower strut mount nut to 106 ft. lbs. (144 Nm).

Strut Bar

REMOVAL & INSTALLATION

LS400

1. Disconnect the negative battery cable. Raise and support the vehicle safely.
2. Remove the steering knuckle with the axle hub.
3. Remove the strut assembly or the pneumatic cylinder, depending on what the vehicle is equipped with.
4. Remove the strut assembly lower bracket.
5. Place matchmarks on the screw part and nut of the strut bar.
6. Remove the nut and washer from the front side of the strut bar.
7. Remove the 2 nuts and remove the strut bar from the lower arm.

8. Remove the 3 nuts and remove the strut bar cushion and strut bar.

To install:

9. Installation is the reverse order of the removal procedure. Be sure to align the match marks on the strut bar when installing it.
10. Use the following torque specifications during installation:

   a. Torque the 2 strut bar bolts to 59 ft. lbs. (80 Nm).

   b. Torque the 2 strut bar nuts to 121 ft. lbs. (164 Nm).

   c. Torque the through bolt of the strut assembly bracket to 83 ft. lbs. (113 Nm) and the other bolt to 43 ft. lbs. (59 Nm).

   d. Lower the vehicle and after stabilizing the suspension, torque the front strut bar nut to 87 ft. lbs. (118 Nm).

Stabilizer Bar

REMOVAL & INSTALLATION

ES250

1. Disconnect the negative battery

cable. Raise and support the vehicle safely.
2. Remove the lower suspension crossmember.
3. Remove the nuts and retainers holding the stabilizer bar to the lower suspension arms.
4. Remove the stabilizer bar brackets.
5. Remove the the 2 engine under covers. Remove the control cable clamp bolts from the engine center mounting member, if so equipped. Remove the 10 bolts and engine center mounting member.
6. Pull off the stabilizer bar from the lower suspension arms. Remove the retainers and spacers from the stabilizer bar.

To install:

7. Installation is the reverse order of the removal procedure.
8. Use the following torque specifications during installation:

   a. Torque the stabilizer bar bracket cushion bolts to 94 ft. lbs. (127 Nm).

   b. Torque the 2 bolts on the right hand end and the 2 bolts on the left hand end of the engine center mounting member to 29 ft. lbs. (39 Nm) and the other retaining bolts to 32 ft. lbs. (43 Nm).

   c. Torque the suspension lower crossmember bolts to 112 ft. lbs. (152 Nm).

   d. Torque the stabilizer bar mounting nuts 156 ft. lbs. (212 Nm).

LS400

1. Disconnect the negative battery cable. Raise and support the vehicle safely.
2. Remove the steering knuckle with the axle hub.
3. Remove the strut assembly or the pneumatic cylinder, depending on what the vehicle is equipped with.
4. Remove the strut assembly lower bracket.
5. Place matchmarks on the screw part and nut of the strut bar.
6. Remove the nut and washer from the front side of the strut bar.
7. Remove the 2 nuts and remove the strut bar from the lower arm.
8. Remove the 3 nuts and remove the strut bar cushion and strut bar.
9. Remove the right and left stabilizer bar bushings.
10. Remove the strut bar brackets with the stabilizer bar. Remove the stabilizer bar and inspect the stabilizer bar link as follows.

   a. Flip the ball joint stud back and forth 5 times.

   b. Using a torque gauge, turn the stud continuously 1 turn per 2-4 seconds and take the torque reading on the 5th turn.

c. Turning torque should be 0.4-13 in. lbs.

To install:

11. Installation is the reverse order of the removal procedure. Be sure to install the stabilizer bar through the right and left strut bar brackets. Install the strut bar brackets. Insert the 2 strut bar bolts before hand into the holes in the lower arm. Torque the bolts to 53 ft. lbs. (72 Nm).

12. Use the following torque specifications during installation:

a. Torque the 2 strut bar bolts to 59 ft. lbs. (80 Nm).

b. Torque the 2 strut bar nuts to 121 ft. lbs. (164 Nm).

c. Torque the through bolt of the strut assembly bracket to 83 ft. lbs. (113 Nm) and the other bolt to 43 ft. lbs. (59 Nm).

d. Lower the vehicle and after stabilizing the suspension, torque the front strut bar nut to 87 ft. lbs. (118 Nm).

e. Torque the stabilizer bar links to 70 ft. lbs. (95 Nm).

f. Torque the stabilizer bar bushings to 21 ft. lbs. (28 Nm).

Lower Ball Joints

INSPECTION

1. Raise and safely support the vehicle. Place a wooden block under the tire and wheel assembly.

2. Lower the jacking device until there is about ½ the load on the front spring.

3. Make sure the front wheels are in a straight forward position and block the wheel with wheel chocks.

4. Move the lower arm up and down and check that the ball joint has no excessive play. The ball joint vertical play limit is 0 in. (0mm) on the ES250 or 0.012 in. (0.3mm) on the LS400.

REMOVAL & INSTALLATION

ES250

1. Raise and safely support the vehicle. Remove the tire and wheel assembly.

2. Loosen the nut holding the stabilizer bar to the lower suspension arm.

3. Loosen the nut holding the lower suspension arm to the lower suspension arm shaft.

4. Disconnect the lower ball joint from the lower suspension arm with the proper tool.

5. Remove the mounting bolts and using a suitable prybar, push down the lower suspension arm, the remove the ball joint.

6. The installation is the reverse of the removal procedure. Tighten the

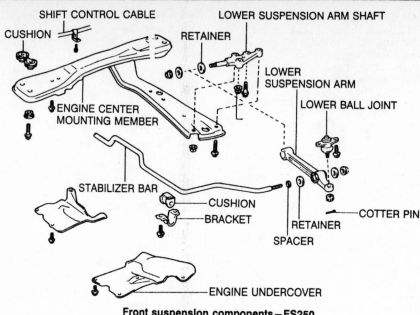

Front suspension components—ES250

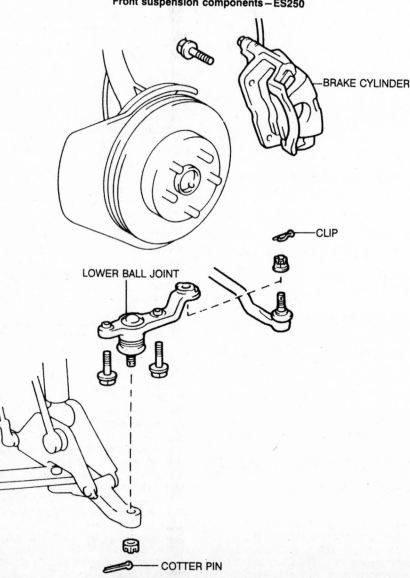

Lower ball joint components—LS400

ball joint mounting bolts to 83 ft. lbs. (113 Nm)., the castle nut to 90 ft. lbs. (12 Nm), the stabilizer bar—lower suspension arm nuts to 156 ft. lbs. (212 Nm) and the lower arm shaft—lower suspension arm nuts to 156 ft. lbs. (212 Nm).

7. Check the front wheel alignment.

LS400

1. If equipped with air suspension, move the height control switch, located in the trunk area to the **OFF** position.

2. Raise and safely support the vehicle. Remove the tire and wheel assembly.

3. Disconnect the brake caliper and support, using a piece of wire.

4. Loosen the lower mounting bolts. Do not remove.

5. Disconnect the tie rod end from the steering arm with the proper tool. Remove the bolts and disconnect the lower ball joint from the steering knuckle.

6. Remove the nut and disconnect the lower ball joint from the lower arm with the proper tool.

7. The installation is the reverse of the removal procedure. Tighten the ball joint castle nut to 112 ft. lbs. (152 Nm), the tie rod end nut to 43 ft. lbs. (58 Nm) and the lower ball joint bolts to 83 ft. lbs. (113 Nm).

Upper Control Arms

REMOVAL & INSTALLATION

LS400

1. Raise and safely support the vehicle. Remove the tire and wheel assembly.

2. Remove the shock absorber or pneumatic cylinder, if equipped.

3. Remove the mounting bolts and the upper suspension arm.

4. Install the upper suspension arm and tighten the mounting bolts to 83 ft. lbs. (113 Nm).

5. Reverse the remainder of the procedures for installation.

Lower Control Arms

REMOVAL & INSTALLATION

ES250

1. Raise and safely support the vehicle.

2. Remove the nut holding the stabilizer to the lower suspension arm. Remove the nut holding the lower suspension arm to the lower suspension arm shaft.

3. Remove the mounting bolts and disconnect the lower ball joint from the steering knuckle with the proper tool.

4. Remove the suspension lower crossmember. Remove the mounting bolt and the lower arm with the shaft.

5. Remove the nut and disconnect the lower ball joint from the lower arm with the proper tool.

6. The installation is the reverse of the removal procedure. Tighten the lower arm shaft-to-body bolts to 153 ft. lbs. (207 Nm), ball joint-to-steering knuckle bolts to 83 ft. lbs. (113 Nm) and the lower suspension arm mounting nuts to 156 ft. lbs. (212 Nm).

LS400

1. Raise and safely support the vehicle. Remove the tire and wheel assembly.

2. Remove the shock absorber or pneumatic cylinder, if equipped.

3. Disconnect the tie rod end from the steering knuckle with the proper tool. Remove the lower shock bracket.

4. Remove the nuts and disconnect the lower strut bar from the lower arm.

5. Place matchmarks on the camber adjusting cam. Remove the nut, the adjusting cam and the lower arm with the lower ball joint.

NOTE: To help remove the adjusting cam, fully pull the steering wheel toward the lower arm being removed.

To install:

6. Insert the camber adjusting cam from the rear side of the vehicle and temporarily tighten the nut. Put 2 strut bar bolts into the holes of the lower arm beforehand.

7. Connect the strut bar to the lower arm and tighten the bolts to 121 ft. lbs. (164 Nm). Install the lower shock bracket.

8. Connect the tie rod to the steering arm and tighten the nut to 43 ft. lbs. (58 Nm). Install a new clip.

9. Install the shock absorber or pneumatic cylinder, if equipped.

10. Install the steering knuckle with the axle hub. Replace the tire and wheel assembly.

11. Lower the vehicle and stabilize the suspension.

12. Support the lower arm with a suitable jacking device and remove the tire and wheel assembly.

13. Align the matchmarks and tighten the nut to 185 ft. lbs. (250 Nm). Replace the tire and wheel assembly. Remove the jacking device.

14. Check the front end alignment.

Sway Bar

REMOVAL & INSTALLATION

ES250

1. Raise and safely support the vehicle. Remove the suspension lower crossmember.

2. Remove the mounting nuts and retainers holding the suspension bar to the lower suspension arms.

3. Remove the stabilizer bar brackets.

4. Remove the 2 engine undercovers and the automatic transaxle control cables from the engine center mounting member.

5. Remove the bolts and lower the engine center mounting member. Pull the stabilizer bar from the suspension arms.

6. Remove the retainers and the spacers from the stabilizer bar.

7. The installation is the reverse of the removal procedure. Tighten the stabilizer bar brackets to 94 ft. lbs. (127 Nm), the engine center mounting member to 29–32 ft. lbs. (40-44 Nm), the lower crossmember to 153 ft. lbs. (207 Nm) and the stabilizer bar mounting nuts to 156 ft. lbs. (212 Nm).

8. Check the front wheel alignment.

Front Wheel Bearings

REMOVAL & INSTALLATION

LS400

1. Raise and safely support the vehicle. Remove the tire and wheel assembly.

2. Remove the steering knuckle with the axle hub. Remove the nut and the speed sensor rotor.

3. Remove the 4 bolts and shift the brake dust cover towards the hub side (outside). Remove the axle shaft from the steering knuckle with the proper tool.

4. Using the proper tool, remove the inner race (outside) from the axle shaft. Pry out the oil seal from the steering knuckle with the proper tool.

5. Remove the snapring and press out the bearing from the steering knuckle with the proper tools.

To install:

6. Using the proper tool, press the new bearing into the steering knuckle. Install the snapring, using suitable snapring pliers.

7. Install the inner race (outside) and press in the new oil seal until it is flush with the end surface of the steering knuckle.

8. Install the brake dust cover to the steering knuckle. Press the axle

hub to the steering knuckle with the proper tool.

9. Install the speed sensor rotor and the steering knuckle.

10. Replace the tire and wheel assembly. Lower the vehicle.

REAR SUSPENSION

Pneumatic Cylinder

REMOVAL & INSTALLATION

LS400

1. Remove the rear seat cushion and seat back. Remove the rear scuff plates and the roof side inner trim panel and the speaker panel.

2. Remove the trunk trim panel. Move the height control switch, located in the trunk area to the **OFF** position.

3. Raise and safely support the vehicle. Remove the tire and wheel assembly.

4. Disconnect the stabilizer links from the stabilizer bar.

5. Disconnect and support the brake caliper, using a piece of wire. Do not disconnect the brake line.

6. Disconnect the height control sensor link from the suspension arm. Remove the nut on the lower side of the shock absorber. Do not remove the bolt.

7. Support the rear axle assembly with a suitable jacking device. Remove the grommet and disconnect the air tube from the shock absorber.

8. Remove the mounting bolts and the actuator cover. Remove the mounting bolts and the actuator.

9. Remove the upper mounting nuts and lower the rear axle assembly. Remove the bolt on the lower side of the shock absorber.

10. Remove the shock absorber.

To install:

11. Install the shock to the vehicle and tighten the upper mounting nuts to 43 ft. lbs. (58 Nm).

12. Install the actuator and replace the actuator cover. Tighten the mounting nuts to 13 ft. lbs. (18 Nm).

13. Install new O-rings and connect the air line to the shock absorber. Tighten the fitting to 13 ft. lbs. (18 Nm).

14. Install the shock to the rear axle carrier. Insert the bolt from the vehicle's rear and temporarily tighten the nut.

15. Connect the height control sen-

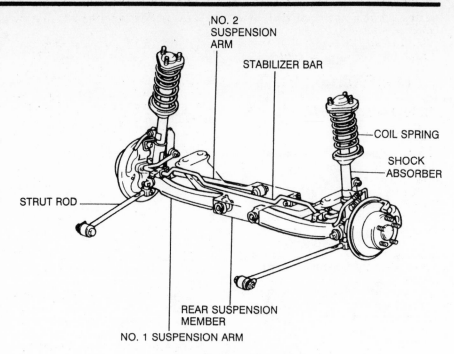

Rear suspension—ES250

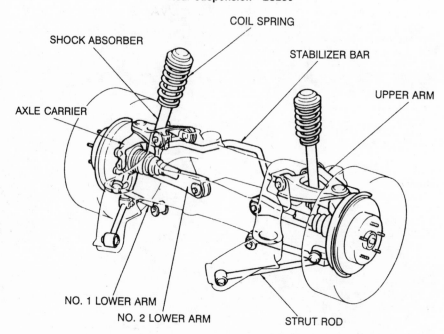

Rear suspension—LS400

sor link to suspension arm and tighten to 48 inch lbs.

16. Install the rear brake caliper to the rear axle carrier and tighten the mounting bolts to 77 ft. lbs. (104 Nm).

17. Connect the stabilizer links and tighten to 26 ft. lbs. (35 Nm).

18. Stabilize the suspension by:

a. Install the tire and wheel assembly and lower the vehicle.

b. Move the height control switch to the **ON** position. Start the engine an fill the pneumatic cylinder with air.

c. Bounce the vehicle up and down several times to stabilize the suspension.

19. Raise and safely support the vehicle. Remove the tire and wheel assembly.

20. Support the rear axle carrier with a suitable jacking device. Tighten the lower shock bolt to 101 ft. lbs. (137 Nm).

21. Replace the tire and wheel assembly. Lower the vehicle.

22. Install the rear seat cushion and seat back. Replace the rear scuff

plates, the roof side inner trim panel and the speaker panel.

23. Install the trunk trim panel. Check the rear wheel alignment.

MacPherson Strut

REMOVAL & INSTALLATION

LS250

1. Raise and safely support the vehicle. Remove the tire and wheel assembly.

2. Disconnect and plug the brake line from the strut assembly.

3. Disconnect the stabilizer link from the strut assembly.

4. Remove the rear seat back and package tray trim. Remove the dust cover from the upper suspension support. Loosen the nut but do not remove.

5. Remove the strut mounting bolts and disconnect the strut assembly. Remove the upper mounting bolts and remove the strut assembly.

6. The installation is the reverse of the removal procedure. Tighten the upper mounting bolts to 29 ft. lbs., lower strut-to-axle bolts to 166 ft. lbs., upper center mounting nut to 36 ft. lbs. and the stabilizer link top 47 ft. lbs.

LS400

1. Remove the rear seat cushion and seat back. Remove the rear scuff plates and the roof side inner trim panel and the speaker panel.

2. Raise and safely support the vehicle. Remove the tire and wheel assembly.

3. Remove the rear halfshaft and disconnect the stabilizer links.

4. Disconnect and support the brake caliper. Do not disconnect the brake line.

5. Remove the nut on the lower side of the strut. Do not remove the bolt.

6. Support the rear axle assembly with a suitable jacking device. Remove the 3 nuts and the strut cap.

7. Loosen the nut in the middle of the suspension support. Do not remove it.

8. Remove the other 3 mounting bolts. Lower the rear axle assembly and remove the bolt on the lower side of the strut assembly.

9. Remove the strut assembly with the coil spring.
To install:
10. Install the strut assembly with the coil spring to the vehicle and tighten the nuts to 47 ft. lbs. (64 Nm). Tighten the nut in the middle of the suspension support to 20 ft. lbs. (27 Nm).

11. Install the strut to the rear axle carrier. Install the bolt from the rear

of the vehicle and temporarily tighten the nut.

12. Install the brake caliper and tighten the mounting bolts to 77 ft. lbs. (104 Nm). Connect the stabilizer links and tighten to 26 ft. lbs. (35 Nm).

13. Install the rear halfshaft. Replace the tire and wheel assembly. Lower the vehicle.

14. Bounce the vehicle up and down to stabilize the suspension.

15. Raise and safely support the vehicle. Remove the tire and wheel assembly.

16. Support the rear axle assembly with a suitable jacking device. Tighten the bolt to 101 ft. lbs. (137 Nm).

17. Install the tire and wheel assembly. Lower the vehicle.

18. Replace the rear scuff plates and the roof side inner trim panel and the speaker panel.

19. Replace the rear seat cushion and seat back.

20. Check the rear wheel alignment.

Rear Control Arms

REMOVAL & INSTALLATION

LS400
UPPER

1. Raise and safely support the ve-

hicle. Remove the tire and wheel assembly.

2. Remove the rear axle carrier with the upper arm assembly.

3. Install the axle carrier in a suitable holding fixture.

4. Disconnect the upper control arm from the axle carrier.

5. The installation is the reverse of the removal procedure. Tighten the upper arm-to-axle bolt to 80 ft. lbs. (108 Nm).

Rear Wheel Bearings

REMOVAL & INSTALLATION

ES250

1. Raise and safely support the vehicle. Remove the tire and wheel assembly.

2. Remove the rear brakes and rotor assembly.

3. Remove the axle hub mounting bolts and remove the axle hub with the parking brake assembly.

4. Using the proper tool, unstake the locknut and remove.

5. Push the rear axle shaft off the axle hub with the proper tool. Remove the bearing inner race (inside) with the proper tool.

6. Using the proper tool, pull off the

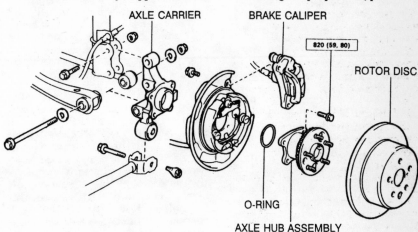

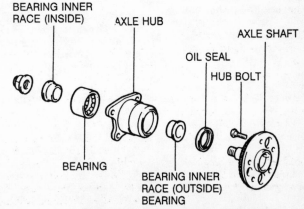

Rear axle bearing assembly—ES250

bearing inner race (outside) from the axle shaft. Remove the oil seal.

7. Install the inner race (outside) of the bearing to be removed, using the proper tool, press out the bearing.

NOTE: Always replace the bearing as an assembly.

To install:

8. Apply a suitable multi-purpose grease to the outer race of a new bearing. Press the bearing into the axle hub with the proper tool.

9. Install a new bearing inner race (outside) and drive in a new oil seal with the proper tool. Apply a suitable multi-purpose grease to the seal lip.

10. Install a new bearing inner race (inside) and press the inner races onto the axle shaft, using the proper tool. Install and tighten the locknut to 90 ft. lbs. (122 Nm). Stake the nut.

11. Install the parking brake assembly and a new oil seal to the axle carrier.

12. Install the axle hub and tighten the bolts to 59 ft. lbs. (80 Nm).

13. Install the tire and wheel assembly. Lower the vehicle.

Rear Axle Assembly

REMOVAL & INSTALLATION

ES250

1. Raise and safely support the vehicle. Remove the tire and wheel assembly.

2. Remove the rear brakes and rotor assembly.

3. Remove the axle hub mounting bolts and remove the axle hub with the parking brake assembly. Remove the O-ring from the axle carrier.

NOTE: Be careful not to damage the sensor control rotor.

4. Remove the strut rod nut and bolt from the axle carrier. Remove the suspension arm mounting bolts from the axle carrier.

5. Supporting the axle carrier, remove the axle carrier mounting bolt and nut from the strut assembly. Remove the axle carrier assembly.

To install:

6. Place the axle carrier into position.

7. Install the axle carrier mounting bolt and nut to the strut assembly. Tighten to 166 ft. lbs. (225 Nm).

8. Temporarily, connect the suspension arms to axle carrier. Temporarily, connect the strut rod to the axle carrier.

9. Install the parking brake assembly and a new oil seal to the axle carrier.

10. Install the axle hub and tighten

the mounting bolts to 59 ft. lbs. (80 Nm).

11. Install the rotor and disc brake assembly. Stabilize the suspension.

12. Tighten the axle carrier-to strut rod mounting bolts to 83 ft. lbs. (113 Nm) and the suspension arm-to-axle carrier bolts to 134 ft. lbs. (182 NM). Tighten with the vehicle weight on the suspension.

13. Check the rear wheel alignment.

Suspension Arms

REMOVAL & INSTALLATION

ES250

1. Disconnect the negative battery cable. Raise and support the vehicle safely.

2. Disconnect the ABS speed sensor wire clamp from the No. 1 suspension arm.

3. Remove the bolt and nut holding the strut rod to the axle carrier and remove the strut rod from the carrier.

4. Place a match mark on the adjusting cam and body of the No. 1 and No. 2 suspension arms. Remove the volt and nut holding the No. 1 and No. 2 suspension arms to the axle carrier.

5. Remove the fuel tank protector by removing the 2 bolts and clips.

6. Remove the cam, cam bolt and No. 2 suspension arm from the body.

7. Remove the bolt, retainer, nut and No. 1 suspension arm.

8. Installation is the reverse order of the removal procedure.

NOTE: The right and left suspension arms have been stamped with an R and L respectively for identification. Install the suspension arm so that the directional hole faces toward the outside of the rear of the vehicle. Temporarily install the No. 1 suspension arm to the body with the retaining bolt and nut. Install the No. 2 suspension arm, so that the bushing with the slit side faces toward the rear. Then install the suspension arm with the small paint spot towards the outside of the vehicle. Place the No. 2 suspension arm, in position and temporarily install the cam bolt and cam to the body.

9. Use the following torque specifications:

    a. Torque the No. 1 suspension arm to the body mounting bolt with the vehicle weight on the suspension to 83 ft. lbs. (113 Nm).

    b. Align the matchmarks on the adjusting cam and body, torque the No. 2 suspension arm to the body mounting bolt with the vehicle

weight on the suspension to 83 ft. lbs. (113 Nm).

    c. Torque the No. 1 and No. 2 suspension arms to the axle carrier mounting bolt and nut with the vehicle weight on the suspension to 134 ft. lbs. (181 Nm).

    d. Torque the strut rod to the axle carrier mounting bolt with the vehicle weight on the suspension to 83 ft. lbs. (113 Nm).

    e. Reconnect the negative battery cable and Check the rear wheel alignment.

Strut Rod

REMOVAL & INSTALLATION

ES250

1. Disconnect the negative battery cable. Raise and support the vehicle safely.

2. Disconnect the ABS speed sensor wire clamp from the No. 1 suspension arm.

3. Remove the bolt and nut holding the strut rod to the axle carrier and remove the strut rod from the carrier. If necessary, remove the rear floor heat upper insulator on the right hand side.

4. Installation is the reverse order of the removal procedure. Position the strut rod to the body and temporarily install the nut and bolt. Be sure the lip of the nut is resting on the flange of the bracket.

5. Torque the strut rod to the body mounting bolt with the vehicle weight on the suspension to 83 ft. lbs. (113 Nm).

6. Torque the strut rod to the axle carrier mounting bolt with the vehicle weight on the suspension to 83 ft. lbs. (113 Nm).

7. Reconnect the negative battery cable and Check the rear wheel alignment.

Stabilizer Bar

REMOVAL & INSTALLATION

ES250

1. Disconnect the negative battery cable. Raise and support the vehicle safely.

2. Remove the tail pipe assembly.

3. Remove the stabilizer links. If the ball joint stud turns together with the nut, use a hexagon wrench to hold the stud.

4. Disconnect the ABS speed sensor wire clamp. Remove the 4 bolts, stabilizer brackets and cushions.

5. Using a suitable jack and a block of wood, support the fuel tank. Remove the 2 fuel tank band installation bolts.

6. Lower the fuel tank approximately 1.57 in. (40 mm) and then remove the stabilizer bar from the body.

7. Rotate the ball joint stud in all directions, If the movement is not smooth and free, replace the stabilizer link.

To install:

8. Installation is the reverse order of the removal installation. Use the following torque specifications:

a. After placing the fuel tank in its proper position, torque the tank strap retaining bolts to 29 ft. lbs. (39 Nm).

b. Torque the stabilizer bar cushions retaining bolts to 14 ft. lbs. (19 Nm).

c. Torque the stabilizer link to 47 ft. lbs. (64 Nm).

d. Torque the tail pipe clamp to 29 ft. lbs. (39 Nm).

Lower Suspension Arm and Strut Rod

REMOVAL & INSTALLATION

LS400

1. Disconnect the negative battery cable. If the vehicle is equipped with air suspension. move the height control ON/OFF switch (located in the luggage compartment) to the **OFF** position.

2. Raise and support the vehicle safely and remove the rear wheel assembly.

3. Disconnect the strut rod from the rear axle carrier. Remove the strut rod.

4. Disconnect the height control sensor link if so equipped, from the No. 1 suspension arm.

5. Place matchmarks on the adjusting cam and body. Remove the adjusting cam. Remove the nut on the axle carrier side of the No. 1 lower suspension arm.

6. Using a tie rod removal tool, remove the No. 1 suspension arm.

7. Disconnect the stabilizer bar link from the No. 2 suspension arm.

8. Place matchmarks on the adjusting cam and body. Remove the adjusting cam. Remove the No. 2 lower suspension arm. Remove the arm cover.

9. Inspect the No. 1 lower suspension arm ball joint as follows.

a. Flip the ball joint stud back and forth 5 times.

b. Using a torque gauge, turn the stud continuously 1 turn per 2–4 seconds and take the torque reading on the 5th turn.

c. Turning torque should be 7.4–30 inch lbs.

d. If not within specifications, replace the No. 1 suspension arm.

To install:

10. Temporarily install the No. 2 and No. 1 lower suspension arms. Install the bolt and nut into the No. 2 suspension arm, connected the No. 2 lower suspension arm to the axle carrier and torque the bolts to 121 ft. lbs. (164 Nm). Install a new nut to the No. 1 suspension arm ball joint and torque it to 43 ft. lbs. (59 Nm).

11. Connect the height control sensor link to the No. 1 lower suspension arm with a new nut. Torque it to 48 inch lbs.

12. Temporarily install the strut rod.

13. Install the rear wheel assemblies and lower the vehicle.

14. Move the height control switch back to the **ON** position, if so equipped. Bounce the vehicle up and down several times to stabilize the suspension.

15. Raise and safely support the vehicle. Remove the rear wheel assemblies.

16. Support the axle carrier with a suitable jack. Torque the bolt and nut of the strut rod to 136 ft. lbs. (184 Nm). Torque the nut on the body side of the No. 1 lower suspension arm to 136 ft. lbs. (184 Nm). Torque the nut on the body side of the No. 2 lower suspension arm to 136 ft. lbs. (184 Nm).

17. Install the rear wheel assemblies, lower the vehicle and torque the rear wheel retaining nuts to 76 ft. lbs. (103 Nm).

18. Reconnect the negative battery cable. Check the rear wheel alignment.

STEERING

Steering Wheel

CAUTION

On vehicles equipped with an air bag, the negative battery cable must be disconnected, before working on the system. Failure to do so may result in deployment of the air bag and possible personal injury. Work must be started after approximately 20 seconds or longer from the time the ignition switch is turned to the LOCK position and the negative battery cable is disconnected from the battery. If the wiring connector of the airbag system is disconnected with the ignition switch at ON or ACC, diagnostic coded will be recorded.

REMOVAL & INSTALLATION

1. Disconnect the negative battery cable. Position the front wheels in a straight ahead position.

2. Loosen the screw until the groove along the screw circumference catches on the screw case.

3. Pull the wheel pad away from the steering wheel and disconnect the air bag connector.

NOTE: When removing the wheel pad, take care not to pull the air bag harness connector. When storing the wheel pad, keep the upper surface of the pad facing upward.

4. Disconnect the wire connector and remove the set nut. Place matchmarks on the steering wheel and main shaft.

5. Remove the steering with a suitable puller.

To install:

6. Check that the front wheels are facing straight ahead. Center the spiral cable. When centering the spiral cable be sure to use the following procedure:

a. Check that the front wheels are pointing straight ahead.

b. Turn the spiral cable counterclockwise by hand until it becomes harder to turn the cable. The spiral cable will rotate approximately 2½ turns to either the left or right of the center.

c. Then rotate the spiral cable clockwise approximately 2½ turns to align the red mark.

7. Align the matchmarks and install the steering wheel. Tighten the set nut to 26 ft. lbs. (35 Nm).

8. Connect the connector.

9. Install the steering wheel pad after confirming that the circumference groove of the screws is caught on the screw case.

NOTE: Make sure that the wheel pad is installed to the specified torque. If the wheel pad has been dropped, or there are cracks, dents or other defects in the case or connector, replace the wheel pad with a new one. When installing the wheel pad, be sure that the wires and connectors do not interfere with other parts and are not pinched between other parts.

10. Tighten the screws to 65 inch lbs. Connect the negative battery cable.

Steering Column

REMOVAL & INSTALLATION

ES250

1. Disconnect the negative battery cable. Remove the steering wheel.

2. Using a centering punch, mark the center of the tapered-head bolts.

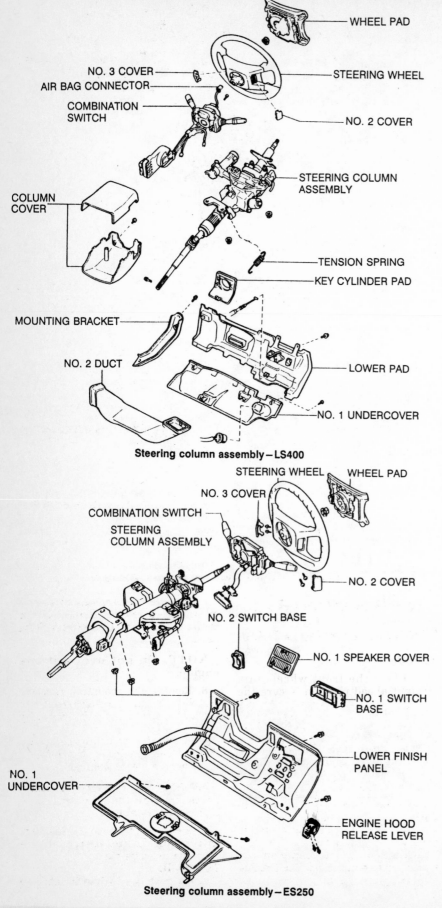

Steering column assembly — LS400

Steering column assembly — ES250

3. Drill into the tapered-head bolts, using a 0.12–0.16 in. (0.3–0.4mm) drill. Remove the tapered-head bolts.

4. Use a screw extractor and remove the bolts and separate the upper bracket and column tube. Remove the thrust stopper set bolts and disconnect the snapring.

5. Remove the main shaft. Using the proper tool, remove the snapring and thrust stopper.

6. The installation is the reverse of the removal procedure. Tighten the thrust stopper bolts to 9 ft. lbs. (12 Nm).

LS400

1. Disconnect the negative battery cable. Remove the steering wheel.

2. Disconnect the ignition key light. Remove the No. 2 intermediate shaft.

3. Remove the lower dust cover and gasket. Remove the tilt position sensor.

4. Remove the turn signal bracket, the lower shield protectors and connector bracket.

5. Using a centering punch, mark the center of the tapered-head bolts.

6. Drill into the tapered-head bolts, using a 0.16–0.20 in. (0.4–0.5mm) drill. Remove the tapered-head bolts.

7. Use a screw extractor and remove the bolts and separate the upper bracket and the break-away bracket.

8. Remove the nuts, spacers, springs, bushings and bond cable. Using the proper tool, remove the 2, No. 2 tilt steering shafts.

9. Remove the tilt steering gear assembly. Disconnect the nut and telescopic lever serration attachment.

10. Install a double nut on the lock bolt and remove the lock bolt. Remove the steering column tube upper stop bolt.

11. Remove the snapring from the mainshaft. Disconnect the break away bracket from the main shaft.

12. Remove the No. 1 and No. 2 telescopic spring seats and compression ring from the lower tube.

13. Remove the steering shaft thrust stopper from the break-away bracket. Remove the lock wedges from the break-away bracket.

14. Remove the upper tube with the main shaft and the tilt housing support.

15. Remove the main shaft using the proper tool to compress the mainshaft spring and remove the snapring.

16. Remove the main shaft from the upper tube. Remove the spring, thrust collar and bearing.

To install:

17. Install the spring, thrust collar and spring to the main shaft. Insert the main shaft in the upper tube.

18. Using the proper tool, compress

the main shaft spring and install the snapring.

19. Install the tilt steering upper housing support. Tighten to 9 ft. lbs. (12 Nm).

20. Insert the upper tube and the mainshaft into the lower tube. Press in the 2 tilt steering shafts.

21. Replace the lock wedges to the break-away bracket. Install the steering shaft thrust stopper to the break-away bracket.

22. Install the No. 1 and No. 2 telescopic spring seats and compression spring. Install the break-away bracket to the main shaft.

23. Install the snapring to the main shaft and the steering column tube stopper bolt. Tighten the bolt to 14 ft. lbs. (19 Nm).

24. Pull the steering column away from the break-away bracket and facing the body installation surface of the break-away bracket upward, temporarily install the lock bolt.

25. Install a double nut on the lock bolt. Tighten to 12 ft. lbs. (16 Nm), loosen once, then tighten again to 65 inch lbs.

26. Install the compression spring and ball, washer, lever, collar and bolt. Tighten the bolt to 19 ft. lbs. (26 Nm), with the depression of the lever matching the depression of the ball.

27. Rotate the telescopic lever until it touches the break-away bracket. Install the serration attachment so the alignment marks on the telescopic lever align with the serration attachment. Tighten to 9 ft. lbs. (12 Nm).

28. Install the tilt steering gear assembly with the motor by:

    a. Install the bushings and support stopper bolts.

    b. Press in the 2, No. 2 tilt steering shafts, using the proper tools.

    c. Install the bushing, spring and spacer to each stopper bolt.

    d. Install the bond cable and the locknuts. Tighten to 26 inch lbs.

29. Install the upper bracket and tighten the tapered-head bolts until the bolt heads break off.

30. Install the connector bracket and tighten to 43 inch lbs. Replace both protective covers.

31. Install the turn signal bracket so the upper surface is parallel with the upper surface of the capsule of the break-away bracket.

32. Install the tilt position sensor. Install the lower dust cover, using the proper tool to press in the dust seal.

33. Install the No. 2 intermediate shaft and tighten to 26 ft. lbs. (35 Nm). Replace the ignition key cylinder lamp.

34. Check that there is no axial play when the lever is turned fully counterclockwise.

35. Check that the main shaft moves smoothly when the lever is turned clockwise

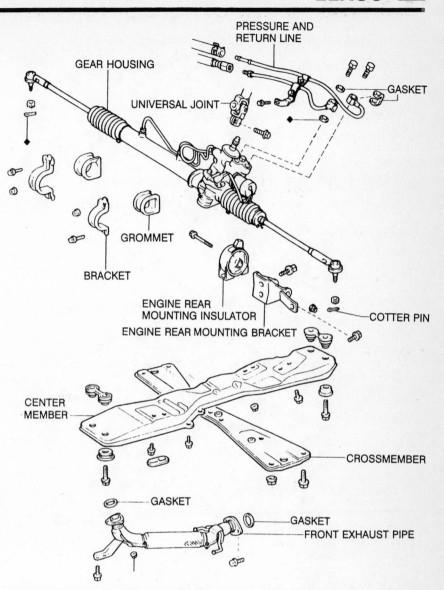

Exploded view of the power rack assembly – ES250

Power Steering Rack

REMOVAL & INSTALLATION

ES250

1. Place the front wheels in a straight ahead position, secure the steering wheel with a suitable device to prevent the wheel from turning.

2. Place matchmarks in the universal joint and control valve shaft. Disconnect the joint.

3. Disconnect and plug the hydraulic lines to rack assembly.

4. Raise and safely support the vehicle. Disconnect the front exhaust pipe.

5. Remove the center support member and the crossmember.

6. Matchmark and disconnect the tie rod ends from the steering knuckle with the proper tool.

7. Remove the rear engine mount and mounting bracket. Raise the front of the engine with a suitable jacking device.

NOTE: Do not over-tilt the engine

8. Remove the mounting brackets and slide the gear housing to the right side to put the tie rod end in the body panel.

9. Pull the gear housing out through the opening in the left side lower side of the vehicle body.

NOTE: Do not damage the turn pressure tube and the transmission control cables.

10. Secure the gear housing in a suitable holding fixture. Remove the return and pressure tubes.

To install:

11. Connect the hydraulic tubes to

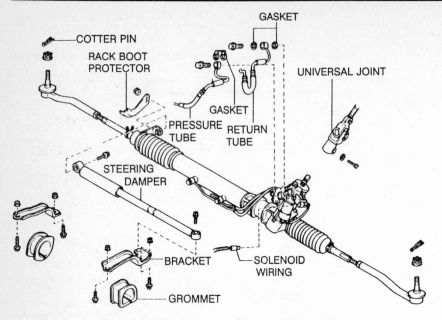

Exploded view of the power rack assembly—LS400

8. Disconnect the solenoid wiring and remove the mounting grommets and brackets. Remove the rack assembly.

9. The installation is the reverse of the removal procedure. Follow the following torque specifications:

   a. Joint connecting bolt—26 ft. lbs. (35 Nm).

   b. Tie rod end nuts—43 ft. lbs. (58 Nm).

   c. Steering damper bolts—20 ft. lbs. (27 Nm).

   d. Mounting bracket bolts—56 ft. lbs. (76 Nm).

Power Steering Pump

REMOVAL & INSTALLATION

ES250

1. Disconnect the negative battery cable. Disconnect the hydraulic lines at the pump assembly.

2. Raise and safely support the vehicle. Disconnect the right side tie rod end with the proper tool.

3. Remove the lower crossmember and the front fender apron seal.

4. Loosen the adjusting and through bolt and push the power steering pump forward. Remove the drive belt.

5. Remove the adjusting bolt and

the rack assembly and tighten to 38 ft. lbs. (51 Nm). Insert the rack assembly into position.

12. Replace the mounting brackets and tighten to 43 ft. lbs. (58 Nm).

13. Replace the rear engine mount and mounting bracket. Tighten the the through bolt to 64 ft. lbs. (85 Nm) (and the mounting bolts to 38 ft. lbs. (52 Nm). Lower the front of the engine with a suitable jacking device. Remove the jacking device.

14. Connect the tie rod ends to the steering knuckle. Tighten to 36 ft. lbs. (49 Nm).

15. Replace the center support member and the crossmember. Tighten the crossmember bolts to 153 ft. lbs. (207 Nm) and the center member support bolts to 29 ft. lbs. (39 Nm).

16. Install the front exhaust pipe and lower the vehicle.

17. Connect the hydraulic lines to the rack assembly.

18. Align the matchmarks on the universal joint and the control valve shaft and connect. Tighten the connecting bolt to 26 ft. lbs. (35 Nm).

19. Check the steering wheel center point and the toe-in. Refill the power steering reservoir.

LS400

1. Disconnect the negative battery cable.

2. Place the front wheels in a straight ahead position, secure the steering wheel with a suitable device to prevent the wheel from turning.

3. Place matchmarks in the universal joint and control valve shaft. Disconnect the joint.

4. Disconnect and plug the hydraulic lines to rack assembly.

5. Raise and safely support the vehicle. Disconnect the brake caliper and support with a piece of wire.

6. Disconnect the tie rod end from the steering knuckle with the proper tool.

7. Remove the steering damper and rack boot protector.

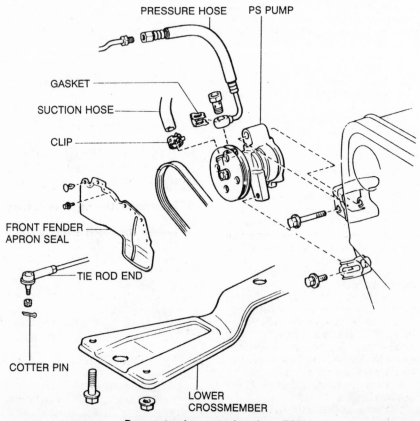

Power steering pump location—ES250

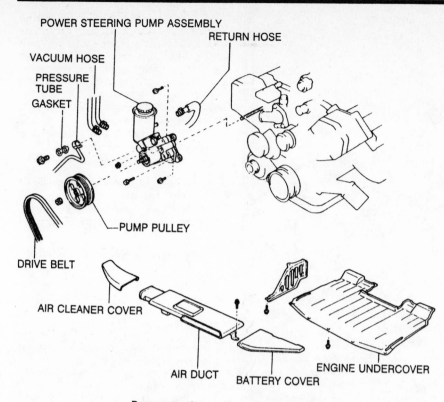

POWER STEERING PUMP ASSEMBLY
RETURN HOSE
VACUUM HOSE
PRESSURE TUBE
GASKET
PUMP PULLEY
DRIVE BELT
AIR CLEANER COVER
AIR DUCT
BATTERY COVER
ENGINE UNDERCOVER

Power steering pump location – LS400

through bolt, then remove the power steering pump.

6. Remove the pressure hose from the pump.

7. The installation is the reverse of the removal procedure. Follow the following torque specifications:

    a. Power steering pump mounting bolts – 29 ft. lbs. (39 Nm).

    b. Lower crossmember bolts – 153 ft. lbs. (207 Nm).

    c. Tie rod end nuts – 36 ft. lbs. (49 Nm).

    d. Pressure line fitting – 27–33 ft. lbs. (37-45 Nm).

LS400

1. Disconnect the negative battery cable.

2. Remove the air cleaner cover, air duct and battery cover.

3. Turn the drive belt tensioner counterclockwise and remove the drive belt. Remove the pump pulley with the proper tool.

4. Raise and safely support the vehicle. Remove the engine undercover.

5. Disconnect the hydraulic and vacuum lines at the pump assembly.

6. Remove the pump mounting bolts and the pump assembly.

7. The installation is the reverse of the removal procedure. Follow the following torque specifications:

    a. Pump mounting bolts – 29 ft. lbs. (40 Nm).

    b. Pump pulley bolt – 32 ft. lbs. (44 Nm).

    c. Pressure line fitting – 36 ft. lbs. (49 Nm).

8. Bleed the power steering system.

NOTE: Note the curvature of the No. 2 and No. 3 power steering return tubes and the rear left hand engine cover have been modified. The new curvature is a 50 degree bend. Previous and new parts are interchangeable as a set only.

BELT ADJUSTMENT

On the LS400, a belt tensioner is used. There is no need for belt adjustment.

ES250

1. Disconnect the negative battery cable.

2. Loosen the power steering pump adjusting bolt and the through bolt.

3. Using the proper tool, move the pump assembly to attain the proper belt tension.

4. Tighten the pump mounting bolts.

5. Connect the negative battery cable.

SYSTEM BLEEDING

1. Check that the fluid level in the reservoir tank is at the maximum level.

2. Start the engine and turn the steering wheel from lock to lock until the air bubbles are removed from the fluid.

3. Stop the engine and measure the fluid level.

4. Make sure the rise of the fluid is not over 0.020 in.

Tie Rod Ends

REMOVAL & INSTALLATION

1. Raise and safely support the vehicle.

2. Place matchmarks on the threads of the tie rod end and the rack end.

3. On LS400, disconnect the brake caliper and suspend with a piece of wire. Do not disconnect the lines.

4. Remove the cotter pin and nut. Disconnect the tie rod end from the steering knuckle with the proper tool.

5. Unscrew the tie rod end from the rack end.

To install:

6. Install the tie rod end to the rack end, counting the same number of threads as were removed.

7. Connect the tie rod end to the steering knuckle. Tighten the nut to 41–43 ft. lbs.

8. Connect the caliper, if removed.

9. Lower the vehicle. Check the toe-in.

BRAKES

For all brake system repair and service procedures not detailed below, please refer to "Brakes" in the Unit Repair section.

Master Cylinder

REMOVAL & INSTALLATION

ES250

1. Disconnect the negative battery cable. Remove the charcoal canister.

2. Remove the fluid from the reservoir, using a suitable syringe.

3. Disconnect and plug the hydraulic lines to the master cylinder.

4. Disconnect the level warning switch connector. Remove the mounting nuts and the union and clamp.

5. Remove the master cylinder from the booster and remove the gasket.

To install:

6. Adjust the length of the booster pushrod before installing the master cylinder.

7. Install the master cylinder with a new gasket. Replace the union, clamp

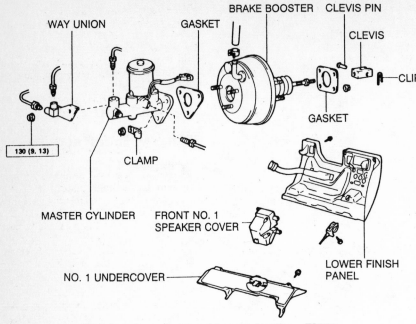

WAY UNION   GASKET   BRAKE BOOSTER   CLEVIS PIN

CLEVIS

CLIP

GASKET

130 (9, 13)

CLAMP

MASTER CYLINDER   FRONT NO. 1 SPEAKER COVER

NO. 1 UNDERCOVER

LOWER FINISH PANEL

Brake master cylinder and booster location—ES250

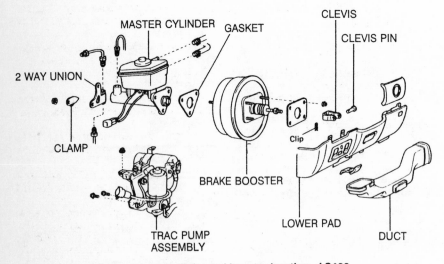

MASTER CYLINDER   GASKET   CLEVIS

CLEVIS PIN

2 WAY UNION

Clip

CLAMP

TRAC PUMP ASSEMBLY

BRAKE BOOSTER

LOWER PAD

DUCT

Brake master cylinder and booster location—LS400

and tighten the mounting nuts to 9 ft. lbs. (12 Nm).

8. Connect the brake lines with the proper tool and tighten the union nuts to 11 ft. lbs. (15 Nm).

9. Connect the level warning switch connector. Fill the reservoir with brake fluid.

10. Bleed the brake system and check for leaks.

LS400

1. Disconnect the negative battery cable.

2. If equipped with a Traction Control System (TRAC), perform the following procedure:

   a. Remove the air cleaner.

   b. Connect a vinyl tube from a container to the bleeder plug of the

TRAC actuator, then loosen the bleeder plug with the ignition in the **OFF** position.

   c. Tighten the plug when the fluid stops flowing out.

NOTE: The fluid is under high pressure and could spray out with great force, use caution.

3. Remove the fluid from the reservoir, using a suitable syringe.

4. If equipped with a TRAC, disconnect the hydraulic lines, mounting bolts and connector. Remove the TRAC pump assembly.

5. Disconnect the brake lines from the master cylinder.

6. Remove the mounting nuts and 2-way union. Remove the master cyl-

inder and gasket from the brake booster.

To install:

7. Adjust the length of the booster pushrod before installation.

8. Install the master cylinder and replace the 2-way union. Tighten the mounting nuts to 9 ft. lbs. (12 Nm).

9. Connect the brake lines to the master cylinder and the 2-way union. Tighten the nuts to 11 ft. lbs. (15 Nm).

10. Install the TRAC pump assembly, if equipped.

11. Bleed the brake system and the TRAC system, if equipped.

12. Check for leaks. Check and adjust the brake pedal, if needed.

Proportioning Valve

REMOVAL & INSTALLATION

1. Disconnect the brake line from the valve union. Use caution if under pressure.

2. Remove the valve mounting bolts and remove the valve assembly.

3. The installation is the reverse of the removal procedure.

4. Bleed the brake system. Check for leaks.

Power Brake Booster

REMOVAL & INSTALLATION

ES250

1. Disconnect the negative battery cable. Remove the master cylinder and the vacuum hose.

2. Remove the No. 1 undercover and the lower finish panel. Disconnect the pedal return spring and the theft deterrent horn.

3. Remove the clip and clevis pin from the operating rod.

4. Remove the pedal bracket mounting bolt, the steering support nuts and the break-away bracket nuts. Lower the steering column.

5. Remove the mounting nuts and pull out the booster and gasket.

To install:

6. Adjust the length of the pushrod by:

   a. Install the gasket on the master cylinder. Place the proper tool on the gasket and lower the pin of the tool until it's tip slightly touches the pin.

   b. Turn the tool upside down and set it on the booster.

   c. Measure the distance pushrod, the clearance is 0 in. (0mm). Adjust the booster pushrod length until the pushrod slightly touches the pin head.

7. Install the booster and gasket.

Replace the clevis pin to the operating rod.

8. Install and tighten the mounting nuts to 9 ft. lbs. (12 Nm).

9. Insert the clevis pin to the clevis and brake pedal. Install the clip to the clevis pin.

10. Lift up the steering column and install the steering support nuts and break-away nuts. Tighten to 19 ft. lbs. (26 Nm).

11. Install the pedal bracket mounting bolt and tighten to 13 ft. lbs. (18 Nm). Replace the pedal return spring.

12. Install the master cylinder and connect the vacuum hose to the brake booster.

13. Fill the reservoir with brake fluid and bleed the brake system. Check for leaks.

14. Check and adjust the brake pedal. Tighten the clevis locknut to 19 ft. lbs. (26 Nm).

15. Install the lower finish panel and No. 1 undercover. Connect the negative battery cable.

LS400

1. Disconnect the negative battery cable.

2. Remove the TRAC pump assembly, if equipped.

3. Remove the master cylinder and the vacuum hose from the booster.

4. Remove the No. 1 undercover and the lower finish panel. Disconnect the pedal return spring and the theft deterrent horn.

5. Remove the clip and clevis pin from the operating rod.

6. Remove the brake line between the TRAC accumulator and the actuator from the clamp.

7. Remove the booster mounting nuts and lift out the booster.

To install:

8. Adjust the length of the pushrod by:

   a. Install the gasket on the master cylinder. Place the proper tool on the gasket and lower the pin of the tool until it's tip slightly touches the pin.

   b. Turn the tool upside down and set it on the booster.

   c. Measure the distance pushrod, the clearance is 0 in. (0mm). Adjust the booster pushrod length until the pushrod slightly touches the pin head.

9. Install the brake booster and gasket. Replace the clevis on the operating rod.

10. Install and tighten the mounting nuts to 9 ft. lbs. (12 Nm).

11. Insert the clevis pin into the clevis and brake pedal and install the clip on the clevis pin.

12. If equipped with a TRAC, install the brake tube between the TRAC ac-

cumulator and actuator to the No. 2 clamp.

13. Install the No. 1 undercover and lower finish panel.

14. Install the master cylinder and connect the vacuum hose to the booster.

15. Replace the TRAC pump, if equipped.

16. Fill the brake reservoir with brake fluid and bleed the brake system.

17. Check and adjust the brake pedal. Check for leaks.

18. Connect the negative battery cable.

Brake Caliper

REMOVAL & INSTALLATION

Front

1. Raise and safely support the vehicle. Remove the front tire and wheel assembly.

2. Disconnect and plug the brake line at the caliper.

3. Remove the mounting bolts and the caliper assembly.

4. Remove the brake pads, shims, springs and indicators from the caliper.

5. The installation is the reverse of the removal procedure.

6. Tighten the mounting bolts to 29 ft. lbs. (40 Nm) on the ES250 or 25 ft. lbs. (34 Nm) on the LS400.

7. Bleed the brake system.

Rear

1. Raise and safely support the vehicle. Remove the rear tire and wheel assembly.

2. Disconnect and plug the brake line at the caliper.

3. Remove the mounting bolts and the caliper assembly.

4. Remove the brake pads, shims, springs and indicators from the caliper.

5. The installation is the reverse of the removal procedure.

6. Tighten the mounting bolts to 29 ft. lbs. (40 Nm) on the ES250 or 25 ft. lbs. (34 Nm) on the LS400.

7. Bleed the brake system.

Disc Brake Pads

REMOVAL & INSTALLATION

Front

1. Raise and safely support the vehicle. Remove the front tire and wheel assembly.

3. Remove the mounting bolts and lift off the caliper assembly. Support the caliper. Do not disconnect the brake line.

4. Remove the brake pads, shims, springs and indicators from the caliper.

To install:

5. Install the pad support plates and the pad wear indicator plate on the inside pad.

6. Apply disc brake grease to both sides of the anti-squeal shims to each pad.

7. Install the inside pad with the wear indicator facing upward. Install the outside pad.

8. Install the anti-squeal springs. Press in the caliper piston with the proper tool and install the caliper. Tighten the mounting bolts to 29 ft. lbs. (40 Nm) on the ES250 or 25 ft. lbs. (34 Nm) on the LS400.

9. Install the front tire and wheel assembly. Lower the vehicle.

Rear

1. Raise and safely support the vehicle. Remove the rear tire and wheel assembly.

3. Remove the mounting bolts and lift off the caliper assembly. Support the caliper. Do not disconnect the brake line.

4. Remove the brake pads, shims, springs and indicators from the caliper.

To install:

5. Install the pad support plates and the pad wear indicator plate on the inside pad.

6. Apply disc brake grease to both sides of the anti-squeal shims to each pad.

7. Install the inside pad with the wear indicator facing upward. Install the outside pad.

8. Install the anti-squeal springs. Press in the caliper piston with the proper tool and install the caliper. Tighten the mounting bolts to 29 ft. lbs. (40 Nm) on the ES250 or 25 ft. lbs. (34 Nm) on the LS400.

9. Install the rear tire and wheel assembly. Lower the vehicle.

Brake Rotor

REMOVAL & INSTALLATION

Front

1. Raise and safely support the vehicle. Remove the front tire and wheel assembly.

2. Remove the mounting bolts and lift off the caliper assembly. Support the caliper. Do not disconnect the brake line.

3. Remove the torque plate from the steering knuckle, if equipped.

4. Remove the hub bolts and pull the brake rotor.

5. The installation is the reverse of

the removal procedure. Tighten the torque plate bolts to 79 ft. lbs. (107 Nm) on the ES250 and 87 ft. lbs. (118 Nm) on the LS400.

Rear

1. Raise and safely support the vehicle. Remove the rear and wheel assembly.
2. Remove the 2 retaining bolts from the rear disc brake caliper. Suspend the caliper with a suitable piece of wire so as not to stretch the brake hose.
3. Place matchmarks on the rotor disc and the rear axle shaft. Remove the 2 rotor retaining screws from the rotor and remove the rotor.

NOTE:If the rotor disc cannot be removed easily, return the shoe adjuster until the wheel turns freely.

4. The installation is the reverse of the removal procedure. Tighten the caliper retaining bolts to 34 ft. lbs. (47 Nm) on the ES250 and 77 ft. lbs. (104 Nm) on the LS400.

Parking Brake Shoes

REMOVAL & INSTALLATION

1. Raise and safely support the vehicle. Remove the rear and wheel assembly. On the ES250 models, remove the axle carrier mounting bolt and nut.
2. Remove the 2 retaining bolts from the rear disc brake caliper. Suspend the caliper with a suitable piece of wire so as not to stretch the brake hose.
3. Place matchmarks on the rotor disc and the rear axle shaft. Remove the 2 rotor retaining screws from the rotor and remove the rotor.

NOTE:If the rotor disc cannot be removed easily, return the shoe adjuster until the wheel turns freely.

4. Remove the parking brake shoe return springs.
5. Slide out the front shoe and remove the shoe adjuster. Remove the shoe strut with spring. Disconnect the tension spring and remove the front shoe.
6. Slide out the rear shoe. Disconnect the tension spring from the rear shoe. Disconnect the parking brake cable from the parking brake shoe lever. Remove the shoe hold down spring cups, springs and pin. Remove the rear shoe.
7. Installation is the reverse order of the removal procedure. On the ES250 models, torque the disc brake caliper bolts to 34 ft. lbs. (47 Nm).

Torque the axle carrier bolt and nut of the upper side to 166 ft. lbs. (226 Nm).
8. On the LS400 models, torque the disc brake caliper bolts to 77 ft. lbs.

Parking Brake Cable

ADJUSTMENT

ES250

1. Pull the parking brake lever all the way up and count the number of clicks, should be 5–8 clicks.
2. Before adjusting the parking brake, make sure the rear parking brake shoe clearance is adjusted. Adjust by:
   a. Raise and safely support the vehicle. Remove the rear tire and wheel assembly.
   b. Remove the hole plug.
   c. Turn the adjuster and expand the shoes until the rotor locks.
   d. Return the adjuster 8 notches.
   e. Install the hole plug. Replace the tire and wheel assembly.
   f. Lower the vehicle.
3. Remove the console box.
4. Loosen the locknut and turn the adjusting nut until the lever travel is correct.
5. Tighten the locknut to 48 inch lbs. Install the console box.

LS400

1. Depress the parking brake all the way and count the number of clicks, should be 5–7 clicks.
2. Before adjusting the parking brake, make sure the rear parking brake shoe clearance is adjusted. Adjust by:
   a. Raise and safely support the vehicle. Remove the rear tire and wheel assembly.
   b. Remove the hole plug.
   c. Turn the adjuster and expand the shoes until the rotor locks.
   d. Return the adjuster 8 notches.
   e. Install the hole plug. Replace the tire and wheel assembly.
   f. Lower the vehicle.
3. Raise and safely support the vehicle.
4. Loosen the lock adjuster locknut and adjuster until the parking brake pedal travel is correct. Tighten the locknut.
5. The installation is the reverse of the removal procedure.

Brake System Bleeding

1. Fill the reservoir to the maximum level with brake fluid.
2. To bleed the master cylinder:

a. Disconnect the brake lines from the master cylinder.
   b. Depress the brake pedal and hold it.
   c. Block off the outer holes with fingers and release the brake pedal. Repeat 2–3 times.
3. To bleed the wheels:
   a. Start bleeding the brakes from the farthest point.
   b. Connect a vinyl tube to the brake cylinder bleeder plug and insert the other end of the tube in a ½ full container of brake fluid.
   c. Press on the brake pedal and loosen the bleeder plug until brake fluid comes out.
   d. Repeat until there is no more air bubbles in the fluid. Tighten the bleeder screw.
   e. Repeat the procedure for each wheel.
4. To bleed TRAC Control System:
   a. Remove the air cleaner, then temporarily reinstall it so the engine can be started.
   b. Connect a vinyl tube to the bleeder plug of the TRAC actuator, then loosen the bleeder plug.
   c. Start the engine, then operate the TRAC pump motor until all the air has been bled out of the fluid.
   d. Tighten the bleeder screw and stop the engine. Install the air cleaner.

Anti-Lock Brake System Service

The ABS system controls the hydraulic pressure of all 4 wheels during sudden braking and braking on slippery road surfaces, preventing the wheels from locking.

ABS Actuator

REMOVAL & INSTALLATION

ES250

1. Disconnect the negative battery cable. Remove brake fluid using a proper syringe.
2. Remove the actuator cover and disconnect the connectors from the actuator.
3. Remove the cover bracket and stud bolt.
4. Disconnect the brake tubes from the actuator, mounting nuts, wave washers and washers.
5. Remove the actuator from the actuator bracket.
To install:
6. Install the actuator to the actuator bracket. Replace the washers, wave washers and tighten the mounting nuts to 48 inch lbs.

7. Connect the brake tubes to the actuator with the proper tool and tighten to 11 ft. lbs. (15 Nm).

8. Connect the connectors to the actuator. Install the stud and bracket.

9. Install the actuator cover.

10. Fill the brake reservoir to the proper level and bleed the brake system.

11. Connect the negative battery cable. Check for leaks.

LS400

1. Disconnect the negative battery cable. Remove brake fluid using a proper syringe.

2. Disconnect the dust cover, air cleaner and the air duct.

3. Disconnect the brake line from the ABS, actuator with the proper tool. Disconnect the brake lines from the TRAC actuator, if equipped.

4. Remove the mounting bolts from the ABS actuator or TRAC actuator, if equipped.

5. Disconnect the connectors and remove the actuator.

To install:

6. Install the ABS actuator or TRAC actuator, if equipped. Tighten the mounting bolts to 9 ft. lbs. (12 Nm). Connect the actuator connectors.

7. Using the proper tool, connect the brake lines to the actuator and tighten to 11 ft. lbs. (15 Nm).

8. Connect the dust cover, air cleaner and the air duct.

9. Bleed the brake system and connect the negative battery cable.

10. Check for leaks.

Front Speed Sensor

REMOVAL & INSTALLATION

ES250

1. Disconnect the negative battery cable.

2. Raise and safely support the vehicle. Remove the tire and wheel assembly.

3. Disconnect the speed sensor connector.

4. Remove the front hub and steering knuckle assembly.

5. Remove the nut and the speed sensor rotor. Do not scratch the serrations of the speed sensor rotor.

6. The installation is the reverse of the removal procedure.

LS400

1. Disconnect the negative battery cable.

2. Raise and safely support the vehicle. Remove the tire and wheel assembly.

---

3. Disconnect the speed sensor connector.

4. Remove the front axle hub from the steering knuckle. Remove the sensor control rotor from the axle hub with the proper tool.

5. The installation is the reverse of the removal procedure.

Rear Speed Sensor

REMOVAL & INSTALLATION

ES250

1. Disconnect the negative battery.

2. Remove the rear seat cushion and rear side seat back cushion.

3. Disconnect the sensor connector and pull out the sensor wire harness. Remove the 2 clamp bolts holding the sensor wire harness to the body and suspension arm.

4. Remove the axle carrier mounting bolt and nut of the upper side. Remove the 2 bolts and remove the disc brake caliper assembly, Be sure to suspend the disc brake caliper with a piece of safety wire.

5. Remove the rotor disc.

6. Remove the 4 axle hub mounting bolts. Remove the rear axle hub. Remove the backing plate with the parking brake assembly and O-ring.

7. Remove the speed sensor bolts and remove the speed sensor from the backing plate.

8. Installation is the reverse order of the removal procedure. Torque speed sensor retaining bolts to 69 in. lbs. Install the rear hub retaining bolts and 59 ft. lbs. (80 Nm). Torque the brake caliper retaining bolts to 34 ft.

---

lbs. (47 Nm). Torque the axle carrier mounting bolt and nut of the upper side to 166 ft. lbs. (226 Nm).

LS400

1. Disconnect the negative battery cable.

2. Raise and safely support the vehicle. Remove the tire and wheel assembly.

3. Disconnect the speed sensor connector.

4. Remove the rear axle hub and backing plate.

5. Remove the mounting bolts and the speed sensor from the backing plate.

6. The installation is the reverse of the removal procedure.

CHASSIS ELECTRICAL

Air Bag

DISARMING

On vehicles equipped with an air bag, the negative battery cable must be disconnected, before working on the system. Failure to do so may result in deployment of the air bag and possible personal injury. Work must be started after approximately 20 seconds or longer from the time the ignition switch is turned to the **LOCK** position and the negative battery cable is dis-

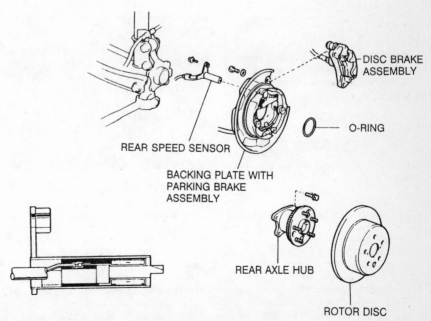

Exploded view of the ABS rear wheel speed sensor

connected from On vehicles equipped with an air bag, the negative battery cable must be disconnected, before working on the system.

Heater Blower Motor

REMOVAL & INSTALLATION

ES250

1. Disconnect the negative battery cable.
2. Disconnect the electrical connection at the blower motor.
3. Remove the mounting bolts and the blower motor.
4. The installation is the reverse of the removal procedure.

LS400

1. Disconnect the negative battery cable. Remove the right side instrument panel cover.
2. Properly discharge the air conditioning system. Remove the cruise control actuator.
3. Remove the bolt and disconnect the actuator tubes.
4. Disconnect both the liquid and the suction tubes for the air conditioner.
5. Remove the cover plate to the control assembly.

NOTE: Cap the open tubes to keep moisture out of the system.

6. Disconnect the drain hose clamp. Remove the mirror control electronic unit.
7. Disconnect the electrical connections and the mounting bracket from the blower control.
8. Disconnect the vehicle side wire harness from the blower unit.
9. Remove the mounting nuts and screws and remove the blower control unit.
10. The installation is the reverse of the removal procedure. Evacuate and recharge the air conditioning system.

Windshield Wiper Motor

REMOVAL & INSTALLATION

ES250

1. Disconnect the negative battery cable. Remove the wiper arm and blade assembly.
2. Disconnect the electrical connections at the wiper motor.
3. Remove the mounting bolts and disconnect the wiper arm.
4. Remove the wiper motor.
5. The installation is the reverse of the removal procedure.

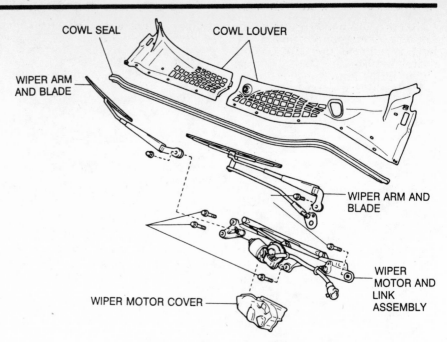

Wiper motor location—LS400

LS400

1. Disconnect the negative battery cable. Remove the wiper arm and blade assembly.
2. Remove the hood to cowl seal and remove the cowl louver.

NOTE: Raise the front side of the cowl louver up to remove the louver.

3. Remove the mounting bolts and disconnect the electrical connection.
4. Raise the front side of the wiper motor and link assembly up. Remove the wiper motor and link assembly.
5. Remove the wiper motor cover.
6. The installation is the reverse of the removal procedure. Tighten the mounting bolts to 48 inch lbs.

NOTE: With the front side of the louver raised, install the protector on the glass, then push the louver down.

Windshield Wiper Switch

REMOVAL & INSTALLATION

1. Disconnect the negative battery cable.
2. Remove the combination switch assembly.
3. Remove the mounting screws and separate the bracket from the switch body.
4. Remove the mounting screws and the switch from the switch body.
5. Remove the boot.
6. The installation is the reverse of the removal procedure.

Instrument Cluster

REMOVAL & INSTALLATION

ES250

1. Disconnect the negative battery cable.

NOTE: Work must be started after approximately 20 seconds or longer from the time the ignition switch is turned to the LOCK position and the negative battery cable is disconnected.

2. Remove the steering wheel.
3. Pry out the clips and pull the front pillar molding upward and remove.
4. Remove the left lower dash panel. Disconnect the hood release lever and remove the panel cover.
5. Pry out the switch bases and the speaker panel with the proper tool. Remove the mounting screws and lower the lower column panel.
6. Remove the mounting screws and the cluster panel. Disconnect the electrical connections.
7. Remove the steering column cover.
8. Disconnect the mounting screws, the speedometer cable and the electrical connections. Remove the instrument cluster and the speedometer.
9. The installation is the reverse of the removal procedure.

LS400

1. Disconnect the negative battery cable.

NOTE: Work must be started after approximately 20 seconds or longer from the time the ignition switch is turned to the LOCK position and the negative battery cable is disconnected.

2. Remove the steering wheel. Remove the left and right front pillar moldings.

3. Remove the steering column cover and the upper console panel, with the proper tool, to prevent damaging the cover.

4. Remove the front ash receptacle and disconnect the electrical connection.

5. Pry out the front side of the lower console cover and remove by sliding the cover forward.

6. Remove the mounting screws and pry out the lower console box. Remove the cup holder.

7. Pry out the rear end panel and disconnect the wire connector. Remove the console box.

8. Remove the left lower trim panel. Remove the hood release lever and disconnect the cable from the lever.

9. Remove the mounting bolts and the left trim pad. Disconnect the wire connections and hose from the pad.

10. Remove the key cylinder pad and disconnect the park brake lever. Remove the outer mirror switch assembly and disconnect the electrical connection.

11. Remove the mounting screws and carefully pry out the instrument cluster. Remove the speedometer from the cluster.

12. The installation is the reverse of the removal procedure.

NOTE: A rattle or squeak noise from the glove box door arear may be apparent when the glove box door is closes and/or the vehicle is driven over rough roads. This condition may be caused by the movement of the glove box door check arm against the glove box door and/or the check arm through hole in the dash.

Speedometer

REMOVAL & INSTALLATION

1. Disconnect the negative battery cable.

2. Remove the combination meter from the instrument cluster.

3. Remove the speedometer from the combination meter.

4. Installation is the reverse order of the removal procedure.

Combination Switch

The combination switch incorporates the headlight switch, turn signal switch, dimmer switch and the windshield wiper switch.

REMOVAL & INSTALLATION

1. Disconnect the negative battery cable.

2. On the ES250 models, remove the instrument panel No. 1 under cover subassembly. Remove the instrument panel lower finish panel and cluster finish panel.

3. On the LS400 models, remove the under cover, lower pad, key cylinder pad. The No. 3 finish panel mounting bracket. The No. 2 heater to register duct.

NOTE: Position the front wheels in a straight ahead position. Work must be started after approximately 20 seconds or longer from the time the ignition switch is turned to the LOCK position and the negative battery cable is disconnected from the battery. If the wiring connector of the airbag system is disconnected with the ignition switch at ON or ACC, diagnostic coded will be recorded.

4. Remove the steering wheel center pad.

NOTE: When removing the wheel pad, take care not to pull the air bag harness connector. When storing the wheel pad, keep the upper surface of the pad facing upward. Since the air bag con-

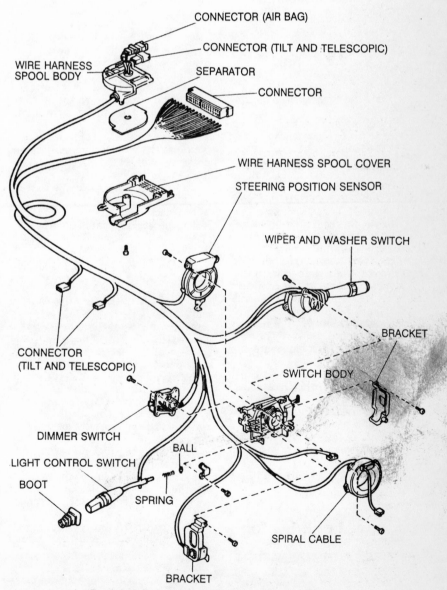

Exploded view of the combination switch—LS400

nector has a 2-stage lock, remove the 1st stage lock and disconnect the connector.

5. Remove the steering wheel assembly.

6. Remove the steering column cover. Remove the 4 combination switch retaining switch. Disconnect the connectors and remove the combination switch assembly from the steering column.

7. To remove the headlight switch:
   a. Remove the mounting screws and separate the bracket from the switch body.
   b. Remove the screws and the ball set plate from the switch body.
   c. Remove the ball and slide out the switch from the switch body with the spring.
   d. Remove the boot.

8. Loosen the mounting screws and remove the dimmer switch and the turn signal switch from the switch body.

9. Separate the bracket from the wiper switch body. Remove the wiper switch from the switch body.

To install:

10. Install the wiper switch to the switch body and connect the mounting bracket.

11. Install the dimmer and turn signal switch to the switch body and tighten the mounting screws.

12. To install the headlight switch:
   a. Slide the switch and install the switch body.
   b. Set the lever in the **HIGH** position. Install the ball and plate.

13. Install the combination switch assembly to the steering column and tighten the mounting screws.

14. Connect the electrical connectors. Push in the terminals until they are securely locked in the connector lug.

15. Replace both steering column covers. Install the steering wheel, using the proper tool.

16. Connect the air bag connector and replace the 1st stage lock.

17. Replace the cluster finish panel. Install the steering wheel center pad and connect the wire connector.

18. Replace the key cylinder trim pad assembly and heater duct register, if necessary.

19. Install the lower instrument trim panel and cover assembly.

20. Connect the negative battery cable.

Ignition Lock/Switch

REMOVAL & INSTALLATION

1. Disconnect the negative battery cable.

2. Remove the lower trim panel and pad assembly, if necessary.

3. Remove the key cylinder trim panel and pad assembly. Disconnect the key cylinder lamp assembly.

4. Remove the trim panel mounting bracket and the heater duct register, if necessary.

5. Remove the mounting screws and disconnect the electrical connections. Remove the ignition switch.

6. The installation is the reverse of the removal procedure.

Stoplight Switch

ADJUSTMENT

1. Disconnect the negative battery cable.

2. Remove the instrument panel undercover, the lower trim panel and air duct, if necessary.

3. Remove the power steering computer, if equipped. Disconnect the connector from the stoplight switch.

4. Loosen the stoplight switch locknut, the switch and the pushrod locknut.

5. Adjust the pedal height by turning the pushrod. Return the stoplight switch until it contacts the pedal stopper.

6. Tighten the stoplight switch locknut, the switch and the pushrod locknut. Connect the electrical connection at the switch.

7. Connect the negative battery cable.

8. Check that the stoplights light when the brake pedal is depressed. It goes off when the pedal is released.

REMOVAL & INSTALLATION

1. Disconnect the negative battery cable.

2. Remove the instrument panel undercover, the lower trim panel and air duct, if necessary.

3. Remove the power steering computer, if equipped. Disconnect the connector from the stoplight switch. Loosen the pushrod locknut.

4. Remove the stoplight switch locknut and the stoplight switch.

To install:

5. Install the stoplight switch and replace the locknut, do not tighten.

6. Adjust the pedal height by turning the pushrod. Return the stoplight switch until it contacts the pedal stopper.

7. Tighten the stoplight switch locknut, the switch and the pushrod locknut. Connect the electrical connection at the switch.

8. Connect the negative battery cable.

9. Check that the stoplights light when the brake pedal is depressed and off when the pedal is released.

Clutch Switch

ADJUSTMENT

ES250

1. Disconnect the negative battery cable.

2. Remove the lower dash trim panel and disconnect the air duct.

3. Loosen the locknut and disconnect the connection at the clutch switch. Turn the clutch switch until the pedal height is 7.52–7.91 in. from stop.

4. Tighten the locknut and connect the electrical connection at the switch.

5. Connect the air duct and replace the lower dash panel.

6. Connect the negative battery cable.

Neutral Safety Switch

ADJUSTMENT

1. Disconnect the negative battery cable. Raise and safely support the vehicle.

2. Loosen the neutral start switch bolt and place the shift lever in the **N** position.

3. Align the groove and the neutral basic line.

4. Hold in position and tighten the bolt to 9 ft. lbs.

5. Lower the vehicle and connect the negative battery cable.

Fuses, Circuit Breakers and Relays

LOCATION

There is a fuse block, relay center and circuit breaker center located on the left lower portion of the instrument panel. There is a fuse block and relay center located on the right lower portion of the instrument panel. There is a fuse block and relay center located in the engine compartment on the driver's side.

Flashers

LOCATION

The turn signal and hazard flashers are located on the left lower section of the instrument panel.

Toyota-Truck

4RUNNER • LAND CRUISER • PICK-UP • PREVIA • VAN

SPECIFICATIONS

ENGINE IDENTIFICATION

Year	Model	Engine Displacement cu. in. (cc/liter)	Engine Series Identification	No. of Cylinders	Engine Type
1988	Pick-Up	144.4 (2366/2.4)	22R	4	OHC
	Pick-Up	144.4 (2366/2.4)	22R-E	4	OHC
	Pick-Up	144.4 (2366/2.4)	22R-TE	4	OHC-Turbo
	Pick-Up	180.5 (2959/3.0)	3VZ-E	6	OHC
	4Runner	144.4 (2366/2.4)	22R	4	OHC
	4Runner	144.4 (2366/2.4)	22R-E	4	OHC
	4Runner	144.4 (2366/2.4)	22R-TE	4	OHC-Turbo
	4Runner	180.5 (2959/3.0)	3VZ-E	6	OHC
	Van	136.5 (2237/2.2)	4Y-EC	4	OHC
	Land Cruiser	241.3 (3956/4.0)	3F-E	6	OHV
1989	Pick-Up	144.4 (2366/2.4)	22R	4	OHC
	Pick-Up	144.4 (2366/2.4)	22R-E	4	OHC
	Pick-Up	180.5 (2959/3.0)	3VZ-E	6	OHC
	4Runner	144.4 (2366/2.4)	22R-E	4	OHC
	4Runner	180.5 (2959/3.0)	3VZ-E	6	OHC
	Land Cruiser	241.3 (3956/4.0)	3F-E	6	OHV
	Van	136.5 (2237/2.2)	4Y-EC	4	OHC
1990	Pick-Up	144.4 (2366/2.4)	22R	4	OHC
	Pick-Up	144.4 (2366/2.4)	22R-E	4	OHC
	Pick-Up	180.5 (2959/3.0)	3VZ-E	6	OHC
	4Runner	144.4 (2366/2.4)	22R-E	4	OHC
	4Runner	180.5 (2959/3.0)	3VZ-E	6	OHC
	Land Cruiser	241.3 (3956/4.0)	3F-E	6	OHV
1991-92	Pick-Up	144.4 (2366/2.4)	22R-E	4	OHC
	Pick-Up	180.5 (2959/3.0)	3VZ-E	6	OHC
	4Runner	144.4 (2366/2.4)	22R-E	4	OHC
	4Runner	180.5 (2959/3.0)	3VZ-E	6	OHC
	Land Cruiser	241.3 (3956/4.0)	3F-E	6	OHV
	Previa	146.4 (2399/2.4)	2TZ-FE	4	DOHC

OHC—Overhead cam   OHV—Overhead valves   DOHC—Dual overhead cam

GENERAL ENGINE SPECIFICATIONS

Year	Model	Engine Displacement cu. in. (cc)	Fuel System Type	Net Horsepower @ rpm	Net Torque @ rpm (ft. lbs.)	Bore × Stroke (in.)	Compression Ratio	Oil Pressure @ rpm
1988	Pick-Up	144.4 (2366)	2 bbl	96 @ 4800	129 @ 2800	3.62 × 3.50	9.3:1	36–71 @ 3000
	Pick-Up	144.4 (2366)	EFI	116 @ 4800	140 @ 2800	3.62 × 3.50	9.3:1	36–71 @ 3000
	Pick-Up	144.4 (2366)	EFI ①	135 @ 4800	173 @ 2800	3.62 × 3.50	7.5:1	36–71 @ 3000
	Pick-Up	180.5 (2959)	EFI	150 @ 4800	180 @ 3400	3.44 × 3.23	9.0:1	36–71 @ 4000
	4Runner	144.4 (2366)	2 bbl	96 @ 4800	129 @ 2800	3.62 × 3.50	9.3:1	36–71 @ 3000
	4Runner	144.4 (2366)	EFI	116 @ 4800	140 @ 2800	3.62 × 3.50	9.3:1	36–71 @ 3000
	4Runner	144.4 (2366)	EFI ①	135 @ 4800	173 @ 2800	3.62 × 3.50	7.5:1	36–71 @ 3000
	Van	136.5 (2237)	EFI	101 @ 4000	132 @ 3000	3.58 × 3.40	8.8:1	50–70 @ 3000
	Land Cruiser	241.3 (3956)	EFI	154 @ 4000	220 @ 3000	3.70 × 3.74	8.1:1	36–71 @ 4000

GENERAL ENGINE SPECIFICATIONS

Year	Model	Engine Displacement cu. in. (cc)	Fuel System Type	Net Horsepower @ rpm	Net Torque @ rpm (ft. lbs.)	Bore × Stroke (in.)	Compression Ratio	Oil Pressure @ rpm
1989	Pick-Up	144.4 (2366)	2 bbl	103 @ 4800	133 @ 2800	3.62 × 3.50	9.3:1	36–71 @ 3000
	Pick-Up	144.4 (2366)	EFI	116 @ 4800	140 @ 2800	3.62 × 3.50	9.3:1	36–71 @ 3000
	Pick-Up	180.5 (2959)	EFI	150 @ 4800	180 @ 3400	3.44 × 3.23	9.0:1	36–71 @ 4000
	4Runner	144.4 (2366)	EFI	116 @ 4800	140 @ 2800	3.62 × 3.50	9.3:1	36–71 @ 3000
	4Runner	180.5 (2959)	EFI	150 @ 4800	180 @ 3400	3.44 × 3.23	9.0:1	36–71 @ 4000
	Land Cruiser	241.3 (3956)	EFI	154 @ 4000	220 @ 3000	3.70 × 3.74	8.1:1	36–71 @ 4000
	Van	136.5 (2237)	EFI	101 @ 4000	132 @ 3000	3.58 × 3.40	8.8:1	50–70 @ 3000
1990	Pick-Up	144.4 (2366)	2 bbl	103 @ 4800	133 @ 2800	3.62 × 3.50	9.3:1	36–71 @ 3000
	Pick-Up	144.4 (2366)	EFI	116 @ 4800	140 @ 2800	3.62 × 3.50	9.3:1	36–71 @ 3000
	Pick-Up	180.5 (2959)	EFI	150 @ 4800	180 @ 3400	3.44 × 3.23	9.0:1	36–75 @ 3000
	4Runner	144.4 (2366)	EFI	116 @ 4800	140 @ 2800	3.62 × 3.50	9.3:1	36–71 @ 3000
	4Runner	180.5 (2959)	EFI	150 @ 4800	180 @ 3400	3.44 × 3.23	9.0:1	36–75 @ 3000
	Land Cruiser	241.3 (3956)	EFI	155 @ 4000	220 @ 3000	3.70 × 3.74	8.1:1	36–71 @ 4000
1991–92	Pick-Up	144.4 (2366)	EFI	116 @ 4800	140 @ 2800	3.62 × 3.50	9.3:1	36–71 @ 3000
	Pick-Up	180.5 (2959)	EFI	150 @ 4800	180 @ 3400	3.44 × 3.23	9.0:1	36–75 @ 3000
	4Runner	144.4 (2366)	EFI	116 @ 4800	140 @ 2800	3.62 × 3.50	9.3:1	36–71 @ 3000
	4Runner	180.5 (2959)	EFI	150 @ 4800	180 @ 3400	3.44 × 3.23	9.0:1	36–75 @ 3000
	Land Cruiser	241.3 (3956)	EFI	155 @ 4000	220 @ 3000	3.70 × 3.74	8.1:1	36–71 @ 4000
	Previa	146.4 (2399)	EFI	138 @ 5000	154 @ 4000	3.74 × 3.39	9.1:1	36–71 @ 3000

EFI—Electronic Fuel Injection
① Turbocharged

GASOLINE ENGINE TUNE-UP SPECIFICATIONS

Year	Model	Engine Displacement cu. in. (cc)	Spark Plugs Type	Gap (in.)	Ignition Timing (deg.) MT	AT	Compression Pressure (psi)	Fuel Pump (psi)	Idle Speed (rpm) MT	AT	Valve Clearance In.	Ex.
1988	Pick-Up①	144.4 (2366)	W16-EXRU	0.031	0④	0④	142–171	2.1–4.3	700	750	0.008	0.012
	Pick-Up②	144.4 (2366)	W16-EXRU	0.031	5B⑤	—	142–171	36–38	750	—	0.008	0.012
	Pick-Up③	144.4 (2366)	W16-EXRU	0.031	5B⑤	—	142–171	36–38	800	—	0.008	0.012
	Pick-Up	180.5 (2959)	Q16-RU	0.031	10B	10B	142–171	38–44	800	800	⑨	⑨
	4Runner①	144.4 (2366)	W16-EXRU	0.031	0④	0④	142–171	2.1–4.3	700	750	0.008	0.012
	4Runner②	144.4 (2366)	W16-EXRU	0.031	5B⑤	—	142–171	36–38	750	—	0.008	0.012
	4Runner③	144.4 (2366)	W16-EXRU	0.031	5B⑤	—	142–171	36–38	800	—	0.008	0.012
	4Runner	180.5 (2959)	Q16-RU	0.031	10B	10B	142–171	38–44	800	800	⑨	⑨
	Van	136.5 (2237)	P-16R	0.043	12B⑤	—	128–178	27–31	700	750	Hyd.	Hyd.
	Land Cruiser	241.3 (3956)	W16-EXRU	0.031	7B④	7B	114–149	37–46	650	650	0.008	0.014
1989	Pick-Up①	144.4 (2366)	W16-EXRU	0.031	0④	0④	142–171	2.1–4.3	700	750	0.008	0.012
	Pick-Up②	144.4 (2366)	W16-EXRU	0.031	5B⑤	—	142–171	36–38	750	—	0.008	0.012
	Pick-Up	180.5 (2959)	Q16-RU	0.031	10B	10B	142–171	38–44	800	800	⑨	⑨
	4Runner②	144.4 (2366)	W16-EXRU	0.031	5B⑤	—	142–171	36–38	750	—	0.008	0.012
	4Runner	180.5 (2959)	Q16-RU	0.031	10B	10B	142–171	38–44	800	800	⑨	⑨
	Land Cruiser	241.3 (3956)	W16-EXRU	0.031	7B④	7B	114–149	37–46	650	650	0.008	0.014
	Van	136.5 (2237)	P-16R	0.043	12B⑤	—	128–178	27–31	700	750	Hyd.	Hyd.

GASOLINE ENGINE TUNE-UP SPECIFICATIONS

Year	Model	Engine Displacement cu. in. (cc)	Spark Plugs Type	Gap (in.)	Ignition Timing (deg.) MT	AT	Compression Pressure (psi)	Fuel Pump (psi)	Idle Speed (rpm) MT	AT	Valve Clearance In.	Ex.
1990	Pick-Up ①	144.4 (2366)	W16-EXRU	0.031	0 ④	0	142–171	2.1–4.3	750	850	0.008	0.012
	Pick-Up ②	144.4 (2366)	W16-EXRU	0.031	5B	5B	142–171	38–44	750	850	0.008	0.012
	Pick-Up	180.5 (2959)	K16-RU	0.031	10B	10B	142–171	38–44	800	800	⑨	⑨
	4Runner	144.4 (2366)	W16-EXRU	0.031	5B	5B	142–171	38–44	750	750	0.008	0.012
	4Runner	180.5 (2959)	K16-RU	0.031	10B	10B	142–171	38–44	800	800	⑨	⑨
	Land Cruiser	241.3 (3956)	W16-EXRU	0.031	—	7B	114–149	37–46	650	650	0.008	0.014 ⑪
1991	Pick-Up ②	144.4 (2366)	W16-EXRU	0.031	5B	5B	142–171	38–44	750	850	0.008	0.012
	Pick-Up	180.5 (2959)	K16-RU	0.031	10B	10B	142–171	38–44	800	800	⑨	⑨
	4Runner ②	144.4 (2366)	W16-EXRU	0.031	5B	5B	142–171	38–44	750	750	0.008	0.012
	4Runner	180.5 (2959)	K16-RU	0.031	10B	10B	142–171	38–44	800	800	⑨	⑨
	Land Cruiser	241.3 (3956)	W16-EXR-U11	0.043	—	7B	114–149	37–46	650	650	0.008	0.014 ⑪
	Previa ⑥	146.4 (2399)	PK16R11	0.043	5B	5B	128–178	38–44	700	750	⑦	⑧
1992	SEE UNDERHOOD STICKER											

Hyd.—Hydraulic
① 2 bbl
② EFI
③ Turbocharged
④ 950 rpm
⑤ "T" terminal shorted
⑥ DOHC
⑦ Intake: 0.006–0.010 (cold)
⑧ Exhaust: 0.010–0.014 (cold)
⑨ Intake: 0.007–0.011 (cold)
Exhaust: 0.009–0.013 (cold)
⑩ Intake: 0.010–0.014 (cold)
Exhaust: 0.011–0.015 (cold)
⑪ Engine hot

FIRING ORDERS

NOTE: To avoid confusion, always replace spark plug wires one at a time.

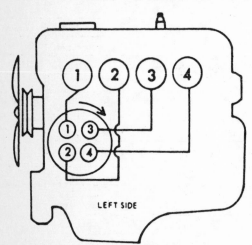

22R, 22R–E, 22R–TE, 2TZ–FE and 4Y–EC Engines
Engine Firing Order: 1–3–4–2
Distributor Rotation: Clockwise

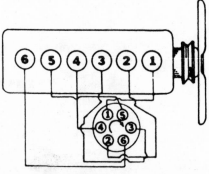

3F–E Engines
Engine Firing Order: 1–5–3–6–2–4
Distributor Rotation: Clockwise

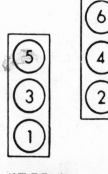

3VZ–E Engine
Engine Firing Order: 1–2–3–4–5–6
Distributor Rotation: Counterclockwise

CAPACITIES

Year	Model	Engine Displacement cu. in. (cc)	Engine Crankcase with Filter	Engine Crankcase without Filter	Transmission (qts.) 4-Spd	Transmission (qts.) 5-Spd	Transmission (qts.) Auto.	Drive Axle (qts.)	Fuel Tank (gal.)	Cooling System (qts.)
1988	Pick-Up①	144.4 (2366)	4.5	4.0	2.5	2.5	⑦	⑧	⑨	8.9
	Pick-Up②	144.4 (2366)	4.5	4.0	—	4.1	⑦	⑧	⑨	8.9
	Pick-Up③	144.4 (2366)	4.5	4.0	—	4.1	⑦	⑧	⑨	8.9
	Pick-Up	180.5 (2959)	4.8	4.4	—	3.2	—	2.5	17.2	⑪
	4Runner①	144.4 (2366)	4.5	4.0	2.5	2.5	⑦	⑧	⑨	8.9
	4Runner②	144.4 (2366)	4.5	4.0	—	4.1	⑦	⑧	⑨	8.9
	4Runner③	144.4 (2366)	4.5	4.0	—	4.1	⑦	⑧	⑨	8.9
	4Runner	180.5 (2959)	4.8	4.4	—	3.2	⑦	2.5	17.2	⑪
	Van	136.5 (2237)	3.7	3.2	—	2.5	6.9	1.5⑩	15.9	7.5
	Land Cruiser	258.1 (4200)	8.2	7.4	—	—	15.9	2.2⑥	22.2④	17.5⑤
1989	Pick-Up①	144.4 (2366)	4.5	4.0	2.5	2.5	⑦	⑧	⑨	8.9
	Pick-Up②	144.4 (2366)	4.5	4.0	—	4.1	⑦	⑧	⑨	8.9
	Pick-Up	180.5 (2959)	4.8	4.4	—	3.2	—	2.5	17.2	⑪
	4Runner	144.4 (2366)	4.5	4.0	—	4.1	⑦	⑧	17.2	⑪
	4Runner	180.5 (2959)	4.8	4.4	—	3.2	⑦	2.5	17.2	⑪
	Van	136.5 (2237)	3.7	3.2	—	2.5	6.9	1.5⑩	15.9	7.5
	Land Cruiser	241.3 (3956)	8.2	7.4	—	—	15.9	2.2⑥	22.2④	17.5⑤
1990	Pick-Up①	144.4 (2366)	4.5	4.0	2.5	⑯	8.7	⑰	⑨	⑮
	Pick-Up②	144.4 (2366)	4.5	4.0	2.5	⑯	8.7	⑰	⑨	⑮
	Pick-Up	180.5 (2959)	4.8	4.4	—	⑯	8.7	⑰	⑨	⑪
	4Runner	144.4 (2366)	4.5	4.0	—	⑯	8.7	⑰	17.2⑭	⑮
	4Runner	180.5 (2959)	4.8	4.4	—	⑯	8.7	⑰	17.2⑭	⑪
	Land Cruiser	241.3 (3956)	8.2	7.4	—	—	5.3	2.6⑬	25.1	18.5⑫
1991–92	Pick-Up①	144.4 (2366)	4.5	4.0	—	⑯	⑱	⑰	⑨	⑮
	Pick-Up	180.5 (2959)	4.8	4.4	—	⑯	⑱	⑰	⑨	⑪
	4Runner	144.4 (2366)	4.5	4.0	—	⑯	⑱	⑰	17.2⑭	⑮
	4Runner	180.5 (2959)	4.8	4.4	—	⑯	⑱	⑰	17.2⑭	⑪
	Land Cruiser	241.3 (3956)	8.2	7.4	—	—	6.3	3.0	25.1	18.5⑫
	Previa	146.4 (2399)	6.1	5.8	—	⑲	2.5	⑳	19.8	12.3

① 2 bbl
② EFI
③ Turbocharged
④ 23.8 gallon optional
⑤ Wagon—18.3 qts.
⑥ Front Axle—1.7 pts. for Pick-Up and 4Runner
   3.2 pts. for Land Cruiser
⑦ A43D Automatic Transmission—6.9 qts.
   A340E Automatic Transmission—7.3 qts.
   A340H Automatic Transmission—10.9 qts.
⑧ 2WD
   7.5 in: 1.42 qts.
   8 in: 1.9 qts.
   4WD
   22R and 22R Engines: 2.3 qts.
   22R-TE Engine: 2.5 qts.

⑨ Short bed—13.7 gallons
   Long bed—17.2 gallons
   Long bed 4WD—19.3 gallons
⑩ 4WD—2.0 qts.
⑪ Manual Transmission—11.0 qts.
   Automatic Transmission—10.8 qts.
⑫ With rear heater: 20.6 qts.
⑬ Front axle—3.2 qts.
⑭ 19.3 gallons optional
⑮ 4WD automatic transmission—9.6 qts.
   Except 4WD automatic transmission—8.9
⑯ 2WD—2.5 qts.
   4WD:
   G58—4.1 qts.
   W56 and R150F—3.2 qts.

⑰ 2WD:
   7.5 in.—1.4 qts.
   8.0 in.—1.9 qts.
   4WD Front
   Standard—1.7 qts.
   A.D.D.—2.0 qts.
   4WD Rear—2.3 qts.
⑱ 2WD:
   A340E—1.7 qts.
   A44D—2.5 qts.
   4WD:
   A340F—1.7 qts.
   A340H—4.8 qts.
⑲ 2WD—2.3 qts.
   4WD—2.7 qts.
⑳ 2WD—1.6 qts.
   4WD:
   Front—1.1 qts.
   Rear—1.6 qts.

CAMSHAFT SPECIFICATIONS

All measurements given in inches.

Year	Engine Displacement cu. in. (cc)	Journal Diameter					Lobe Lift		Bearing Clearance	Camshaft End Play
		1	2	3	4	5	In.	Ex.		
1988	144.4 (2366)	1.2984–1.2992	1.2984–1.2992	1.2984–1.2992	1.2984–1.2992	—	1.6783–1.6891	1.6807–1.6842	0.0004–0.0020	0.0031–0.0071
	136.5 (2237)	1.8291–1.8297	1.8192–1.8199	1.8094–1.8100	1.7996–1.8002	1.7913–1.7929	1.5205–1.5244	1.5208–1.5248	0.0010–0.0032	0.0028–0.0087
	180.5 (2959)	1.3370–1.3376	1.3370–1.3376	1.3370–1.3376	1.3370–1.3376	—	1.6783–1.6891	1.6807–1.6842	0.0010–0.0030	0.0031–0.0075
	241.3 (3956)	1.8880–1.8888	1.8289–1.8297	1.7699–1.7707	1.7108–1.7116	—	1.5102–1.5142	1.5059–1.5098	0.0010–0.0030	0.0079–0.0103
1989	144.4 (2366)	1.2984–1.2992	1.2984–1.2992	1.2984–1.2992	1.2984–1.2992	—	1.6783–1.6891	1.6807–1.6842	0.0004–0.0020	0.0031–0.0071
	136.5 (2237)	1.8291–1.8297	1.8192–1.8199	1.8094–1.8100	1.7996–1.8002	1.7913–1.7929	1.5205–1.5244	1.5208–1.5248	0.0010–0.0032	0.0028–0.0087
	180.5 (2959)	1.3370–1.3376	1.3370–1.3376	1.3370–1.3376	1.3370–1.3376	—	1.6783–1.6891	1.6807–1.6842	0.0010–0.0030	0.0031–0.0075
	241.3 (3956)	1.8880–1.8888	1.8289–1.8297	1.7699–1.7707	1.7108–1.7116	—	1.5102–1.5142	1.5059–1.5098	0.0010–0.0030	0.0079–0.0103
1990	144.4 (2366)	1.2984–1.2992	1.2984–1.2992	1.2984–1.2992	1.2984–1.2992	—	1.6783–1.6891	1.6807–1.6842	0.0004–0.0020	0.0031–0.0071
	180.5 (2959)	1.3370–1.3376	1.3370–1.3376	1.3370–1.3376	1.3370–1.3376	—	1.8701–1.8870	1.8701–1.8870	0.0010–0.0039	0.0031–0.0098
	241.3 (3956)	1.8880–1.8888	1.8289–1.8297	1.7699–1.7707	1.7108–1.7116	—	1.5102–1.5142	1.5059–1.5098	0.0010–0.0030	0.0079–0.0130
1991–92	144.4 (2366)	1.2984–1.2992	1.2984–1.2992	1.2984–1.2992	1.2984–1.2992	—	1.6783–1.6891	1.6807–1.6842	0.0004–0.0020	0.0031–0.0071
	146.4 (2399)	1.0614–1.0620	1.0614–1.0620	1.0614–1.0620	1.0614–1.0620	—	1.7839–1.7878	1.7740–1.7779	0.0010–0.0031	0.0016–0.0047
	180.5 (2959)	1.3370–1.3376	1.3370–1.3376	1.3370–1.3376	1.3370–1.3376	—	1.8701–1.8870	1.8701–1.8870	0.0010–0.0039	0.0031–0.0098
	241.3 (3956)	1.8880–1.8888	1.8289–1.8297	1.7699–1.7707	1.7108–1.7116	—	1.5102–1.5142	1.5059–1.5098	0.0010–0.0030	0.0079–0.0130

CRANKSHAFT AND CONNECTING ROD SPECIFICATIONS

All measurements are given in inches.

Year	Engine Displacement cu. in. (cc)	Crankshaft				Connecting Rod		
		Main Brg. Journal Dia.	Main Brg. Oil Clearance	Shaft End-play	Thrust on No.	Journal Diameter	Oil Clearance	Side Clearance
1988	144.4 (2366)	2.3616–2.3622	0.0010–0.0022	0.0008–0.0087	3	2.0861–2.0866	0.0010–0.0022	0.0008–0.0087
	136.5 (2237)	2.2829–2.2835	0.0008–0.0020	0.0008–0.0087	3	1.8892–1.8898	0.0008–0.0020	0.0063–0.0123
	180.5 (2959)	2.5195–2.5197	0.0009–0.0017	0.0008–0.0098	3	2.1648–2.1654	0.0009–0.0021	0.0059–0.0130
	241.3 (3956)	①	0.0008–0.0017	0.0024–0.0063	3	2.1252–2.1260	0.0008–0.0024	0.0043–0.0091

CRANKSHAFT AND CONNECTING ROD SPECIFICATIONS

All measurements are given in inches.

Year	Engine Displacement cu. in. (cc)	Crankshaft Main Brg. Journal Dia.	Crankshaft Main Brg. Oil Clearance	Crankshaft Shaft End-play	Crankshaft Thrust on No.	Connecting Rod Journal Diameter	Connecting Rod Oil Clearance	Connecting Rod Side Clearance
1989	144.4 (2366)	2.3616–2.3622	0.0010–0.0022	0.0008–0.0087	3	2.0861–2.0866	0.0010–0.0022	0.0008–0.0087
	136.5 (2237)	2.2829–2.2835	0.0008–0.0020	0.0008–0.0087	3	1.8892–1.8898	0.0008–0.0020	0.0063–0.0123
	180.5 (2959)	2.5195–2.5197	0.0009–0.0017	0.0008–0.0098	3	2.1648–2.1654	0.0009–0.0021	0.0059–0.0130
	241.3 (3956)	①	0.0008–0.0017	0.0006–0.0080	3	2.0861–2.0866	0.0008–0.0020	0.0063–0.0118
1990	144.4 (2366)	2.3616–2.3622	0.0010–0.0022	0.0008–0.0087	3	2.0861–2.0866	0.0010–0.0022	0.0008–0.0087
	180.5 (2959)	2.5195–2.5197	0.0009–0.0017	0.0008–0.0098	3	2.1648–2.1654	0.0009–0.0021	0.0059–0.0130
	241.3 (3956)	①	0.0008–0.0017	0.0006–0.0080	3	2.0861–2.0866	0.0008–0.0020	0.0063–0.0118
1991–92	144.4 (2366)	2.3616–2.3622	0.0010–0.0022	0.0008–0.0087	3	2.0861–2.0866	0.0010–0.0022	0.0008–0.0087
	146.4 (2399)	2.3617–2.3622	0.0008–0.0019	0.0008–0.0087	3	2.0861–2.0866	0.0012–0.0023	0.0063–0.0123
	180.5 (2959)	2.5195–2.5197	0.0009–0.0017	0.0008–0.0098	3	2.1648–2.1654	0.0009–0.0021	0.0059–0.0130
	241.3 (3956)	①	0.0008–0.0017	0.0006–0.0080	3	2.0861–2.0866	0.0008–0.0020	0.0063–0.0118

① No. 1—2.6367–2.6376
No. 2—2.6957–2.6967
No. 3—2.7548–2.7557
No. 4—2.8139–2.8148

VALVE SPECIFICATIONS

Year	Engine Displacement cu. in. (cc)	Seat Angle (deg.)	Face Angle (deg.)	Spring Test Pressure (lbs.)	Spring Installed Height (in.)	Stem-to-Guide Clearance (in.) Intake	Stem-to-Guide Clearance (in.) Exhaust	Stem Diameter (in.) Intake	Stem Diameter (in.) Exhaust
1988	144.4 (2366)	45①	44.5	66	1.909②	0.0010–0.0024	0.0012–0.0026	0.3138–0.3144	0.3136–0.3142
	136.5 (2237)	45①	44.5	64–77	1.850②	0.0010–0.0024	0.0012–0.0026	0.3138–0.3144	0.3136–0.3142
	180.5 (2959)	45①	44.5	57	1.594	0.0010–0.0024	0.0012–0.0026	0.3138–0.3144	0.3136–0.3142
	241.3 (3956)	45	44.5	71.6	1.693	0.0012–0.0024	0.0016–0.0028	0.3140	0.3137
1989	144.4 (2366)	45①	44.5	66	1.909②	0.0010–0.0024	0.0012–0.0026	0.3138–0.3144	0.3136–0.3142
	136.5 (2237)	45①	44.5	64–77	1.850②	0.0010–0.0024	0.0012–0.0026	0.3138–0.3144	0.3136–0.3142
	180.5 (2959)	45①	44.5	57	1.594	0.0010–0.0024	0.0012–0.0026	0.3138–0.3144	0.3136–0.3142
	241.3 (3956)	45	44.5	71.6	1.693	0.0012–0.0024	0.0016–0.0028	0.3140	0.3137

VALVE SPECIFICATIONS

Year	Engine Displacement cu. in. (cc)	Seat Angle (deg.)	Face Angle (deg.)	Spring Test Pressure (lbs.)	Spring Installed Height (in.)	Stem-to-Guide Clearance (in.)		Stem Diameter (in.)	
						Intake	Exhaust	Intake	Exhaust
1990	144.4 (2366)	45①	44.5	63–66	1.909②	0.0010–0.0024	0.0012–0.0026	0.3138–0.3144	0.3136–0.3142
	180.5 (2959)	45①	44.5	54–57	1.850②	0.0010–0.0024	0.0012–0.0026	0.3138–0.3144	0.3136–0.3142
	241.3 (3956)	45③	45③	60–72	2.028②	0.0010–0.0024	0.0014–0.0028	0.3138–0.3144	0.3134–0.3140
1991–92	144.4 (2366)	45①	44.5	63–66	1.909②	0.0010–0.0024	0.0012–0.0026	0.3138–0.3144	0.3136–0.3142
	146.4 (2399)	45①	44.5	57–63	1.406②	0.0010–0.0024	0.0012–0.0026	0.2350–0.2356	0.2348–0.2354
	180.5 (2959)	45①	44.5	54–57	1.850②	0.0010–0.0024	0.0012–0.0026	0.3138–0.3144	0.3136–0.3142
	241.3 (3956)	45③	45③	60–72	2.028②	0.0010–0.0024	0.0014–0.0028	0.3138–0.3144	0.3134–0.3140

① Blend the seat with 30° and 60° cutters to center the 45° portion on the valve face.
② Free length
③ Intake:
Blend the seat with 25° and 70° cutters to center the 45° portion of the valve face.
Exhaust:
Blend the seat with 60° cutters to center the 45° portion of the valve face.

PISTON AND RING SPECIFICATIONS

All measurements are given in inches.

Year	Engine Displacement cu. in. (cc)	Piston Clearance	Ring Gap			Ring Side Clearance		
			Top Compression	Bottom Compression	Oil Control	Top Compression	Bottom Compression	Oil Control
1988	144.4 (2366)	0.0008–0.0016①	0.0098–0.0421	0.0236–0.0559	0.0079–0.0461	0.0012–0.0080	0.0012–0.0080	0.0012–0.0080
	136.5 (2237)	0.0026–0.0033	0.0091–0.0189	0.0063–0.0173	0.0051–0.0185	0.0012–0.0028	0.0012–0.0028	0.0012–0.0028
	180.5 (2959)	0.0031–0.0039	0.0091–0.0327	0.0150–0.0366	0.0059–0.0354	0.0012–0.0028	0.0012–0.0028	0.0012–0.0028
	241.3 (3956)	0.0011–0.0019	0.0079–0.0165	0.0197–0.0283	0.0079–0.0323	0.0012–0.0028	0.0020–0.0035	snug
1989	144.4 (2366)	0.0008–0.0016①	0.0098–0.0421	0.0236–0.0559	0.0079–0.0461	0.0012–0.0080	0.0012–0.0080	0.0012–0.0080
	136.5 (2237)	0.0026–0.0033	0.0091–0.0189	0.0063–0.0173	0.0051–0.0185	0.0012–0.0028	0.0012–0.0028	0.0012–0.0028
	180.5 (2959)	0.0031–0.0039	0.0090–0.0327	0.0150–0.0366	0.0060–0.0354	0.0012–0.0028	0.0012–0.0028	0.0012–0.0028
	241.3 (3956)	0.0011–0.0019	0.0079–0.0165	0.0197–0.0283	0.0079–0.0323	0.0012–0.0028	0.0020–0.0035	snug
1990	144.4 (2366)	0.0006–0.0014	0.0098–0.0421	0.0236–0.0559	0.0079–0.0461	0.0012–0.0080	0.0012–0.0080	0.0012–0.0080
	180.5 (2959)	0.0031–0.0039	0.0091–0.0327	0.0150–0.0366	0.0059–0.0354	0.0012–0.0028	0.0012–0.0028	0.0012–0.0028
	241.3 (3956)	0.0011–0.0019	0.0079–0.0165	0.0197–0.0283	0.0079–0.0323	0.0012–0.0028	0.0020–0.0035	snug

PISTON AND RING SPECIFICATIONS

All measurements are given in inches.

Year	Engine Displacement cu. in. (cc)	Piston Clearance	Ring Gap			Ring Side Clearance		
			Top Compression	Bottom Compression	Oil Control	Top Compression	Bottom Compression	Oil Control
1991–92	144.4 (2366)	0.0006–0.0014	0.0098–0.0421	0.0236–0.0559	0.0079–0.0461	0.0012–0.0080	0.0012–0.0080	0.0012–0.0080
	146.4 (2399)	0.0012–0.0020	0.0118–0.0406	0.0177–0.0472	0.0051–0.0386	0.0008–0.0028	0.0012–0.0028	snug
	180.5 (2959)	0.0031–0.0039	0.0091–0.0327	0.0150–0.0366	0.0059–0.0354	0.0012–0.0028	0.0012–0.0028	0.0012–0.0028
	241.3 (3956)	0.0011–0.0019	0.0079–0.0402	0.0197–0.0520	0.0079–0.0559	0.0012–0.0028	0.0020–0.0035	snug

① 22R-TE Engine—0.0022–0.0030 ② Maximum

TORQUE SPECIFICATIONS

All readings in ft. lbs.

Year	Engine Displacement cu. in. (cc)	Cylinder Head Bolts	Main Bearing Bolts	Rod Bearing Bolts	Crankshaft Pulley Bolts	Flywheel Bolts	Manifold	
							Intake	Exhaust
1988	144.4 (2366)	53–63	69–83	40–47	102–130	73–86	13–19	29–36
	136.5 (2237)	②	58	36	116	61①	36	36
	180.5 (2959)	③	⑪	⑩	181	65①	13	29
	241.3 (3956)	87–93	④	40–46	247–259	60–68	⑤	⑤
1989	144.4 (2366)	53–63	69–83	40–47	102–130	73–86	13–19	29–36
	136.5 (2237)	②	58	36	116	61①	36	36
	180.5 (2959)	③	43–47	16–20	176–186	63–67	11–15	25–33
	241.3 (3956)	87–93	④	40–46	247–259	60–68	⑤	⑤
1990	144.4 (2366)	58⑨	76	51	116	80①	14	33
	180.5 (2959)	③	⑪	⑩	181	65①	13	29
	241.3 (3956)	90⑫	④	43	253	64	⑤	⑤
1991–92	144.4 (2366)	58⑨	76	51	116	80①	14	33
	146.4 (2399)	⑥	⑨	⑦	192	65①	15	36
	180.5 (2959)	③	⑪	⑩	181	65①	13	29
	241.3 (3956)	90⑫	④	43	253	64	⑤	⑤

① Drive plate—54 ft. lbs.
② 12 mm bolt—14 ft. lbs.
   14 mm bolt—65 ft. lbs.
③ Step 1—27
   Step 2—33
   Step 3—90 degree turn
   Step 4—90 degree turn
④ 19 mm bolt—99 ft. lbs.

   17 mm bolt—85 ft. lbs.
⑤ 14 mm bolt—37 ft. lbs.
   17 mm bolt—51 ft. lbs.
   Nut—41 ft. lbs.
⑥ Step 1—29 ft. lbs.
   Step 2—90 degree turn
   Step 3—90 degree turn
⑦ Step 1—22 ft. lbs.

⑧ Step 1—29 ft. lbs.
   Step 2—90 degree turn
⑨ Step 1—20 ft. lbs.
   Step 2—35 ft. lbs.
   Step 3—58 ft. lbs.
⑩ Step 1—18 ft. lbs.
   Step 2—90 degree turn

⑪ Step 1—45 ft. lbs.
   Step 2—90 degree turn
⑫ Torque in 3 steps

BRAKE SPECIFICATIONS

All measurements in inches unless noted.

Year	Model	Lug Nut Torque (ft. lbs.)	Master Cylinder Bore	Brake Disc		Standard Brake Drum Diameter	Minimum Lining Thickness	
				Minimum Thickness	Maximum Runout		Front	Rear
1988	Pick-Up	65–86	NA	①	0.0059	②	0.039	0.039
	4Runner	65–86	NA	①	0.0059	②	0.039	0.039
	Van	65–86	NA	0.748	0.0059	10.00	0.039	0.039
	Land Cruiser	65–86	NA	0.748	0.0059	11.61	0.040	0.059

BRAKE SPECIFICATIONS
All measurements in inches unless noted.

Year	Model	Lug Nut Torque (ft. lbs.)	Master Cylinder Bore	Brake Disc		Standard Brake Drum Diameter	Minimum Lining Thickness	
				Minimum Thickness	Maximum Runout		Front	Rear
1989	Pick-Up	65–86	NA	①	0.0059	②	0.039	0.039
	4Runner	65–86	NA	①	0.0059	②	0.039	0.039
	Van	65–86	NA	0.748	0.0059	10.00	0.039	0.039
	Land Cruiser	65–86	NA	0.748	0.0059	11.61	0.040	0.059
1990	Pick-Up	65–86	NA	⑤	⑥	⑦	0.039	0.039
	4Runner	65–86	NA	0.709	0.0035	11.61	0.039	0.039
	Land Cruiser	65–86	NA	0.748	0.0059	11.61	0.039	0.059
1991–92	Pick-Up	65–86	NA	⑤	⑥	⑦	0.039	0.039
	4Runner	65–86	NA	0.709	0.0035	11.61	0.039	0.039
	Land Cruiser	65–86	NA	0.906	0.0059	11.61	0.039	0.059
	Previa	65–86	NA	0.827③	0.0028④	10.00	0.039	0.039

NA—Not available
① 2WD with PD60 Brake Disc—0.945
   2WD with FS17 Brake Disc—0.827
   4WD—0.748
② 2WD—10.00
   4WD—11.64

③ Rear disc—0.669
④ Rear disc—0.0039
⑤ PD60—0.907
   PD66—1.102
   FS17, 18—0.787
   S12 & 12—0.709

⑥ PD60—0.0035
   PD66—0.0047
   FS17, 18—0.0035
   S12 & 12—0.0035
⑦ 2WD—10.00
   4WD—11.70

WHEEL ALIGNMENT

Year	Model	Caster		Camber		Toe-in (in.)	Steering Axis Inclination (deg.)
		Range (deg.)	Preferred Setting (deg.)	Range (deg.)	Preferred Setting (deg.)		
1988	Pick-Up 2WD① :						
	Extra long bed	½P–1½P	1P	0–1P	½P	⅛–3/16	10
	Cab & Chassis	1/16P–1 1/16P	9/16P	0–1P	½P	⅛–3/16	10
	Dual rear wheels	1/16P–1 1/16P	9/16P	0–1P	½P	⅛–3/16	10
	Extra long cab	5/16P–1 5/16P	13/16P	0–1P	½P	⅛–3/16	10
	1 ton	1/16P–1 1/16P	9/16P	0–1P	½P	⅛–3/16	10
	Pick-Up 4WD	13/16P–1 13/16P	1 5/16P	⅛P–1 ⅛P	⅝P	5/64–5/32	11 5/16
	4Runner	1 ⅝P–2 ⅝P	2 ⅛P	⅛P–1 ⅛P	⅝P	5/64–5/32	11 15/16
	Van 2WD	2P–3P	2½P	9/16N–7/16P	1/16N	3/64–3/64	10 9/16
	Van 4WD	1 11/16P–2 9/16P	2 13/16P	5/16N–11/16P	3/16P	3/64–3/64	12 7/16
	Land Cruiser	3/16N–1 13/16P	13/16P	¼P–1 ¾P	1P	0–5/64	9½
1989	Pick-Up 2WD① :						
	RN80LTRKR, TRTR	3/16P–1 3/16P	11/16P	1/16N–15/16P	7/16P	3/64–7/64	10
	RN80LTRLD, TRMR	¼P–1 ¼P	¾P	0–1P	½P	1/32–3/32	10
	RN85LTRLD, TRMD	½P–1½P	1P	1/16N–15/16P	7/16P	1/64–⅛	10 1/32
	RN90LCRLD, CRMD	⅝P–1 ⅝P	1 ⅛P	⅛N–7/8P	⅜P	⅛–3/16	10 ⅛
	RN80LTRMD, TRSD	¼P–1 ¼P	¾P	0–1P	½P	1/32–3/32	10
	RN85LTRMD, TRSD	½P–1½P	1P	1/16N–15/16P	7/16P	1/16–⅛	10 1/32
	RN85LTRMS, TRSS	11/16P–1 11/16P	1 3/16P	⅛N–7/8P	⅜P	⅛–3/16	10 ⅛
	RN90LCRMD, CRDS	10/16P–1 11/16P	1 3/16P	⅛N–7/8P	⅜P	⅛–3/16	10 ⅛
	RN90LCRMS, CRSS	7/8P–1 7/8P	1 ⅜P	5/16N–11/16P	3/16P	7/32–9/32	10 5/16
	VZ85LTRMD, TRSD	⅝P–1 ⅝P	1 ⅛P	1/16N–15/16P	7/16P	1/16–⅛	10

WHEEL ALIGNMENT

Year	Model	Caster Range (deg.)	Caster Preferred Setting (deg.)	Camber Range (deg.)	Camber Preferred Setting (deg.)	Toe-in (in.)	Steering Axis Inclination (deg.)
1989	VZ85LTHMD, THSD	1/16P–1 1/16P	9/16P	1/16P–1 1/16P	9/16P	3/16–1/4	9 29/32
	VZ85LTRMR, TRSR	7/16N–9/16P	1/16P	1/16P–1 1/16P	9/16P	11/64–15/64	9 15/16
	VZN90LCRMD, CRSD	13/16P–1 13/16P	1 5/16P	1/8N–7/8P	3/8P	3/32–5/32	10 3/32
	VZN90LCRMG, CRPG	15/16P–1 15/16P	1 7/16P	3/16N–13/16P	5/16P	5/32–7/32	10 3/16
	VZN95LTWMR, TWSR	3/8N–5/8P	1/8P	1/16P–1 1/16P	9/16P	5/32–7/32	9 15/16
	Pick-Up 4WD	2P–3P	2 1/2P	1/4P–1 1/4P	3/4P	0–1/16	11 13/16
	4Runner	1 5/8P–2 5/8P	2 1/8P	1/8P–1 1/8P	5/8P	5/64–5/32	11 15/16
	Van 2WD	2P–3P	2 1/2P	9/16N–7/16P	1/16N	3/64–3/64	10 9/16
	Van 4WD	1 11/16P–2 9/16P	2 13/16P	5/16N–11/16P	3/16P	3/64–3/64	12 7/16
	Land Cruiser	1/16P–1 9/16P	13/16P	1/4P–1 3/4P	1P	3/64–9/64	9 1/2
1990	Pick-Up 2WD①:						
	RN80LTRTR	1/16N–15/16P	7/16P	0–1P	1/2P	1/32–3/32	10
	RN80LTRMR	1/8P–1 1/8P	5/8P	0–1P	1/2P	1/32–3/32	10
	RN85LTRLD	3/16P–1 3/16P	11/16P	0–1P	1/2P	1/64–5/64	10
	RN85LTRMD, TRLD	7/16P–1 7/16P	15/16P	1/16N–15/16P	7/16P	1/16–1/8	10
	RN90LCRLDL	3/4P–1 3/4P	1 1/4P	1/8N–7/8P	3/8P	3/32–5/32	10 3/32
	RN80LTRKR	3/16P–1 3/16P	11/16P	0–1P	1/2P	1/32–3/32	10
	RN80LTRSD, TRMD	3/16P–1 3/16P	11/16P	0–1P	1/2P	1/64–5/64	10
	RN85LTRMD, TRSD	7/16P–1 7/16P	15/16P	1/16N–15/16P	7/16P	1/16–1/8	10
	RN85LTRMS, TRSS	5/8P–1 5/8P	1 1/8P	1/8N–7/8P	3/8P	1/4–3/8	10 1/8
	RN90LCRMD, CRSD	3/4P–1 3/4P	1 1/4P	1/8N–7/8P	3/8P	3/32–5/32	10 3/32
	RN90LCRMJ, CRSS	7/8P–1 7/8P	1 3/8P	5/16N–11/16P	3/16P	3/16–1/4	10 1/4
	VZN85LTRMD, TRSD	3/8P–1 3/8P	7/8P	1/16N–15/16P	7/16P	1/16–1/8	10
	VZN85LTHMD, THSD	1/16P–1 1/16P	9/16P	0–1P	1/2P	5/32–7/32	10
	VZN85LTRMR, TRSR	7/16N–9/16P	1/16P	0–1P	1/2P	5/32–7/32	10
	VZN85LTWMR, TWSR	3/4P–1 3/4P	1 1/4P	0–1P	1/2P	3/16–1/4	10
	VZN90LCRMD, CRSD	11/16P–1 11/16P	1 3/16P	1/16N–15/16P	7/16P	3/32–5/32	10 1/16
	VZN90LCRMG, CRPG	11/16P–1 11/16P	1 3/16P	1/16N–15/16P	7/16P	1/16–1/8	10 1/16
	VZN95LTWMR, TWSR	1/2N–1/2P	0	0–1P	1/2P	3/16–1/4	10
	Pick-Up 4WD	2P–3P	2 1/2P	1/4P–1 1/4P	3/4P	0–1/16	11 13/16
	4Runner	2P–3P	2 1/2P	1/4P–1 1/4P	3/4P	1/32–3/32	11 13/16
	Land Cruiser	1/16P–1 9/16P	13/16P	1/4P–1 3/4P	1P	3/64–9/64	9 1/2
1991–92	Pick-Up 2WD①:						
	RN80LTRSD, TRMD	0–1 1/2P	3/4P	1/4N–1 1/4P	1/2P	1/32–5/32	10
	RN80LTRMR	1/16N–1 7/16P	11/16P	5/16N–1 3/16P	7/16P	1/32–5/32	10
	RN85LTRMD, TRSD	7/32P–1 23/32P	31/32P	5/16N–1 3/16P	7/16P	0–3/16	10 1/32
	RN90LCRSD, CRMD	1/2P–2P	1 1/4P	3/8N–1 1/8P	3/8P	1/32–7/32	10 3/32
	VZN85LTHMD	3/16N–15/16P	9/16P	1/4N–1 1/4P	1/2P	1/8–5/16	10
	VZN85LTHSD	3/16N–15/16P	9/16P	1/4N–1 1/4P	1/2P	3/32–9/32	10
	VZN85LTWMR, TWSR	1P–2 1/2P	1 3/4P	1/4N–1 1/4P	1/2P	3/32–9/32	10
	VZN90LCRMD, CRSD	7/16P–1 15/16P	1 3/16P	3/8N–1 1/8P	3/8P	1/32–7/32	10 3/32
	VZN90LCRMG, CRPG	7/16P–1 15/16P	1 3/16P	5/16N–1 3/16P	7/16P	1/32–7/32	10 1/16
	VZN95LTWMR, TWSR	1P–2 1/2P	1 3/4P	1/4N–1 1/4P	1/2P	1/8–5/16	10

WHEEL ALIGNMENT

Year	Model	Caster Range (deg.)	Caster Preferred Setting (deg.)	Camber Range (deg.)	Camber Preferred Setting (deg.)	Toe-in (in.)	Steering Axis Inclination (deg.)
1991–92	Pick-Up 4WD	2P–3P	2½P	¼P–1¼P	¾P	0–1/16	11 13/16
	4Runner	2P–3P	2½P	¼P–1¼P	¾P	1/32–3/32	11 13/16
	Land Cruiser with 10.5R tire	11/16P–2 11/16P	1 11/16P	¼P–1¾P	1P	0–3/16	13
	W/O 10.5R tire	2P–4P	3P	¼P–1¾P	1P	0–3/16	13
	Previa 2WD	4¾P–6¼P	5½P	21/32N–27/32P	3/32P	0–3/16	10 19/32
	Previa 4WD	4 19/32P–6 3/32P	5 11/32P	½N–1P	¼P	1/32–7/32	10 11/32

N—Negative
P—Positive
① **NOTE:** Front end alignment specifications are given according to Vehicle Identification Number. Before using this information, verify that the Vehicle Identification Number is correct for the alignment data that is used.

ENGINE MECHANICAL

NOTE: Disconnecting the negative battery cable on some vehicles may interfere with the functions of the on board computer systems and may require the computer to undergo a relearning process, one the negative battery cable is reconnected.

Engine Assembly

REMOVAL & INSTALLATION

22R, 22R–E and 22R–TE Engines

1. Disconnect the negative battery cable.
2. Remove the engine undercover.
3. Disconnect the windshield washer hose and then remove the hood. Scribe matchmarks around the hinges for easy installation.
4. Drain the engine oil. Drain the engine coolant from the radiator and the cylinder block.
5. Drain the automatic transmission fluid on models so equipped.
6. Disconnect the air cleaner hose and then remove the air cleaner.
7. Remove the radiator and shroud. Disconnect the No. 1 turbocharger water line on the 22R–TE engine.
8. Remove the coupling fan.

9. Disconnect the heater hoses at the engine.
10. If equipped with automatic transmissions, disconnect the accelerator and throttle cables at their bracket.
11. Disconnect the following:
    a. No. 1 and No. 2 PCV hoses
    b. Brake booster hose
    c. Air control valve hoses
    d. EVAP hose at the canister
    e. Actuator hose on vehicles with cruise control
    f. Vacuum modulator hose at the EGR valve
    g. Air valve hoses at the throttle body and chamber
    h. Two water bypass hoses at the throttle body
    i. Air control valve hose at the actuator
    j. Pressure regulator hose at the chamber
    k. Cold start injector pipe
    l. BVSV hose.
12. Tag and disconnect the cold start injector wire and the throttle position sensor wire.
13. Remove the EGR valve from the throttle chamber.
14. Disconnect the throttle chamber at the stay. Remove the chamber-to-intake manifold mounting bolts and lift off the throttle chamber.
15. Tag and disconnect the following wires:
    a. Cold start injector time switch wire
    b. Water temperature sensor wire
    c. If equipped with air condition-

ing: VSV and air conditioning compressor wires
    d. Oxygen sensor wire for 22R–TE engine only
    e. OD temperature switch wire (with automatic transmission)
    f. Injector wires
    g. Knock sensor connector
    h. Air valve wire
    i. Oil pressure switch wire
    j. Starter wire.
16. Remove the power steering pump from its bracket, if equipped. Disconnect the ground strap from the bracket.
17. If equipped with air conditioning, loosen the drive belt and remove the air conditioning compressor. Position it aside with the refrigerant lines attached.
18. Disconnect the engine ground straps at the rear and right side of the engine.
19. If equipped with a manual transmission, remove the shift lever from inside the vehicles.
20. Raise and safely support the vehicle. Drain the engine oil. Remove the rear driveshaft.
21. If equipped with automatic transmission, disconnect the manual shift linkage at the neutral start switch. On 4WD vehicles with automatic transmission, disconnect the transfer shift linkage.
22. Disconnect the speedometer cable. Be sure not to lose the felt dust protector and washers.
23. Remove the transfer case undercover on 4WD vehicles.

24. Remove the stabilizer bar on 4WD vehicles.
25. Remove the front driveshaft on 4WD vehicles.
26. Remove the No. 1 frame crossmember.
27. Disconnect the front exhaust pipe at the manifold and tail pipe and remove the exhaust pipe.
28. If equipped with manual transmission, remove the clutch release cylinder and its bracket from the transmission.
29. Remove the No. 1 front floor heat insulator and the brake tube heat insulator on 4WD vehicles.
30. On 2WD vehicles, remove the rear engine mount bolts, raise the transmission slightly with a floor jack and then remove the support member mounting bolts.
31. On 4WD vehicles, remove the 4 rear engine mount bolts, raise the transmission slightly with a floor jack and then remove the bolts from the side member and remove the No. 2 frame crossmember.
32. Lower the vehicle. Attach an engine hoist chain to the lifting brackets on the engine. Remove the engine mount nuts and bolts and slowly lift the engine/transmission out of the truck.

To install:
33. Slowly lower the engine assembly into the engine compartment.
34. Raise the transmission onto the crossmember with a floor jack.
35. Align the holes in the engine mounts and the frame, install the bolts and then remove the engine hoist chain.
36. On 2WD vehicles, raise the transmission slightly and align the rear engine mount with the support member and tighten the bolts to 9 ft. lbs. (13 Nm). Lower the transmission until it rests on the extension housing and then tighten the bracket mounting bolts to 19 ft. lbs. (25 Nm).
37. On 4WD vehicles, raise the transmission slightly and tighten the No. 2 frame crossmember-to-side frame bolts to 70 ft. lbs. (95 Nm). Lower the transmission and tighten the rear engine mount bolts to 9 ft. lbs. (13 Nm).
38. On 4WD vehicles, install the brake tube and front floor heat insulators.
39. Install the clutch release cylinder and its bracket to the manual transmission. Tighten the bracket bolts to 29 ft. lbs. (39 Nm) and the cylinder bolts to 9 ft. lbs. (13 Nm).
40. Reconnect the exhaust pipe. Install the No. 1 frame crossmember.
41. On 4WD vehicles, install the front driveshaft, stabilizer bar and the transfer case undercover.
42. Connect the speedometer cable. Connect the transfer shift linkage on

4WD vehicles with automatic transmission.
43. Connect the manual shift linkage to the neutral start switch (automatic transmission only).
44. Install the rear driveshaft. Install the shift lever for manual transmission only.
45. Connect the engine ground straps. Install the air conditioning compressor.
47. Install the power steering pump and connect the ground strap.
48. Connect all of the following wires:
    a. Cold start injector time switch wire
    b. Water temperature sensor wire
    c. If equipped with air conditioning: VSV and air conditioning compressor wires
    d. Oxygen sensor wire for 22R–TE engine only
    e. OD temperature switch wire for automatic transmission
    f. Injector wires
    g. Knock sensor connector
    h. Air valve wire
    i. Oil pressure switch wire
    j. Starter wire
49. Connect all of the following parts:
    a. No. 1 and No. 2 PCV hoses
    b. Brake booster hose
    c. Air control valve hoses
    d. EVAP hose at the canister
    e. Actuator hose on vehicles with cruise control
    f. Vacuum modulator hose at the EGR valve
    g. Air valve hoses at the throttle body and chamber
    h. Two water bypass hoses at the throttle body
    i. Air control valve hose at the actuator
    j. Pressure regulator hose at the chamber
    k. Cold start injector pipe
    l. BVSV hose.
50. Connect the accelerator and throttle cables to the bracket for automatic transmission only.
51. Connect the heater hoses and install the coupling fan. Install the radiator and shroud.
52. Install the air cleaner. Refill the engine with oil and the radiator with coolant. Install the engine undercover.
53. Install and adjust the hood.
54. Connect the battery cable, start the engine and road test the vehicle.

4Y–EC Engine

1. Disconnect the negative battery terminal.
2. Remove the right seat and the engine service hole cover.
3. Drain the coolant from the radiator. Remove the reservoir tank, the heater hoses and the radiator.

4. Remove the air cleaner, the breather tube, the brake booster, the charcoal canister and the fuel hoses from the engine.
5. If equipped with power steering, remove the drive belt, the pulley, the Woodruff key and the pump from the engine, then, move the pump aside.

NOTE: When removing the power steering pump, do not disconnect the pressure lines unless it is absolutely necessary.

6. Disconnect the accelerator cable with the bracket from the throttle body.
7. Disconnect the following wiring connectors from the: water temperature sender, oil pressure switch, IIA unit, air conditioning compressor, idle-up solenoid, VSV, water temperature switch (automatic transmission), alternator connector and wire, air flow meter and solenoid resistor.
8. Remove the fan shroud, the fan, the fluid coupling and the water pump pulley.
9. From inside the vehicle, remove the center pillar cover, the seat belt retractor and cover, then, disconnect the electrical connectors from the ECU.
10. If equipped with air conditioning, remove the drive belt and the compressor mounting bolts, then, move the compressor aside.
11. Raise and safely support the vehicle.
12. Drain the engine oil. Remove the driveshaft and the front exhaust pipe.
13. Remove the transmission selector and shift cables, then, the clutch release cylinder for manual transmission.
14. Disconnect the starter wires, the mounting bolts and the starter from the engine.
15. Remove the speedometer cable, the bond cable and the back-up light switch connector.
16. If equipped with a rear heater, disconnect the mode selector and the air mix damper cable from the damper. Disconnect the heater hoses to the rear heater unit.
17. Disconnect the bond cable(s) from the engine mount(s). Remove the engine under cover.
18. Disconnect the oil level sensor and the oil cooler hoses, if equipped with automatic transmission.
19. Place matchmarks on the front strut bar and the rear mounting nut. Remove the rear nut, the strut bar-to-lower control arm bolts and the strut.
20. Using an engine saddle, place it under the engine and support it. Place a floor jack under the transmission and support it.

NOTE: If equipped with a manual transmission, remove the en-

gine rear mounting bracket from the body. If equipped with automatic transmission, remove the engine mounting member-to-transmission through bolt.

21. Remove the engine mounts-to-body nuts/bolts and lower the engine/transmission assembly, then, remove the engine mounting member from the engine.

22. Remove the transmission from the engine.

To install:

23. Install the engine to the transmission.

24. Install the engine mounts-to-body nuts/bolts and lower the engine/transmission assembly.

25. Remove the engine support and floor jack.

26. Install the front strut bar and the rear mounting nut. Install the rear nut, the strut bar-to-lower control arm bolts and the strut.

27. Connect the oil level sensor and the oil cooler hoses, if equipped with automatic transmission.

28. Connect the bond cable(s) to the engine mount(s). Install the engine under cover.

29. If equipped with a rear heater, connect the mode selector and the air mix damper cable to the damper. Connect the heater hoses to the rear heater unit.

30. Install the speedometer cable, the bond cable and the back-up light switch connector.

31. Connect the starter wires, the mounting bolts and the starter to the engine.

32. Install the transmission selector and shift cables, then, the clutch release cylinder for manual transmission.

33. Refill the engine oil. Install the driveshaft and the front exhaust pipe.

34. Lower the vehicle.

35. If equipped with air conditioning, install the drive belt and the compressor mounting bolts.

36. From inside the vehicle, install the center pillar cover, the seat belt retractor and cover.

37. Install the fan shroud, the fan, the fluid coupling and the water pump pulley.

38. Connect the following wiring connectors to the: water temperature sender, oil pressure switch, IIA unit, air conditioning compressor, idle-up solenoid, VSV, water temperature switch for automatic transmission, alternator connector and wire, air flow meter and solenoid resistor.

39. Connect the accelerator cable with the bracket to the throttle body.

40. If equipped with power steering, install the drive belt, the pulley, the woodruff key and the pump to the engine.

41. Install the air cleaner, the breather tube, the brake booster, the charcoal canister and the fuel hoses to the engine.

42. Refill the coolant and install the reservoir tank, the heater hoses and the radiator.

43. Install the right seat and the engine service hole cover.

44. Connect the negative battery terminal to the battery and check for leaks.

3F–E Engine

1. Disconnect the negative battery cable. Drain the engine coolant.

2. Scribe matchmarks around the hood hinges and then remove the hood.

3. Remove the battery and its tray.

4. Disconnect the accelerator and throttle cables.

5. Remove the air intake hose, air flow meter and air cleaner assembly.

6. Remove the coolant reservoir tank.

7. Remove the radiator.

8. Tag and disconnect the following wires and connectors:
   a. Oil pressure connector
   b. High tension cord at the coil
   c. Neutral start switch and transfer connectors near the starter
   d. Front differential lock connector
   e. Starter wire and connector
   f. Starter ground strap
   g. O_2 sensor connectors
   h. Alternator wire and connector
   i. Cooling fan connector
   j. Check connector.

9. Disconnect the following hoses:
   a. Heater hoses
   b. Fuel hoses
   c. Transfer case hose
   d. Brake booster hose
   e. Air injection hoses
   f. Distributor hose
   g. Emission control hoses

10. Remove the glove box, pull out the 4 connectors and then pull the EFI wiring harness from the cowl.

11. Unbolt the power steering pump and position it aside with the hoses still connected.

12. Do the same with the air conditioning compressor.

13. Raise the vehicle and remove the transfer case undercover. Drain the engine oil.

14. Remove the front and rear driveshafts.

15. Disconnect the speedometer cable.

16. Disconnect the engine ground strap.

17. Disconnect the 2 vacuum hoses at the diaphragm cylinder under the transfer case.

18. Remove the clip and pin and then disconnect the shift rod at the transfer case. Remove the nut, disconnect the washers and the shift lever at the shift rod.

19. Disconnect the transmission control rod.

20. Disconnect the exhaust pipe at the manifold.

21. With a floor jack under the transmission, remove the bolts and nuts that attach the frame crossmember and then remove the crossmember. Lower the vehicle.

22. Attach an engine hoist to the lifting brackets on the engine. Remove the engine mount nuts and bolts and slowly lift the engine/transmission out of the vehicle.

To install:

23. Slowly lower the engine assembly into the engine compartment.

24. Raise the transmission onto the crossmember with a floor jack.

25. Align the holes in the engine mounts and the frame, install the bolts and then remove the engine hoist chain.

26. Raise the transmission slightly and tighten the frame crossmember-to-chassis bolts to 29 ft. lbs. (39 Nm). Tighten the 2 nuts to 43 ft. lbs. (59 Nm).

27. Install the exhaust pipe with a new gasket and tighten the nuts to 46 ft. lbs. (62 Nm).

28. Connect the transmission control rod. Connect the transfer case shift lever.

29. Connect the engine ground strap. Connect the speedometer cable.

30. Install the front and rear driveshafts. Tighten the nuts to 65 ft. lbs. (88 Nm).

31. Install the transfer case undercover. Install the air conditioning compressor. Install the power steering pump and tighten the pulley nut to 35 ft. lbs. (47 Nm).

32. Connect the EFI wiring harness at the ECU. Connect all of the following hoses:
   a. Heater hoses
   b. Fuel hoses
   c. Transfer case hose
   d. Brake booster hose
   e. Air injection hoses
   f. Distributor hose
   g. Emission control hoses.

33. Connect all of the following the wires and connectors:
   a. Oil pressure connector
   b. High tension cord at the coil
   c. Neutral start switch and transfer connectors near the starter
   d. Front differential lock connector
   e. Starter wire and connector
   f. Starter ground strap
   g. Oxygen sensor connectors
   h. Alternator wire and connector
   i. Cooling fan connector
   j. Check connector

34. Install the radiator and the coolant reservoir tank.

35. Install the air intake hose, air flow meter and air cleaner. Connect the accelerator and throttle cables.

36. Refill the engine with oil and the radiator with coolant. Install the engine undercover. Install the battery.

37. Install and adjust the hood.

38. Install the battery, start the engine and road test the vehicle.

3VZ–E Engine

1. Disconnect the battery cables and remove the battery.

2. Remove the engine undercover.

3. Disconnect the windshield washer hose and then remove the hood. Scribe matchmarks around the hinges for easy installation.

4. Drain the engine coolant from the radiator and the cylinder block.

5. Raise and safely support the vehicle. Drain the engine oil. Drain the automatic transmission fluid, if equipped.

6. Lower the vehicle. Disconnect the air cleaner hose and then remove the air cleaner.

7. Remove the radiator.

8. Remove all drive belts and then remove the fluid coupling and fan pulley.

9. Tag and disconnect the following wires and connectors:
   a. Left side and rear ground straps
   b. Alternator connector and wire
   c. Igniter connector
   d. Oil pressure switch connector
   e. ECU connectors
   f. VSV connectors
   g. Starter relay connector for manual transmission only
   h. Solenoid resistor connector
   i. Check connector
   j. Air conditioning compressor connector

10. Tag and disconnect the following hoses:
   a. Power steering hoses at the gas filter and air pipe
   b. Brake booster hose
   c. Cruise control vacuum hose, if equipped
   d. Charcoal canister hose at the canister
   e. VSV vacuum hoses.

11. Disconnect the accelerator, throttle and cruise control cables where applicable.

12. Unbolt the power steering pump and position it aside with the hydraulic lines still connected.

13. Properly discharge the air conditioning system. Remove the air conditioning compressor if equipped.

14. Disconnect the clutch release cylinder hose for manual transmission only.

15. Disconnect the 2 heater hoses.

16. Disconnect and plug the fuel inlet and outlet lines.

17. Remove the shift levers for manual transmission only.

18. Raise and safely support the vehicle. Remove the rear driveshaft.

19. Disconnect the manual shift linkage for automatic transmission only.

20. Disconnect the speedometer cable, don't lose the felt dust protector and washers.

21. Remove the transfer case undercover. Remove the stabilizer bar.

22. Remove the front driveshaft. Remove the front exhaust pipe.

23. Remove the No. 1 front floor heat insulator and the brake tube heat insulator.

24. Remove the rear engine mount bolts, raise the transmission slightly with a floor jack and then remove the 4 bolts from the side member and remove the No. 2 frame crossmember. Lower the vehicle.

25. Attach an engine hoist chain to the lifting brackets on the engine. Remove the engine mount nuts and bolts and slowly lift the engine/transmission out of the vehicle.

To install:

26. Slowly lower the engine assembly into the engine compartment.

27. Raise the transmission onto the crossmember with a floor jack.

28. Align the holes in the engine mounts and the frame, install the bolts and then remove the engine hoist chain.

29. Raise the transmission slightly and tighten the No. 2 frame crossmember-to-side frame bolts to 70 ft. lbs. (95 Nm). Lower the transmission and tighten the 4 rear engine mount bolts to 9 ft. lbs. (13 Nm).

30. Install the brake tube and front floor heat insulators. Reconnect the exhaust pipe.

31. Install the No. 1 frame crossmember. Install the front driveshaft, stabilizer bar and the transfer case undercover.

32. Connect the speedometer cable. Connect the manual shift linkage for automatic transmission only.

33. Install the rear driveshaft. Install the shift levers for manual transmission only.

34. Install the fuel inlet and outlet lines. Connect the heater hoses.

35. Connect the clutch release cylinder hose. Install the air conditioning compressor.

36. Install the power steering pump and connect the ground strap.

37. Connect the throttle, cruise control and accelerator cables.

38. Connect the following hoses:
   a. Power steering hoses at the gas filter and air pipe
   b. Brake booster hose

   c. Cruise control vacuum hose, if equipped
   d. Charcoal canister hose at the canister
   e. VSV vacuum hoses.

39. Install the fan pulley, belt guide, fluid coupling and drive belt. Connect the following wires:
   a. Left side and rear ground straps
   b. Alternator connector and wire
   c. Igniter connector
   d. Oil pressure switch connector
   e. ECU connectors
   f. VSV connectors
   g. Starter relay connector for manual transmission only
   h. Solenoid resistor connector
   i. Check connector
   j. Air conditioning compressor connector

40. Install the air conditioning belt. Install the power steering pump and connect the ground strap.

41. Install the radiator and shroud. Install the air cleaner. Refill the engine with oil and the radiator with coolant.

42. Install the engine undercover. Install the battery. Install and adjust the hood.

43. Install the battery, start the truck and road test it.

2TZ–FE Engine

1. Disconnect the negative battery cable and drain the engine coolant and oil.

2. Raise the vehicle and support safely. Remove the engine under covers.

3. Remove the accessory drive belt and service bolts and nut at the front end of the driveshaft. Matchmark and disconnect the equipment driveshaft from the crankshaft pulley.

4. Label and disconnect all hoses, electrical connectors, vacuum lines and cables from the engine and position aside.

5. Disconnect the front driveshaft for 4WD only.

6. Remove the air intake connector and disconnect the engine wiring from the floor pan.

7. Remove the exhaust pipe from the manifold and disconnect the oxygen sensor connector.

8. Remove the rear driveshaft.

9. Place a engine/transmission jack under the engine and remove the engine and transmission mount bolts.

9. With the vehicle on a hoist, lower the engine with the transmission from the bottom of the vehicle.

To install:

10. Raise the engine/transmission into the vehicle and torque the mount bolts and nuts to 27 ft. lbs. (34 Nm).

11. Connect all electrical connectors,

vacuum hoses, coolant hoses and cables to the engine.

12. Install the exhaust pipe and torque to 32 ft. lbs. (43 Nm).

13. Install the front (4WD) and rear driveshafts and torque to 14 ft. lbs. (21 Nm).

14. Install the equipment drive shaft to the pulley and torque the bolts to 38 ft. lbs. (54 Nm). Install the belt and adjust.

15. Refill the engine with oil and coolant.

16. Connect the battery cable, start the engine and check for leaks.

Engine Mounts

REMOVAL & INSTALLATION

1. Raise the vehicle and support safely.

2. Remove the engine-to-mount bolt or nut.

3. Position a suitable jack under the engine block.

4. Raise the engine far enough to take the weight off of the mount. Wedge a piece of wood between the engine and chassis in case the engine jack slips.

5. Remove the mount-to-body bolts and remove the mount.

To install:

6. Install the mount and torque the bolts to 34 ft. lbs. (46 Nm).

7. Remove the wood and lower the jack to engage the mount.

8. Torque the mount nut to 25 ft. lbs. (34 Nm).

9. Lower the vehicle and check for proper operation.

Cylinder Head

REMOVAL & INSTALLATION

22R Engine

1. Disconnect the negative battery cable.

2. Drain the cooling system.

3. Mark all vacuum hoses and disconnect them.

4. Remove the air cleaner assembly, complete with hoses, from the carburetor.

NOTE: Cover the carburetor with a clean shop cloth so nothing can fall into it.

5. Remove all linkages, fuel lines, etc., from the carburetor, cylinder head, and manifolds. Remove the wire supports.

6. Mark the spark plug leads and disconnect them from the plugs.

7. Matchmark the distributor housing and block. Disconnect the primary lead and remove the distributor.

8. Unfasten the 14mm nuts which

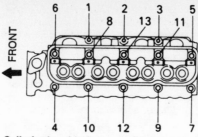

Cylinder head bolt removal sequence— 22R, 22R-E and 22R-TE engines

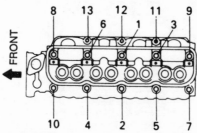

Cylinder head bolt installation sequence —22R, 22R-E and 22R-TE engines

secure the cam cover. Remove the cam cover.

9. Remove the rubber camshaft seals. Turn the crankshaft until the No. 1 piston is at TDC on its compression stroke. Matchmark the timing sprocket to the cam chain, and remove

the semi-circular lug. Using a 19mm wrench, remove the cam sprocket bolt. Slide the distributor drive gear and spacer off the cam and wire the cam sprocket in place.

10. Remove the timing chain cover 14mm bolt at the front of the head. This must be done before the head bolts are removed. Remove the exhaust pipe flange nuts and disconnect the pipe.

11. Remove the cylinder head bolts in the correct order. Improper removal could cause head damage.

12. Using prybars applied evenly at the front and the rear of the valve rocker assembly, pry the assembly off its mounting dowels.

13. Lift the head off its dowels. Do not pry it off.

14. Drain the engine oil from the crankcase after the head has been removed, because the oil will become contaminated with coolant while the head is being removed.

To install:

15. Apply liquid sealer to the front corners of the block and install the head gasket.

16. Lower the head over the locating dowels; do not attempt to slide it into place.

17. Rotate the camshaft so the sprocket aligning pin is at the top. Remove the wire and hold the cam

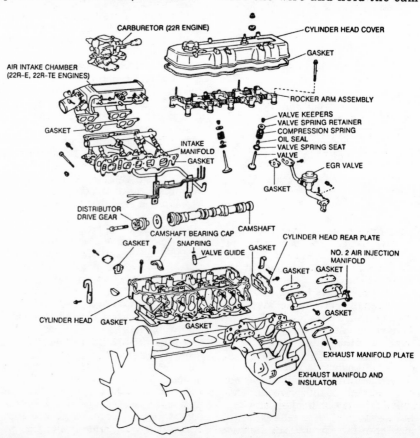

Exploded view of the cylinder head assembly—22R, 22R-E and 22R-TE engines

sprocket. Manually rotate the engine so the sprocket hole is also at the top. Wire the sprocket in place.

18. Install the rocker arm assembly over its positioning dowels.

19. Tighten the cylinder head bolts evenly, in 3 stages and in the correct order.

20. Install the timing chain cover bolt and tighten it to 7–11 ft. lbs. (9.5–14 Nm).

21. Remove the wire and fit the sprocket over the camshaft dowel. If the chain won't allow the sprocket to reach, rotate the crankshaft back and forth, while lifting up on the chain and sprocket.

22. Install the distributor drive gear and tighten the crankshaft bolt to 51–65 ft. lbs. (68–88 Nm).

23. Set the No. 1 piston at TDC of its compression stroke and adjust the valves.

24. After completing valve adjustment, rotate the crankshaft 352 degrees, so the 8 BTDC mark on the pulley aligns with the pointer.

25. Install the distributor.

26. Install the spark plugs and leads.

27. Make sure the oil drain plug is installed. Fill the engine with oil after installing the rubber cam seals. Pour the oil over the distributor drive gear and the valve rockers.

28. Install the rocker cover and tighten the bolts to 8–11 ft. lbs. (11–15 Nm).

29. Connect all the vacuum hoses and electrical leads which were removed during disassembly. Install the spark plug lead supports. Fill the cooling system. Install the air cleaner.

30. Tighten the exhaust pipe-to-manifold flange bolts to 25–33 ft. lbs. (34–45 Nm).

31. Reconnect the battery. Start the engine and allow it to reach normal operating temperature. Check and adjust the timing and valve clearance. Adjust the idle speed and mixture. Road test the vehicle.

22R–E and 22R–TE Engines

1. Relieve the fuel system pressure. Disconnect the negative battery cable.

2. Drain the coolant from the radiator and the cylinder block. Raise and safely support the vehicle. Drain the engine oil.

3. Remove the turbocharger on the 22R–TE engine.

4. Disconnect and remove the air cleaner hose on the 22R–E engine.

5. Disconnect the oxygen sensor wire. Remove the nuts attaching the manifold to the exhaust pipe and then separate them.

6. Remove the oil dipstick. Remove the distributor with the spark plug leads attached.

7. Disconnect the upper radiator hose and the heater hoses where they attach to the engine and then position them aside.

8. Disconnect the actuator cable, the accelerator cable and the throttle cable for the automatic transmission at their bracket.

9. Tag and disconnect the following:
   a. Both PCV vacuum hoses
   b. Brake booster hose
   c. Actuator hose, if equipped with cruise control
   d. Air control valve hoses
   e. Air control valve.

10. Tag and disconnect the EGR vacuum modulator hoses and then remove the modulator itself along with the bracket.

11. Tag and disconnect the following:
   a. Green and brown BVSV hoses
   b. Vacuum advance hoses
   c. The 2 air valve hoses; one at the throttle body, the other at the air chamber
   d. Air control valve hose, if equipped with air conditioning
   e. Pressure regulator hose at the air chamber
   f. Cold start injector pipe and wire
   g. Throttle position sensor wire.

12. Remove the bolt holding the EGR valve to the air chamber. Disconnect the chamber from the stay. Remove the air chamber-to-intake manifold bolts and then lift off the chamber with the throttle body.

13. Disconnect the fuel return hose.

14. Tag and disconnect the following:
   a. Water temperature sender gauge wire
   b. Temperature sensor wire
   c. Start injection time switch wire
   d. Fuel injector wires.

15. Remove the pulsation damper. Remove the bolt holding the fuel hose to the delivery pipe and then disconnect and remove the fuel hose.

16. Disconnect the wire and hose and then remove the air valve from the intake manifold.

17. Disconnect the bypass hose at the intake manifold on the 22R–E engine. On the 22R–TE engine, disconnect the oil cooler hose at the manifold.

18. If equipped with power steering, remove the pump and position it aside without disconnecting the hydraulic lines.

19. Remove the 4 nuts and then remove the cylinder head cover.

20. Remove the rubber camshaft seals. Turn the crankshaft until the No. 1 piston is at TDC of its compression stroke. Matchmark the timing sprocket to the timing chain and then remove the semi-circular plug. Using a 19mm wrench, remove the camshaft sprocket bolt. Slide the distributor

drive gear and spacer off the camshaft and wire the cam sprocket in place.

21. Remove the timing chain cover bolt in front of the cylinder head.

NOTE: This must be done before the cylinder head bolts are removed.

22. Remove the cylinder head bolts gradually, in 2–3 stages, in the correct order.

23. Using prybars applied evenly at the front and rear of the rocker arm assembly, pry the assembly off of its mounting dowels.

24. Lift the cylinder head off of its mounting dowels.

To install:

25. Apply liquid sealer to the front corners of the block and install the head gasket.

26. Lower the head over the locating dowels. Do not attempt to slide it into place.

27. Rotate the camshaft so the sprocket aligning pin is at the top. Remove the wire and hold the cam sprocket. Manually rotate the engine so the sprocket hole is also at the top. Wire the sprocket in place.

28. Install the rocker arm assembly over its positioning dowels.

29. Tighten the cylinder head bolts evenly, in 3 stages and in the correct order.

30. Install the timing chain cover bolt and tighten it to 7–11 ft. lbs. (11–15 Nm).

31. Remove the wire and fit the sprocket over the camshaft dowel. If the chain won't allow the sprocket to reach, rotate the crankshaft back and forth, while lifting up on the chain and sprocket.

32. Install the distributor drive gear and tighten the crankshaft bolt to 51–65 ft. lbs. (68–88 Nm).

33. Set the No. 1 piston at TDC of its compression stroke and adjust the valves.

34. After completing valve adjustment, rotate the crankshaft 352 degrees, so the 8 BTDC mark on the pulley aligns with the pointer.

35. Install the distributor.

36. Install the spark plugs and leads.

37. Make sure the oil drain plug is installed. Fill the engine with oil after installing the rubber cam seals. Pour the oil over the distributor drive gear and the valve rockers.

38. Install the rocker cover and tighten the bolts to 8–11 ft. lbs. (11–15 Nm).

39. Connect all the vacuum hoses and electrical leads which were removed during disassembly. Install the spark plug lead supports. Fill the cooling system. Install the air cleaner.

40. Tighten the exhaust pipe-to-

manifold flange bolts to 25–33 ft. lbs. (34–45 Nm).

41. Reconnect the battery. Start the engine and allow it to reach normal operating temperature. Check and adjust the timing and valve clearance. Adjust the idle speed and mixture. Road test the vehicle.

4Y–EC Engine

1. Disconnect the negative battery terminal from the battery.
2. Remove the right-front seat and the engine service hole cover.
3. Raise and safely support the vehicle. Drain the engine coolant at the radiator. Drain the engine oil. Lower the vehicle.
4. If equipped with power steering, perform the following:
   a. Remove the air hoses from the air control valve
   b. Drain the fluid from the reservoir tank
   c. Disconnect the return hose from the pump
   d. Disconnect the pressure hose from the pump
   e. Remove the drive belt, the pulley nut, the pulley, the Woodruff key, the mounting bolts and the pump
5. Remove the exhaust pipe and the bracket. Remove the air cleaner pipe and the hoses.
6. Disconnect the accelerator cable with the bracket from the throttle body.
7. Disconnect the water temperature sender gauge connector from the cylinder head.
8. Disconnect the following EFI connectors from the:
   a. The water thermo sensor
   b. The cold start injector time switch
   c. The cold start injector
   d. The air valve
   e. The throttle position sensor
   f. The oxygen sensor
   g. The water temperature switch
9. Disconnect the following hoses from the:
   a. The radiator inlet
   b. The radiator breather
   c. The reserve tank
   d. The heater outlet
   e. The PCV valve
   f. The water bypass
   g. The brake booster vacuum
   h. The charcoal canister
   i. Label and disconnect the emission control
10. Remove the throttle body from the air intake chamber.
11. Remove the EGR valve nuts from the intake chamber and the exhaust manifold-to-EGR valve union.
12. Disconnect the cold start injector pipe, the water bypass hoses and the pressure regulator hose.

13. Using a 12mm offset box wrench, remove the air intake chamber brackets, then, the chamber with the air valve.
14. Remove the wire clamp bolts and the injector connectors from the injectors.
15. Remove the exhaust manifold bracket, the heater pipe bracket, the fuel inlet pipe union bolt from the fuel filter and the fuel outlet hose.
16. Remove the spark plugs and the tubes.
17. Remove the cap nuts, the seal washers, the cylinder head cover and the gasket.
18. Remove the rocker arm shaft assembly nuts/bolts a little at a time, in 3–4 steps. Remove the pushrods, keeping them in order.
19. Remove the cylinder head bolts, a little at a time, in 3 passes. Lift the cylinder head off of the engine.
20. Remove the valve lifters from the cylinder block.

To install:

21. Clean the gasket mounting surfaces.
22. To install, use new gaskets and reverse the removal procedures. Torque the cylinder head bolts, in 3 passes, to 65 ft. lbs. (88 Nm) for 14mm bolt or 14 ft. lbs. (19 Nm) for 12mm bolt, the rocker arm shaft-to-cylinder head bolts, in 3 passes, to 17 ft. lbs. (24 Nm), the spark plugs to 13 ft. lbs. (18 Nm), the air intake chamber bolts to 9 ft. lbs. (12 Nm), the throttle body-to-intake chamber to 9 ft. lbs. (12 Nm) and the exhaust pipe-to-exhaust manifold to 29 ft. lbs. (41 Nm).
23. Adjust the drive belts.
24. Refill the cooling system and the engine with oil.
25. Check and/or adjust the timing.

NOTE: Use special tool 09270–71010 or equivalent, to hold the pushrods in position when installing the pushrods.

3F–E Engine

1. Drain the coolant.
2. Disconnect the negative battery cable.
3. Scribe matchmarks around the hood hinges and then remove the hood.
4. Disconnect the accelerator and throttle cables.
5. Remove the air intake hose, air flow meter and air cleaner cap.
6. Unbolt the power steering pump and position it aside without disconnecting the hydraulic lines.
7. Unbolt the air conditioning compressor and position it aside without disconnecting the refrigerant lines.
8. Remove the power steering pump and air conditioning compressor brackets.

9. Disconnect the high tension leads from the spark plugs and the coil.
10. Disconnect and remove the heater water (oil cooler) pipe.
11. Disconnect the upper radiator hose.
12. Disconnect and plug the fuel lines.
13. Disconnect the exhaust pipe at the manifold.
14. Remove the air pump.
15. Remove the fuel delivery pipe along with the fuel injectors.
16. Remove the air injection manifold.
17. Remove the intake and exhaust manifolds.
18. Disconnect the water bypass hose at the water outlet and then remove the outlet.
19. Remove the spark plugs.
20. Remove the cylinder head cover and its gasket.
21. Loosen the bolts and nuts that attach the rocker shaft assembly in several stages and then remove the rocker shaft.
22. Remove the pushrods.
23. Remove the cylinder head bolts in the reverse of the tightening sequence. Remove the air pump bracket and engine hanger.
24. Lift the cylinder head off of its mounting dowels.

To install:

25. Install the cylinder head on the cylinder block using a new gasket.
26. Lightly coat the threads of the cylinder head bolts with engine oil and then install them into the head. Tighten in several stages, to the correct torque.
27. Install the pushrods in the order that they were removed.
28. Position the rocker shaft assembly on the cylinder head and align the rocker arm adjusting screws with the heads of the pushrods. Tighten the mounting bolts with a 12mm head to 17 ft. lbs. (24 Nm); tighten the bolts with a 14mm head to 25 ft. lbs. (33 Nm).
29. Adjust the valve clearance and install the spark plugs. Install the cylinder head cover and tighten the cap nuts to 78 inch lbs. (8.8 Nm).
30. Install the water outlet and connect the bypass hose. Tighten the bolts to 18 ft. lbs. (25 Nm).
31. Install the intake and exhaust manifolds using a new gasket. Make sure the front mark on the gasket is towards the front of the engine.
32. Install the heat insulators and the manifold stay.
33. Install the air injection manifold and tighten the union nuts and clamp bolts to 15 ft. lbs. (21 Nm).
34. Install the fuel injector/delivery pipe assembly.

35. Install the air pump and connect the air hose.

36. Connect the exhaust pipe to the manifold. Use a new gasket and tighten the bolts to 46 ft. lbs. (62 Nm).

37. Connect the fuel lines and the upper radiator hose.

38. Install the heater water pipe.

39. Connect the high tension cords.

40. Install the air conditioning compressor and the power steering pump. Remember to adjust the belt tension later.

41. Install the air intake hose, the air flow meter and the air cleaner cap.

42. Connect and adjust the accelerator and throttle cables.

43. Connect the battery cable, fill the engine with coolant, start the engine and check for any leaks. Road test the vehicle.

3VZ–E Engine

1. Disconnect the negative battery cable.

2. Remove the air cleaner hose and case.

3. Drain the engine coolant.

4. Remove the radiator.

5. Unbolt the power steering pump and position it aside with the hoses still attached.

6. Remove all drive belts and then remove the fluid coupling and fan pulley.

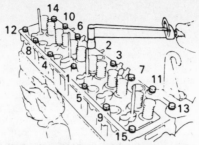

Cylinder head bolt installation sequence —3F–E engine

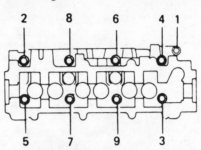

Cylinder head bolt removal sequence– 3VZ–E engine

7. Tag and disconnect all wires and connectors that will interfere with cylinder head removal.

8. Disconnect the following hoses:
   a. Power steering air hoses
   b. Brake booster hose

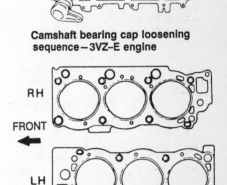

Camshaft bearing cap loosening sequence–3VZ–E engine

RH

FRONT

LH

Cylinder head gasket installation— 3VZ–E engine

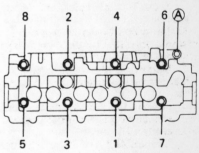

Cylinder head bolt tightening sequence —3VZ–E engine

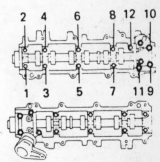

Camshaft bearing cap bolt tightening sequence–3VZ–E engine

c. Cruise control vacuum hose

d. Charcoal canister has at the canister

e. VSV vacuum hose.

9. Disconnect the accelerator, throttle and cruise control cables.

10. Disconnect the clutch release cylinder hose for manual transmission only.

11. Disconnect the heater hoses and the fuel lines.

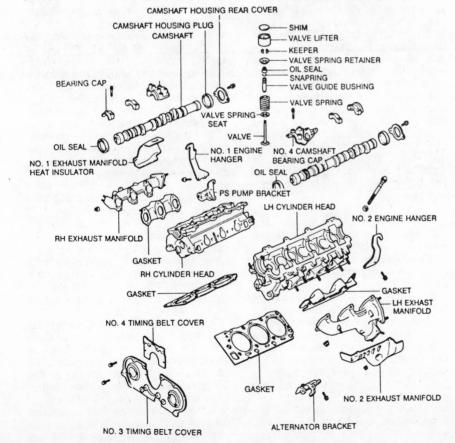

CAMSHAFT HOUSING REAR COVER
CAMSHAFT HOUSING PLUG
CAMSHAFT
SHIM
VALVE LIFTER
KEEPER
VALVE SPRING RETAINER
OIL SEAL
SNAPRING
VALVE GUIDE BUSHING
BEARING CAP
VALVE SPRING
VALVE SPRING SEAT
OIL SEAL
VALVE
NO. 1 ENGINE HANGER
NO. 4 CAMSHAFT BEARING CAP
NO. 1 EXHAUST MANIFOLD HEAT INSULATOR
OIL SEAL
PS PUMP BRACKET
LH CYLINDER HEAD
NO. 2 ENGINE HANGER
RH EXHAUST MANIFOLD
GASKET
RH CYLINDER HEAD
GASKET
LH EXHAUST MANIFOLD
NO. 4 TIMING BELT COVER
GASKET
GASKET
NO. 2 EXHAUST MANIFOLD
NO. 3 TIMING BELT COVER
ALTERNATOR BRACKET

Exploded view of the cylinder head assembly–3VZ–E engine

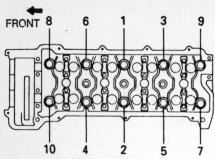

Cylinder head torque sequence— 2TZ–FE engine

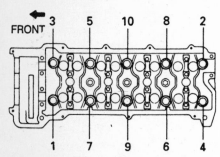

Cylinder head loosengine sequence— 2TZ–FE engine

12. Remove the left side scuff plate and disconnect the O_2 sensor and then remove the front exhaust pipe.

13. Remove the timing belt as detailed later in this chapter.

14. Remove the distributor with the spark plug leads attached; position it aside.

15. Remove the air intake chamber.

16. Disconnect the connectors and then remove the engine wire.

17. Remove the Nos. 2 and 3 fuel pipes.

18. Remove the No. 4 timing belt cover.

19. Remove the No. 2 idler pulley and the No. 3 timing belt cover.

20. Disconnect the hose and remove the water bypass outlet.

21. Remove the intake manifold.

22. Remove the exhaust crossover pipe.

23. For the right side, remove the following:

a. Remove the reed valve with the No. 1 air injection manifold.

b. Remove the water bypass pipe mounting bolt.

c. Remove the cylinder head cover.

d. Remove the camshaft.

e. Loosen the cylinder head bolts in several stages, in the opposite order of the tightening sequence. Remove the air pump bracket and engine hanger.

f. Lift the cylinder head off of its mounting dowels, do not pry it off.

24. For the left side, remove the following:

a. Remove the alternator.

b. Remove the oil dipstick guide tube.

c. Remove the cylinder head cover.

e. Remove the camshaft.

f. Loosen the cylinder head bolts in several stages, in the opposite order of the tightening sequence. Remove the air pump bracket and engine hanger.

g. Lift the cylinder head off of its mounting dowels, do not pry it off.

To install:

25. Install the cylinder head on the cylinder block using a new gasket.

26. Lightly coat the threads of the cylinder head bolts with engine oil and then install them into the head. Tighten them in several stages, in the correct order. After the initial tightening, mark the front side the the top of the bolt with paint. Tighten the bolts an additional 90 degrees (¼ turn) and check that the mark is now facing the side of the head. Tighten the bolts an additional 90 degrees and check that the mark is now facing the rear of the head. Install the bolt (A) and tighten it to 27 ft. lbs. (37 Nm).

27. Install the camshaft.

28. Install the alternator and the water bypass pipe mounting bolt.

29. Install the reed valve with the No. 1 injection manifold.

30. Install the oil dipstick tube.

31. Install the crossover pipe and tighten it to 29 ft. lbs. (39 Nm).

32. Connect the oxygen sensor wire.

33. Install the intake manifold with new gaskets and tighten the mounting bolts to 29 ft. lbs. (39 Nm).

34. Install the water bypass outlet and tighten the bolts to 13 ft. lbs. (18 Nm).

35. Install the fuel delivery pipes and injectors.

36. Install the No. 2 idler pulley. Install the Nos. 3 and 4 timing belt covers and tighten the bolts to 74 inch lbs. (8.3 Nm).

37. Install the fuel pipes and tighten the union bolts to 22 ft. lbs. (29 Nm).

38. Install the timing belt. Install the cylinder head covers.

39. Install the air intake chamber and tighten the nuts and bolts to 13 ft. lbs. (18 Nm).

40. Install the EGR valve and connect all hoses and lines. Install the distributor and the front exhaust pipe.

41. Connect the fuel lines and heater hoses. Connect the clutch release cylinder hose.

42. Install the power steering pump. Connect all cables and adjust, hoses and wires previously removed.

43. Install the fan pulley, fluid coupling and drive belts. Install the radiator.

44. Install the air cleaner hose, refill the engine with coolant and connect the battery cable.

2TZ–FE Engine

1. Disconnect the negative battery cable.

2. Remove the engine/transmission assembly from the vehicle.

3. Remove the engine wiring from the engine and move aside.

4. Remove the No. 2 valve cover.

5. Mark the spark plug wires and disconnect. Matchmark the distributor, rotor and cylinder head. Remove the distributor and disconnect the wiring.

6. Remove the EGR valve.

7. Remove the fuel rail and water outlet.

8. Remove the intake and exhaust manifold assembly.

9. Remove the exhaust manifold heat insulator and oil return pipe.

10. Remove the No. 1 valve cover.

11. Place matchmarks on the timing sprocket and chain. Hold the camshaft with a wrench and remove the sprocket bolt.

12. Remove the chain tensioner and gasket.

13. Remove the No. 6 camshaft bearing cap.

14. Set the knock pin hole of the exhaust camshaft at the 5–30 degree BTDC. Uniformly loosen the camshaft bearing caps and remove the exhaust camshaft from the head.

15. Set the knock pin hole of the intake camshaft at the 75–100 degree BTDC. Uniformly loosen the camshaft bearing caps and remove the exhaust camshaft from the head.

16. Remove the 2 bolts in front of the head before removing the cylinder haed retaining bolts.

17. Using a 12 sided socket wrench, remove the 10 cylinder head retaining bolts in sequence and remove the head from the engine.

To install:

18. Clean the gasket mating surfaces and check for warpage.

19. Install the head gasket and install the cylinder head.

20. Oil the bolts and torque in 3 steps, in sequence. If any of the bolts break, deform or do not meet the torque specification replace them.

21. Install and torque the 2 front bolts to 15 ft. lbs. (21 Nm).

22. Grease to all camshaft journals and caps.

23. Place the intake camshaft at 75–100 degrees BTDC. Install the bearing caps with the marking arrows facing forward. Uniformly torque the bearing cap bolts to 12 ft. lbs. (16 Nm).

24. Apply sealant to the bearing cap next to the timing chain sprocket. Place the intake camshaft at 5–30 degrees BTDC. Install the bearing caps

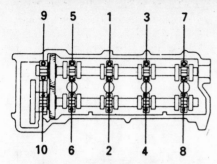

Intake camshaft bearing cap torque sequence — 2TZ-FE engine

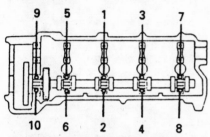

Exhaust camshaft bearing cap torque sequence — 2TZ-FE engine

with the marking arrows facing forward. Uniformly torque the bearing cap bolts to 12 ft. lbs. (16 Nm). Make sure the exhaust and intake camshaft gear alignment marks are facing each other. The one gear has 2 dots and the other has 1 dot.

25. Install the timing chain sprocket and torque the bolt to 54 ft. lbs. (74 Nm).

26. Release the ratchet pawl and fully push the plunger and apply the hook on the tensioner so it can not spring out.

27. Install the tensioner and torque the bolts to 15 ft. lbs. (21 Nm). Turn the crankshaft to the left so the hook of the tensioner is released from the pin. If it does not spring out, pull the slipper into the tensioner to release the hook.

28. Install the cylinder head covers and torque to 69 inch lbs. (7.8 Nm).

29. Install the intake and exhaust manifolds.

30. Install the remaining components onto the engine.

31. Install the engine/transmission assembly into the vehicle.

32. Refill the engine coolant and oil. Connect the battery cable and check for leaks.

Valve Lifters

REMOVAL & INSTALLATION

Always replace the camshaft and lifters as a set. If not replacing, label all components for exact reinstallation.

4Y-EC and 3F-E Engines

1. Disconnect the negative battery cable.

2. Remove the spark plugs and tubes.

3. Remove the valve cover.

4. Uniformly loosen and remove the rocker arm shaft bolts and nuts. Remove the rocker shaft and pushrods. Label all components for installation.

5. Using a piece of wire or magnetic finger, remove the 8 valve lifters for 4Y-EC engine.

NOTE: Always keep the lifters upright and in correct order.

6. Remove the pushrod cover. Remove the 12 lifters for 3F-E engine.
To install:

7. Lubricate the lifters with prelube and install into their original location.

8. Install the pushrod cover and torque to 35 inch lbs. (3.9 Nm) for 3F-E engines.

9. Install the rocker arms and torque to 17 ft. lbs. (24 Nm) and 25 ft. lbs. (33 Nm) for the bolt/nut combination.

10. Install the rocker arm cover and torque to 69 inch lbs. (7.8 Nm).

11. Install the remaining components and check for leaks.

Valve Lash

ADJUSTMENT

22R, 22R-E and 22R-TE Engines

1. Start the engine and allow it to reach normal operating temperatures above 175° F.

2. Stop the engine. Remove the air cleaner assembly, the hoses and the bracket, then any cables, hoses, wires and etc., which are attached to the valve cover. Remove the valve cover.

3. Set the No. 1 cylinder to TDC of the compression stroke. Place a wrench on the crankshaft pulley bolt and turn the engine until the notch on the crankshaft pulley is aligned with the 0 degree mark on the timing plate; the engine is at TDC.

NOTE: The rocker arms on cylinder No. 1 should be loose and the rocker arms on cylinder No. 4 should be tight.

4. With the engine hot, the valve clearances are 0.008 in. for intake or 0.012 in. for exhaust.

NOTE: The clearance is measured with a feeler gauge between the valve stem and the adjusting screw.

5. To adjust the valve clearance, loosen the locknut and turn the ad-

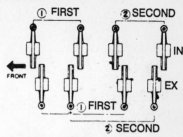

Valve adjustment sequence — 22R, 22R-E and 22R-TE engines

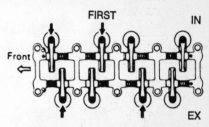

First step of the valve adjustment procedure — 22R, 22R-E and 22R-TE engines

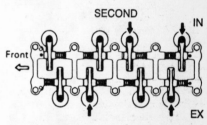

Second step of the valve adjustment procedure — 22R, 22R-E and 22R-TE engines

justing screw until the specified clearance is obtained. Tighten the locknut and check the clearance again. Adjust the intake valves of No. 1 and 2 cylinders; the exhaust valves of No. 1 and 3 cylinders.

6. Turn the crankshaft one full revolution, 360 degrees. Adjust the intake valves of No. 3 and 4 cylinders; the exhaust valves of No. 2 and 4 cylinders.

7. To install the components, reverse the removal procedures.

4Y-EC Engine

The valve tappets of this engine are hydraulic; no adjustment is necessary.

2TZ-FE Engine

Check the valve clearance with the engine cold.

1. Remove the front seat and engine service hole cover.

2. Remove the valve cover.

3. Install a service bolt and nut into the equipment driveshaft.

4. Set the No. 1 cylinder to TDC/compression stoke.

5. Measure the clearance of the first set of valves and record the measurements. The clearance should be 0.006–

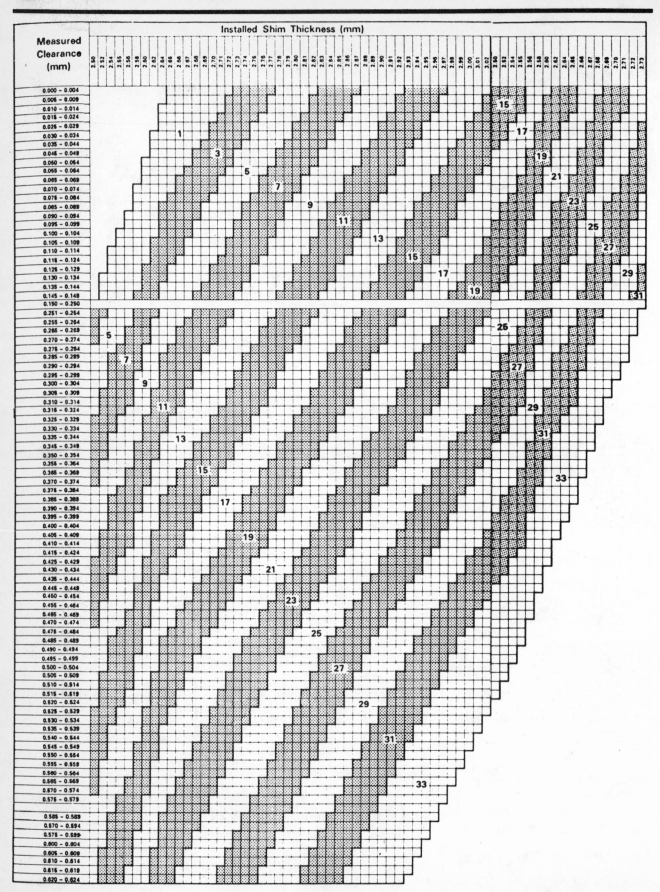

Intake shim selection chart (cont) — 2TZ-FE engine

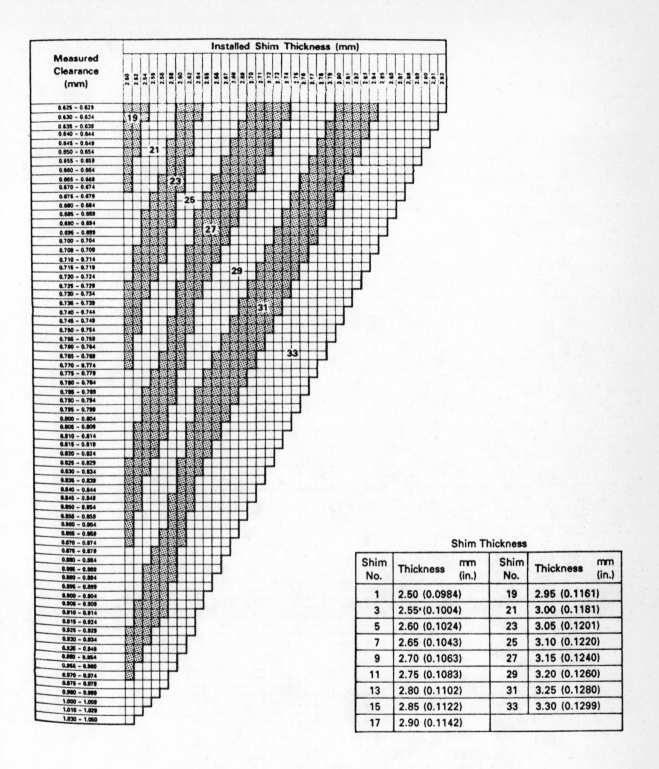

Intake shim selection chart (cont) — 2TZ–FE engine

Shim Thickness

Shim No.	Thickness mm (in.)	Shim No.	Thickness mm (in.)
1	2.50 (0.0984)	19	2.95 (0.1161)
3	2.55 (0.1004)	21	3.00 (0.1181)
5	2.60 (0.1024)	23	3.05 (0.1201)
7	2.65 (0.1043)	25	3.10 (0.1220)
9	2.70 (0.1063)	27	3.15 (0.1240)
11	2.75 (0.1083)	29	3.20 (0.1260)
13	2.80 (0.1102)	31	3.25 (0.1280)
15	2.85 (0.1122)	33	3.30 (0.1299)
17	2.90 (0.1142)		

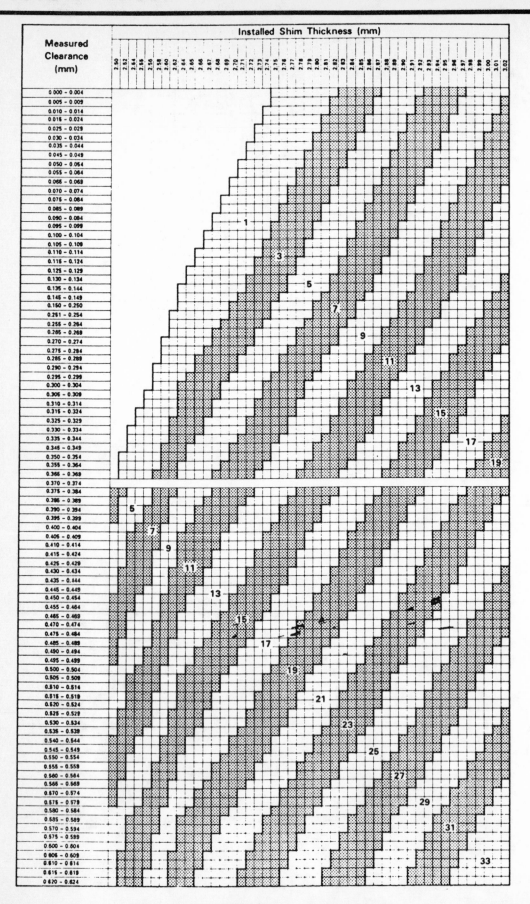

Exhaust shim selection chart (cont) — 2TZ–FE engine

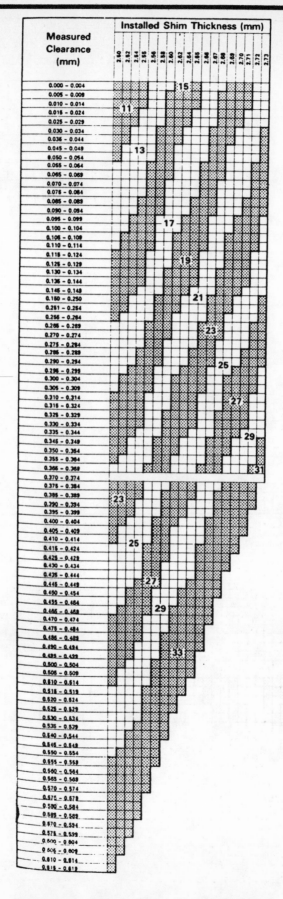

Exhaust shim selection chart (cont) – 2TZ-FE engine

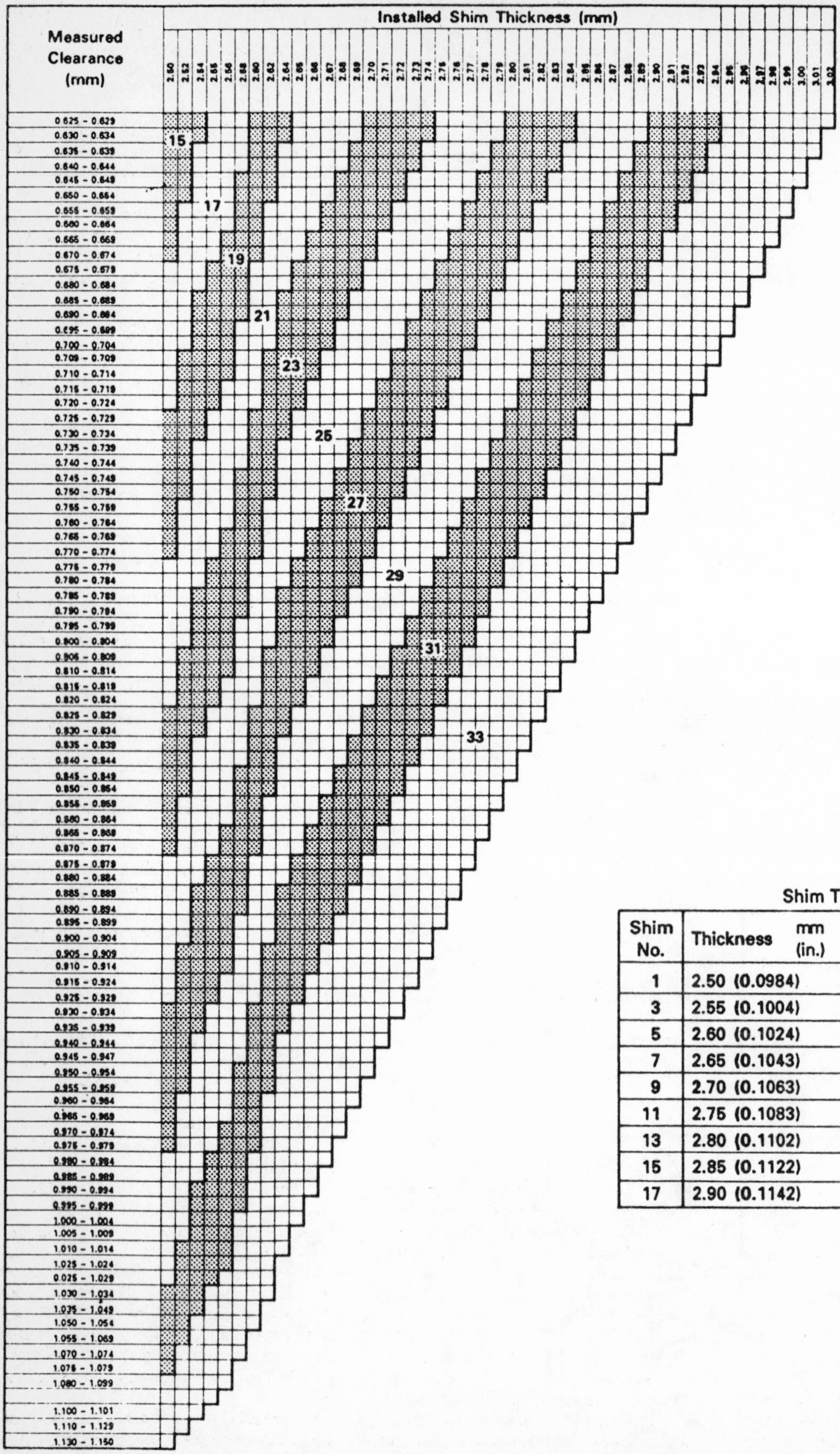

Installed Shim Thickness (mm)

Measured Clearance (mm)

Column headers (Installed Shim Thickness):
2.50, 2.52, 2.54, 2.55, 2.56, 2.58, 2.60, 2.62, 2.64, 2.65, 2.66, 2.67, 2.68, 2.69, 2.70, 2.71, 2.72, 2.73, 2.74, 2.76, 2.77, 2.78, 2.79, 2.80, 2.81, 2.82, 2.83, 2.84, 2.86, 2.87, 2.88, 2.89, 2.90, 2.91, 2.92, 2.94, 2.95, 2.97, 2.98, 2.99, 3.01, 3.02

Measured Clearance rows (mm):
0.625 – 0.629
0.630 – 0.634
0.635 – 0.639
0.640 – 0.644
0.645 – 0.649
0.650 – 0.654
0.655 – 0.659
0.660 – 0.664
0.665 – 0.669
0.670 – 0.674
0.675 – 0.679
0.680 – 0.684
0.685 – 0.689
0.690 – 0.694
0.695 – 0.699
0.700 – 0.704
0.705 – 0.709
0.710 – 0.714
0.715 – 0.719
0.720 – 0.724
0.725 – 0.729
0.730 – 0.734
0.735 – 0.739
0.740 – 0.744
0.745 – 0.749
0.750 – 0.754
0.755 – 0.759
0.760 – 0.764
0.765 – 0.769
0.770 – 0.774
0.775 – 0.779
0.780 – 0.784
0.785 – 0.789
0.790 – 0.794
0.795 – 0.799
0.800 – 0.804
0.805 – 0.809
0.810 – 0.814
0.815 – 0.819
0.820 – 0.824
0.825 – 0.829
0.830 – 0.834
0.835 – 0.839
0.840 – 0.844
0.845 – 0.849
0.850 – 0.854
0.855 – 0.859
0.860 – 0.864
0.865 – 0.868
0.870 – 0.874
0.875 – 0.879
0.880 – 0.884
0.885 – 0.889
0.890 – 0.894
0.895 – 0.899
0.900 – 0.904
0.905 – 0.909
0.910 – 0.914
0.915 – 0.924
0.925 – 0.929
0.930 – 0.934
0.935 – 0.939
0.940 – 0.944
0.945 – 0.947
0.950 – 0.954
0.955 – 0.959
0.960 – 0.964
0.965 – 0.969
0.970 – 0.974
0.975 – 0.979
0.980 – 0.984
0.985 – 0.989
0.990 – 0.994
0.995 – 0.999
1.000 – 1.004
1.005 – 1.009
1.010 – 1.014
1.025 – 1.024
0.025 – 1.029
1.030 – 1.034
1.035 – 1.049
1.050 – 1.054
1.055 – 1.069
1.070 – 1.074
1.075 – 1.079
1.080 – 1.099
1.100 – 1.101
1.110 – 1.129
1.130 – 1.150

Shim numbers shown on chart diagonal: 15, 17, 19, 21, 23, 25, 27, 29, 31, 33

Shim Thickness

Shim No.	Thickness mm (in.)	Shim No.	Thickness mm (in.)
1	2.50 (0.0984)	19	2.95 (0.1161)
3	2.55 (0.1004)	21	3.00 (0.1181)
5	2.60 (0.1024)	23	3.05 (0.1201)
7	2.65 (0.1043)	25	3.10 (0.1220)
9	2.70 (0.1063)	27	3.15 (0.1240)
11	2.75 (0.1083)	29	3.20 (0.1260)
13	2.80 (0.1102)	31	3.25 (0.1280)
15	2.85 (0.1122)	33	3.30 (0.1299)
17	2.90 (0.1142)		

Exhaust shim selection chart—2TZ–FE engine

FIRST ➡ FRONT

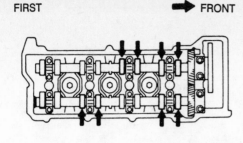

SECOND ➡ FRONT

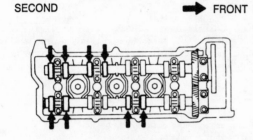

Valve clearance check—2TZ–FE engine

FIRST

FRONT ⬅

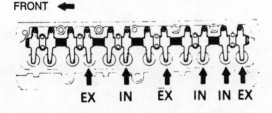

EX IN IN EX IN EX

SECOND

FRONT ⬅

EX IN EX IN IN EX

Adjusting valve clearance—3F–E engine

0.010 in. (0.15–0.25mm) for intake and 0.010–0.014 in. (0.25–0.35mm) for exhaust.

6. Turn the equipment driveshaft 1 full revolution and measure the second set of valves.

7. Using a shim removing tool 0924855010 or equivalent, press down the lifter and remove the shim with a small pick.

8. Determine the replacement shim size by measuring the old shim using a micrometer and calculate the thickness of the new shim using the following formula:

T = thickness of the used shim
A = valve clearance measured
N = thickness of new shim
Intake: $N = T + (A - 0.008$ in. $(0.20$mm$)$
Exhaust: $N = T + (A - 0.012$ in. $(0.30$mm$)$

9. Select a shim with the thickness as close as possible to the calculated values. Shims are available in 17 sizes, in increments of 0.002 in. (0.050mm). The thickness is stamped on the shim.

10. Install the shims and recheck the clearance.

11. Install the valve cover and torque the bolts to 69 inch lbs. (7.8 Nm).

12. Install the remaining components and check for leaks.

3F–E Engine

Check the valve clearance with the engine at normal operating temperature.

1. Remove the air cleaner and valve cover.

2. Set the No. 1 cylinder to TDC. Make sure the No. 1 rocker arms are loose and the No. 6 are tight. If not turn the crankshaft 1 turn and align the timing marks to 0.

3. Adjust the first set of valves to 0.008 in. (0.20mm) for intake and 0.014 in. (0.35mm) for exhaust valves.

4. After tightening the lock nut, recheck the valve clearance.

5. Install the valve cover and air cleaner.

3VZ–E Engine

Check the valve clearance with the engine cold.

1. Remove the valve cover.

2. Set the No. 1 cylinder to TDC/compression.

3. Measure the clearance of the lifters with the camshaft lobe at the base circle and record the measurements. The clearance should be as follows:

1988—0.010–0.014 in. (0.25–0.35mm) intake
1988—0.011–0.015 in. (0.27–0.37mm) exhaust
1989–92—0.007–0.011 in. (0.18–0.28mm) intake
1989–92—0.009–0.013 in. (0.22–0.32mm) exhaust

4. Using a shim removing tool 0924855010 or equivalent, press down the lifter and remove the shim with a small pick.

5. Determine the replacement shim size by measuring the old shim using a micrometer and calculate the thickness of the new shim using the following formula:

T = thickness of the used shim
A = valve clearance measured
N = thickness of new shim
1988 Intake: $N = T + (A - 0.0118$ in. $(0.30$mm$)$
1988 Exhaust: $N = T + (A - 0.0126$ in. $(0.32$mm$)$
1989–92 Intake: $N = T + (A - 0.0091$ in. $(0.23$mm$)$
1989–92 Exhaust: $N = T + (A - 0.0126$ in. $(0.32$mm$)$

6. Select a shim with the thickness as close as possible to the calculated values. Shims are available in 25 sizes, in increments of 0.002 in. (0.050mm). The thickness is stamped on the shim.

7. Install the shims and recheck the clearance.

8. Install the valve cover and torque the bolts to 69 inch lbs. (7.8 Nm).

9. Install the remaining components and check for leaks.

Measured Clearance (mm)	Installed Shim Thickness (mm)																																															
	2.200	2.225	2.250	2.275	2.300	2.325	2.350	2.375	2.400	2.425	2.450	2.475	2.500	2.525	2.550	2.575	2.600	2.625	2.650	2.675	2.700	2.725	2.750	2.775	2.800	2.825	2.850	2.875	2.900	2.925	2.950	2.975	3.000	3.025	3.050	3.075	3.100	3.125	3.150	3.175	3.200	3.225	3.250	3.275	3.300	3.325	3.350	3.375	3.400
0.000-0.025											01	01	01	03	03	05	05	07	07	09	09	11	11	13	13	15	15	17	17	19	19	21	21	23	23	25	25	27	27	29	29	31	31	33	33	35	35	37	37
0.026-0.050										01	01	01	03	03	05	05	07	07	09	09	11	11	13	13	15	15	17	17	19	19	21	21	23	23	25	25	27	27	29	29	31	31	33	33	35	35	37	37	39
0.051-0.075									01	01	01	03	03	05	05	07	07	09	09	11	11	13	13	15	15	17	17	19	19	21	21	23	23	25	25	27	27	29	29	31	31	33	33	35	35	37	37	39	39
0.076-0.100								01	01	01	03	03	05	05	07	07	09	09	11	11	13	13	15	15	17	17	19	19	21	21	23	23	25	25	27	27	29	29	31	31	33	33	35	35	37	37	39	39	41
0.101-0.125							01	01	01	03	03	05	05	07	07	09	09	11	11	13	13	15	15	17	17	19	19	21	21	23	23	25	25	27	27	29	29	31	31	33	33	35	35	37	37	39	39	41	41
0.126-0.150						01	01	01	03	03	05	05	07	07	09	09	11	11	13	13	15	15	17	17	19	19	21	21	23	23	25	25	27	27	29	29	31	31	33	33	35	35	37	37	39	39	41	41	43
0.151-0.175					01	01	01	03	03	05	05	07	07	09	09	11	11	13	13	15	15	17	17	19	19	21	21	23	23	25	25	27	27	29	29	31	31	33	33	35	35	37	37	39	39	41	41	43	43
0.176-0.200				01	01	01	03	03	05	05	07	07	09	09	11	11	13	13	15	15	17	17	19	19	21	21	23	23	25	25	27	27	29	29	31	31	33	33	35	35	37	37	39	39	41	41	43	43	45
0.201-0.225			01	01	01	03	03	05	05	07	07	09	09	11	11	13	13	15	15	17	17	19	19	21	21	23	23	25	25	27	27	29	29	31	31	33	33	35	35	37	37	39	39	41	41	43	43	45	45
0.226-0.249		01	01	01	03	03	05	05	07	07	09	09	11	11	13	13	15	15	17	17	19	19	21	21	23	23	25	25	27	27	29	29	31	31	33	33	35	35	37	37	39	39	41	41	43	43	45	45	47
0.250-0.350																																																	
0.351-0.375	03	05	05	07	07	09	09	11	11	13	13	15	15	17	17	19	19	21	21	23	23	25	25	27	27	29	29	31	31	33	33	35	35	37	37	39	39	41	41	43	43	45	45	47	47	49	49	49	
0.376-0.400	05	05	07	07	09	09	11	11	13	13	15	15	17	17	19	19	21	21	23	23	25	25	27	27	29	29	31	31	33	33	35	35	37	37	39	39	41	41	43	43	45	45	47	47	49	49	49
0.401-0.425	05	07	07	09	09	11	11	13	13	15	15	17	17	19	19	21	21	23	23	25	25	27	27	29	29	31	31	33	33	35	35	37	37	39	39	41	41	43	43	45	45	47	47	49	49	49
0.426-0.450	07	07	09	09	11	11	13	13	15	15	17	17	19	19	21	21	23	23	25	25	27	27	29	29	31	31	33	33	35	35	37	37	39	39	41	41	43	43	45	45	47	47	49	49	49
0.451-0.475	07	09	09	11	11	13	13	15	15	17	17	19	19	21	21	23	23	25	25	27	27	29	29	31	31	33	33	35	35	37	37	39	39	41	41	43	43	45	45	47	47	49	49	49
0.476-0.500	09	09	11	11	13	13	15	15	17	17	19	19	21	21	23	23	25	25	27	27	29	29	31	31	33	33	35	35	37	37	39	39	41	41	43	43	45	45	47	47	49	49	49
0.501-0.525	09	11	11	13	13	15	15	17	17	19	19	21	21	23	23	25	25	27	27	29	29	31	31	33	33	35	35	37	37	39	39	41	41	43	43	45	45	47	47	49	49	49
0.526-0.550	11	11	13	13	15	15	17	17	19	19	21	21	23	23	25	25	27	27	29	29	31	31	33	33	35	35	37	37	39	39	41	41	43	43	45	45	47	47	49	49	49
0.551-0.575	11	13	13	15	15	17	17	19	19	21	21	23	23	25	25	27	27	29	29	31	31	33	33	35	35	37	37	39	39	41	41	43	43	45	45	47	47	49	49	49
0.576-0.600	13	13	15	15	17	17	19	19	21	21	23	23	25	25	27	27	29	29	31	31	33	33	35	35	37	37	39	39	41	41	43	43	45	45	47	47	49	49	49
0.601-0.625	13	15	15	17	17	19	19	21	21	23	23	25	25	27	27	29	29	31	31	33	33	35	35	37	37	39	39	41	41	43	43	45	45	47	47	49	49	49
0.626-0.650	15	15	17	17	19	19	21	21	23	23	25	25	27	27	29	29	31	31	33	33	35	35	37	37	39	39	41	41	43	43	45	45	47	47	49	49	49
0.651-0.675	15	17	17	19	19	21	21	23	23	25	25	27	27	29	29	31	31	33	33	35	35	37	37	39	39	41	41	43	43	45	45	47	47	49	49	49
0.676-0.700	17	17	19	19	21	21	23	23	25	25	27	27	29	29	31	31	33	33	35	35	37	37	39	39	41	41	43	43	45	45	47	47	49	49	49
0.701-0.725	17	19	19	21	21	23	23	25	25	27	27	29	29	31	31	33	33	35	35	37	37	39	39	41	41	43	43	45	45	47	47	49	49	49
0.726-0.750	19	19	21	21	23	23	25	25	27	27	29	29	31	31	33	33	35	35	37	37	39	39	41	41	43	43	45	45	47	47	49	49	49
0.751-0.775	19	21	21	23	23	25	25	27	27	29	29	31	31	33	33	35	35	37	37	39	39	41	41	43	43	45	45	47	47	49	49	49
0.776-0.800	21	21	23	23	25	25	27	27	29	29	31	31	33	33	35	35	37	37	39	39	41	41	43	43	45	45	47	47	49	49	49
0.801-0.825	21	23	23	25	25	27	27	29	29	31	31	33	33	35	35	37	37	39	39	41	41	43	43	45	45	47	47	49	49	49
0.826-0.850	23	23	25	25	27	27	29	29	31	31	33	33	35	35	37	37	39	39	41	41	43	43	45	45	47	47	49	49	49
0.851-0.875	23	25	25	27	27	29	29	31	31	33	33	35	35	37	37	39	39	41	41	43	43	45	45	47	47	49	49	49
0.876-0.900	25	25	27	27	29	29	31	31	33	33	35	35	37	37	39	39	41	41	43	43	45	45	47	47	49	49	49
0.901-0.925	25	27	27	29	29	31	31	33	33	35	35	37	37	39	39	41	41	43	43	45	45	47	47	49	49	49
0.926-0.950	27	27	29	29	31	31	33	33	35	35	37	37	39	39	41	41	43	43	45	45	47	47	49	49	49
0.951-0.975	27	29	29	31	31	33	33	35	35	37	37	39	39	41	41	43	43	45	45	47	47	49	49	49
0.976-1.000	29	29	31	31	33	33	35	35	37	37	39	39	41	41	43	43	45	45	47	47	49	49	49
1.001-1.025	29	31	31	33	33	35	35	37	37	39	39	41	41	43	43	45	45	47	47	49	49	49
1.026-1.050	31	31	33	33	35	35	37	37	39	39	41	41	43	43	45	45	47	47	49	49	49
1.051-1.075	31	33	33	35	35	37	37	39	39	41	41	43	43	45	45	47	47	49	49	49
1.076-1.100	33	33	35	35	37	37	39	39	41	41	43	43	45	45	47	47	49	49	49
1.101-1.125	33	35	35	37	37	39	39	41	41	43	43	45	45	47	47	49	49	49
1.126-1.150	35	35	37	37	39	39	41	41	43	43	45	45	47	47	49	49	49
1.151-1.175	35	37	37	39	39	41	41	43	43	45	45	47	47	49	49	49
1.176-1.200	37	37	39	39	41	41	43	43	45	45	47	47	49	49	49
1.201-1.225	37	39	39	41	41	43	43	45	45	47	47	49	49	49
1.226-1.250	39	39	41	41	43	43	45	45	47	47	49	49	49
1.251-1.275	39	41	41	43	43	45	45	47	47	49	49	49
1.276-1.300	41	41	43	43	45	45	47	47	49	49	49
1.301-1.325	41	43	43	45	45	47	47	49	49	49
1.326-1.350	43	43	45	45	47	47	49	49	49
1.351-1.375	43	45	45	47	47	49	49	49
1.376-1.400	45	45	47	47	49	49	49
1.401-1.425	45	45	47	49	49	49
1.426-1.450	47	47	49	49	49
1.451-1.475	47	49	49	49
1.476-1.500	49	49	49
1.501-1.525	49	49
1.526-1.550	49

Shim Thickness

Shim No.	Thickness mm (in.)	Shim No.	Thickness mm (in.)
01	2.20 (0.0866)	27	2.85 (0.1122)
03	2.25 (0.0886)	29	2.90 (0.1142)
05	2.30 (0.0906)	31	2.95 (0.1161)
07	2.35 (0.0925)	33	3.00 (0.1181)
09	2.40 (0.0945)	35	3.05 (0.1201)
11	2.45 (0.0965)	37	3.10 (0.1220)
13	2.50 (0.0984)	39	3.15 (0.1240)
15	2.55 (0.1004)	41	3.20 (0.1260)
17	2.60 (0.1024)	43	3.25 (0.1280)
19	2.65 (0.1043)	45	3.30 (0.1299)
21	2.70 (0.1063)	47	3.35 (0.1319)
23	2.75 (0.1083)	49	3.40 (0.1339)
25	2.80 (0.1102)		

Intake shim selection chart—1988 3VZ-E engine

Measured Clearance (mm)	Installed Shim Thickness (mm)																																																
	2.200	2.225	2.250	2.275	2.300	2.325	2.350	2.375	2.400	2.425	2.450	2.475	2.500	2.525	2.550	2.575	2.600	2.625	2.650	2.675	2.700	2.725	2.750	2.775	2.800	2.825	2.850	2.875	2.900	2.925	2.950	2.975	3.000	3.025	3.050	3.075	3.100	3.125	3.150	3.175	3.200	3.225	3.250	3.275	3.300	3.325	3.350	3.375	3.400
0.000-0.020													01	01	03	03	05	05	07	07	09	09	11	11	13	13	15	15	17	17	19	19	21	21	23	23	25	25	27	27	29	29	31	31	33	33	35	35	37
0.021-0.045												01	01	03	03	05	05	07	07	09	09	11	11	13	13	15	15	17	17	19	19	21	21	23	23	25	25	27	27	29	29	31	31	33	33	35	35	37	37
0.046-0.070											01	01	03	03	05	05	07	07	09	09	11	11	13	13	15	15	17	17	19	19	21	21	23	23	25	25	27	27	29	29	31	31	33	33	35	35	37	37	39
0.071-0.095										01	01	03	03	05	05	07	07	09	09	11	11	13	13	15	15	17	17	19	19	21	21	23	23	25	25	27	27	29	29	31	31	33	33	35	35	37	37	39	39
0.096-0.120									01	01	03	03	05	05	07	07	09	09	11	11	13	13	15	15	17	17	19	19	21	21	23	23	25	25	27	27	29	29	31	31	33	33	35	35	37	37	39	39	41
0.121-0.145								01	01	03	03	05	05	07	07	09	09	11	11	13	13	15	15	17	17	19	19	21	21	23	23	25	25	27	27	29	29	31	31	33	33	35	35	37	37	39	39	41	41
0.146-0.170							01	01	03	03	05	05	07	07	09	09	11	11	13	13	15	15	17	17	19	19	21	21	23	23	25	25	27	27	29	29	31	31	33	33	35	35	37	37	39	39	41	41	43
0.171-0.195						01	01	03	03	05	05	07	07	09	09	11	11	13	13	15	15	17	17	19	19	21	21	23	23	25	25	27	27	29	29	31	31	33	33	35	35	37	37	39	39	41	41	43	43
0.196-0.220					01	01	03	03	05	05	07	07	09	09	11	11	13	13	15	15	17	17	19	19	21	21	23	23	25	25	27	27	29	29	31	31	33	33	35	35	37	37	39	39	41	41	43	43	45
0.221-0.245				01	01	03	03	05	05	07	07	09	09	11	11	13	13	15	15	17	17	19	19	21	21	23	23	25	25	27	27	29	29	31	31	33	33	35	35	37	37	39	39	41	41	43	43	45	45
0.246-0.269			01	01	03	03	05	05	07	07	09	09	11	11	13	13	15	15	17	17	19	19	21	21	23	23	25	25	27	27	29	29	31	31	33	33	35	35	37	37	39	39	41	41	43	43	45	45	47
0.270-0.370																																																	
0.371-0.395	03	05	05	07	07	09	09	11	11	13	13	15	15	17	17	19	19	21	21	23	23	25	25	27	27	29	29	31	31	33	33	35	35	37	37	39	39	41	41	43	43	45	45	47	47	49	49		
0.396-0.420	05	05	07	07	09	09	11	11	13	13	15	15	17	17	19	19	21	21	23	23	25	25	27	27	29	29	31	31	33	33	35	35	37	37	39	39	41	41	43	43	45	45	47	47	49	49
0.421-0.445	05	07	07	09	09	11	11	13	13	15	15	17	17	19	19	21	21	23	23	25	25	27	27	29	29	31	31	33	33	35	35	37	37	39	39	41	41	43	43	45	45	47	47	49	49
0.446-0.470	07	07	09	09	11	11	13	13	15	15	17	17	19	19	21	21	23	23	25	25	27	27	29	29	31	31	33	33	35	35	37	37	39	39	41	41	43	43	45	45	47	47	49	49
0.471-0.495	07	09	09	11	11	13	13	15	15	17	17	19	19	21	21	23	23	25	25	27	27	29	29	31	31	33	33	35	35	37	37	39	39	41	41	43	43	45	45	47	47	49	49
0.496-0.520	09	09	11	11	13	13	15	15	17	17	19	19	21	21	23	23	25	25	27	27	29	29	31	31	33	33	35	35	37	37	39	39	41	41	43	43	45	45	47	47	49	49
0.521-0.545	09	11	11	13	13	15	15	17	17	19	19	21	21	23	23	25	25	27	27	29	29	31	31	33	33	35	35	37	37	39	39	41	41	43	43	45	45	47	47	49	49
0.546-0.570	11	11	13	13	15	15	17	17	19	19	21	21	23	23	25	25	27	27	29	29	31	31	33	33	35	35	37	37	39	39	41	41	43	43	45	45	47	47	49	49
0.571-0.595	11	13	13	15	15	17	17	19	19	21	21	23	23	25	25	27	27	29	29	31	31	33	33	35	35	37	37	39	39	41	41	43	43	45	45	47	47	49	49
0.596-0.620	13	13	15	15	17	17	19	19	21	21	23	23	25	25	27	27	29	29	31	31	33	33	35	35	37	37	39	39	41	41	43	43	45	45	47	47	49	49
0.621-0.645	13	15	15	17	17	19	19	21	21	23	23	25	25	27	27	29	29	31	31	33	33	35	35	37	37	39	39	41	41	43	43	45	45	47	47	49	49
0.646-0.670	15	15	17	17	19	19	21	21	23	23	25	25	27	27	29	29	31	31	33	33	35	35	37	37	39	39	41	41	43	43	45	45	47	47	49	49
0.671-0.695	15	17	17	19	19	21	21	23	23	25	25	27	27	29	29	31	31	33	33	35	35	37	37	39	39	41	41	43	43	45	45	47	47	49	49
0.696-0.720	17	17	19	19	21	21	23	23	25	25	27	27	29	29	31	31	33	33	35	35	37	37	39	39	41	41	43	43	45	45	47	47	49	49
0.721-0.745	17	19	19	21	21	23	23	25	25	27	27	29	29	31	31	33	33	35	35	37	37	39	39	41	41	43	43	45	45	47	47	49	49
0.746-0.770	19	19	21	21	23	23	25	25	27	27	29	29	31	31	33	33	35	35	37	37	39	39	41	41	43	43	45	45	47	47	49	49
0.771-0.795	19	21	21	23	23	25	25	27	27	29	29	31	31	33	33	35	35	37	37	39	39	41	41	43	43	45	45	47	47	49	49
0.796-0.820	21	21	23	23	25	25	27	27	29	29	31	31	33	33	35	35	37	37	39	39	41	41	43	43	45	45	47	47	49	49
0.821-0.845	21	23	23	25	25	27	27	29	29	31	31	33	33	35	35	37	37	39	39	41	41	43	43	45	45	47	47	49	49
0.846-0.870	23	23	25	25	27	27	29	29	31	31	33	33	35	35	37	37	39	39	41	41	43	43	45	45	47	47	49	49
0.871-0.895	23	25	25	27	27	29	29	31	31	33	33	35	35	37	37	39	39	41	41	43	43	45	45	47	47	49	49
0.896-0.920	25	25	27	27	29	29	31	31	33	33	35	35	37	37	39	39	41	41	43	43	45	45	47	47	49	49
0.921-0.945	25	27	27	29	29	31	31	33	33	35	35	37	37	39	39	41	41	43	43	45	45	47	47	49	49
0.946-0.970	27	27	29	29	31	31	33	33	35	35	37	37	39	39	41	41	43	43	45	45	47	47	49	49
0.971-0.995	27	29	29	31	31	33	33	35	35	37	37	39	39	41	41	43	43	45	45	47	47	49	49
0.996-1.020	29	29	31	31	33	33	35	35	37	37	39	39	41	41	43	43	45	45	47	47	49	49
1.021-1.045	29	31	31	33	33	35	35	37	37	39	39	41	41	43	43	45	45	47	47	49	49
1.046-1.070	31	31	33	33	35	35	37	37	39	39	41	41	43	43	45	45	47	47	49	49
1.071-1.095	31	33	33	35	35	37	37	39	39	41	41	43	43	45	45	47	47	49	49
1.096-1.120	33	33	35	35	37	37	39	39	41	41	43	43	45	45	47	47	49	49
1.121-1.145	33	35	35	37	37	39	39	41	41	43	43	45	45	47	47	49	49
1.146-1.170	35	35	37	37	39	39	41	41	43	43	45	45	47	47	49	49
1.171-1.195	35	37	37	39	39	41	41	43	43	45	45	47	47	49	49
1.196-1.220	37	37	39	39	41	41	43	43	45	45	47	47	49	49
1.221-1.245	37	39	39	41	41	43	43	45	45	47	47	49	49
1.246-1.270	39	39	41	41	43	43	45	45	47	47	49	49
1.271-1.295	39	41	41	43	43	45	45	47	47	49	49
1.296-1.320	41	41	43	43	45	45	47	47	49	49
1.321-1.345	41	43	43	45	45	47	47	49	49
1.346-1.370	43	43	45	45	47	47	49	49
1.371-1.395	43	45	45	47	47	49	49
1.396-1.420	45	45	47	47	49	49
1.421-1.445	45	47	47	49	49
1.446-1.470	47	47	49	49
1.471-1.495	47	49	49
1.496-1.520	49	49
1.521-1.545	49

Exhaust shim selection chart—1988 3VZ-E engine

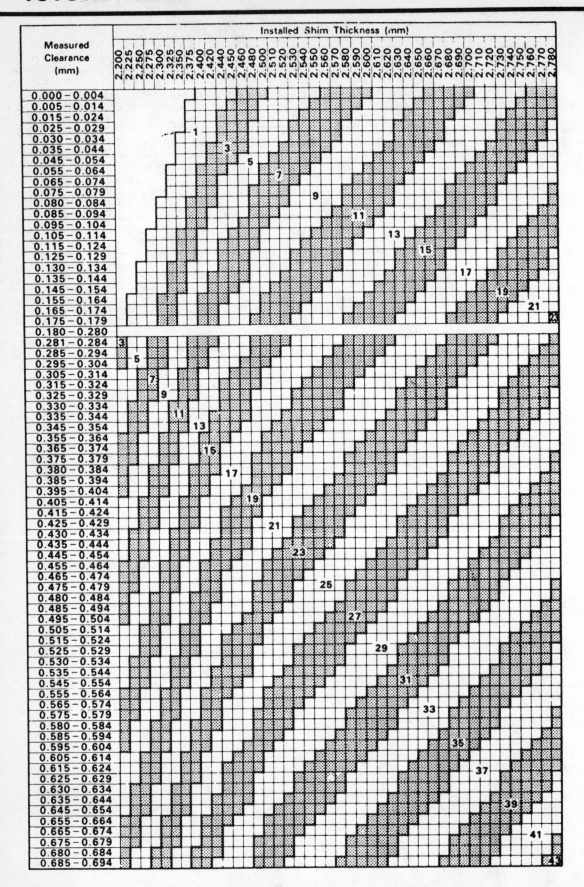

Intake shim selection chart (cont) –1989–92 3VZ–E engine

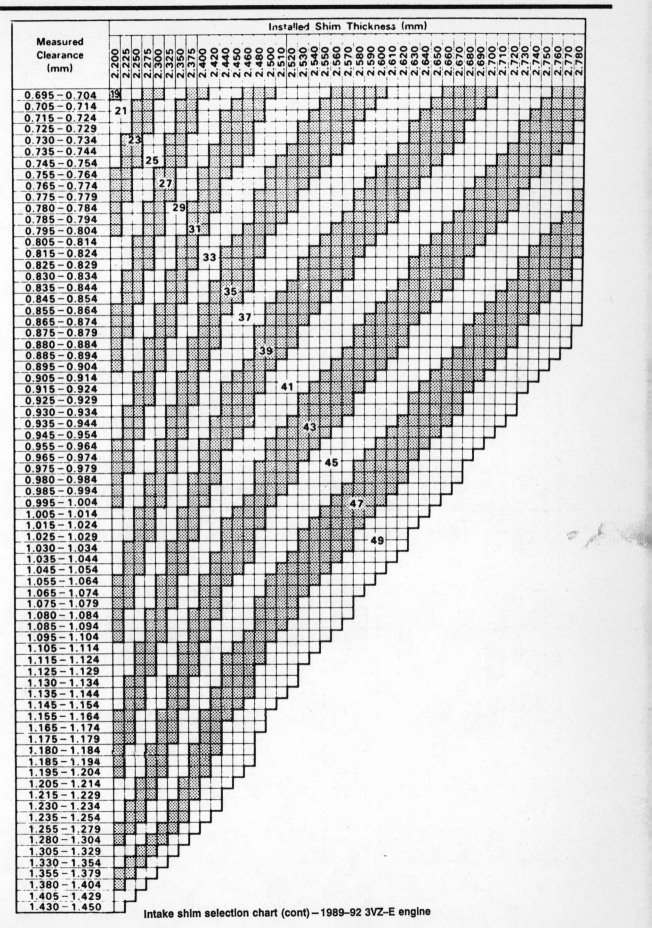

Intake shim selection chart (cont) – 1989–92 3VZ–E engine

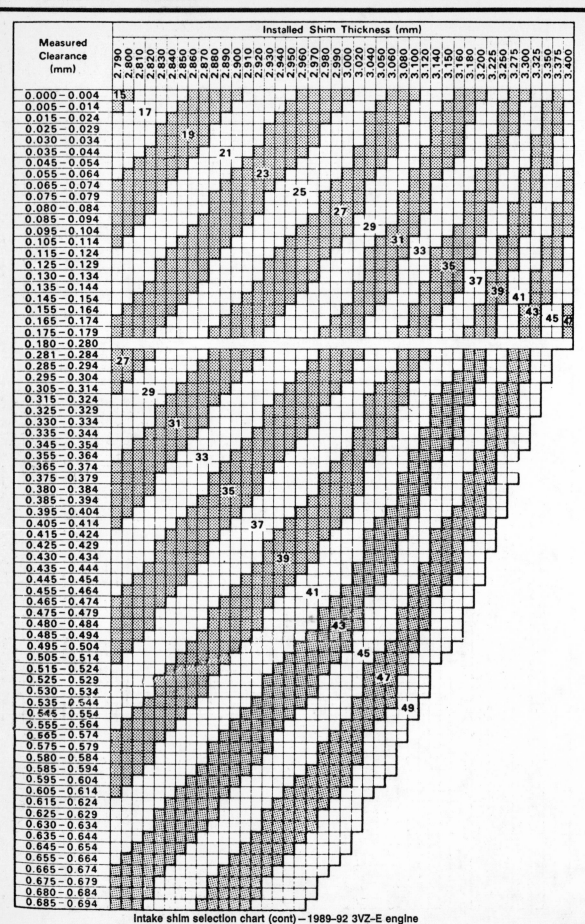

Intake shim selection chart (cont) — 1989–92 3VZ–E engine

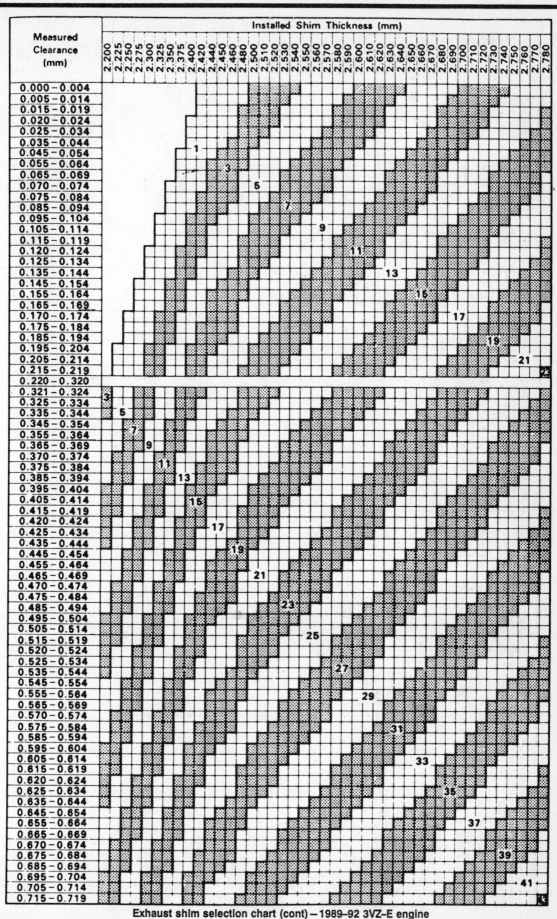

Exhaust shim selection chart (cont) – 1989–92 3VZ–E engine

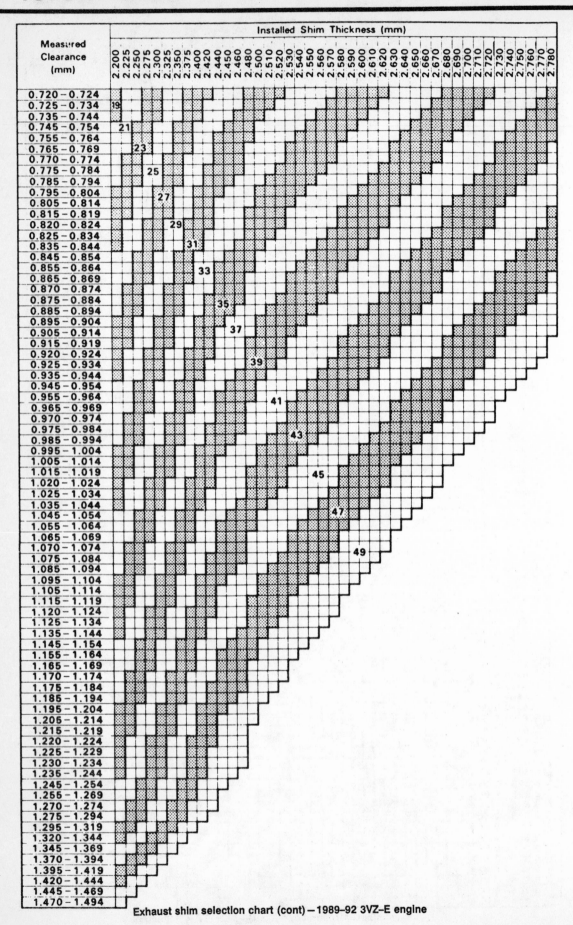

Exhaust shim selection chart (cont) – 1989–92 3VZ–E engine

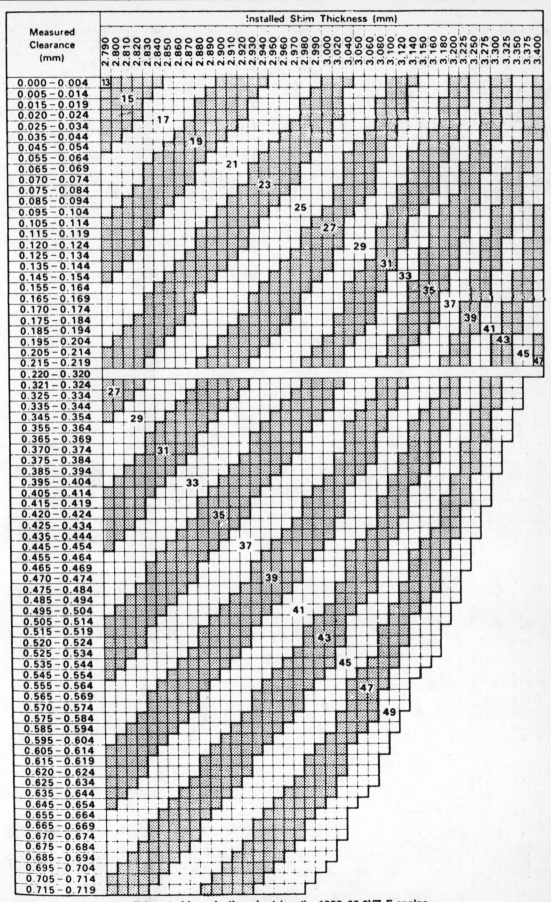

Exhaust shim selection chart (cont) — 1989–92 3VZ–E engine

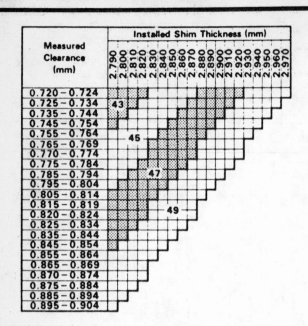

Shim No.	Thickness mm (in.)	Shim No.	Thickness mm (in.)
01	2.20 (0.0866)	27	2.85 (0.1122)
03	2.25 (0.0886)	29	2.90 (0.1142)
05	2.30 (0.0906)	31	2.95 (0.1161)
07	2.35 (0.0925)	33	3.00 (0.1181)
09	2.40 (0.0945)	35	3.05 (0.1201)
11	2.45 (0.0965)	37	3.10 (0.1220)
13	2.50 (0.0984)	39	3.15 (0.1240)
15	2.55 (0.1004)	41	3.20 (0.1260)
17	2.60 (0.1024)	43	3.25 (0.1280)
19	2.65 (0.1043)	45	3.30 (0.1299)
21	2.70 (0.1063)	47	3.35 (0.1319)
23	2.75 (0.1083)	49	3.40 (0.1339)
25	2.80 (0.1102)		

Exhaust shim selection chart—1989–92 3VZ–E engine

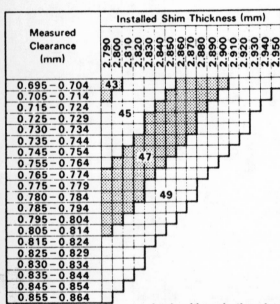

Shim No.	Thickness mm (in.)	Shim No.	Thickness mm (in.)
01	2.20 (0.0866)	27	2.85 (0.1122)
03	2.25 (0.0886)	29	2.90 (0.1142)
05	2.30 (0.0906)	31	2.95 (0.1161)
07	2.35 (0.0925)	33	3.00 (0.1181)
09	2.40 (0.0945)	35	3.05 (0.1201)
11	2.45 (0.0965)	37	3.10 (0.1220)
13	2.50 (0.0984)	39	3.15 (0.1240)
15	2.55 (0.1004)	41	3.20 (0.1260)
17	2.60 (0.1024)	43	3.25 (0.1280)
19	2.65 (0.1043)	45	3.30 (0.1299)
21	2.70 (0.1063)	47	3.35 (0.1319)
23	2.75 (0.1083)	49	3.40 (0.1339)
25	2.80 (0.1102)		

Intake shim selection chart—1989–92 3VZ–E engine

Rocker Arm Shafts

REMOVAL & INSTALLATION

22R, 22R–E and 22R–TE Engines

1. Remove the valve cover.
2. Remove the timing chain sprocket and secure the sprocket and chain to the engine with wire to ensure correct valve timing.
3. Remove the distributor drive gear and fuel pump for 22R engine.
4. Uniformly loosen the 10 shaft bolts in the opposite sequence of tightening.
5. Remove the rocker arm shaft as-

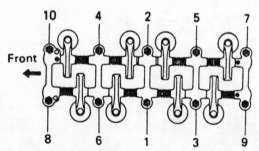

Rocker arm shaft torque sequence—22R, 22R–E, 22R–TE engines

sembly from the cylinder head. Keep the assembly together for installation.
To install:
6. Install the rocker arm shaft as-

sembly and torque the bolts in 3 steps, in sequence to 58 ft. lbs. (78 Nm).
7. Install the chain cover bolt and torque to 9 ft. lbs. (13 Nm).

8. Install timing chain sprocket, distributor drive gear and thrust plate. Torque the bolt to 58 ft. lbs. (78 Nm).

9. Adjust the valve clearance.

10. Install the valve cover and torque to 69 ft. lbs. (7.8 Nm).

11. Install the remaining components and check for leaks.

3F–E and 4Y–EC Engines

1. Remove the valve cover.

2. Uniformly loosen the shaft bolts starting from the ends and working inward.

3. Remove the rocker arm shaft assembly from the cylinder head. Keep the assembly together for installation.

To install:

4. Install the rocker arm shaft assembly and torque the bolts in 3 steps, start in the middle and move to the ends and torque in sequence to 17 ft. lbs. (24 Nm) and 25 ft. lbs. (33 Nm) for the bolt and nut combination.

5. Adjust the valve clearance.

6. Install the valve cover and torque to 69 ft. lbs. (7.8 Nm).

7. Install the remaining components and check for leaks.

Intake Manifold

REMOVAL & INSTALLATION

22R Engine

1. Disconnect the negative battery cable.

2. Drain the cooling system.

3. Remove the air cleaner assembly, complete with hoses, from the carburetor.

4. Disconnect the vacuum lines from the EGR valve and carburetor. Mark them first to aid in the installation.

5. Remove the fuel lines, electrical leads, accelerator linkage, and water hose from the carburetor.

6. Remove the water bypass hose from the manifold.

7. Unbolt and remove the intake manifold, complete with carburetor and EGR valve.

8. Cover the cylinder head ports with clean shop cloths to keep anything from falling into the cylinder head or block.

To install:

9. When installing the manifold, replace the gasket with a new one. Torque the mounting bolts to 14 ft. lbs. (19 Nm). Tighten the bolts in several stages working from the inside bolts outward.

10. Connect the bypass line and all other lines and hoses.

11. Install the air cleaner and refill the cooling system.

22R–E and 22R–TE Engines

1. Relieve the fuel system pressure. Disconnect the negative battery cable.

2. Drain the cooling system.

3. Disconnect the air intake hose from both the air cleaner assembly on one end and the air intake chamber on the other.

4. Tag and disconnect all vacuum lines attached to the intake chamber and manifold.

5. Tag and disconnect the wires to the cold start injector, throttle position sensor and the water hoses from the throttle body.

6. Remove the EGR valve from the intake chamber.

7. Tag and disconnect the actuator cable, accelerator cable and throttle valve cable, if equipped, from the cable bracket on the intake chamber.

8. Unbolt the air intake chamber from the intake manifold and remove the chamber with the throttle body attached.

9. Disconnect the fuel hose from the fuel delivery pipe.

10. Tag and disconnect the air valve hose from the intake manifold.

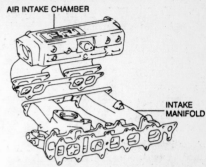

Intake manifold assembly—22R–E and 22R–TE engines

11. Make sure all hoses, lines and wires are tagged for later installation and disconnected from the intake manifold. Unbolt the manifold from the cylinder head, removing the delivery pipe and injection nozzle with the manifold.

To install:

12. Clean the gasket mating surfaces and check for warpage.

13. Install the gasket and manifold. Torque the bolts to 13 ft. lbs. (18 Nm) starting from the middle and move outward.

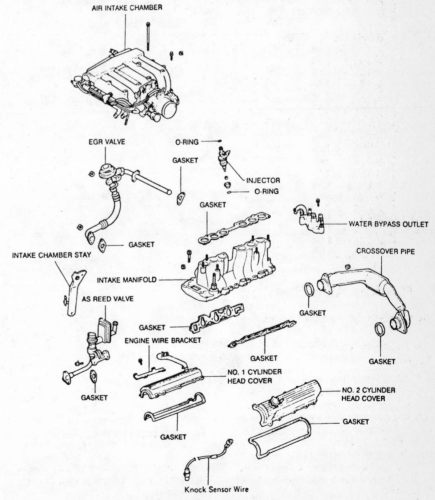

Intake manifold assembly—3VZ–E engine

14. Install the remaining components and check for leaks.

3VZ–E Engine

1. Relieve the fuel system pressure. Disconnect the negative battery cable.
2. Drain the cooling system.
3. Disconnect the air intake hose from both the air cleaner assembly on one end and the air intake chamber on the other.
4. Tag and disconnect all vacuum lines attached to the intake chamber and manifold.
5. Disconnect the throttle position sensor connector at the air chamber. Disconnect the PCV hose at the union.
6. Disconnect the No. 4 water bypass hose at the manifold. Remove the No. 5 bypass hose at the water bypass pipe.
7. Disconnect the cold start injector and the vacuum hose at the fuel filter.
8. Remove the union bolt and gaskets, then remove the cold start injector tube.
9. Disconnect the EGR gas temperature sensor and the EGR vacuum hoses from the air pipe and the vacuum modulator.
10. Remove the EGR valve.
11. Disconnect the No. 1 air hose at the reed valve.
12. Remove the air intake chamber and then remove the engine wire.
13. Remove the union bolts and then remove the No. 2 and 3 fuel pipes.
14. Remove the No. 4 timing belt cover. Remove the the No. 2 idler pulley and the No. 3 timing belt cover.
15. Remove the fuel delivery pipes with their injectors.
16. Remove the water bypass outlet and then remove the intake manifold.

To install:

17. Install the intake manifold with new gaskets and tighten the mounting bolts to 29 ft. lbs. (39 Nm).
18. Install the water bypass outlet and tighten the 2 bolts to 13 ft. lbs. (18 Nm).
19. Install the fuel delivery pipes and injectors.
20. Install the No. 2 idler pulley. Install the No. 3 and 4 timing belt covers and tighten the bolts to 74 inch lbs. (8.3 Nm).
21. Install the fuel pipes and tighten the union bolts to 22 ft. lbs. (29 Nm).
22. Install the engine wire.
23. Install the air intake chamber and tighten the nuts and bolts to 13 ft. lbs. (18 Nm).
24. Install the EGR valve and connect all hoses and lines.
25. Install the air cleaner hose, refill the engine with coolant and connect the battery cable.

2TZ–FE Engine

1. Disconnect the negative battery cable.
2. Remove the engine from the vehicle.
3. Disconnect the fuel pipes and remove the fuel rail.
4. Label and disconnect all hoses, wiring and cables from the intake manifold.
5. Remove the water outlet and disconnect the PCV hose.
6. Remove the intake manifold stays.
7. Remove the retaining bolts, intake manifold and gasket.

To install:

8. Clean the gasket mating surfaces and check for warpage.
9. Install the gasket, manifold and bolts. Torque the bolts to 15 ft. lbs. (21 Nm), starting from the inside and work outward.
10. Install the manifold stays and torque the bolts to 27 ft. lbs. (37 Nm).
11. Connect the water bypass pipe and fuel rail.
12. Reconnect all wiring, hoses and cables to the manifold.
13. Install the engine into the vehicle.
14. Install the remaining components and check for leaks.

Exhaust Manifold

REMOVAL & INSTALLATION

22R, 22R–E, 22R–TE and 3VZ–E Engines

1. Remove the 3 exhaust pipe flange bolts for turbocharger flange bolts on the 22R–TE engine and disconnect the exhaust pipe from the manifold.
2. Tag and disconnect the spark plug leads.
3. Position the spark plug wires aside.
4. Remove the air cleaner tube from the heat stove, on carbureted engines, and remove the outer part of the heat stove.
5. Use a 14mm wrench to remove the manifold securing nuts.
6. Remove the manifold(s), complete with air injection tubes and the inner portion of the heat stove.

To install:

7. Separate the inner portion of the heat stove from the manifold.
8. When installing the manifold(s), torque the retaining nuts to 29–36 ft. lbs. (41–49 Nm), working from the inside out, and in several stages. Install the distributor and set the timing. Tighten the exhaust pipe flange nuts to 25–32 ft. lbs. (34–45 Nm).
9. Install the remaining components and check for leaks.

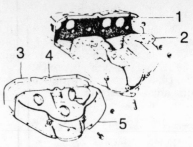

1. Inner heat stove
2. Exhaust manifold
3. Gasket
4. Gasket
5. Outer heat stove

Exhaust manifold—22R, 22R–E and 22R–TE engines

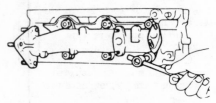

Right exhaust manifold—3VZ–E engine

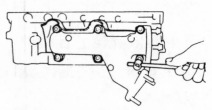

Left exhaust manifold—3VZ–E engine

Combinaton Manifold

REMOVAL & INSTALLATION

3F–E Engine

1. Relieve the fuel system pressure on the 3F–E engine. Disconnect the negative battery cable. Remove the air cleaner assembly, complete with hoses.
2. Disconnect the accelerator and choke linkages, as well as the fuel and vacuum lines. Remove the throttle linkage.
3. Remove or move aside, any of the emission control system components which are in the way.
4. Disconnect the oil filter lines and remove the oil filter assembly from the intake manifold. Unfasten the solenoid valve wire from the ignition coil terminal. Remove the EGR pipes from the exhaust gas cooler, if equipped.
5. Unfasten the retaining bolts and remove the carburetor from the manifold. On the fuel injected engine, disconnect the throttle chamber from the manifold.
6. Loosen the manifold retaining

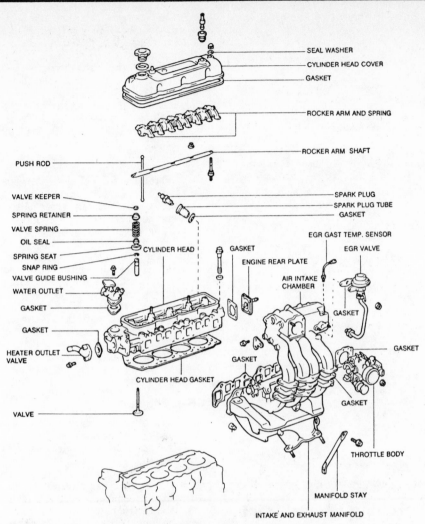

Intake manifold assembly—4YE-C engine

Labels on diagram: SEAL WASHER, CYLINDER HEAD COVER, GASKET, ROCKER ARM AND SPRING, ROCKER ARM SHAFT, PUSH ROD, VALVE KEEPER, SPRING RETAINER, VALVE SPRING, OIL SEAL, SPRING SEAT, SNAP RING, VALVE GUIDE BUSHING, WATER OUTLET, GASKET, GASKET, HEATER OUTLET VALVE, CYLINDER HEAD, CYLINDER HEAD GASKET, VALVE, SPARK PLUG, SPARK PLUG TUBE, GASKET, GASKET, ENGINE REAR PLATE, EGR GAST TEMP. SENSOR, EGR VALVE, AIR INTAKE CHAMBER, GASKET, GASKET, GASKET, THROTTLE BODY, GASKET, MANIFOLD STAY, INTAKE AND EXHAUST MANIFOLD

nuts, working from the inside out, in 2–3 stages.

7. Remove the intake/exhaust manifold assembly from the cylinder head as a complete unit.

To install:

8. Clean the gasket mating surfaces and check for warpage.

9. When installing the manifolds, always use new gaskets. Torque the bolts to 36 ft. lbs. (49 Nm) working from the inside out.

4Y–EC Engine

1. Relieve the fuel system pressure. Disconnect the negative battery terminal from the battery. Remove the right seat and the engine service hole cover.

2. Drain the engine coolant to a level below the throttle body.

3. Remove the air cleaner-to-throttle body hose.

4. Remove the accelerator cable with the bracket from the throttle body.

5. Disconnect the air valve connector, the throttle position sensor connector and the O_2 sensor connector.

6. Disconnect the PCV hose from the air intake chamber, the water by-pass hoses from the throttle body, the booster vacuum hose and the charcoal canister hose, then label and disconnect the emission control hoses.

7. Remove the throttle body from the air intake chamber, the EGR tube union nut from the exhaust manifold, the EGR valve from the air intake chamber.

8. Disconnect the cold start injector tube, the water bypass hoses and the pressure regulator hose from the intake manifold.

9. Remove the air intake chamber brackets and the air intake chamber with the air valve from the intake manifold.

10. Remove the wire clamp bolt from the fuel injector rail, then, the fuel injector rail from the fuel injectors.

11. Remove the exhaust manifold-to-intake manifold bracket, the exhaust

manifold-to-engine bracket, the exhaust pipe from the exhaust manifold, the fuel inlet and outlet tubes union nut from the fuel rail.

12. Remove the spark plug wires, the spark plugs and the tubes.

13. Remove the retaining bolts, then, the intake/exhaust manifolds as an assembly.

To install:

14. Clean the gasket mounting surfaces and check for leakage.

15. Use new gaskets and reverse the removal procedures. Torque the manifold-to-cylinder head bolts to 36 ft. lbs. (48 Nm), the air intake chamber-to-intake manifold bolts to 9 ft. lbs. (14 Nm) and the throttle body bolts to 9 ft. lbs. (14 Nm).

16. Refill the cooling system. Start the engine, allow it to reach operating temperatures and check for leaks.

Turbocharger

REMOVAL & INSTALLATION

22R–TE Engine

1. Disconnect the negative battery cable.

2. Drain the coolant.

3. Disconnect the oxygen sensor wire clamp and connector.

4. Disconnect the No. 1 and 3 PCV hoses.

5. Disconnect the No. 1 and 2 turbocharger water hoses.

6. Loosen the clamp on the throttle body, remove the nuts and lift off the air tube assembly.

7. Remove the No. 1 air cleaner hose assembly and the No. 2 air cleaner hose.

8. Remove the exhaust manifold and turbocharger heat insulators.

9. Disconnect the No. 3 turbocharger water hose.

10. Raise and safely support the vehicle. Disconnect the exhaust pipe from the turbine outlet.

11. Remove the turbocharger bracket stay.

12. Disconnect the turbocharger oil pipe.

13. Remove the turbocharger and exhaust manifold as an assembly.

14. Remove the No. 2 turbocharger water pipe.

15. Remove the oil pipe.

16. Remove the No. 1 water pipe.

17. Remove the turbine outlet elbow with the O_2 sensor attached.

18. Disconnect the turbocharger from the manifold.

To install:

19. Pour approximately 20cc of new oil into the oil inlet and then turn the

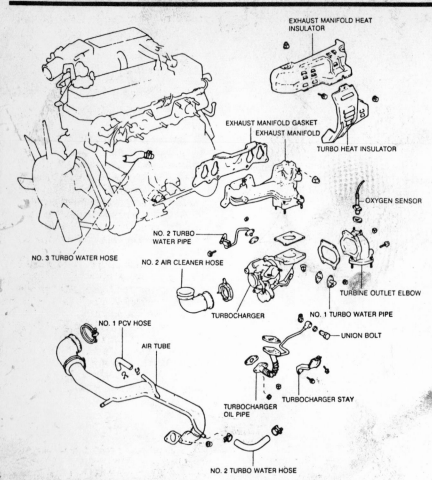

Exhaust manifold and turbocharger assembly—22R-TE engine

turbocharger impeller wheel so as to wet the bearing.

20. Using a new gasket, attach the turbocharger to the manifold and tighten the nuts to 29 ft. lbs. (39 Nm).

21. Install the turbine outlet elbow and tighten the nuts to 19 ft. lbs. (25 Nm).

22. Install the No. 1 water pipe. Install the turbo oil pipe and tighten the nuts to 14 ft. lbs. (19 Nm).

23. Install the No. 2 water pipe and then mount the assembly to the cylinder head.

24. Install the oil pipe and tighten the union bolt to 20 ft. lbs. (27 Nm). Tighten the nuts to 14 ft. lbs. (19 Nm).

7. Install the turbocharger bracket stay.

25. Connect the exhaust pipe to the turbine outlet elbow and tighten the nuts to 32 ft. lbs. (43 Nm). Lower the truck.

26. Connect the No. 3 water hose and install the heat insulators.

27. Install the No. 2 air cleaner hose with the arrow facing the turbocharger and fasten the clip as shown.

28. Install the air tube assembly. Connect the O₂ sensor and clamp.

29. Refill the engine with coolant, start it and check for leaks.

Timing Chain Front Cover

REMOVAL & INSTALLATION

22R, 22R-E and 22R-TE Engines

1. Disconnect the negative battery cable. Remove the cylinder head.

2. Remove the radiator.

3. Remove the alternator. Remove the oil pan.

4. On engines equipped with air pumps, unfasten the adjusting link bolts and the drive belt. Remove the hoses from the pump; remove the pump and bracket from the engine.

5. Remove the fan and water pump as a complete assembly.

NOTE: To prevent the fluid from running out of the fan coupling, do not tip the assembly over on its side.

6. Unfasten the crankshaft pulley securing bolt and remove the pulley with a suitable puller.

7. Remove the water bypass pipe.

8. Remove the fan belt adjusting bar.

9. Disconnect and remove the heater water outlet pipe. On the 22R-TE engine, remove the No. 3 turbocharger water pipe.

10. Remove the bolts securing the timing chain cover. Remove the cover.

To install:

11. Install the cover and tighten the 8mm bolts to 9 ft. lbs. (13 Nm). Tighten the 10mm bolts to 29 ft. lbs. (39 Nm). Apply sealer to the gaskets for both the timing chain cover and the oil pan.

12. Install the fan belt adjusting bar and tighten it to 9 ft. lbs. (13 Nm).

13. Install the heater water outlet pipe and the No. 3 turbo water hose. Install the water bypass pipe.

14. Install the crankshaft pulley and tighten the bolt to the proper torque.

15. Install the water pump and fluid coupling. Install the air conditioning compressor and then adjust the tension on all drive belts.

16. Install the oil pan. Install the radiator and then install the cylinder head.

17. Refill the engine with oil and coolant. Road test the vehicle and check for leaks.

3F-E Engine

1. Disconnect the negative battery cable and drain the coolant.

2. Disconnect the accelerator and throttle cables.

3. Remove the air intake hose, air flow meter and air cleaner as an assembly.

4. Loosen the power steering pump drive pulley nut.

5. Remove the fluid coupling with the fan and water pump pulley.

6. Remove the power steering pump and the air conditioning compressor. Remove their brackets. Remove the power steering pump idler pulley and its bracket.

7. Remove the cylinder head cover. Remove the rocker shaft assembly.

8. Remove the distributor.

9. Remove the pushrod cover and then remove the valve lifters. Be certain that they are kept in order.

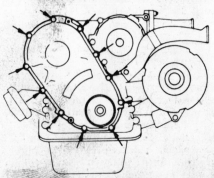

Timing chain cover removal—3F-E engine

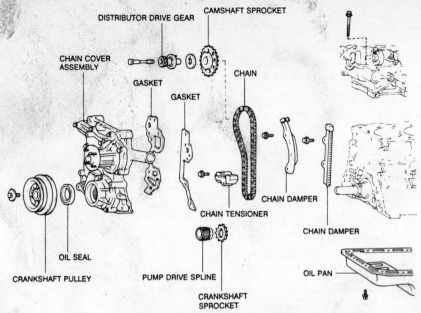

Timing chain and sprockets—22R, 22R-E and 22R-TE engines

10. Loosen the 6 bolts and then slide the power steering pump pulley off the crankshaft.

11. Using special tool 09213-58011 or equivalent, and a 46mm socket wrench, remove the crankshaft pulley bolt. Remove the pulley.

12. Remove the oil cooler pipe and its hose.

13. Remove the timing gear cover and gasket.

To install:

14. There are 3 sizes of timing gear cover bolts. Apply adhesive to the two **A** bolts. Install a new gasket and then position the cover. Finger tighten all bolts. Align the crankshaft pulley set key with the groove of the pulley; gently tap the pulley onto the crankshaft. Tighten the cover bolts marked **A** to 18 ft. lbs. (25 Nm). Tighten those marked **B** or **C** to 43 inch lbs. (5 Nm). Tighten the pulley bolt to 253 ft. lbs. (343 Nm).

15. Position the power steering pulley on the crankshaft and tighten the bolts to 13 ft. lbs. (18 Nm).

16. Insert the valve lifters into their bores and install the pushrod cover. Tighten the bolts to 35 inch lbs. (4 Nm). Make sure the valve lifters are installed in the same bore that they were removed from.

17. Install the rocker shaft assembly, the cylinder head cover and the distributor.

18. Install the water pump pulley, fluid coupling and fan.

19. Install the power steering pump idler pulley and bracket. Install the power steering pump and air conditioning compressor. Adjust the drive belts.

20. Install the air cleaner assembly

and then connect and adjust the accelerator and throttle cables.

21. Fill the engine with coolant and connect the battery cable. Start the engine and check for leaks. Check the ignition timing.

4Y–EC Engine

1. Disconnect the negative battery terminal from the battery. Drain the cooling system.

2. If equipped with an automatic transmission, remove and plug the oil cooler lines at the radiator.

3. Remove the fan shroud, the radiator hoses, the coolant reservoir hose and the upper radiator bolt. Raise and safely support the vehicle. Remove the engine under cover, the mounting bolts and the radiator. Lower the vehicle.

4. Remove the drive belts, the fan, the fluid coupling and the water pump pulley from the water pump.

5. Using the appropriate tools, remove the crankshaft pulley center bolt. Using an appropriate puller, pull the crankshaft pulley from the crankshaft.

6. Remove the front cover mounting bolts. Using a small prybar, lift the front cover from the engine.

To install:

7. Using a putty knife, clean the gasket mounting surfaces.

8. Use a new gasket, sealant and reverse the removal procedures.

9. Using a soft faced hammer, drive the crankshaft pulley onto the crankshaft. Torque the crankshaft pulley bolt to 80 ft. lbs. (108 Nm).

10. Adjust the drive belts and refill the cooling system.

11. Start the engine, allow it to reach

normal operating temperatures and check for leaks.

2TZ–FE Engine

1. Disconnect the negative battery cable.

2. Remove the engine from the vehicle.

3. Remove the cylinder head from the engine.

4. Remove the crankshaft pulley bolt by using a holding tool 0921358012 and 0933000021 or equivalent. Using a pulley pulling tool 0995020017 or equivalent, remove the crankshaft pulley.

5. Remove the left engine mounting.

6. Remove the oil pressure switch and engine ventilation case.

7. Remove the oil pan using a pan removing tool 0903200100 or equivalent. Remove the oil baffle.

8. Remove the 3 bolts and oil filter bracket from the timing cover.

9. Remove the 12 bolts, 2 nuts and timing cover. 3 bolts are in the back of the cover. Be careful not to damage the mating surfaces during removal.

To install:

10. Clean the gasket surfaces and check for warpage.

11. Install new gaskets and torque the (A) bolts to 14 ft. lbs. (21 Nm), (B) bolts to 21 ft. lbs. (28 Nm) and the (C) bolts to 32 ft. lbs. (43 Nm).

12. Install the oil filter bracket and torque to 14 ft. lbs. (21 Nm).

13. Install the crankcase baffle plate, oil pan and ventilation case.

14. Install the oil pressure switch and left engine mount.

15. Install the crankshaft pulley and torque the bolt to 192 ft. lbs. (260 Nm).

16. Install the cylinder head and engine assembly into the vehicle.

17. Install the remaining components and check for leaks.

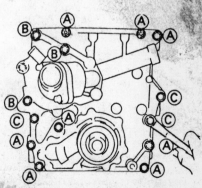

Timing chain cover torque sequence—2TZ-FE engine

Front Cover Oil Seal

REPLACEMENT

22R, 22R–E and 22R–TE Engines

1. Disconnect the negative battery cable. Remove the crankshaft pulley.

2. Using a small prybar, pry the oil seal from the oil pump housing.

3. Using the appropriate seal installer, drive the new seal into the oil pump housing. Apply multi-purpose grease to the lip of the new seal.

4. To complete the installation, reverse the removal procedures.

3F–E Engine

1. Disconnect the negative battery cable. Remove the crankshaft pulley.

2. Using a small prybar, pry the oil seal from the front cover.

3. Using the appropriate seal installer, drive the new seal into the front cover. Apply multi-purpose grease to the lip of the new seal.

4. To complete the installation, reverse the removal procedures.

4Y–EC Engine

FRONT COVER REMOVED

1. Using a drift punch and a hammer, drive the oil seal from the front cover.

2. Using an appropriate seal installer and a hammer, drive the new oil seal into the front cover.

3. Apply grease to the lip of the new seal.

FRONT COVER INSTALLED

1. Disconnect the negative battery cable. Remove the crankshaft pulley.

2. Using the appropriate seal remover, pull the oil seal from the front cover.

3. Apply multi-purpose grease to the lip of the new seal.

4. Using the appropriate seal installer and a hammer, drive the new oil seal into the front cover.

5. To complete the installation, reverse the removal procedures.

Timing Chain and Tensioner

REMOVAL & INSTALLATION

22R, 22R–E and 22R–TE Engines

1. Disconnect the negative battery cable. Remove the cylinder head and timing chain cover.

2. Separate the chain from the damper and remove the chain, complete with the camshaft sprocket.

3. Remove the crankshaft sprocket and the oil pump drive with a puller.

4. Inspect the chain for wear or damage. Replace it if necessary.

5. Inspect the chain tensioner for wear. If it measures less than 11mm, replace it.

6. Check the dampers for wear. If their measurements are below the following specifications, replace them. The specification for the upper damper is 5.0mm and the lower damper is 4.5mm.

To install:

7. Rotate the crankshaft until its key is at TDC. Slide the sprocket in place over the key.

8. Place the chain over the sprocket so its single bright link aligns with the mark on the camshaft sprocket.

9. Install the cam sprocket so the timing mark falls between the two bright links on the chain.

10. Fit the oil pump drive spline over the crankshaft key.

11. Install the timing cover gasket on the front of the block.

12. Rotate the camshaft sprocket counterclockwise to remove the slack from the chain.

13. Install the timing chain cover and cylinder head.

4Y–EC Engine

1. Disconnect the negative battery cable. Remove the front cover.

NOTE: Using a tension gauge, measure the slack of the timing chain, it should be 0.531 in. at 22 lbs. (29 N) pressure.

2. Remove the mounting bolts and the timing chain tensioner.

3. Install the crankshaft pulley on the crankshaft. Using the proper tools, secure the crankshaft pulley, remove the camshaft mounting bolt with a socket wrench and remove the crankshaft pulley.

4. Using an appropriate puller tool, uniformly remove the camshaft sprocket with the crankshaft sprocket and chain.

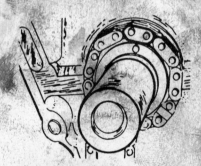

Aligning the crankshaft gear with the single bright link of the timing chain— 22R, 22R–E and 22R–TE engines

Aligning the camshaft sprocket mark between the 2 bright links of the timing chain—22R, 22R–E and 22R–TE engines

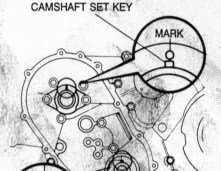

Aligning the timing chain and sprockets —4Y–EC engine

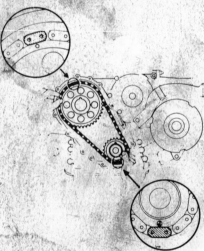

Installing the timing chain on the sprockets—4Y–EC engine

5. Clean the gasket mounting surfaces.

6. Upon installation, align the timing chain with the timing marks on the sprockets, then install the sprockets on their respective shafts.

7. To complete the installation, use new gaskets, sealant and reverse the removal procedures. Torque the camshaft mounting bolt to 67 ft. lbs. (90

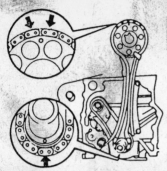

Timing chain alignment marks—2TZ–FE engine

Nm), the timing chain tensioner bolts to 13 ft. lbs. (18 Nm), the crankshaft pulley bolt to 80 ft. lbs. (108 Nm). Adjust the drive belt tension and refill the cooling system. Check and/or adjust the engine timing.

2TZ–FE Engine

1. Remove the timing chain cover. There are 3 bolts in back of the cover.
2. Remove the cylinder head assembly.
3. Remove the chain slipper, damper and oil nozzle.
4. Remove the oil pump drive chain and idle gear.
5. Remove the crankshaft sprocket using a puller 0921336020 or equivalent.

To install:

6. Install the crankshaft sprocket using an installer tool 0960806040 or equivalent.
7. Install the idle sprocket, chain and tensioner. Torque the bolts to 14 ft. lbs. (21 Nm).
8. Check the spring is operating normally against the chain guide by pressing on the chain with a finger and then release. With the guide against the chain, torque the tensioner bolt to 14 ft. lbs. (20 Nm).
9. Install the oil nozzle and torque to 13 ft. lbs. (18 Nm).
10. Install the timing chain damper and slipper. Torque the bolts to 20 ft. lbs. (26 Nm).
11. Place the timing chain on the camshaft sprocket so the timing mark is between the bright chain links and at 12 o'clock.
12. Place the timing chain on the crankshaft sprocket with the single bright link indicated aligned with the timing mark on the crankshaft sprocket at 6 o'clock.
13. Turn the camshaft sprocket counterclockwise to take the slack out of the chain.
14. Tie the timing chain with a cord.
15. Install the timing chain cover and cylinder head.
16. Install the remaining components and check for leaks.

Timing Gears

REMOVAL & INSTALLATION

3F–E Engine

1. Disconnect the negative battery cable. Remove the cylinder head and the front cover from the engine.
2. Remove the oil slinger from the crankshaft. Remove the camshaft thrust plate retaining bolts, by working through the holes provided in the camshaft timing gear.
3. Remove the camshaft through the front of the cylinder block. Support the camshaft while removing it, so the bearings or the lobes do not become damaged.

NOTE: The timing gear is a press-fit and cannot be removed without removing the camshaft.

4. Inspect the crankshaft timing gear. Replace it if it has worn or damaged teeth.
5. Remove the sliding key, then, pull the crankshaft timing gear from the crankshaft with a gear puller.
6. Use a large piece of pipe to drive the timing gear onto the crankshaft. Lightly and evenly tap the end of the pipe until the gear is in its original position.

To install:

7. Apply a coat of engine oil to the camshaft journals and bearings, then, insert the camshaft into the block.
8. Align the mating marks on the timing gears. Slip the camshaft into position. Torque the camshaft thrust plate bolts to 14.5 ft. lbs. (20 Nm).
9. Using a feeler gauge, check the gear backlash, inserted between the crankshaft and the camshaft timing gears. The maximum backlash should be 0.002–0.005 in.; if it exceeds this, replace one or both of the gears, as required.
10. Using a dial indicator, check the gear run-out. Maximum run-out, for both gears, is 0.008 in.; if not, replace the gear.
11. Install the oil nozzle, if removed, by screwing it in place with a screwdriver and punching it in 2 places, to secure it.

NOTE: Be sure the oil hole in the nozzle is pointed toward the timing gear before securing it.

12. To complete the installation, use new gaskets, sealant and reverse the removal procedures.

Timing Belt Front Cover

REMOVAL & INSTALLATION

3VZ–E Engine

1. Disconnect the negative battery cable and drain the coolant.
2. Remove the radiator and shroud.
3. Remove the power steering belt and pump.
4. Remove the spark plugs.
5. Disconnect the No. 2 and 3 air hoses at the air pipe.
6. Disconnect the No. 1 water bypass hose at the air pipe and then remove the water outlet.
7. Remove the air conditioning belt. Remove the alternator drive belt, fluid coupling, guide and fan pulley.
8. Disconnect the high tension cords and their clamps at the No. 2

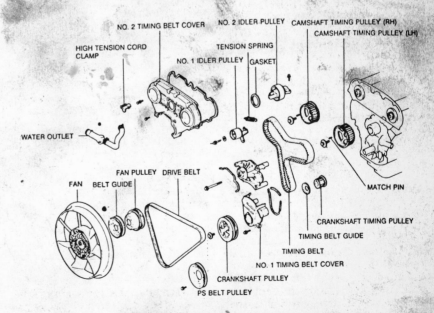

Timing belt components—3VZ–E engine

(upper) timing belt cover and then remove the cover and its gaskets.

9. Rotate the crankshaft pulley until the groove on its lip is aligned with the **0** on the No. 1 (lower) timing belt cover, this should set the No. 1 cylinder at TDC of its compression stroke. The matchmarks on the camshaft timing pulleys must be in alignment with those on the No. 3 (upper rear) timing cover. If not, rotate the engine 360 degrees (1 complete revolution).

10. Remove the crankshaft pulley using a puller.

11. Remove the fan pulley bracket and then remove the No. 1 timing belt cover.

To install:

12. Install the No. 1 cover with the 2 gaskets and tighten the bolts to 48 inch lbs. (5.4 Nm).

13. Install the fan pulley bracket and tighten it to 30 ft. lbs. (41 Nm).

14. Install the No. 2 cover and tighten the bolts 48 inch lbs. (5.4 Nm).

15. Position the crankshaft pulley so the groove in the pulley is aligned with the Woodruff key in the crankshaft. Tighten the bolt to 181 ft. lbs. (245 Nm).

16. Install the fan pulley, guide, fluid coupling and alternator drive belt. Adjust the belt tension.

17. Install the power steering pump and belt. Install the air conditioning belt. Adjust the belt tension.

18. Install the water outlet and connect the bypass hose. Connect the No. 2 and 3 air hoses.

19. Install the spark plugs. Install the radiator, fill with coolant and road test the vehicle. Check for leaks and check the ignition timing.

Timing Belt and Tensioner

REMOVAL & INSTALLATION

3VZ–E Engine

1. Disconnect the negative battery cable. Remove the timing belt covers.

2. Draw a directional arrow on the timing belt and matchmark the belt to each of the pulleys. Remove the timing belt guide and then remove the tension spring.

3. Loosen the idler pulley bolt and shift it left as far as it will go. Tighten the set bolt and relieve the tension on the timing belt. Remove the belt.

4. Remove the crankshaft and camshaft sprocket timing pulleys. Remove the No. 1 idler pulley.

To install:

5. Align the groove in the crankshaft pulley with the key on the crankshaft and press the pulley onto the shaft.

Aligning the timing marks with the rear cover—3VZ–E engine

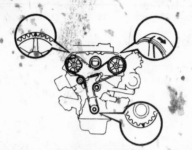

Installing the timing belt—3VZ–E engine

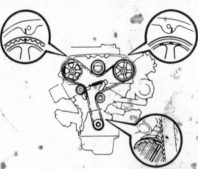

Rechecking the sprocket alignment—3VZ–E engine

6. Install the idler pulley. Align the groove on the pulley with the cavity of the oil pump and then force it to the left as far as it will go. Temporarily tighten it to 27 ft. lbs. (37 Nm).

7. Position the camshaft pulleys on the camshafts so the match holes in each pulley are in alignment with those on the No. 3 (upper rear) timing cover. Align the pulley matchmark with the one on the cover.

NOTE: Do not install the match pin. Check that the bolt head is not touching the pulley.

8. Install the timing belt around the timing pulleys. If reusing the old belt, make sure the arrow and matchmarks all line up with those made earlier on the pulleys.

9. Move the idler pulley to the right as far as it will go. Install the tension spring and then loosen the pulley bolt until the pulley moves lightly with the tension spring force.

10. Check the valve timing and belt tension by turning the crankshaft 2 complete revolutions clockwise. Check that each pulley aligns with its timing marks. Retighten the idler pulley bolt to 27 ft. lbs. (37 Nm).

11. Remove the camshaft timing pulley bolts and align the match pin hole with the match pin hole in the camshaft. Install the pin and bolt and tighten to 80 ft. lbs. (108 Nm).

12. Remove the crankshaft timing pulley bolt and position the belt guide over the crankshaft pulley so the cupped side is out.

13. Install the timing covers. Connect the negative battery cable.

Oil Seal Replacement

Remove the crankshaft pulley, timing belt cover and pry the seal from the retainer. Be careful not to damage the crankshaft. Install the seal with a seal installer 0930937010 or equivalent.

Camshaft

REMOVAL & INSTALLATION

22R, 22R–E and 22R–TE Engines

1. Disconnect the negative battery cable. Remove the cylinder head cover.

2. Remove the rocker arm assembly from the cylinder head.

NOTE: It may be necessary to use a small prybar to lift the rocker arm assembly from the cylinder head.

3. Using a feeler gauge, measure the thrust bearing clearance at the front of the camshaft; the standard clearance is 0.003–0.007 in., it should not exceed 0.0098 in.

4. Remove the camshaft bearing caps and lift out the camshaft. Keep the bearings in order so they may be installed in their original position.

5. Check the camshaft journal caps for damage. Clean all of the bearing surfaces, including the caps, cam journal and the cylinder head.

To install:

6. With the camshaft in place on the cylinder head, lay small strips of plastigage® on each of the camshaft journals (at the tops of the journals, facing front-to-rear).

7. Reinstall the journal caps in their original locations, arrows facing forward, and torque the caps to 13–16 ft. lbs. (18–22 Nm).

8. Remove the journal caps and gauge the width of the plastigage® against the chart on the plastigage® package. Maximum journal clearance is 0.004 in. If the journal clearance is greater than specified, measure the cam journal diameters with a micrometer. If the diameter of any cam journal is less than specified, obtain a new camshaft and recheck the journal clearance. If the clearance is still ex-

cessive, the cylinder head must be replaced.

9. To complete the installation, use new gaskets, sealant and reverse the removal procedures. Refill the cooling system. Torque the camshaft bearing cap bolts to 14 ft. lbs. (19 Nm), the cylinder head-to-engine block to 58 ft. lbs. (79 Nm), the timing chain cover-to-cylinder head bolt to 9 ft. lbs. (12 Nm), the camshaft sprocket-to-camshaft bolt to 58 ft. lbs. (79 Nm), the intake manifold bolts to 14 ft. lbs. (19 Nm), the exhaust manifold bolts to 33 ft. lbs. (45 Nm) and the rocker arm cover to 7–12 ft. lbs. (10–15 Nm). Replace the cooling system fluid and the engine oil. Adjust the valves, the drive belts, then, check and/or adjust the timing.

NOTE: If a new cam is installed, use an assembly lube on the cam lobes and engine oil on the journals. Change the engine oil and filter.

3F–E Engine

1. Disconnect the negative battery cable. Remove the cylinder head and the front cover from the engine.

2. Remove the oil slinger from the crankshaft. Remove the camshaft thrust plate retaining bolts, by working through the holes provided in the camshaft timing gear.

3. Remove the camshaft through the front of the cylinder block. Support the camshaft while removing it, so the bearings or the lobes do not become damaged.

4. Inspect the crankshaft timing gear. Replace it if it has worn or damaged teeth.

5. Remove the sliding key, then, pull the crankshaft timing gear from the crankshaft with a gear puller.

6. Use a large piece of pipe to drive the timing gear onto the crankshaft. Lightly and evenly tap the end of the pipe until the gear is in its original position.

To install:

7. Apply a coat of engine oil to the camshaft journals and bearings, then, insert the camshaft into the block.

8. Align the mating marks on the timing gears. Slip the camshaft into position. Torque the camshaft thrust plate bolts to 14.5 ft. lbs. (20 Nm).

9. Using a feeler gauge, check the gear backlash, inserted between the crankshaft and the camshaft timing gears. The maximum backlash should be 0.002–0.005 in.; if it exceeds this, replace one or both of the gears, as required.

10. Using a dial indicator, check the gear run-out. Maximum run-out, for both gears, is 0.008 in.; if not, replace the gear.

11. Install the oil nozzle, if removed, by screwing it in place and punching it in 2 places, to secure it.

NOTE: Be sure the oil hole in the nozzle is pointed toward the timing gear before securing it.

12. To complete the installation, use new gaskets, sealant and reverse the removal procedures.

4Y–EC Engine

1. Disconnect the negative battery cable. Remove the timing chain from the engine.

2. Remove the right-front seat, the service hole cover and the distributor.

3. Disconnect the cold start injector connector, place a shop towel under the injector tube, then, remove the cold start injector union bolts, the injector and the gaskets. Remove the valve cover, the mounting bolts and the rocker arm assembly.

4. Remove the pushrods, keeping them in order. Using a wire hook or a magnetic finger, remove the valve lifters, keeping them in order.

5. Remove the thrust plate mounting bolts and the plate.

6. While turning the camshaft, slowly pull it out through the front of the engine, making sure not to damage the bearings, the camshaft lobes or the camshaft bearing surfaces.

To install:

7. Install the thrust plate, the camshaft sprocket and bolt onto the camshaft. Using a feeler gauge, measure the thrust bearing clearance, it should be 0.0028–0.0087 in.; if the clearance exceeds 0.012 in., replace the thrust plate.

8. Using a micrometer, check the bearing diameters of the camshaft. Using an internal micrometer, check the camshaft bearing diameters on the engine block.

9. Clean the gasket surfaces.

NOTE: Before installing the valve lifters, coat them with oil.

10. Use new gaskets, sealant and reverse the removal procedures. Torque the camshaft thrust bearing plate bolts to 13 ft. lbs. (17 Nm), the camshaft sprocket bolt to 67 ft. lbs. (93 Nm), the timing chain tensioner bolts to 13 ft. lbs. (17 Nm) and the crankshaft pulley bolt to 80 ft. lbs. (108 Nm). Adjust the drive belts and refill the cooling system. Check and/or adjust the timing.

3VZ–E Engine

1. Disconnect the negative battery cable.

2. Remove the timing belt covers and remove the timing belt.

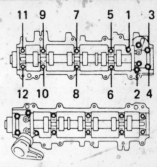

Removing the camshaft bearing bolts

3. Disconnect all wires and hoses to the air intake chamber.

4. Remove the bolts retraining the air intake chamber and remove the air intake chamber from the intake manifold.

5. Remove the rear timing belt cover.

6. Remove the idler pulley and timing cover.

7. Remove the fuel rail and injectors from the intake manifold.

8. Remove the cylinder head cover.

9. Remove the camshaft housing rear cover. Loosen the camshaft retaining bolts a little at a time in the correct sequence.

10. Remove the camshaft from the cylinder head.

11. Installation is the reverse of the removal procedure. When installing the bearing caps, make sure the arrow faces the front of the engine. Torque the caps to 12 ft. lbs. (16 Nm) in the correct sequence.

2TZ–FE Engine

1. Disconnect the negative battery cable.

2. Remove the engine from the vehicle.

3. Remove the valve cover.

4. Place the matchmarks on the camshaft sprocket and chain facing 12 o'clock with the engine upright.

5. Hold the camshaft with a wrench and remove the sprocket bolt.

6. Remove the chain tensioner, timing chain and sprocket. Hold to the side with a wire to maintain camshaft timing.

7. Remove the No. 6 camshaft bearing cap.

8. To remove the exhaust camshaft:
   a. Set the knock pin hole of the camshaft at 5–30 degrees BTDC.
   b. Secure the camshaft sub-gear to the main gear with a service bolt.
   c. Uniformly loosen and remove the bearing caps No. 1, No. 2, No. 3 and No. 5., in that order. Leave No. 4 tight.
   d. Loosen and remove the No. 4 bearing cap.

9. To remove the intake camshaft:

a. Set the knock pin hole of the camshaft at 75–100 degrees BTDC.

b. Uniformly loosen and remove the bearing caps No. 1, No. 2, No. 4 and No. 5., in that order. Leave No. 3 tight.

c. Loosen and remove the No. 3 bearing cap.

To install:

10. To install the intake camshaft:

a. Set the knock pin hole of the camshaft at 75–100 degrees BTDC.

b. Apply prelube to the camshaft journals and lobes.

c. Tighten caps No. 1 and No. 3 to draw the camshaft to the cylinder head.

d. Uniformly torque the bearing caps in several steps to ensure the camshaft does not bend.

e. Torque the bolts to 12 ft. lbs. (16 Nm).

11. To install the exhaust camshaft:

a. Set the knock pin hole of the camshaft at 5–30 degrees BTDC.

b. Apply prelube to the camshaft journals and lobes.

c. Tighten caps No. 2 and No. 4 to draw the camshaft to the cylinder head.

d. Uniformly torque the bearing caps in several steps to ensure the camshaft does not bend.

e. Torque the bolts to 12 ft. lbs. (16 Nm).

12. Apply sealer to the bottom of No. 6 bearing cap and install. Torque the cap to 12 ft. lbs. (16 Nm).

13. Install the camshaft sprocket and chain. Torque the bolt to 54 ft. lbs. (74 Nm).

14. Release the chain tensioner ratchet pawl. Fully push in the plunger and apply the hook to the pin so the plunger can not spring out and install the tensioner. Torque the bolts to 15 ft. lbs. (21 Nm).

15. Set the tensioner by pulling back on the slipper to release the hook.

16. Check the valve clearances.

17. Install the remaining components and check for leaks.

Piston and Connecting Rod

POSITIONING

FRONT MARK

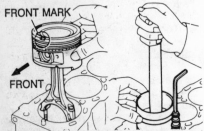
FRONT

Piston and connecting rod positioning—22R, 22R-E and 22R-TE, 3F-E and 4Y-EC engines

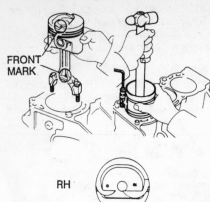
FRONT MARK

RH

FRONT ←

LH

Piston and connecting rod positioning—3VZ-E engine, 2TZ-FE similar

ENGINE LUBRICATION

Oil Pan

REMOVAL & INSTALLATION

Pick-Up, Van and 4Runner

1. Raise the hood and disconnect the negative battery cable.

2. Raise and safely support the vehicle.

3. Drain the engine oil.

4. Remove the steering relay rod and the tie rods from the idler arm, pitman arm, and steering knuckles.

5. Remove the engine stiffening plates.

6. Remove the splash pans from under the engine.

7. Position a floor jack under the transmission and raise the engine/transmission assembly slightly.

8. Remove the front motor mount attaching bolts.

9. Remove the oil pan bolts and remove the oil pan.

10. Scrape the cylinder block and oil pan mating surfaces clean of any old sealing material. Apply gasket sealer to the oil pan when installing a new gasket. If equipped with the 22R, 22R-E, 22R-TE or 4Y-EC engine, apply 5mm bead of gasket sealer; if equipped with the 3F-E or 3VZ-E engine, use 3mm bead. The parts should be installed within 5 minutes of applying the sealer.

11. The oil pan bolts should be tightened to 9 ft. lbs. (12 Nm) on the 22R, 22R-E, 22R-TE or 4Y-EC engines; 6 ft. lbs. (10 Nm) (bolt), 52 inch lbs. (70 Nm) for the 3F-E or 3V-ZE engines. Tighten the bolts in a circular pattern, starting in the middle of the pan and working out towards the ends.

12. Lower the engine and tighten the motor mount bolts. Install the splash shields and stiffening plates.

13. Install any steering arms removed in Step 4 and then lower the vehicle; tighten all suspension components and the motor mounts to their final torque with the vehicle resting on the ground.

14. Fill the engine with oil, road test the vehicle and check for leaks.

Land Cruiser

1. Raise and safely support the vehicle. Remove the engine skid plates.

2. Remove the flywheel side cover and skid plate.

3. Disconnect the front driveshaft from the engine.

4. Drain the engine oil.

5. Remove the bolts which secure the oil pan. Remove the pan and its gasket.

6. Scrape away any old gasket material and then apply gasket sealer to the cylinder block mating surface and the No. 1 and No. 4 main bearing caps.

7. Install the oil pan and tighten the bolts to 69 inch lbs. (7.8 Nm). Always use a new pan gasket.

8. Connect the driveshaft, skid plate and flywheel side cover.

9. Lower the vehicle, fill the engine with oil and check for any leaks.

Previa

This engine has 2 oil pans. If the crankshaft is going to be serviced, the side crankcase pan has to be removed. If the oil pump sump is going to be serviced, the bottom oil pan has to be removed.

1. Drain the engine oil and disconnect the battery cable.

2. Remove the oil level sensor and gasket. Be careful not to drop the sensor when removing.

3. Remove the 14 bolts and 2 nuts.

4. Using a pan removing tool 0903200100 or equivalent, pry the pan from the engine, being careful not to damage the flange.

To install:

5. Clean the gasket mating surfaces and apply gasket sealer No. 0882600080 or equivalent, to the pan and assembly within 5 minutes.

6. Install the pan and torque the bolts and nuts to 48 inch lbs. (5.4 Nm).

7. Install the gasket, oil sensor and torque to 9 ft. lbs. (13 Nm).

8. Install the remaining components.

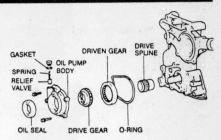

Oil pan and pump assembly—3VZ-E engine

9. Refill the engine with oil and check for leaks.

Oil Pump

REMOVAL & INSTALLATION

22R, 22R-E and 22R-TE Engines

1. Raise and safely support the vehicle. Drain the oil, and remove the oil pan and the oil strainer and pick-up tube.

2. Remove the drive belts from the crankshaft pulley.

3. Remove the crankshaft bolt, and remove the pulley with a gear puller.

4. Remove the 5 bolts from the oil pump and remove the oil pump assembly.

5. Inspect the drive spline, driven gear, pump body, and timing chain cover for excessive wear or damage. If necessary, replace the gears or pump body or cover. Unbolt the relief valve (the vertical bolt on the pump body) when attached to the engine) and check the pistons, oil passages, and sliding surfaces for burrs or scoring. Inspect the crankshaft front oil seal and replace if worn or damaged.

6. When installing, use a new O-ring if necessary.

7. Apply a sealer to the upper bolt and install the 5 bolts.

8. Install the crankshaft pulley and use a new gasket on the oil strainer and oil pan. Be sure to apply sealer to the corners of the oil pan gasket before installing the pan.

3F-E Engine

1. Disconnect the negative battery

cable. Raise and safely support the vehicle. Remove the oil pan.

2. Remove the oil strainer and unfasten the union nuts on the oil pump pipe.

3. Remove the lock wire and the oil pump retaining bolt and pipe from the engine.

4. Remove the oil pump cover and inspect the following parts for nicks, scoring, grooving, etc.:
    a. pump cover
    b. drive and driven gears
    c. pump body

5. Replace either the damaged parts or the complete pump if damage is excessive.

To install:

6. Install the oil pump so the slot in the oil pump shaft is in alignment with the protrusion on the governor shaft of the distributor. Tighten the mounting bolts to 13 ft. lbs. (18 Nm).

7. Install the outlet pipe and tighten the union bolt to 33 ft. lbs. (44 Nm); use new gaskets.

8. Install the oil pan, fill the engine with oil and check for leaks.

NOTE: Be sure to check all of the gaskets and replace if necessary.

4Y-EC Engine

1. Disconnect the negative battery cable. Raise and safely support the vehicle. Remove the oil pan.

2. Remove the oil pump mounting bolts, then pull out the pump assembly.

3. Using a gasket scraper, clean the gasket mounting surfaces.

4. To install, use new gaskets, sealant and reverse the removal procedures. Torque the oil pump bolts to 13 ft. lbs. (17 Nm).

3VZ-E Engine

1. Disconnect the negative battery cable.

2. Remove the timing belt.

3. Raise and safely support the vehicle. Remove the engine under cover.

4. Remove the front differential.

5. Drain the oil.

6. Remove the crankshaft timing pulley.

7. Raise the engine slightly and remove the oil pan.

8. Remove the oil strainer. Insert a drift between the cylinder block and the oil pan baffle plate, cut off the sealer and remove the baffle plate.

NOTE: When removing the baffle plate with the drift, do not damage the baffle plate flange.

9. Remove the oil pump and O-ring.

To install:

10. Apply sealer to the oil pump mating surface running the bead on the inside of the bolts holes. Position a new O-ring in the groove in the cylinder block and install the pump so the spline teeth of the drive gear engage the large teeth on the crankshaft. Tighten the mounting bolts to 14 ft. lbs. (20 Nm).

11. Remove any old sealer and install the baffle plate with new sealer.

12. Install the oil strainer and tighten the bolts to 61 inch lbs. (7 Nm).

13. Install the oil pan, crankshaft pulley and the timing belt.

14. Install the front differential and the undercovers.

15. Fill the engine with oil and check for leaks.

2TZ-FE Engine

1. Remove the oil level sensor. Be careful not to drop the sensor when removing.

2. Remove the pan retaining bolts and pry the pan from the engine using a prying tool 0903200100 or equivalent. Be careful not to damage the flange.

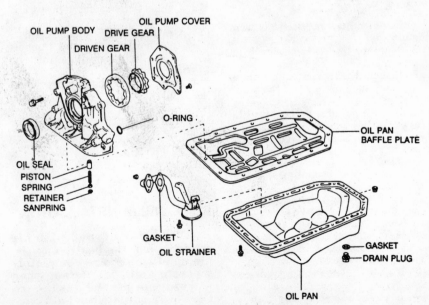

Oil pump assembly—22R, 22R-E and 22R-TE engines

To install:

3. Clean the gasket mating surfaces with a scraper and solvent.

4. Apply a bead of sealer 0882600080 or equivalent to the pan surface and install within 5 minutes.

5. Torque the pan bolts and nuts to 48 inch lbs. (5.4 Nm).

6. Install the oil sensor and torque the bolts to 9 ft. lbs. (13 Nm).

7. Install the remaining components and refill the with engine oil.

Rear Main Bearing Oil Seal

REMOVAL & INSTALLATION

22R, 22R–E and 22R–TE Engines

1. Raise and safely support the vehicle. Remove the transmission and the clutch assembly, if equipped. Remove the transfer case, if equipped.

2. Remove the flywheel or the flexplate from the crankshaft. Remove the cover plate from the rear of the engine.

3. Remove oil pan-to-oil seal retaining plate bolts, the oil seal retaining plate-to-engine bolts and oil seal retaining plate.

4. Carefully pry or drive the old seal from the retaining plate. Be careful not to damage the retaining plate.

5. Using an oil seal driver tool, drive the new seal into the oil seal retaining plate, until the surface is flush.

6. Lubricate the lips of the seal with multi-purpose grease.

7. Clean the gasket mounting surfaces.

8. To install, use new gaskets and reverse the removal procedures. Adjust the clutch.

3F–E Engine

1. Raise and safely support the vehicle. Remove the transfer case, the transmission and the clutch assembly.

2. Remove the flywheel from the crankshaft.

3. Using a small prybar, carefully pry the oil seal from the rear of the crankshaft.

4. Lubricate the lips of the seal with multipurpose grease.

5. Using an oil seal driver tool, drive the new seal into the rear of the crankshaft.

6. Clean the gasket mounting surfaces.

7. To install, reverse the removal procedures. Adjust the clutch.

4Y–EC and 3VZ–E Engines

1. Raise and safely support the vehicle. Remove the transmission and the clutch assembly, if equipped. Remove the transfer case, if equipped.

2. Remove the flywheel or the flexplate from the crankshaft. Remove the cover plate from the rear of the engine.

3. To replace the oil seal with the retaining plate removed:

  a. Remove oil pan-to-oil seal retaining plate bolts, the oil seal retaining plate-to-engine bolts and oil seal retaining plate.

  b. Carefully pry or drive the old seal from the retaining plate. Be careful not to damage the retaining plate.

  c. Using an oil seal driver tool, drive the new seal into the oil seal retaining plate, until the surface is flush.

  d. Lubricate the lips of the seal with multipurpose grease.

4. To replace the oil seal with the retaining plate installed:

  a. Cut off the oil seal lip.

  b. Using a small prybar, pry the oil seal from the retaining plate.

  c. Apply multi-purpose grease to the new oil seal.

  d. Using an oil seal driver tool, drive the new seal into the oil seal retaining plate until the surface is flush.

5. Clean the gasket mounting surfaces.

6. To complete the installation, reverse the removal procedures. Adjust the clutch.

2TZ–FE Engine

1. Remove the transmission from the vehicle.

2. Separate the transmission from the engine.

3. Remove the flywheel or flexplate.

4. Using a knife, cut off the lip of the oil seal.

5. Using a suitable prybar, pry the seal out of the seal carrier. Be careful not to damage the crankshaft.

To install:

6. Clean the seal mating surfaces with solvent.

7. Apply grease to the seal lip.

8. Using an installer tool 0922356010 or equivalent, install the seal. Make sure the seal is seated properly. If not, the seal will leak and the engine will have to be removed again.

9. Install the engine/transmission assembly.

10. Install the remaining components and check for leaks.

ENGINE COOLING

Radiator

REMOVAL & INSTALLATION

1. Disconnect the negative battery cable. Drain the cooling system.

2. Unfasten the hose clamps and disconnect the hoses from the radiator. On Land Cruiser equipped with air conditioning, properly discharge the system.

3. Disconnect the transmission cooling lines, if equipped with an automatic transmission. Disconnect the air conditioning air intake duct on Previa vehicles.

4. On the 22R–TE engine, disconnect the No. 1 turbocharger cooling line.

5. Raise the vehicle, support safely and remove the engine under cover on Van.

6. Remove the fan shrouds, if equipped.

7. Remove the grille assembly and remove the hood lock from the radiator support. On Land Cruiser with air conditioning, remove the condenser to radiator bolts.

8. Remove the coolant recovery tank. Unbolt the radiator and remove it from the vehicle.

To install:

9. Install the radiator into position.

10. On Land Cruiser with air conditioning, install the condensor-to-radiator bolts.

11. Install the hood lock assembly and install the grille.

12. Connect the radiator hoses and the transmission cooler lines.

13. On the 22R–TE engine, connect the turbocharger water line. Connect the air conditioning intake duct on Previa vehicles.

14. Install the coolant recovery bottle. Refill the cooling system to the correct level.

15. Run the engine and check for leaks.

Heater Core

REMOVAL & INSTALLATION

NOTE: If equipped with air conditioning, the heater and the air conditioner are completely separate units. Be certain when working under the dashboard that only the heater hoses are disconnected.

───── **CAUTION** ─────

The air conditioning hoses are under pressure; if disconnected, the escaping refrigerant will freeze any surface with which it comes in contact, including skin and eyes.

Pick-Up, Van, 4Runner and Previa

1. Disconnect the negative battery terminal from the battery.
2. Drain the cooling system.
3. Remove the glove box, the defroster hoses, the air damper, the air duct and the 2 side defroster ducts.
4. Remove the control unit from the instrument panel.
5. Disconnect the heater hoses from the core tubes.
6. Remove the retaining bolts and lift out the heater unit. At this point, the core may be pulled from the case.
7. To install, reverse the removal procedures. Refill the cooling system.

Land Cruiser

FRONT HEATER

NOTE: The entire heater unit must be removed to gain access to the heater core. This procedure requires almost complete disassembly of the instrument panel and lowering of the steering column.

1. Disconnect the negative battery terminal. Remove the glove box and the glove box door.
2. Remove the lower heater ducts. Remove the large heater duct from the passenger side of the heater unit.
3. Remove the ductwork from behind the instrument panel. Remove the radio.
4. Disconnect the wiring connector from the right-side inner portion of the glove opening.
5. Remove the instrument panel pad. Remove the hood release lever. Disconnect the hand throttle control cable.
6. Remove the retaining screw from the left-side of the fuse block.
7. Remove the steering column-to-instrument panel attaching nuts and carefully lower the steering column. Tag and disconnect the wiring as necessary in order to lower the column assembly.
8. Disconnect the electrical connector from the rheostat located to the left of the steering column opening.
9. Remove the center dual outlet duct which is attached to the upper portion of the heater unit.
10. Remove the lower instrument panel.
11. Tag and disconnect the hoses from the heater unit. Remove the heater unit-to-firewall fasteners and the heater unit.

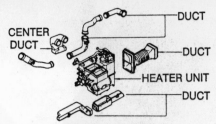

Typical heater unit and ducts— Land Cruiser

12. Remove the heater core-to-heater unit pipe clamps and the heater core retaining clamp, then, withdraw the heater core from the heater unit.
13. To install, reverse the removal procedures. Torque the steering column-to-instrument panel fasteners to 14–15 ft. lbs. (18–20 Nm). Refill the cooling system.

REAR HEATER

1. Turn off the water valve and disconnect both hoses from the rear heater core.
2. Disconnect the wiring from the rear heater.
3. Remove the mounting bolts and lift out the core.
4. To install, reverse the removal procedures. Refill the cooling system.

Water Pump

REMOVAL & INSTALLATION

22R, 22R–E and 22R–TE Engines

1. Disconnect the negative battery cable. Drain the cooling system.
2. If equipped, remove the fan shroud bolts and the shroud.
3. Loosen the alternator adjusting link bolt and remove the drive belt, then, swing the alternator toward the engine.
4. If equipped with an air pump, air conditioning compressor or power steering pump drive belts, it may be necessary to loosen the adjusting bolt, remove the drive belt(s) and move the component(s) aside.
5. Remove the fan from the fluid coupling, the fluid coupling and pulley from the water pump, then, the water pump-to-engine bolts and the pump.
6. Clean the gasket mounting surfaces.
7. To install, use a new gasket, sealant and reverse the removal procedures. Adjust the drive belt(s) tension. Refill the cooling system.

4Y–EC Engine

1. Disconnect the negative battery cable. Drain the cooling system. Disconnect the drive belt from the water pump.

2. Remove the fan from the fluid coupling and the fluid coupling/pulley from the water pump.
3. Remove the drive belt adjusting bar from the water pump, the water pump-to-engine bolts and the water pump.
4. Clean the gasket mounting surfaces.
5. To install, use a new gasket, sealant and reverse the removal procedures.
6. Torque the water pump nuts/bolts to 13 ft. lbs. (17 Nm), the drive belt adjusting bar to 29 ft. lbs. (40 Nm), the pulley/fluid coupling-to-water pump nuts to 10 ft. lbs. (14 Nm) and the fan-to-fluid coupling nuts to 10 ft. lbs. (14 Nm). Adjust the drive belt tension. Refill the cooling system.

3F–E Engine

1. Disconnect the negative battery cable.
2. Drain the engine coolant.
3. Remove the accessory drive belt. Loosen the power steering pump mount, idler pulley and adjusting bolts.
4. Disconnect the overflow tank hose.
5. Disconnect the radiator inlet hose and remove the fan shroud.
6. Remove the fan, fluid coupling and the water pump pulley.
7. Remove the alternator. Disconnect the hoses from the water pump.
8. Remove the water pump, power steering idler pulley and bracket as an assembly.
9. Installation is the reverse of the removal procedure. Torque the water pump mounting bolts to 27 ft. lbs. (37 Nm).

3VZ–E Engine

1. Disconnect the negative battery cable.
2. Remove the timing belt assembly.
3. Remove the thermostat.
4. Remove the idler pulley.
5. Remove the water pump mounting bolts and remove the water pump.

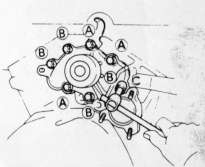

Water pump mounting bolt locations— 3VZ–E engine

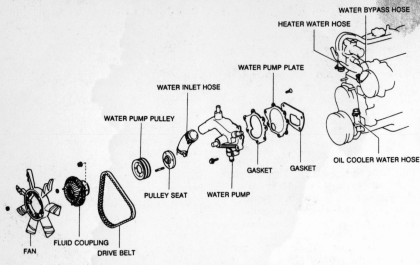

Water pump assembly – 3F-E engine

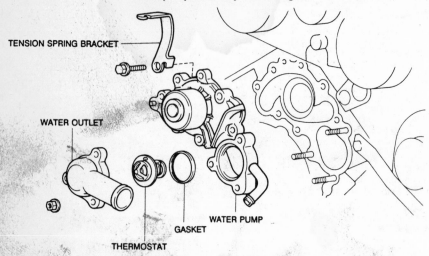

Water pump assembly – 3VZ-E engine

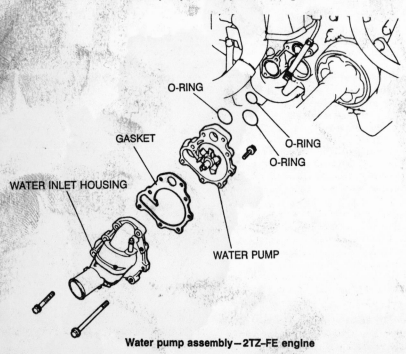

Water pump assembly – 2TZ-FE engine

To install:

6. Clean the mounting surface.

7. Use new seal packing on the water pump and install it in position on the engine.

8. Torque the bolts marked **A** to 13 ft. lbs. (18 Nm) and the bolts marked **B** to 14 ft. lbs. (20 Nm).

9. Install the thermostat and the idler pulley.

10. Install the timing belt assembly.

11. Connect the negative battery cable and refill the cooling system.

2TZ-FE Engine

1. Disconnect the negative battery cable and drain the engine coolant.

2. Raise the vehicle and support safely.

3. Disconnect the heater hose and radiator outlet hoses.

4. Remove the oil filter bracket.

5. Remove the pump retaining bolts and pump from the timing cover.

To install:

6. Install the water pump with new O-rings.

7. Torque the bolts to 21 ft. lbs. (28 Nm).

8. Reconnect the coolant hoses and oil filter bracket.

9. Refill the engine coolant and check for leaks.

Thermostat

REMOVAL & INSTALLATION

22R, 22R-E and 22R-TE Engines

1. Disconnect the negative battery cable. Partially drain the cooling system to a level below the thermostat.

NOTE: Unless the upper radiator hose is positioned over one of the thermostat housing (water outlet) bolts, it is not necessary to detach the hose.

2. Remove the mounting bolts, the water outlet and the thermostat from the intake manifold.

3. Clean the gasket mounting surfaces.

4. To install, use a new gasket, sealant and reverse the removal procedures. Refill the cooling system.

NOTE: When installing a new thermostat, be sure the thermostat is positioned with the spring down.

5. Bleed the cooling system.

4Y-EC Engine

1. Disconnect the negative battery cable. Drain the cooling system to a level below the thermostat.

2. Disconnect the radiator outlet hose from the thermostat housing.

3. Remove the mounting bolts, the thermostat housing and the thermostat.

4. Clean the gasket mounting surfaces.

5. To install, use a new gasket, sealant and reverse the removal procedures, making sure the jiggle valve is placed at the upper-left position. Torque the thermostat housing to 9 ft. lbs. (12 Nm). Refill the cooling system.

NOTE: When installing a new thermostat, be sure the thermostat is positioned with the spring facing the engine block.

6. Bleed the cooling system.

3F–E Engine

1. Disconnect the negative battery cable.

2. Drain the cooling system.

3. Disconnect the cold start injector wire and the BVSV vacuum lines.

4. Remove the thermostat housing bolts and remove the housing.

5. Remove the thermostat housing.

6. Installation is the reverse of removal. Torque the housing bolts to 13 ft. lbs. (18 Nm).

7. Refill the cooling system. Bleed the cooling system.

3VZ–E Engine

1. Disconnect the negative battery cable.

2. Drain the coolant.

3. Remove the radiator outlet hose from the housing.

4. Remove the thermostat housing and thermostat from the engine.

5. Installation is the reverse of the removal procedure. Use a new gasket for installation.

6. Torque the housing bolts to 14 ft. lbs. (20 Nm). Refill and bleed the cooling system.

2TZ–FE Engine

1. Raise the vehicle and support safely. Drain the engine coolant.

2. Disconnect the radiator outlet hose.

3. Remove the retaining bolts, water inlet and thermostat.

4. Installation is the reverse of removal. Torque the retaining bolts to 14 ft. lbs. (20 Nm). Refill the engine with coolant.

Cooling System bleeding

After working on the cooling system, even to replace the thermostat, it must be bled. Air trapped in the system will prevent proper filling and leave the radiator coolant level low, causing a risk of overheating.

1. To bleed the system, start with the system cool, the radiator cap off

and the radiator filled to about an inch below the filler neck.

2. Start the engine and run it at slightly above normal idle speed. This will insure adequate circulation. If air bubbles appear and the coolant level drops, fill the system with an antifreeze/water mixture to bring the level back to the proper level.

3. Run the engine this way until the thermostat opens. When this happens, coolant will move abruptly across the top of the radiator and the temperature of the radiator will suddenly rise.

4. At this point, air is often expelled and the level may drop quite a bit. Keep refilling the system until the level is near the top of the radiator and remains constant.

5. If the vehicle has an overflow tank, fill the radiator right up to the filler neck. Replace the radiator filler cap.

ENGINE ELECTRICAL

NOTE: Disconnecting the battery cable on some vehicles may interfere with the functions of the on board computer systems and may require the computer to undergo a relearning process, once the negative battery cable is disconnected.

Distributor

REMOVAL

3F–E, 22R, 22R–E, 22R–TE and 3VZ–E Engines

1. Disconnect the negative battery cable. Label and disconnect the high tension cables from the spark plugs. Remove the high tension cable from the coil.

2. Remove the primary wire or the electrical connector and the vacuum line, if equipped, from the distributor. Remove the distributor cap spring clips or screws, then the cap.

3. Using a piece of chalk, matchmark the rotor-to-distributor housing and the distributor-to-engine block. This will aid in correct positioning of the distributor during installation.

4. Remove the distributor holddown clamp bolt and the distributor from the engine.

NOTE: It is easier to install the distributor if the engine timing is not disturbed while it is removed.

4Y–EC Engine

1. Disconnect the negative battery terminal from the battery.

2. Remove the front-right seat from the vehicle.

3. Remove the engine service hole cover.

4. Disconnect the distributor vacuum advance hoses.

5. Disconnect the high tension cables from the spark plugs.

6. Using a piece of chalk, matchmark the rotor-to-distributor housing and the distributor housing-to-engine.

7. Remove the distributor-to-engine bolt and the distributor from the engine.

2TZ–FE Engine

1. Disconnect the negative battery cable.

2. Label and disconnect the spark plug wires.

3. Disconnect the distributor connector and ventilation hoses.

4. Remove the cap and packing.

5. Set the No. 1 cylinder to TDC of the compression stroke. Install the service bolt and nut into the equipment driveshaft to turn the crankshaft pulley. Turn the crankshaft 1 turn if the rotor is not facing No. 1 spark plug wire.

6. Remove the distributor holddown and pull the distributor out of the engine.

INSTALLATION

Timing Not Disturbed
ALL ENGINES

1. Insert the distributor into the engine block by aligning the matchmarks made during removal.

2. Engage the distributor drive with the oil pump drive shaft.

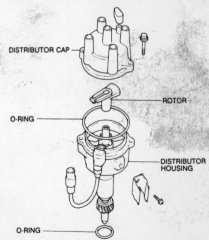

Distributor assembly—22R, 22R–E and 22R–TE engines

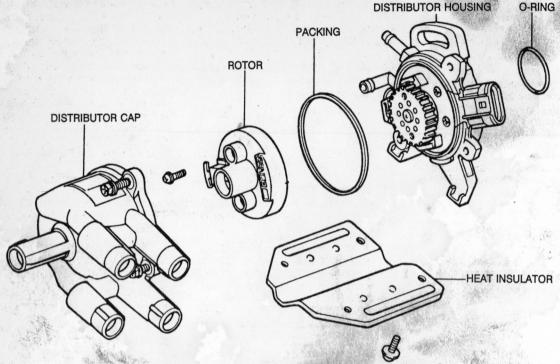

DISTRIBUTOR CAP

ROTOR

PACKING

DISTRIBUTOR HOUSING

O-RING

HEAT INSULATOR

Distributor assembly—2TZ–FE engine

3. Install the distributor hold-down clamp, the cap, the high tension wire, the primary wire or the electrical connector and the vacuum line(s).

4. Install the spark plugs cables.

5. Connect the negative battery cable.

Timing Disturbed

3F–E, 22R, 22R–E, 22R–TE AND 3VZ–E ENGINES

If the engine has been cranked, dismantled or the timing otherwise lost, proceed as follows:

1. Determine the Top Dead Center (TDC) of the No. 1 cylinder's compression stroke by removing the spark plug from the No. 1 cylinder and placing a finger or a compression gauge over the spark plug hole.

NOTE: Using a wrench, turn the crankshaft until the compression pressure starts to build up. Continue cranking the engine until the timing marks indicate TDC (0 degrees).

2. Turn the crankshaft to align the timing marks on the 22R–E and 22R–TE engines to 5 degrees BTDC or on the 22R and 3VZ–E engines to 0 degree TDC.

3. Temporarily install the rotor on the distributor shaft so the rotor is pointing toward the No. 1 terminal of the distributor cap.

4. Using a small prybar, align the slot on the distributor drive (oil pump driveshaft) with the key on the bottom of the distributor shaft.

5. Install the distributor in the block by rotating it slightly (no more than a gear tooth in either direction) until the driven gear meshes with the drive.

NOTE: Oil the distributor drive gear and the oil pump driveshaft end before installation.

6. Temporarily tighten the lock bolt.

7. Remove the rotor, then install the dust cover, the rotor and the distributor cap.

8. Install the primary wire or the electrical connector and the vacuum line(s).

9. Install the No. 1 cylinder spark plug. Connect the cables to the spark plugs in the proper order. Install the high tension wire on the coil.

10. Start the engine and adjust the ignition timing.

4Y–EC ENGINE

1. Remove the No. 1 spark plug, place a finger over the opening and rotate the crankshaft, using a turning tool, in the clockwise direction, until pressure is felt, then replace the spark plug.

NOTE: Make sure the notch on the crankshaft pulley is aligned with the 0 degree mark on the timing plate.

2. Position the oil pump drive rotor slot 30 degrees from the centerline.

3. On the distributor, align the groove on the housing with the pin of the driven gear (the drill mark side).

4. Insert the distributor by aligning the flange center with the bolt hole in the engine block.

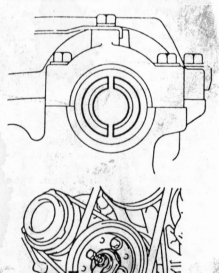

Exhaust camshaft positioning—2TZ–FE engine

5. Lightly tighten the hold-down bolt.

6. To complete the installation, reverse the removal procedures. Adjust the ignition timing.

2TZ–FE ENGINE

1. Remove the No. 1 spark plug, place a finger over the opening and rotate the equipment driveshaft, using a turning tool, in the clockwise direction, until pressure is felt, then replace the spark plug.

NOTE: Make sure the slit in the exhaust camshaft is in the proper position.

2. Remove the service bolt and nut.

3. Align the cut out portion of the coupling with the groove on the housing.

4. Install the distributor and align the center of the flange with the bolt hole on the cylinder head.

5. Install the hold-down bolt loosely.

6. Install the seal packing, distributor cap and connect the wiring.

7. Adjust the timing to specifications and torque the hold-down bolt to 14 ft. lbs. (19 Nm).

Ignition Timing

ADJUSTMENT

Except 2TZ–FE Engine

NOTE: The timing mark locations differ between the engines used in the Pick-Up and 4Runner for 22R, 22R-E, 22R-TE and 3VZ-E engines; Van for 4Y-EC engine and Land Cruiser for 3F-E engine. On the 22R, 22R-E, 22R-TE, 4Y-EC and 3VZ-E engines, the timing marks are located on the crankshaft pulley (painted notch) and the timing cover (plate). On the 3F-E engine the timing marks are located on the flywheel (ball) and the bell housing (pointer).

1. Set the parking brake and block the wheels.

2. Clean off the timing marks and mark them with chalk or paint. The crankshaft may have to rotated to find the marks.

3. Warm the engine to operating temperatures. Connect a tachometer to the engine, then, check and/or adjust the engine idle speed.

NOTE: On the 22R, 22R-E, 22R-TE and 3VZ-E engines, connect the positive (+) tachometer terminal either to the negative (−) ignition coil terminal or to the yellow service connector. On the 4Y-EC engine, connect the positive (+) tachometer to the service

connector on the ignition coil/igniter assembly. Do not connect it to the distributor side. Improper connections will damage the transistorized igniter. On the 4Y-EC and 3VZ-E engines, use a service wire to short the engine check connector

4. Turn off the engine and connect a timing light according to the manufacturer's directions.

5. On the 22R-E and 22R-TE engines, disconnect and short the **T** and the **E₁** connector of the engine check harness (near the front of the vehicle). On all other engines, disconnect and plug the vacuum hose(s) from the distributor vacuum unit.

NOTE: If equipped with a High Altitude Compensation (HAC) system there are 2 vacuum hoses which connect to the distributor. Both must be disconnected and plugged. These systems require an extra step in the timing procedure.

6. Be sure the timing light wires are clear of the fan and pulleys, then start the engine.

7. Allow the engine to run at the

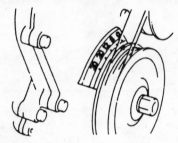

Timing mark location—4Y-EC engine

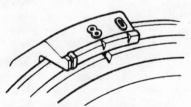

Timing mark location—22R, 22R-E and 22R-TE engines

Timing mark location—3VZ-E engine

specified idle speed with the shift selector in neutral for manual transmission or D for automatic transmission.

8. Point the timing light at the marks. With the engine at the specified idle, the marks should align.

9. If the timing is incorrect, loosen the bolt at the base of the distributor just enough so the distributor can be turned. Hold the distributor by its base and turn it slightly to advance or retard the timing as required. Once the marks are seen to align properly, tighten the bolt.

10. After tightening the distributor bolt or adjusting the octane selector, recheck the timing. Turn off the engine, then, disconnect the timing light and connect the vacuum line(s) at the distributor or the electrical **T** and **E₁** connector, except on engines with HAC.

11. On engines with HAC after setting the initial timing, reconnect the vacuum hoses at the distributor. Recheck the timing.

12. If the advance is still low, pinch the hose between the HAC valve and the 3 way connector; it should now be to specifications. If not, the HAC valve should be checked for proper operation.

2TZ–FE Engine

NOTE: On the 2TZ-FE engine the timing mark is on the equipment driveshaft U-joint and front cover.

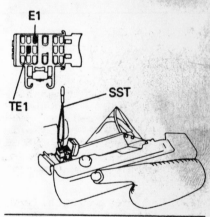

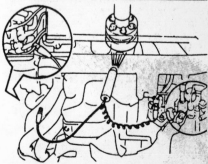

Grouding the ckeck connector and timing location—2TZ-FE engine

1. Ground terminals **TE1** and **E1** of the check connector using tool 0984318020 or equivalent. The check connector is next to the emergency brake lever.

2. Connect a timing light terminal to terminal **30** of the starter and test probe to the No. 1 spark plug wire (light blue).

3. Start the engine and warm up. Slowly turn the distributor unitl the timing mark on the crankshaft pulley is aligned with the **5** mark. Tighten the distributor bolt and recheck timing.

4. Remove the grounding tool from the check connector and timing light.

Alternator

PRECAUTIONS

Several precautions must be observed with alternator equipped vehicles to avoid damage to the unit.

• If the battery is removed for any reason, make sure it is reconnected with the correct polarity. Reversing the battery connections may result in damage to the one-way rectifiers.

• When utilizing a booster battery as a starting aid, always connect the positive to positive terminals and the negative terminal from the booster battery to a good engine ground on the vehicle being started.

• Never use a fast charger as a booster to start vehicles.

• Disconnect the battery cables when charging the battery with a fast charger.

• Never attempt to polarize the alternator.

• Do not use test lights of more than 12 volts when checking diode continuity.

• Do not short across or ground any of the alternator terminals.

• The polarity of the battery, alternator and regulator must be matched and considered before making any electrical connections within the system.

• Never separate the alternator on an open circuit. Make sure all connections within the circuit are clean and tight.

• Disconnect the battery ground terminal when performing any service on electrical components.

• Disconnect the battery if arc welding is to be done on the vehicle.

BELT TENSION ADJUSTMENT

Inspection and adjustment to the alternator drive belt should be performed every 30,000 miles or if the alternator has been removed.

1. Inspect the drive belt to see that it is not cracked or worn. Be sure its surfaces are free of grease or oil.

2. If not using a belt tension gauge, push down on the belt halfway between the fan and the alternator pulleys, (or crankshaft pulley) with thumb pressure; belt deflection should be ⅜–½ in.

3. If using the an appropriate belt tension gauge, position it in the middle of the drive belt and check the belt tension; a new belt should be 170–180 lbs. (230–244 N) for Van or 100–150 lbs. (135–202 N) for Pick-Up and 4Runner, a used belt should be 95–135 lbs. (130–182 N) for Van or 60–100 lbs. (81–136 N) for Pick-Up and 4Runner.

4. If the belt tension requires adjustment, loosen the adjusting link bolt and move the alternator until the proper belt tension is obtained.

5. Tighten the adjusting link bolt.

REMOVAL & INSTALLATION

NOTE: On some engines the alternator is mounted very low. On these engines it may be necessary to remove the gravel shield and work from under the vehicle in order to gain access to the alternator.

Except 2TZ–FE Engine

1. Disconnect the negative battery cable.

2. Remove the air cleaner, if necessary, to gain access to the alternator.

3. On the 22R remove the vane pump pulley.

4. On the 22R–E and 22R–TE engines, drain the engine coolant. If necessary, remove the under engine cover.

5. If equipped with power steering, remove the water inlet pipe bolts and the water inlet hose from the engine.

NOTE: If equipped with air conditioning, it may be necessary to remove the No. 2 fan shroud.

6. Remove the nut or the wiring connector and the wire(s) from the alternator.

7. Remove the adjusting lock, the pivot and the adjusting bolt(s), then the drive belt from the alternator.

8. Remove the alternator attaching bolt and then withdraw the alternator from its bracket.

9. To install, reverse the removal procedures. Rotate the drive belt 8 revolutions for new belt or 5 revolutions for used belt. Adjust the drive belt tension.

10. Refill the cooling system, if it was drained.

11. Connect the negative battery cable.

2TZ–FE Engine

NOTE: To service many engine components, the engine service hole covers may have to be removed. Remove the front seats, scuff plate and seat legs to access the service hole covers.

1. Disconnect the negative battery cable.

2. Raise the vehicle and safely support it.

3. Disconnect the alternator connectors.

4. Loosen the lock bolt, adjusting bolt and pivot bolt. Remove the drive belt.

5. Remove the pivot and lock bolts. Remove the alternator.

To install:

6. Install the alternator and bolts.

7. Install the belt and adjust. Do not pry against the housing.

8. Connect the alternator wiring and battery cable.

9. Lower the vehicle and check operation.

Starter

REMOVAL & INSTALLATION

Except 2TZ–FE Engine

1. If necessary, raise and support the vehicle safely.

2. Disconnect the negative battery terminal from the battery.

NOTE: On some 22R, 22R–E and 22R–TE engines, equipped with an automatic transmission, it may be necessary to remove the transmission oil filler tube.

3. Disconnect the wiring connectors and the wiring from the starter.

4. Remove the starter-to-engine bolts and the starter from the engine.

5. To install, reverse the removal procedures.

2TZ–FE Engine

1. Disconnect the negative battery cable.

2. Raise the vehicle and safely support it.

3. Disconnect the starter connectors.

4. Remove the front driveshaft for 4WD only.

5. Remove the clutch release cylinder (manual transmission only).

6. Remove the 3 bolts and starter for 2WD. Remove the 4 bolts, 2 nuts, center support bracket and starter for 4WD.

To install:

7. Install the starter and torque the long bolts to 41 ft. lbs. (54 Nm) and short bolts to 30 ft. lbs. (41 Nm).

8. Connect the starter connectors.

9. Install the front driveshaft and clutch release cylinder, if removed.

10. Connect the battery cable and lower the vehicle.

FUEL SYSTEM

Fuel System Service Precaution

When working with the fuel system certain precautions should be taken; always work in a well ventilated area, keep a dry chemical (Class B) fire extinguisher near the work area. Always disconnect the negative battery cable and do not make any repairs to the fuel system until all the necessary steps for repair have been reviewed.

RELIEVING FUEL SYSTEM PRESSURE

1. Disconnect the negative battery terminal from the battery.

2. Allow the system enough time to bleed off the fuel pressure through the fuel return line.

3. Before disconnecting any fuel line component, place a rag under the item to catch any excess fuel.

4. After installation, install the negative battery terminal, turn **ON** the ignition switch and check for fuel leaks.

Fuel Tank

REMOVAL & INSTALLATION

1. Disconnect the negative battery cable. Relieve the fuel pressure.

2. This procedure should be done with the tank less than ¼ full. Place a suitable drain pan under fuel tank and remove the drain plug. If the tank is more than ¼ full, use an approved syphon to remove the fuel from the tank.

3. Disconnect all hoses, wiring and tubes that can be accessed at this point.

4. Place a jack under the tank and remove the tank protector.

5. Remove the tank straps or retaining bolts.

6. Lower the tank far enough to disconnect the remaining hoses, wiring and tubes. Be careful not to damage the nylon fuel lines when removing the tank.

7. Drain the remaining fuel from the tank.

To install:

8. Raise the tank and connect the hoses, wiring and tubes.

9. Install the retaining bolts or straps and torque to 29 ft. lbs. (39 Nm).

10. Connect the remaining hoses or wiring.

11. Refill the fuel tank and check for leaks.

12. Check the battery cable and check operation.

Fuel Filter

REMOVAL & INSTALLATION

Except 22R Engine

The fuel filter is located in the engine compartment, at the inlet line to the fuel rail.

1. Disconnect the negative battery cable.

2. Relieve the fuel system pressure.

3. Disconnect and plug the inlet and outlet lines from the filter.

4. Remove the filter retaining bolts and remove the filter.

5. Installation is the reverse of the removal procedure.

6. Use new O-rings and tighten the lines to 22 ft. lbs. (29 Nm).

7. Connect the negative battery cable. Run the engine and check for leaks.

22R Engine

The fuel filter is located under the rear of the vehicle, next to the fuel tank.

1. Disconnect the negative battery cable.

2. Raise and safely support the vehicle.

3. Remove the protective cover from the filter.

4. Disconnect and plug the inlet and outlet lines from the filter.

5. Remove the filter retaining bolts and remove the filter.

6. Installation is the reverse of the removal procedure.

7. Use new O-rings and tighten the lines to 22 ft. lbs. (29 Nm).

8. Connect the negative battery cable. Run the engine and check for leaks.

Mechanical Fuel Pump

PRESSURE TESTING

1. Attach a pressure gauge to the pressure side of the fuel line. If equipped with a vapor return system, squeeze off the return hose.

2. Run the engine at idle and note the reading on the gauge. Stop the engine and compare the reading with the specification. If the pump is operating properly, the pressure will be as specified and will be constant at idle speed. If the pressure varies or is too high or low, the pump should be repaired or replaced.

3. Remove the pressure gauge.

FLOW TEST

1. Disconnect the fuel line from the carburetor. Run the fuel line into a suitable measuring container.

2. Run the engine at idle until there is one pint of fuel in the container. One pint should be pumped in 30 seconds or less.

3. If the flow is below minimum, check for a restriction in the line.

REMOVAL & INSTALLATION

22R Engine

1. Disconnect the negative battery terminal from the battery.

2. Drain the cooling system to a level below the upper radiator hose and remove the upper radiator hose.

3. Remove all 3 lines from the fuel pump, the mounting bolts, the fuel pump and the gasket.

NOTE: The fuel pump is not repairable. It must be replaced as a complete unit.

4. Clean the gasket mounting surfaces.

5. To install, use a new gaskets and reverse the removal procedures. Refill the cooling system. Start the engine and check for leaks.

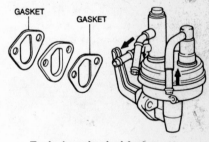

Typical mechanical fuel pump

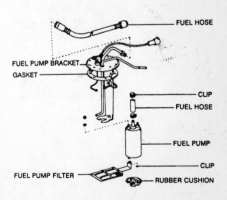

Typical electric fuel pump

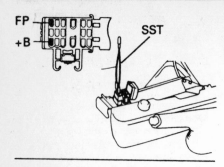

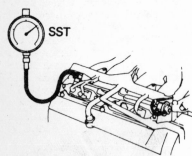

Testing the fuel pump pressure

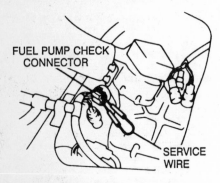

Fuel pump check connector—Van

Electric Fuel Pump

An electric fuel pump is used on all fuel injected engines. The fuel pump is wired into the ignition switch and oil pressure switch circuits. In the event of an oil pressure loss, the fuel pump is turned **OFF** so the engine will stall, thus preventing engine damage due to the oil pressure loss. The fuel pump will operate only when the ignition switch is turned to the **START** position and when the oil pressure is normal.

OPERATION TESTING

1. Disconnect the electrical clip from the oil pressure switch.
2. Turn the ignition switch to the **ON** position; do not start the engine.
3. Short the **Fp** and the **+B** terminals of the check connector. Check the cold start injector hose for pressure.
4. Check for a smooth flow of gasoline from the fuel filter outlet. If the pump is noisy, it is probably defective.

If the pump does not run, check the pump resistor and relay.
5. Disconnect the jumper wire and reconnect the check connector. Turn the ignition switch **OFF**.

PRESSURE TESTING

1. Disconnect the negative battery terminal from the battery and the wiring connector from the cold start injector.
2. Place a container or a shop towel near the end of the delivery tube.
3. Slowly loosen the cold start injector union bolt, then, remove the bolt and the gaskets. Drain the fuel line.
4. Using pressure gauge tool 09268–45011 or equivalent, connect it in line with the cold start injector. Reconnect the battery cable.
5. Short the **Fp** and **+B** terminals of the check connector wire. Turn the ignition switch to the **ON** position and measure the fuel pump pressure. It should be as follows:

22R–E, 2TZ–FE, 3VZ–E, 4Y–E engines—38–44 psi (265–304 kPa)

22R–TE engine—33–38 psi (226–265 kPa)

3F–E engine—37–46 psi (255–314 kPa)
6. Turn the ignition switch **OFF**.

NOTE: If the pressure is high, replace the pressure regulator; if the pressure is low, check the hoses, the connections, the fuel pump, the fuel filter or the pressure regulator.

6. Remove the jumper wire from the check connector. Start the engine. Disconnect and plug the vacuum sensing hose at the pressure regulator, then, measure the fuel pressure at idle. It should be as follows:

22R–E, 2TZ–FE, 3VZ–E, 4Y–E engines—38–44 psi (265–304 kPa)

22R–TE engine—33–38 psi (226–265 kPa)

3F–E engine—37–46 psi (255–314 kPa)
7. Reconnect the vacuum sensing hose to the pressure regulator. The pressure should now be as follows; if not, check the vacuum hose and/or the pressure regulator.

22R–E, 2TZ–FE, 3VZ–E, 4Y–E engines—33–38 psi (226–265 kPa)

22R–TE engine—27–31 psi (186–216 kPa)

3F–E engine—33–37 psi (226–265 kPa)
8. Stop the engine and check that the fuel pressure remains at 21 psi. for 5 minutes. If not, check the fuel pump, the pressure regulator and/or the injectors.

REMOVAL & INSTALLATION

1. Disconnect the negative battery cable. Drain the fuel tank.
2. Disconnect the electrical connector and the fuel lines from the fuel tank.
3. Remove the inlet tube and mounting bolts/straps, then, the fuel tank from the vehicle.
4. Remove the access plate-to-fuel tank bolts, then, pull out the plate/fuel pump assembly.
5. Disconnect the electrical connectors from the fuel pump. Pull the bracket from the lower-side of the fuel pump, then, remove the fuel pump from the fuel hose.
6. Remove the rubber cushion, the clip and the fuel filter from the bottom of the fuel pump.
To install:
7. Install the fuel pump and use new gaskets.
8. Install the fuel tank and connect all electrical and fuel connections.
9. Torque the fuel pump bracket-to-fuel tank to 43 inch lbs. (5.3 Nm). Refill the fuel tank and check for leaks.

Carburetor
REMOVAL & INSTALLATION

1. Disconnect the negative battery terminal from the battery.
2. Label and disconnect the emission control hoses. Disconnect the air intake hose. Remove the mounting and butterfly nuts, then, lift the air cleaner from the carburetor.
3. If equipped with an automatic transmission, disconnect the throttle cable or rod. Disconnect the fuel hose, the emission control hose, the PCV hose and the wiring connector(s) from the carburetor.
4. Disconnect the accelerator linkage and the choke pipe, if equipped.
5. Remove the carburetor-to-manifold nuts/bolts and lift it from the manifold.
6. Cover the open manifold with a clean cloth to prevent dirt and small objects from entering into the engine. Clean the gasket mounting surfaces.
To install:
7. Install a new gasket and carburetor. Torque the retaining nuts to 56 inch lbs. (6.4 Nm).
8. Reconnect all electrical, vacuum and fuel connectors.
9. Install the remaining components. Adjust the idle speed.

IDLE SPEED AND MIXTURE ADJUSTMENT

22R Engine

NOTE: The idle mixture screw

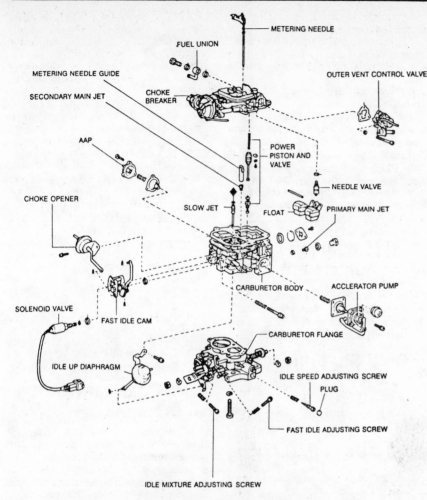

Carburetor components—22R engine

IDLE SPEED ADJUSTMENT

22R–E, 22R–TE, 4Y–EC and 3VZ–E Engines

The engines are equipped with a computer activated, electronic fuel injection system. Prior to adjusting the idle speed, make sure the air cleaner is installed. All vacuum hoses are connected. All pipes and hoses in the air intake system are connected and in good condition. All fuel injection system wiring is connected and in good condition. The engine is at normal operating temperature. All accessories are **OFF**. Transmission selector lever in **N**.

1. Connect the tachometer positive (+) lead to the coil's (−) negative terminal or to the igniter's service connector, if provided.
2. Run the engine at 2500 rpm for 2 minutes.
3. Run the engine at idle and turn the idle speed adjusting screw to obtain the correct idle speed.
4. Disconnect and remove the tachometer.

3F–E and 2TZ–FE Engines

These engines are equipped with electronic fuel injection. The idle speed is controlled by a Idle Speed Control (ISC) valve. The idle speed is preset at the factory and requires no adjust-

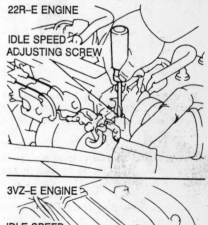

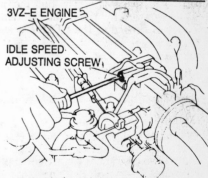

Idle speed adjusting screw—22R–E, 22R–TE and 3VZ–E engines shown, 4Y–EC engine similar

---

is preset at the factory and adjustment should not be necessary.

The idle speed should be adjusted under the following conditions: the air cleaner must be installed, the choke fully opened, the transmission should be in **N**, all accessories should be turned **OFF**, all vacuum lines should be connected and the ignition timing should be set to specification.

1. Start the engine and allow it to reach normal operating temperatures.
2. Check the float setting; the fuel level should be just about even with the spot on the sight glass. If the fuel level is too high or low, adjust the float level.

NOTE: Do not connect the tachometer to the distributor side; damage to the transistorized ignition could result. Never allow the tachometer terminal to touch ground for damage to the igniter or the ignition coil could result.

3. Connect a tachometer in accordance with its manufacturer's instructions. However, connect the tachometer positive (+) lead to the coil's (−)

negative terminal or to the igniter's service connector, if provided.

NOTE: If the idle mixture caps have been removed, turn the idle mixture screws to the fully closed position, then, open them 3½ turns.

4. Using a pair of pliers, remove the caps from the idle speed adjusting screw. Turn the idle speed adjusting screw to obtain the correct idle speed: 700 rpm for a manual transmission or 750 rpm for an automatic transmission.
5. Disconnect the tachometer and install new idle speed adjusting screw cap.

SERVICE ADJUSTMENTS

For all carburetor service adjustment procedures and specifications, please refer to "Carburetor Service" in the Unit Repair section.

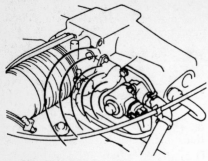

Checking Idle Speed Control (ISC) valve – 3F–E engine

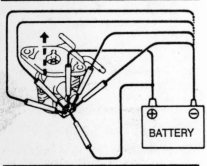

Checking ISC valve – 3F–E engine

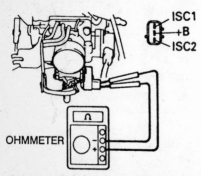

Checking ISC valve – 2TZ–FE engine

ment. There is no idle speed adjusting screws.

1. 3F–E engine: check that there is a clicking sound immediately after stopping the engine. Check the valve

using an ohmmeter. Measure the resistance between the terminals.

Terminals **B1–S1** or **S3** – 10–30 ohms

Terminals **B2–S2** or **S4** – 10–30 ohms

2. If not within specifications, replace the ISC valve.

3. 2TZ–FE engine: using an ohmmeter, measure the resistance between terminals **B+** and the other 2. The resistance should be 18.8–22.8 ohms.

4. If the valve check OK, remove the valve and clean the mounting area with carburetor cleaner. Install new gaskets and valve

Idle Mixture Adjustment

The idle mixture adjustment is preset at the factory and controlled by the electronic control unit. There is no adjustment necessary or possible.

Cold Start Injector

The EFI engines have a cold start injector located in the intake air chamber which aids in cold weather starting.

REMOVAL & INSTALLATION

1. Disconnect the negative battery cable and the cold start injector wire.

2. Place a shop towel or a container under the fuel delivery pipe and drain the fuel from the pipe.

3. Disconnect the fuel pipe from the cold start injector.

4. Remove the mounting bolts and the cold start injector from the intake air chamber.

5. To install, use new gaskets and reverse the removal procedures. Torque the injector bolts to 44–60 inch lbs. (5.4–6.8 Nm).

Fuel Pressure Regulator

The fuel pressure regulator is located on the fuel delivery pipe of the fuel system, it maintains a constant fuel pressure in the injection system.

REMOVAL & INSTALLATION

1. Relieve the fuel pressure. Disconnect the vacuum sensing hose from the pressure regulator.

NOTE: On the 22R–E and 22R–TE engines, remove the No. 1 EGR pipe.

2. Place a shop towel or a container

under the fuel hose connection and disconnect the fuel return hose from the regulator.

3. Remove the locknut for 22R–E, 22R–TE and 3F–E engines, the mounting bolts for 4Y–EC and 3VZ–E engines and the pressure regulator from the fuel delivery pipe.

4. To install, reverse of removal. Torque the locknut to 22 ft. lbs. (30 Nm) for 22R–E, 22R–TE and 3F–E engines or bolts to 44–60 inch lbs. (5.4–6.8 Nm) for 4Y–EC and 3VZ–E engines. Start the engine and check for fuel leaks.

Fuel Injector

TESTING

Except 2TZ–FE Engine

Each injector may be tested for operation while on the engine, in 2 ways.

1. Listen for a clicking at the injector.

2. Using an ohmmeter, check the continuity at each injector's terminal; the resistance should be 1.5–3.0 ohms.

REMOVAL & INSTALLATION

Except 2TZ–FE Engine

1. Disconnect the negative battery terminal from the battery and the ground strap from the rear side of the engine.

2. Disconnect the accelerator wire. If equipped with an automatic transmission, disconnect the throttle cable from the bracket and the clamp.

3. Disconnect the No. 1 and No. 2 PCV hoses.

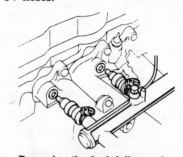

Removing the fuel delivery pipe

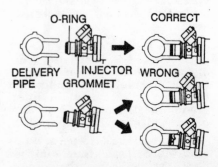

Injector installation into delivery pipe

4. If equipped with a 22R–TE engine, disconnect the No. 2 PCV hose and the Vacuum Control Valve (VCV).

5. Disconnect the following items:
   a. Brake power booster hose
   b. Air control valve hoses
   c. Vacuum Switching Valve (VSV)
   d. Evaporative emission control hose
   e. EGR vacuum hose and modulator
   f. Pressure regulator hose for 2WD
   g. Fuel pressure-up (VSV) and hose
   h. No. 1 and No. 2 air valve hose from the throttle body
   i. No. 2 and No. 3 water bypass hoses from the throttle body
   j. Cold start injector wire
   k. Throttle position wire

6. Remove the following items:
   a. Cold start injector-to-plenum chamber bolt.
   b. No. 1 EGR pipe-to-plenum chamber bolts.
   c. Manifold stay-to-plenum chamber bolts.
   d. Fuel hose clamp, 4 bolts, 2 nuts and the bond strap.
   e. The plenum chamber with the throttle body and gaskets.

7. Disconnect the fuel return hose.

8. Disconnect the following wires:
   a. Auxiliary air valve wire
   b. Knock sensor wire
   c. Oil pressure sender gauge/switch
   d. Starter wire (terminal 50)
   e. Transmission wires
   f. Air conditioning compressor wires
   g. Injector wires
   h. Water temperature sender gauge wire
   i. Overdrive temperature switch wire (air conditioning)
   j. Oxygen sensor and igniter wire
   k. Vacuum Switching Valve (VSV) wire (air conditioning)
   l. Cold start injector time switch wire
   m. Water temperature sensor wire

9. Disconnect the fuel hose from the delivery pipe with the pulsation damper and gaskets.

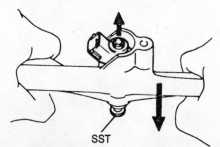

SST

Injector removal – 2TZ–FE engine

10. Remove the injectors from the engine. Take care in handling the injectors.

NOTE: Injector performance tests are possible but special tools are required. If these tools are unavailable, use the test procedures above.

11. To install, use new O-rings and reverse the removal procedures. Torque the hold-down bolts to 14 ft. lbs. (20 Nm). Check for fuel leakage.

NOTE: Each injector should have 4 insulators. Prior to installation, coat the O-rings with clean gasoline. Prior to tightening the hold-down bolts, make sure the injector rotates smoothly in its bore. If not, the O-rings are twisted.

2TZ–FE Engine

1. Relieve the fuel pressure and disconnect the negative battery cable.
2. Remove the right engine service hole.
3. Disconnect the PCV hose, engine wiring harness-to-injectors, vacuum hoses and fuel rail pipes.
4. Remove the fuel rail and spacers.
5. Remove the 4 bolts from the injector covers.
6. Using an injector remover tool 0926874010 and remove the injector from the fuel rail.
To install:
7. Lubricate the O-rings with spindle oil or gasoline. Install new O-rings.
8. Push the injector into the fuel rail. Check the connectors are along the center line of the fuel rail.
9. Install the insulator onto each injector.
10. Install the injector covers and fuel rail onto the engine.
11. Torque the fuel rail retaining bolts to 14 ft. lbs. (20 Nm).
12. Install the remaining components, connect the battery cable and check for leaks.

DRIVE AXLE

Front Halfshaft

REMOVAL & INSTALLATION

4WD Pick-Up, 4Runner and Van

1. Remove the 4WD hub (with the flange) from the axle hub.
2. Raise and safely support the vehicle. Remove the wheel and tire assembly.

3. Disconnect and plug the brake line from the caliper. Remove the caliper from the axle hub.
4. Using a drift punch and a hammer, drive the lock washer tabs away from the locknut.
5. Remove the locknut from the halfshaft. Remove the lock washer, the adjusting nut, the thrust washer, the outer bearing and the axle hub/disc assembly from the vehicle.
6. Remove the knuckle spindle bolts, the dust seal and the dust cover. Using a brass bar and a hammer, tap the steering spindle from the steering knuckle.
7. Turn the halfshaft until a flat spot on the outer shaft is in the upper position, then pull the halfshaft from the steering knuckle.
8. Using a slide hammer, pull the oil seal from the axle housing.
9. Using a clean shop towel, wipe the grease from inside the steering knuckle housing and the halfshaft.
To install:
10. Using an oil seal installation tool, drive a new oil seal into the axle housing until it seats.
11. Install the halfshaft into the axle housing.
12. Using multi-purpose grease, fill the steering knuckle cavity to about ¾ full.
13. To complete the installation, use new seals/gaskets and reverse the removal procedures.
14. Torque the steering spindle-to-steering knuckle bolts to 38 ft. lbs. (52 Nm), the axle hub adjusting nut to 18 ft. lbs. (25 Nm), the axle hub locknut to 33 ft. lbs. (44 Nm), the free wheel/locking hub nuts to 23 ft. lbs. (33 Nm) and the brake caliper to 65 ft. lbs. (88 Nm).

NOTE: To install the wheel bearings with the axle hub, torque the adjusting nut to 43 ft. lbs. (58 Nm), turn the axle hub (back and forth, several times), loosen the nut and retorque the adjusting nut to 18 ft. lbs. (24 Nm).

15. Install the wheel and tire assembly. Lower the vehicle.

4WD Previa

1. Raise and safely support the vehicle. Remove the wheel and tire assembly.
2. Using a drift punch and a hammer, drive the lock washer tabs away from the locknut.
3. Remove the locknut from the halfshahft. Remove the lock washer, the adjusting nut and the thrust washer.
4. Using a tie rod remover 0962810011 or equivalent, remove the tie rod end from the knuckle.
5. Remove the 6 bolts from the inner shaft joint.

6. Disconnect the knuckle from the lower ball joint by removing the 2 bolts.

7. Remove the halfshaft by pulling the knuckle outward and remove the halfshaft from the wheel hub. If the outer shaft will not come out of the hub, soak the splines with penetrating lube, install the nut and tap on the halfshaft with a soft hammer. Be careful not to damage the shaft threads.

To install:

NOTE: Coat the halfshaft splines with anti-seize compound to prevent spline secure. This will help for future halfshaft removal.

8. Connect the lower ball joint and torque the bolts to 94 ft. lbs. (127 Nm).

9. Install the inner halfshaft joint-to-drive axle and torque the bolts to 51 ft. lbs. (61 Nm).

10. Install the tie rod end, torque the nut to 36 ft. lbs. (49 Nm) and install a new cotter pin.

11. Install the halfshaft nut and torque to 76 ft. lbs. (103 Nm).

12. Install the remaining components and check operation.

CV-Boot

REMOVAL & INSTALLATION

Inner Boot

1. Remove the halfshaft from the vehicle.

2. Clean the halfshaft with soap and water. Remove the 2 inner boot clamps and slide the inner boot off the the joint.

3. Place matchmarks on the inner joint outer race and shaft. Remove the outer race from the shaft. Some joints have a large snapring on the inner side of the race that has to be removed.

4. Remove the snapring from the end of the inner joint. Pull the inner joint and boot from the shaft splines.

To install:

5. Install the boot, inner joint and snapring.

6. Fill the boot and joint with special CV-boot grease. The grease ussualy comes with replacement boot kits.

7. Slide the boot into position and install the clamps.

Outer Boot

1. Remove the inner joint and boot from the shaft.

2. Tape the inner splines to protect the boots.

3. Remove the 2 outer boot clamps and slide the outer boot off the the joint.

4. Installation is the reverse of removal. Clean the outer joint from all old grease and repack with special CV-boot grease.

Driveshaft and U-Joints

REMOVAL & INSTALLATION

4WD Vehicles Except Long Bed Pick-Up

REAR

1. Raise and safely support the vehicle and place a drain pan under the transmission.

2. Paint a mating mark on the halves of the rear universal joint flange.

3. Remove the bolts which hold the rear flange together.

4. Remove the splined end of the driveshaft from the transmission. The Previa rear shaft is bolted at both ends.

NOTE: Plug the end of the transmission with a rag or dummy flange to avoid losing transmission oil.

5. Remove the driveshaft from under the vehicle.

6. To install, reverse the removal procedures. Grease the splined end of the shaft before installing. Torque bolts to 31 ft. lbs. (42 Nm) for Van and Previa or 54 ft. lbs. (75 Nm) for Pick-up and 4Runner.

All 4WD Vehicles

FRONT

1. Raise and safely support the vehicle.

2. Matchmark the driveshaft flange at the front axle housing and the transfer case.

3. On Vans, remove the differential support bracket. On Previa, remove the center support bracket.

4. On Pick-Up and 4Runner, remove the flange dust cover.

5. Remove the bolts retaining the driveshaft and remove the driveshaft from the vehicle.

6. Install the driveshaft by aligning the matchmarks made during removal.

7. Torque the retaining bolts to 54–58 ft. lbs. (74–78 Nm). Torque the center support bracket bolts to 35 ft. lbs. (48 Nm). Make sure the drain hole is facing downward.

2WD Long Bed Pick-Up

REAR

1. Raise and safely support the vehicle.

2. Paint mating marks on all 6 flange halves.

3. Remove the bolts attaching the rear universal joint flange to the drive pinion flange.

4. Drop the rear section of the shaft slightly and pull the unit out of the center bearing sleeve yoke.

5. Remove the center bearing support from the crossmember.

6. Unbolt the driveshaft flange from the rear of the transmission and remove driveshaft along with center bearing support.

7. To install, align the matchmarks and reverse the removal procedures. Torque the flange bolts to 54 ft. lbs. (75 Nm).

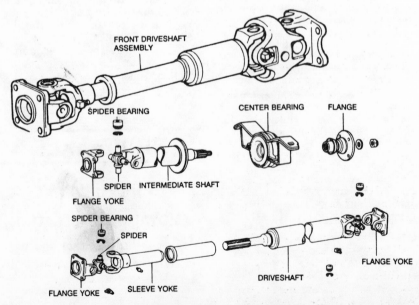

Driveshaft assemblies – Pick-Up and 4Runner

Front Axle Shaft

REMOVAL & INSTALLATION

Land Cruiser

1. Raise and safely support the vehicle. Remove the wheel and tire assembly.
2. Plug the brake master cylinder reservoir to prevent brake fluid leakage from the disconnected brake flexible hose.
3. Remove the outer axle shaft flange cap for automatic locking hub, the hub cover bolts and the cover for free wheel locking hub and the shaft snapring from the axle hub.
4. Remove the outer axle shaft flange for automatic locking hub or the hub ring-to-axle hub bolts, then alternately, screw 2 service bolts into the shaft flange or hub ring and remove the shaft flange or the hub ring with it's gasket.
5. Remove the caliper and disc.
6. Straighten the lock washer and remove the front wheel bearing adjusting nuts with front wheel adjusting nut wrench or similar tool.
7. Remove the front axle hub together with its claw washer, bearings and oil seal.
8. Remove the clip and disconnect the brake flexible hose from the brake tube.
9. Cut and remove the lock wire.
10. Using a soft mallet, lightly, tap the steering knuckle spindle and remove the spindle with it's gasket.

NOTE: When removing the steering knuckle spindle, if equipped with the ball joint type axle shaft joint, be prepared for the disconnection of the outer axle shaft from the joint. Prevent the shaft joint ball from falling from the joint.

11. If equipped with the ball type axle shaft joint, slide the inner front axle shaft out of the axle housing. If equipped with the Birfield constant velocity joint type of axle shaft joint, remove the entire axle shaft assembly from the axle housing.
12. Using a bearing puller, remove the bushing from inside of knuckle spindle and the axle housing oil seal. Using a metal tube as a seating tool, drive oil seal into the axle housing and the new bushing into the knuckle spindle.

NOTE: If equipped with the ball joint type axle joint, install the inner axle with its proper spacer in position until the splines are fully meshed with the differential. If equipped with the Birfield constant velocity joint

axle joint, install the axle into the housing and rotate the axle shaft until its splines mesh with the differential. Fill the steering knuckle about ¾ full with grease and place the joint ball on the inner shaft end.

13. To complete the installation, reverse the removal procedures. Adjust the wheel bearing preload.

Front Axle Bearing

REMOVAL & INSTALLATION

Except 4WD Previa

1. Raise and safely support the vehicle. Remove the 4WD hubs.
2. Using a small prybar, pry the grease seal from the rear of the disc/hub assembly, then remove the inner bearing from the assembly.
3. Using a shop cloth, wipe the grease from inside the disc/hub assembly.
4. Using a brass drift, drive the outer bearing races from each side of the disc/hub assembly.
5. Using solvent, clean all of the parts and blow dry with compressed air.

To install:

6. Using a bearing installation tool, drive the outer races into the disc/hub assembly until they seat against the shoulder.
7. Using multi-purpose grease, coat the area between the races and pack the bearings.
8. Place the inner bearing into the rear of the disc/hub assembly. Using a bearing installation tool, drive a new grease seal into the rear of the disc/hub assembly until it is flush with the housing.

9. Install the disc/hub assembly onto the axle shaft, the outer bearing, the thrust washer and the adjusting nut.
10. To adjust the bearing preload, perform the following:
   a. Torque the adjusting nut to 43 ft. lbs. (58 Nm).
   b. Turn the disc/hub assembly 2–3 times, from the left to the right.
   c. Loosen the adjusting nut until it can be turned by hand.
   d. Retorque the adjusting nut to 18 ft. lbs. (25 Nm).
   e. Install the lock washer and the locknut. Torque the locknut to 33 ft. lbs. (44 Nm).
   f. Check that the bearing has no play.
   g. Using a spring gauge, connect it to a wheel stud, the gauge should be held horizontal, then measure the rotating force, it should be 6–12 lbs. (8–15 Nm).
11. Lower the vehicle.

4WD Previa

1. Remove the steering knuckle from the vehicle. Place the assembly in a vise.
2. Using a hub remover tool (slide hammer) 0952000031 or equivalent, remove the wheel hub from the knuckle.
3. Remove the bearing from the hub using a press and arbor tool 0955010012 or equivalent.
4. Remove the dust protector and oil seal.
5. Remove the bearing snapring from the knuckle and press the inner bearing from the knuckle.

To install:

6. Using a press and arbor tool 0960810010 or equivalent, press the bearing into the knuckle.
7. Install the outer oil seal and dust cover.

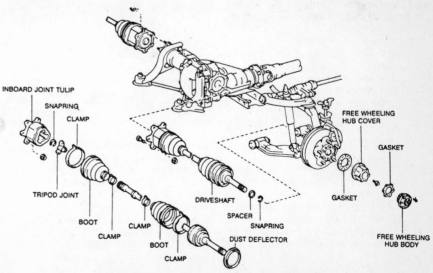

Front drive axle assembly—Pick-Up and 4Runner

8. Press the axle hub onto the knuckle.

9. Install the knuckle assembly onto the vehicle.

Rear Axle Shaft and Bearings

REMOVAL & INSTALLATION

Pick-Up and 4Runner

1. Loosen the rear wheel lug nuts, then raise and safely support the vehicle. Remove the wheel and tire assembly.

2. Place a pan under the axle, remove the plug and drain the axle housing.

3. For 2WD vehicles, remove the clip/clamp-to-frame bolts and disconnect the parking brake cable from the equalizer. For 4WD vehicles, remove the pin and disconnect the rear parking brake cable from the bell crank.

4. Remove the brake drum securing screw and the drum.

5. Disconnect the brake line from the wheel cylinder and plug it, being careful not to damage the fitting.

6. Remove the brake backing plate-to-axle housing nuts and pull the backing plate with the axle from the axle housing.

NOTE: When removing the axle shaft, be careful not to damage the oil seal.

7. Using a pair of snapring pliers, remove the snapring from the axle shaft.

8. Slip tool 09521–25011 or equivalent, over the axle shaft and fasten it to the backing plate. Using 2 metal blocks and a press, press the axle from the backing plate assembly.

9. If necessary to remove the bearing from backing plate, perform the following:

    a. Remove the brake spring, the retracting spring clamp bolt, the lower springs, the shoe strut, the brake shoes and the parking brake lever.

    b. Using a slide hammer puller, pull the outer oil seal from the backing plate.

    c. Press the bearing from the backing plate.

    d. Using the proper installation tools, press the new bearing into the backing plate.

    e. Using the proper seal installation tool, press the new oil seal into the backing plate.

    f. Reassemble the brake components to the backing plate.

10. Using a slide hammer, pull the oil seal from the axle housing.

To install:

11. Using the installation tool and a hammer, drive a new oil seal into the axle housing.

12. Using a press, press the axle shaft into the backing plate and the bearing retainer. Using snapring pliers, install the snapring onto the axle shaft.

13. Clean the gasket mounting surfaces.

14. To complete the installation, reverse the removal procedures. Torque the backing plate-to-axle housing nuts to 51 ft. lbs. (69 Nm). Adjust the brake shoe clearance and bleed the brake system. Refill the axle housing with SAE 90W GL5 gear oil.

Van and Previa

1. Loosen the rear wheel nuts. Raise and safely support the vehicle. Remove the wheel and tire assembly.

2. Remove the brake drum or caliper and rotor.

3. Working through the hole in the axle flange, remove the backing plate-to-axle housing bolts.

4. Using a slide hammer puller, pull the axle shaft from the housing.

5. Using a grinder, grind down the inner bearing retainer on the axle shaft. Using a chisel and a hammer, cut off the retainer and remove it from the shaft.

6. Using a arbor press, press the bearing from the axle shaft.

7. Using a slide hammer puller, pull the oil seal from the axle housing.

To install:

8. Lubricate the new oil seal with multi-purpose grease. Using the proper seal installation tool and a hammer, drive the new oil seal into the axle housing to a depth of 0.236 in.

9. To install, use new gaskets and reverse the removal procedures. Torque the axle retainer-to-housing bolts to 48 ft. lbs. (65 Nm).

Land Cruiser

SEMI-FLOATING TYPE

1. Loosen the rear wheel nuts. Raise and safely support the vehicle. Remove the wheel and tire assembly.

2. Place a pan under the axle, remove the plug and drain the oil from the differential.

3. Remove the brake drum and related parts, as follows:

    a. Remove the cover from the back of the differential housing.

    b. Remove the pin from the differential pinion shaft.

    c. Withdraw the pinion shaft and it's spacer from the case.

    d. Use a mallet to tap the rear axle shaft toward the differential, then remove the C-lock from the axle shaft.

    e. Withdraw the axle shaft from the housing.

4. Using a bearing puller, remove axle bearing and oil seal together from the axle housing. Using a metal tube and a hammer, drive the bearing and the seal into the housing until they seat.

NOTE: Do not mix the parts of the left and right axle shaft assemblies.

5. To complete the installation, reverse the removal procedures. Refill the axle housing with SAE 90W GL5 gear oil.

NOTE: After installing the axle shaft, C-lock, spacer and pinion shaft, measure the clearance between the axle shaft and the pinion shaft spacer with a feeler gauge. The clearance should fall between 0.0024–0.0181 in. If the clearance is not within specifications, use one of the following spacers to adjust it:

    a. 1.172–1.173 in.
    b. 1.188–1.189 in.
    c. 1.204–1.205 in.

FULL FLOATING TYPE

1. Loosen the rear wheel nuts. Raise and safely support the vehicle. Remove the wheel and tire assembly.

2. Place a pan under the axle, remove the plug and drain the oil from the differential.

3. Remove the rear axle shaft plate nuts.

4. Remove the cone washers from the mounting studs by tapping the slits of the washers with a tapered punch.

5. Install bolts into the 2 unused holes of the axle shaft plate.

6. Tighten the bolts to draw the axle shaft assembly out of the housing.

7. To install, use a new gasket, sealant and reverse the removal procedures. Torque the axle shaft nuts to 21–25 ft. lbs. (27–34 Nm).

Front Wheel Hub, Knuckle and Bearing

REMOVAL & INSTALLATION

2WD Pick-Up and Van

1. Raise and safely support the vehicle. Remove the wheel and tire assembly.

2. Remove the brake caliper. Do not disconnect the brake hose from the caliper. Suspend it on a wire.

3. Remove axle hub dust cap, the cotter pin, the nut lock, the adjusting nut, the thrust washer and the outer

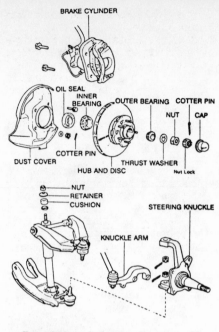

Front hub and steering knuckle— typical 2WD vehicle

bearing, then pull the hub/disc assembly from the axle spindle.

4. Remove the backing plate cotter pins and the mounting nuts or bolts, then the backing plate.

5. Remove steering knuckle arm from the back of the steering knuckle.

6. Remove the nuts, the retainers and the bushings, then the shock absorber from the lower control arm.

7. Support the lower arm with a jack and raise to put pressure on spring.

NOTE: Be careful not to unbalance vehicle support stands when jacking up lower arm.

8. Remove cotter pins, then the upper and lower ball joint nuts. Using a ball joint removal tool, separate the ball joints from the steering knuckle.

9. Remove the steering knuckle from the vehicle.

NOTE: Whenever the hub/disc assembly is removed from the vehicle, it is good practice to replace the grease seal.

10. To install, reverse the removal procedures. Torque the upper ball joint nut to 80 ft. lbs. (109 Nm) for Pick-Up or 58 ft. lbs. (78 Nm) for Van, the lower ball joint nut to 105 ft. lbs. (142 Nm) for Pick-Up or 76 ft. lbs. (109 Nm), the steering knuckle arm-to-steering knuckle bolts to 80 ft. lbs. (109 Nm) for Pick-Up or 61 ft. lbs. (90 Nm) for Van, the shock absorber-to-lower control arm nuts to 19 ft. lbs. (27 Nm) and the backing plate-to-steering knuckle bolts to 80 ft. lbs. (109 Nm) for Pick-Up or 61 ft. lbs. (90 Nm) for Van. Adjust the wheel bearing.

4WD Vehicles

1. Remove the front axle shaft assembly from the vehicle.

2. Remove the oil seal retainer and the oil seal set from the rear of the steering knuckle.

3. At the drag link end of the steering knuckle arm, remove the cotter pin. Using the proper tool, remove the plug from the drag link, then disconnect the drag link from the steering knuckle arm.

4. Remove the tie rod-to-steering knuckle, cotter pin and nut. Using the proper ball joint removal tool, separate the tie rod from the steering knuckle arm.

5. Remove the steering knuckle arm-to-steering knuckle (top) nuts and the steering knuckle-to-bearing cap (bottom) nuts. Using a tapered punch, tap the cone washers slits and remove the washers.

NOTE: Do not mix or lose the upper and lower bearing cap shims.

6. Using a bearing removal tool (without a collar), press the steering knuckle arm with the shims from the steering knuckle.

7. Using a bearing removal tool (without a collar), press the bearing cap with the shims from the steering knuckle.

8. Remove the steering knuckle from the vehicle.

To install:

9. To install the steering knuckle, use a suitable tool to support the upper inner bearing. Using a hammer, tap the steering knuckle arm into the bearing inner race.

10. Using the proper tool, support the lower bearing inner race. Using a hammer, tap the bearing cap into the bearing inner race.

NOTE: When installing the drag link-to-steering knuckle arm, torque the plug all the way, then loosen it 1⅓ turns and secure it with the cotter pin.

11. To install, use gaskets, seals,

Front hub and steering knuckle—typical 4WD vehicle

pack the steering knuckle with multi-purpose grease and reverse the removal procedures for the Land Cruiser.

   a. Torque the steering knuckle arm-to-steering knuckle nuts to 71 ft. lbs. (95 Nm).

   b. Torque the bearing cap-to-steering knuckle nuts to 71 ft. lbs. (95 Nm).

   c. Torque the tie rod-to-steering knuckle arm nut to 67 ft. lbs. (91 Nm).

   d. Torque the axle spindle-to-steering knuckle bolts to 38 ft. lbs. (51 Nm).

12. To install, reverse the removal procedures for the Pick-Up, 4Runner and Van.

   a. Torque the upper ball joint nut to 80 ft. lbs. (109 Nm) for Pick-Up and 4Runner or 58 ft. lbs. (78 Nm) for Van.

   b. Torque the lower ball joint nut to 105 ft. lbs. (142 Nm) for Pick-Up and 4Runner or 76 ft. lbs. (109 Nm) for Van.

   c. Torque the steering knuckle arm-to-steering knuckle bolts to 80 ft. lbs. (109 Nm) for Pick-Up and 4Runner or 61 ft. lbs. (90 Nm) for Van.

   d. Torque the shock absorber-to-lower control arm nuts to 19 ft. lbs. (27 Nm).

   e. Torque the backing plate-to-steering knuckle bolts to 80 ft. lbs. (109 Nm) for Pick-Up and 4Runner or 61 ft. lbs. (90 Nm) for Van.

13. Adjust the wheel bearing preload.

NOTE: To test the knuckle bearing preload, attach a spring scale to the tie rod end hole (at a right angle) in the steering knuckle arm. The force required to move the knuckle from side to side should be 6.6–13 lbs. (10–17 N) except for Land Cruiser or 4–5 lbs. (5–7 N) for Land Cruiser. If the preload is not correct, adjust by replacing shims.

Previa

1. Loosen the halfshaft nut. Raise the vehicle and support safely.

2. Remove the front wheels and disconnect the ABS sensor, if equipped.

3. Remove the front caliper and rotor. Secure the caliper to the vehicle with wire.

4. Remove the the halfshaft nut for 4WD.

5. Remove the MacPherson strut-to-knuckle bolts and lower ball joint bolts.

6. Separate the tie rod end using a remover tool 0962810011 or equivalent.

7. Make sure the halfshaft splines are loose and remove the halfshaft from the knuckle.

8. Remove the steering knuckle/hub assembly.

9. 2WD only, remove the grease cap. Using a chisel, release the nut stake and nut.

10. Using a hub remover tool (slide hammer) 0952000031 or equivalent, remove the wheel hub from the knuckle.

11. Remove the bearing from the hub using a press and arbor tool 0955010012 or equivalent.

12. Remove the dust protector and oil seal.

13. Remove the bearing snapring from the knuckle and press the inner bearing from the knuckle.

To install:

14. Using a press and arbor tool 0960810010 or equivalent, press the bearing into the knuckle.

15. Install the outer oil seal and dust cover.

16. Press the axle hub onto the knuckle.

17. 2WD only, install the lock nut and torque to 147 ft. lbs. (199 Nm). Stake the nut and install the dust cover.

18. Install the knuckle assembly onto the vehicle. Torque the tie rod ends to 36 ft. lbs. (49 Nm), strut bolts to 231 ft. lbs. (314 Nm), lower ball joint bolts to 94 ft. lbs. (147 Nm) and 4WD halfshaft nut to 137 ft. lbs. (186 Nm).

Locking Hubs

REMOVAL & INSTALLATION

1. If equipped with free-wheeling hubs, turn the hub control handle to the **FREE** position.

2. Remove the hub cover bolts and pull off the cover.

3. If equipped with automatic locking hubs, remove the axle bolt with the washer.

4. Using snapring pliers, remove the snapring from the axle shaft.

5. Remove the hub body mounting nuts.

6. Remove the cone washers from the hub body mounting studs by tapping on the washer slits with a tapered punch.

7. Remove the hub body from the axle hub.

8. Apply multi-purpose grease to the inner hub splines.

9. To install, use new gaskets and reverse the removal procedures. Torque the hub body-to-axle hub nuts to 23 ft. lbs. (31 Nm), the plate washer/bolt to 13 ft. lbs. (17 Nm) for auto. locking hub and the hub cover-to-hub body bolts to 7 ft. lbs. (10 Nm).

NOTE: To install the snapring onto the axle shaft, install a bolt into the axle shaft, pull it out and install the snapring.

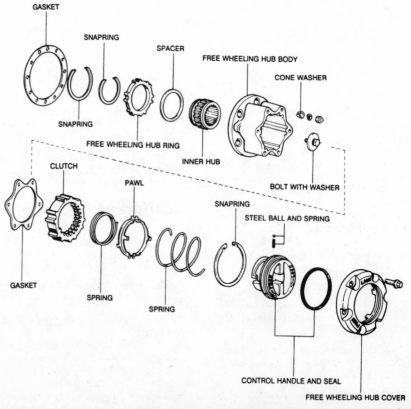

Manual 4WD hub assembly

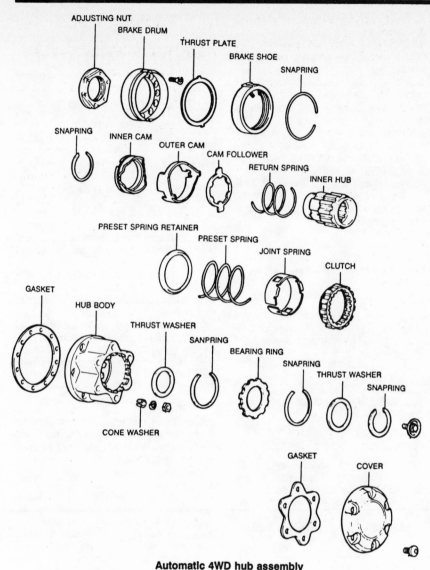

Automatic 4WD hub assembly

Pinion Seal

REMOVAL & INSTALLATION

1. Raise and safely support the vehicle.
2. Matchmark and remove the driveshaft.
3. Remove the companion flange from the differential.
4. Using puller 09308–10010 or equivalent, remove the oil seal from the housing.
5. Remove the oil slinger.
6. Remove the bearing and spacer.
To install:
7. Install the bearing spacer and bearing.

NOTE: Lubricate the seal lips with multi-purpose grease before installing it.

8. Install the oil slinger and using seal installer 09554–30011 or equivalent, install the oil seal.

9. Drive the seal into place, to a depth of 0.59 in below the housing lip for 7.5 in. axles and 0.39 in. below the lip for 8 in. axles.
10. Install the companion flange and install the driveshaft.
11. Lower the vehicle.

Differential Carrier

REMOVAL & INSTALLATION

Rear and Land Cruiser Front

1. Raise and safely support the vehicle. Drain the lubricant from the differential.
2. Remove the axle shafts from the axle housing.
3. Matchmark the driveshaft flange to the differential flange. Remove the mounting bolts and separate the driveshaft from the differential.
4. Remove the differential retaining nuts and pull the assembly out of the differential housing.

5. Place matchmarks on the bearing caps, carrier housing and adjusting nuts.
6. Remove the 4 bolts and bearing caps. Lift the carrier assembly with the side bearings out of the housing. Tag the parts to show the location for reassembly.
To install:
7. Place the bearing outer races on their respective bearings. Install the carrier into the differential housing.
8. Install the bearing adjusting nuts, bearing caps and bolts to the original position. Torque the cap bolts to 58 ft. lbs. (78 Nm).
9. Use new gaskets and reverse the removal procedures. Torque the differential-to-axle nuts to 23 ft. lbs. (31 Nm) and the driveshaft flange-to-differential flange nuts/bolts to 31 ft. lbs. (43 Nm). Refill the axle with 80W–90 gear oil to a level of ¼ in. below the fill hole.

NOTE: Before installing the carrier, apply a thin coat of liquid or silicone sealer to the carrier housing gasket and to the carrier side face of each carrier retaining nut.

Front

1. Raise and safely support the vehicle. Drain the lubricant from the differential.
2. Remove the front axle shafts or halfshafts from the axle housing.
3. Matchmark the front driveshaft flange to the differential flange. Remove the mounting bolts and separate the driveshaft from the differential.
4. Remove the differential carrier cover.
5. Using needle nose pliers or equivalent, remove the snapring from the side gear shafts. Pull the side gear shafts from the housing. A slide hammer may have to be used to remove the side shafts.
6. Remove the A.D.D actuator from the left side of the differential assembly. Remove the intermediate shaft using a puller 0935020015 or equivalent, for A.D.D equipped vehicles.
7. Place matchmarks on the bearing caps, carrier housing and adjusting nuts.
8. Remove the 4 bolts and bearing caps. Using carrier removing tool 0950422011 or equivalent, and a hammer. Dislodge the carrier assembly and lift the carrier assembly with the side bearings out of the housing. Tag the parts to show the location for reassembly.

To install:
9. Install the carrier, bearing races and adjusting shims into the housing.

10. Install the bearing caps and torque the bolts to 58 ft. lbs. (78 Nm).

11. Install the side gear shaft oil seals and side shafts.

12. Install new snaprings to the end of the shafts.

13. Apply sealer to the A.D.D actuator and install. Torque the bolts to 15 ft. lbs. (21 Nm) for A.D.D only.

14. Apply sealer to the differential cover and torque the bolts to 34 ft. lbs. (47 Nm).

15. Torque the differential-to-axle nuts to 19 ft. lbs. (27 Nm) and the front driveshaft flange-to-differential flange nuts/bolts to 54 ft. lbs. (74 Nm). Refill the axle with 80W–90 gear oil to a level of ¼ in. below the fill hole.

Axle Housing

REMOVAL & INSTALLATION

Front

4WD VEHICLES, EXCEPT PREVIA

1. Raise and safely support the vehicle.

2. Matchmark and remove the front driveshaft.

3. Disconnect the axle shafts from the axle assembly.

4. Disconnect vacuum hoses, if equipped with automatic locking hubs.

5. Disconnect the 4WD indicator. Remove the front differential mounting bolt.

6. Support the axle housing with an suitable jack and remove the rear mounting bolts.

To install:

7. Install the differential in position under the vehicle.

8. Install the rear mounting bolts and torque to 123 ft. lbs. (167 Nm).

9. Install the front mounting bolt and torque to 108 ft. lbs. (147 Nm).

10. Connect vacuum hoses and the 4WD indicator.

11. Install the axle shafts and the driveshaft, aligning matchmarks made during removal.

12. Refill the axle with the correct oil and lower the vehicle.

4WD PREVIA

1. Raise the vehicle and safely support it and drain the axle housing.

2. Disconnect the driveshaft and halfshafts. Hange the shafts aside with wire. Be careful not to damage the rubber shaft boots.

3. Remove the left engine undercover and differential support protectors.

4. Place a jack under the axle housing and remove the 3 support bolts and

6 cushions. Lower the jack and remove the assembly.

To install:

5. Raise the assembly and install the collars, cushions and support bolts. Torque the bolts to 54 ft. lbs. (73 Nm).

6. Install the support protectors and torque the bolts to 9 ft. lbs. (12 Nm).

7. Install the under cover.

8. Install the halfshafts and torque the bolts to 51 ft. lbs. (69 Nm).

9. Install the front driveshaft and torque the bolts to 27 ft. lbs. (37 Nm).

10. Refill the axle housing with hypoid gear oil.

Rear

EXCEPT VAN AND PREVIA

1. Raise and safely support the vehicle.

2. Remove the tire and wheel assemblies.

3. Support the axle housing with a suitable jack.

4. Disconnect the shock absorber lower bolts.

5. Disconnect the stabilizer bar and lateral rod. Disconnect the brake lines from the axle housing.

6. Remove the leaf spring U-bolts and carefully lower the axle housing from the vehicle.

7. Installation is the reverse of removal procedure. Torque the shock absorber lower bolts to 19 ft. lbs (25 Nm) on 2WD vehicles or to 47 ft. lbs (64 Nm) on 4WD vehicles. Tighten the U-bolt nuts to 90 ft. lbs (123 Nm).

VAN AND PREVIA

1. Raise and safely support the vehicle.

2. Remove the tire and wheel assemblies.

3. Support the axle housing with a suitable jack.

4. Disconnect the shock absorber lower bolts.

5. Disconnect the stabilizer bar and lateral rod. Also, disconnect the upper and lower control arms.

6. Remove the brake lines from the axle housing.

7. Slowly lower the axle housing from the vehicle.

8. Installation is the reverse of the removal procedure. Observe the following torques:

a. Shock absorber bottom bolts—27 ft. lbs (37 Nm) for 2WD, 94 ft. lbs. (127 Nm) for 4WD vehicles.

b. Upper control arm bolts—105 ft. lbs. (142 Nm).

c. Lower control arm bolts—105 ft. lbs. (142 Nm).

d. Stabilizer bar bolts—19 ft. lbs. (25 Nm).

MANUAL TRANSMISSION

Transmission Assembly

REMOVAL & INSTALLATION

2WD Pick-Up and Van

1. Disconnect the negative battery terminal.

2. Perform the following:

a. Remove the center floor console, if equipped.

b. Remove the shift lever handle, then the floor mat or carpet along with the shift lever boot in order to gain access to the shift lever.

c. Using an shift lever removal tool, remove the shift lever.

NOTE: On the Pick-Up, remove the boot and the shift lever from inside the vehicle.

3. Raise and safely support the vehicle. Drain the transmission fluid.

4. Make matchmarks on the driveshaft flange and the differential pinion flange to indicate their relationships; these marks must be aligned during installation.

5. Remove the driveshaft flange bolts and the center support bearing-to-frame bolts, if equipped with a 2-piece driveshaft. Lower the driveshaft out of the vehicle. Using an appropriate tool, insert it into the end of the transmission to prevent oil leakage.

6. On the Van, disconnect the shift and the select cables from the select outer levers, the clips and the cables.

7. Disconnect the back-up lamp switch electrical connector and the speedometer cable from the transmission, then tie the cable aside.

8. Disconnect the wiring at the starter. Remove the starter mounting bolts and lower the starter out of the vehicle.

9. Remove the exhaust pipe clamp and the exhaust pipe.

10. If the hydraulic line from the clutch release cylinder is clamped to the frame, remove the clamp retaining bolt. Remove the release cylinder mounting bolts and the fork spring, if equipped. Tie the release cylinder aside.

NOTE: It is not necessary to disconnect the hydraulic line from the release cylinder.

11. On column shift vehicles, disconnect the shift selector linkage at the

transmission and remove the transmission cross shafts.

12. Support the rear of the transmission with a jack and remove the transmission-to-crossmember bolts, the crossmember-to-frame bolts and the crossmember from the vehicle.

NOTE: When removing the crossmember, raise the rear of the transmission, just enough to take the weight off of the crossmember.

13. Place a support under the engine with a wooden block (¾ in. thick) between the support and the engine oil pan.

NOTE: The wooden block and support should be no more than about ¼ in. away from the engine so when the engine is lowered, damage will not occur to any underhood components. If possible, shim the support so the wooden block touches the engine.

14. Remove the transmission-to-engine bolts, draw the transmission rearward and down, away from the engine.
To install:
15. Raise the transmission into position under the vehicle.
16. Install the transmission-to-engine bolts.
17. Torque transmission-to-engine bolts to 53 ft. lbs. (70 Nm), the stiffener plate bolts to 27 ft. lbs. (38 Nm), the transmission mount/bracket bolts to 19 ft. lbs. (28 Nm), the rear engine mount bracket-to-crossmember bolts to 9 ft. lbs. (12 Nm).
18. Connect the exhaust pipes and brackets. Install the starter and the clutch release cylinder.
19. Tighten the exhaust pipe-to-manifold bolts to 29 ft. lbs. (40 Nm), the upper exhaust pipe bracket-to-clutch housing bolts to 27 ft. lbs. (38 Nm), the lower exhaust pipe bracket-to-clutch housing bolts to 51 ft. lbs. (68 Nm), the lower starter bolt/release cylinder tube bracket bolt to 29 ft. lbs. (41 Nm), the clutch release cylinder bolts to 9 ft. lbs. (12 Nm).
20. Connect the remaining linkages and connect the driveshaft, aligning the matchmarks made during removal.
21. Refill the transmission to the correct level.
22. Install the shift lever and the center console.
23. Connect the negative battery cable.

4WD Pick-Up, 4Runner and Van

1. Disconnect the negative battery terminal. Remove the starter upper mounting bolt.
2. Working inside the vehicle, pull up the shift lever boot and pull out the shift lever. If equipped with a 22R–E engine, pull up the shift lever boot, then, remove the mounting bolts and pull out the shift lever.
3. Using needle nose pliers, remove the transfer case shift lever snapring and the shift lever.
4. Raise and safely support the vehicle.
5. Drain the lubricant from both the transmission and the transfer case.
6. Make matchmarks on the driveshaft flanges and the differential pinion flanges to indicate their relationships. These marks must be aligned during installation.
7. Remove the driveshaft mounting bolts and remove the front driveshaft assembly.

NOTE: Do not disassemble the front driveshaft to remove it.

8. Using a piece of chalk, place matchmarks on the rear driveshaft and the slip yoke to indicate their relationships; these marks must be aligned during installation.
9. Remove the mounting bolts from the rearward flange of the rear driveshaft. Lower the driveshaft out of the vehicle. Remove the mounting bolts from the slip yoke flange, then, remove the flange and yoke assembly.
10. Unbolt the clutch release cylinder and tie it aside.

NOTE: It is not necessary to disconnect the hydraulic line from the clutch release cylinder.

11. Disconnect the starter motor electrical connectors. Remove the starter bolts and lower the starter from the vehicle.
12. At the transfer case, disconnect the speedometer cable (tie it aside), the back-up light switch connector and the 4WD indicator switch connector.
13. Disconnect the exhaust pipe clamp and the exhaust pipe from the transmission housing.
14. Remove the clutch release cylinder and the tube bracket, then, move the cylinder aside.

NOTE: When removing the clutch release cylinder, do not disassemble the hydraulic line from the cylinder.

15. Remove the crossmember-to-transfer case mounting bolts. Using a jack, raise the transmission and transfer case assembly off of the crossmember. Remove the crossmember-to-frame attaching bolts and remove the crossmember.
16. Place a support under the engine oil pan, with a wooden block (¾ in. thick) between the support and the engine oil pan.

NOTE: The wooden block and support should be no more than about ¼ in. away from the engine so when the engine is lowered, damage will not occur to any underhood components. If possible, shim the support so the wooden block touches the engine.

17. Lower the jack until the engine rests on the support.
18. Remove the exhaust pipe bracket and the stiffener plate bolts.
19. Remove the transmission-to-engine bolts, draw the transmission/transfer case assembly rearward and down away from the engine.
20. Remove the transmission-to-transfer case adapter bolts and pull the transfer case from the transmission.
To install:
21. Raise the transmission into position under the vehicle.
22. Install the transmission-to-engine bolts.
23. Torque transmission-to-engine bolts to 53 ft. lbs. (73 Nm), the stiffener plate bolts to 27 ft. lbs. (38 Nm), the transmission mount/bracket bolts to 19 ft. lbs. (27 Nm), the rear engine mount bracket-to-crossmember bolts to 9 ft. lbs. (12 Nm).
24. Connect the exhaust pipes and brackets. Install the starter and the clutch release cylinder.
25. Tighten the exhaust pipe-to-manifold bolts to 29 ft. lbs. (40 Nm), the upper exhaust pipe bracket-to-clutch housing bolts to 27 ft. lbs. (38 Nm), the lower exhaust pipe bracket-to-clutch housing bolts to 51 ft. lbs. (69 Nm), the lower starter bolt/release cylinder tube bracket bolt to 29 ft. lbs. (40 Nm), the clutch release cylinder bolts to 9 ft. lbs. (12 Nm).
26. Connect the remaining linkages and connect the driveshaft, aligning the matchmarks made during removal.
27. Refill the transmission to the correct level.
28. Install the shift lever and the center console.
29. Connect the negative battery cable.

Previa

1. Disconnect the negative battery cable and drain the transmission fluid.
2. Raise the vehicle and support safely.
3. Remove the starter motor. Matchmark the driveshafts-to-flange and remove the front (4WD) and rear driveshafts.
4. Remove the clutch release cylinder, hose and bracket.
5. Remove the exhaust pipe and bracket.

6. Disconnect the control cables/bracket and speed sensor connector.

7. Remove the engine-to-transmission stiffener plate.

8. Place a suitable transmission jack under the transmission.

9. Remove the engine rear mounting bolts and raise the rear side of the engine.

10. Remove the engine-to-transmission bolts, pull the transmission toward the rear and remove.

To install:

11. Align the input shaft with the clutch disc and push the transmission fully into position.

12. Install the transmission bolts and torque to 53 ft. lbs. (72 Nm).

13. Install the rear engine mounts and stiffener plate. Torque the bolt to 27 ft. lbs. (37 Nm).

14. Connect the speed sensor and control cables.

15. Install the exhaust pipe and torque the bracket to 37 ft. lbs. (51 Nm).

16. Install the clutch release cylinder, starter and driveshafts. Torque the starter to 41 ft. lbs. (56 Nm) and driveshaft bolts to 20 ft. lbs. (25 Nm).

17. Lower the vehicle.

18. Connect the battery cable and refill with transmission fluid.

CLUTCH

Clutch Assembly

REMOVAL & INSTALLATION

1. Raise and safely support the vehicle. Remove the transmission from the vehicle.

2. Make matchmarks on the clutch cover and flywheel, indicating their relationship.

3. Loosen the clutch cover-to-flywheel retaining bolts a turn at a time. The pressure on the clutch disc must be released gradually.

4. Remove the clutch cover-to-flywheel bolts. Remove the clutch cover and the clutch disc.

5. If the clutch release bearing is to be replaced, perform the following:

  a. Remove the bearing retaining clip(s), the bearing and hub.

  b. Remove the release fork and the boot.

  c. The bearing is press fitted to the hub.

  d. Clean all parts and lightly grease the input shaft splines and all of the contact points.

  e. Install the bearing/hub assembly, the fork, the boot and the re-

taining clip(s) in their original locations.

To install:

6. Inspect the flywheel surface for cracks, heat scoring (blue marks) and warpage. Replace or resurface the flywheel, if any damage is present.

NOTE: Before installing any new parts, make sure they are clean. During installation, do not get grease or oil on any of the components, as this will shorten clutch life considerably.

7. Using an clutch alignment tool, position the clutch disc against the flywheel. The raised center section of the disc faces the transmission.

8. Install the clutch cover over the disc and install the bolts loosely. Align the pressure plate-to-flywheel matchmarks. If a new or rebuilt clutch cover assembly is installed, use the matchmark on the old cover assembly as a reference. Torque the pressure plate-to-flywheel bolts to 14 ft. lbs. (19 Nm), using a criss-cross pattern.

9. Install the transmission into the vehicle.

PEDAL HEIGHT/FREE-PLAY ADJUSTMENT

The pedal height measurement is gauged from the angled section of the floorboard to the center of the clutch pedal pad. The correct pedal height is 6.7–7.3 in. for Van, 6.12 for Pick-Up and 4Runner or 6.46 in. for Previa.

If necessary, adjust the pedal height by loosening the locknut and turning the pedal stop bolt which is located above the pedal towards the drivers seat. Tighten the locknut after the adjustment.

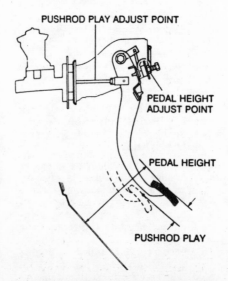

PUSHROD PLAY ADJUST POINT

PEDAL HEIGHT ADJUST POINT

PEDAL HEIGHT

PUSHROD PLAY

Clutch pedal adjustment points

Clutch Master Cylinder

REMOVAL & INSTALLATION

Pick-Up and 4Runner

1. Disconnect the negative battery cable. Disconnect the master cylinder pushrod pin from the top of the clutch pedal.

2. Remove the hydraulic line from the master cylinder, being careful not to damage the compression fitting.

3. Remove the master cylinder-to-cowl nuts/ bolts.

4. To install, reverse the removal procedures. Partially tighten the hydraulic line before tightening the master cylinder mounting nut(s). Torque the nuts/bolts to 9 ft. lbs. (12 Nm). Bleed the clutch system. Adjust the push rod play clearance.

Van

1. Disconnect the negative battery terminal from the battery.

2. Remove the reservoir cap from the cluster finish panel, the mounting screws, then, pull the cluster finish panel forward and remove it.

3. Remove the mounting screws and pull the instrument panel forward, then, disconnect the speedometer and the electrical connectors from it.

4. Remove the No. 3, the No. 1 and the No. 2 air ducts.

5. Disconnect and plug the reservoir hose at the master cylinder. Disconnect the clutch line union.

6. Remove the mounting bolts and the master cylinder.

7. To install, reverse the removal procedures. Bleed the clutch system. Adjust the clutch pedal.

Previa

1. Disconnect the negative battery cable.

2. Remove the instrument panel lower finish panel and steering column cover.

3. Remove the clip and clevis pin.

4. Disconnect the reservoir hose and clutch union line.

5. Remove the mounting bolts and pull out the master cylinder.

6. Installation is the reverse of removal. Torque the mounting bolts to 9 ft. lbs. (12 Nm).

7. Bleed the clutch hydraulic system.

8. Connect the battery cable and check operation.

ADJUSTMENT

Pushrod

The pedal pushrod play is the distance

between the clutch master cylinder piston and the pedal pushrod located above the pedal towards the firewall. Since it is nearly impossible to measure this distance at the source, it must be measured at the pedal pad, preferably with a dial indicator gauge. The pushrod play specification is: 0.040–0.200 for Land Cruiser and Previa or 0.039–0.197 for Vans, Pick-Ups and 4Runner.

If necessary, adjust the pedal play by loosening the pedal pushrod locknut and turning the pushrod. Tighten the locknut after the adjustment.

Free-Play

The free-play measurement is the total travel of the clutch pedal from the fully released position to where resistance is felt as the pedal is pushed downward. The free-play specification is: 0.20–0.59 for all vehicles.

Clutch Slave Cylinder

REMOVAL & INSTALLATION

1. Raise and safely support the vehicle.
2. If equipped, remove the tension spring on the clutch fork.
3. Remove the hydraulic line from the release cylinder. Be careful not to damage the fitting.
4. Turn the release cylinder pushrod in sufficiently to gain clearance from the fork.
5. Remove the mounting bolts and withdraw the cylinder.
6. To install, reverse the removal procedures. Bleed the clutch system. Adjust the fork tip clearance.

BLEEDING THE HYDRAULIC CLUTCH SYSTEM

1. Fill the master cylinder reservoir with brake fluid.
2. Remove the cap and loosen the bleeder screw on the clutch release cylinder. Cover the hole with a finger.
3. Pump the clutch pedal several times. Take the finger off the hole while the pedal is being depressed so the air in the system can be released. Put a finger back on the hole and release the pedal.
4. When fluid pressure can be felt tighten the bleeder screw.
5. Place a short length of hose over the bleeder screw and the other end in a jar half full of clean brake fluid.
6. Depress the clutch pedal and loosen the bleeder screw. Allow the fluid to flow into the jar.
7. Tighten the plug, then release the clutch pedal.
8. Repeat this procedure until no

air bubbles are visible in the bleeder tube.
9. When there are no more air bubbles in the system, tighten the plug fully with the pedal depressed. Replace the plastic cap.
10. Refill the master cylinder to the correct level with brake fluid. Check the system for leaks.

AUTOMATIC TRANSMISSION

Transmission Assembly

REMOVAL & INSTALLATION

1. Disconnect the negative battery terminal from the battery. On the Pick-Up and 4Runner, remove the air cleaner assembly.
2. Disconnect the transmission throttle cable from the carburetor linkage or the throttle body.
3. Raise and safely support the vehicle. Drain the transmission fluid.
4. Disconnect the wiring connectors (near the starter) for the neutral start switch and the back-up light switch. If equipped, disconnect the solenoid (overdrive) switch wiring at the same location. Disconnect the oil level gauge for Previa.
5. Disconnect the starter wiring at the starter. Remove the mounting bolts and the starter from the engine.
6. Make matchmarks on the rear driveshaft flange and the differential pinion flange. These marks must be aligned during installation.
7. Unbolt the rear driveshaft flange. If the vehicle has a 2 piece driveshaft, remove the center bearing bracket-to-frame bolts. Remove the driveshaft from the vehicle.
8. Disconnect the speedometer cable (tie it aside) and the shift linkage from the transmission.
9. Disconnect the transmission oil cooler lines at the transmission.
10. Disconnect the exhaust pipe clamp and remove the oil filler tube.
11. Support the transmission, using a jack with a wooden block placed between the jack and the transmission pan. Raise the transmission, just enough to take the weight off of the rear mount.
12. On the Pick-Up and 4Runner, remove the rear engine mount with the bracket and the engine under cover, to gain access to the engine crankshaft pulley. On the Van, remove the fuel tank mounting bolts and support the

fuel tank; remove the transmission mount through bolt. Remove the stiffener plates on 2WD Previa.
13. Place a wooden block (or blocks) between the engine oil pan and the front frame crossmember.
14. Slowly, lower the transmission until the engine rests on the wooden block.
15. Remove the rubber plug(s) from the service holes located at the rear of the engine in order to gain access to the torque convertor bolts.
16. Rotate the crankshaft (to remove the torque convertor bolts) to access the bolts through the service holes.
17. Obtain a bolt of the same dimensions as the torque convertor bolts. Cut the head off of the bolt and hacksaw a slot in the bolt opposite the threaded end.

NOTE: This modified bolt is used as a guide pin. Two guides pins are needed to properly install the transmission.

18. Thread the guide pin into one of the torque convertor bolt holes. The guide pin will help keep the convertor with the transmission.
19. Remove the stiffener plates from the transmission.
20. Remove the transmission-to-engine bolts, then carefully move the transmission rearward by prying on the guide pin through the service hole.
21. Pull the transmission rearward and lower it (front end down) out of the vehicle.

To install:

22. Apply a coat of multi-purpose grease to the torque convertor stub shaft and the corresponding pilot hole in the flywheel.
23. Install the torque convertor into the front of the transmission. Push inward on the torque convertor while rotating it to completely couple the torque convertor to the transmission.
24. To make sure the convertor is properly installed, measure the distance between the torque convertor mounting lugs and the front mounting face of the transmission. The proper distance is 0.080 in.
25. Install guide pins into 2 opposite mounting lugs of the torque convertor.
26. Raise the transmission to the engine, align the transmission with the engine alignment dowels and position the convertor guide pins into the mounting holes of the flywheel.
27. Install and tighten the transmission-to-engine mounting bolts. Torque the bolts to 47 ft. lbs. (63 Nm).
28. Remove the convertor guide pins and install the convertor mounting bolts. Rotate the crankshaft as necessary to gain access to the guide pins and bolts through the service holes. Evenly, tighten the convertor mount-

ing bolts to 13 ft. lbs. (17 Nm). Install the rubber plugs into the access holes.

29. Install the engine undercover. Raise the transmission slightly and re-move the wood block(s) from under the engine oil pan.

30. Install the transmission cross-member. Torque the crossmember-to-frame bolts to 26–36 ft. lbs. (34–48 Nm).

31. Lower the transmission onto the crossmember and install the transmis-sion mounting bolts. Torque the bolts to 19 ft. lbs. (26 Nm).

32. Install the oil filler tube and con-nect the exhaust pipe clamp.

33. Connect the oil cooler lines to the transmission and torque the fittings to 25 ft. lbs. (34 Nm).

34. To complete the installation, re-verse the removal procedures. Adjust the transmission throttle cable. Refill the transmission with Dexron®II flu-id. Road test the vehicle and check for leaks.

SHIFT LINKAGE ADJUSTMENT

1. Loosen the adjustment nut on the transmission shift cable.

2. Push the manual lever of the transmission fully rearward except for Previaor downward for Previa.

3. Move the manual lever back 2 notches, which is the N position except for Previa or 3 notches for Previa.

4. Set the gearshift selector lever in the N position.

5. Apply a slight amount of forward pressure on the selector lever and tighten the shift cable adjustment nut.

THROTTLE LINKAGE ADJUSTMENT

1. Remove the air cleaner assembly.

2. Push the accelerator to the floor and check that the throttle valve opens fully; if not, adjust the accelerator link, so it does.

3. Push back the rubber boot from the throttle cable which runs down to the transmission. Loosen the throttle cable adjustment nuts so the cable housing can be adjusted.

4. Fully open the carburetor throt-tle by pressing the accelerator all the way to the floor.

5. Adjust the cable housing so, with the throttle wide open, the distance between the outer cable end rubber cap to the inner cable stopper is 0–0.04 in. (0–1mm).

6. Tighten the nuts and double check the adjustment. Install the rub-ber boot and the air cleaner.

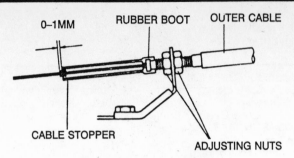

Throttle linkage adjustment

NEUTRAL START SWITCH ADJUSTMENT

The neutral safety switch prevents the vehicle from starting unless the gear-shift selector is in either the P or N po-sitions. If the vehicle will start in these positions, adjustment of the switch is required.

1. Loosen the neutral start switch bolt.

2. Place the selector lever in the N position.

3. Disconnect the wires from the neutral start switch.

4. Connect an ohmmeter between the terminals of the switch.

5. Adjust the switch until there is continuity between the N and B terminals.

6. Reconnect the wires. Torque the bolt to 48 inch lbs. (65 Nm).

TRANSFER CASE

Transfer Case Assembly

REMOVAL & INSTALLATION

The transfer case and transmission are connected together. It is recom-mended by the manufacturer that they be removed from the vehicle as an assembly and then separated for repairs.

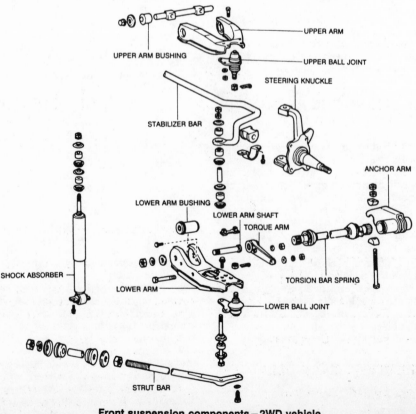

Front suspension components—2WD vehicle

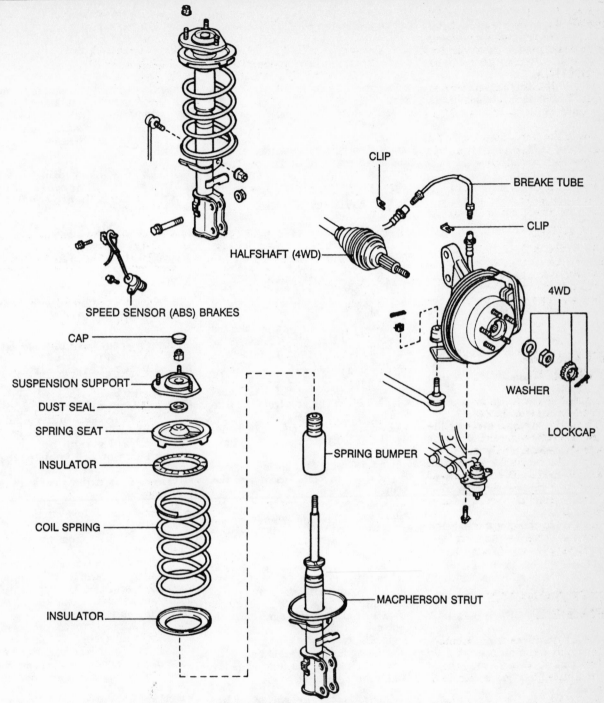

CLIP

BREAKE TUBE

CLIP

HALFSHAFT (4WD)

4WD

SPEED SENSOR (ABS) BRAKES

CAP

SUSPENSION SUPPORT

DUST SEAL

SPRING SEAT

WASHER

INSULATOR

LOCKCAP

SPRING BUMPER

COIL SPRING

INSULATOR

MACPHERSON STRUT

Front suspension components—Previa

FRONT SUSPENSION

Shock Absorbers

REMOVAL & INSTALLATION

2WD Pick-Up and Van

1. Raise and safely support the ve-hicle. Remove the wheel and tire assembly.

2. Unfasten the double nuts at the top end of the shock absorber. Remove the cushions and the cushion retainers.

3. Remove the shock absorber-to-lower control arm bolts.

4. Compress the shock absorber and remove it from the vehicle.

5. To install, reverse the removal procedures. Torque the shock absorb-er-to-lower control arm bolts to 13 ft. lbs. (18 Nm) and the shock absorber-to-body nuts to 19 ft. lbs. (26 Nm).

4WD Pick-Up, 4Runner and Van

1. Raise and safely support the ve-hicle. Remove the wheel and tire assembly.

2. Unfasten the double nuts at the top end of the shock absorber. Remove the cushions and the cushion retainers.

3. Remove the shock absorber-to-axle housing bolt.

4. Compress the shock absorber and remove it from the vehicle.

5. To install, reverse the removal procedures. Torque the shock absorber-to-suspension arm nut/bolt to 101 ft. lbs. (137 Nm) for Pick-Up and 4Runner or 70 ft. lbs. (95 Nm) for Van and the shock absorber-to-body nuts to 19 ft. lbs. (27 Nm).

Land Cruiser

1. Raise and safely support the vehicle. Remove the wheel and tire assembly.

2. Remove mounting bolts from the top and the bottom of the shock and remove shock.

3. To install, reverse the removal procedures.

MacPherson Strut

REMOVAL & INSTALLATION

Previa

1. Raise the vehicle and support safely.

2. Remove the halfshaft cotter pin and nut for 4WD only.

3. Disconnect the brake hose from the strut by removing the clips.

4. Loosen the 2 strut-to-knuckle bolts, do not remove.

5. Remove the stabilizer bar from the strut.

6. Remove the 3 upper strut mount nuts and remove the strut from the vehicle.

To install:

7. Install the strut and torque the upper nuts to 47 ft. lbs. (64 Nm).

8. Install the strut-to-knuckle bolts and torque to 231 ft. lbs. (314 Nm).

9. Install the stabilizer bar and torque the bolt to 76 ft. lbs. (103 Nm).

10. Install the halfshaft nut and torque to 136 ft. lbs. (187 Nm). Install a new cotter pin.

11. Install the remaining components and align the front end.

Coil Springs

REMOVAL & INSTALLATION

1991–92 Land Cruiser

1. Raise the vehicle and support the body and frame.

2. Raise the front axle assembly.

3. Hold the shock piston rod with a wrench and remove the upper mounting nut.

4. Disconnect the stabilizer bar with the cushion and bracket.

5. Lower the jack and install a spring compressor tool 0972730020 or equivalent. Compress the spring to take the load off the the upper mount.

6. Remove the upper spring retaining nuts and remove the spring.

To install:

7. Compress the coil spring and position into the vehicle. Align the spring end with the lower seat. Torque the upper mount nuts to 30 ft. lbs. (40 Nm).

8. Remove the spring compressor and raise the jack.

9. Connect the stabilizer bar and shock absorber. Torque the stabilizer and shock nut to 51 ft. lbs. (69 Nm).

Leaf Springs

REMOVAL & INSTALLATION

1988–90 Land Cruiser

1. Raise and safely support the vehicle. Support the axle housing with a floor jack. Remove the wheel and tire assembly.

2. Lower the floor jack to take the tension off of the spring. Remove the shock absorber mounting nuts/bolts and the shock absorber.

3. Remove the cotter pins and the nuts from the lower end of the stabilizer link. Detach the link from the axle housing.

4. Remove the spring-to-axle housing U-bolt nuts, the spring bumper and the U-bolt.

5. At the front of the spring, remove the hanger pin bolt. Disconnect the spring from the bracket.

6. Remove the spring shackle retaining nuts and the spring shackle inner plate, then carefully pry out the spring shackle with a pry bar.

7. Remove the spring from the vehicle.

To install:

8. To install, perform the following procedure:

   a. Install the rubber bushings in the eye of the spring.

   b. Align the eye of the spring with the spring hanger bracket and drive the pin through the bracket holes and rubber bushings.

NOTE: Use soapy water as lubricant, if necessary, to aid in pin installation. Never use oil or grease.

   c. Finger-tighten the spring hanger nuts/bolts.

   d. Install the rubber bushings in the spring eye at the opposite end of the spring.

   e. Raise the free end of the spring. Install the spring shackle through the bushings and the bracket.

   f. Install the shackle inner plate and finger-tighten the retaining nuts.

   g. Center the bolt head in the hole

which is provided in the spring seat on the axle housing.

   h. Fit the U-bolts over the axle housing. Install the spring bumper and the nuts.

9. To complete the installation, reverse the removal procedures. Torque the U-bolt nuts to 90 ft. lbs. (122 Nm), the hanger pin-to-frame nut to 67 ft. lbs. (93 Nm), the shackle pin nuts to 67 ft. lbs. (93 Nm), the shock absorber bolts to 47 ft. lbs. (65 Nm).

NOTE: When installing the U-bolts, tighten the nuts so the length of the bolts are equal.

Torsion Bars

REMOVAL & INSTALLATION

2WD Pick-Up and Van

The vehicles are equipped with torsion bar front springs.

NOTE: Great care must be taken to make sure springs are not mixed after removal. It is strongly suggested that before removal, each spring be marked with paint, showing front and rear of spring and from which side of the vehicle it was taken. If the springs are installed backwards or on the wrong sides of the vehicle, they could fracture. If replacing the springs, it is not necessary to mark them.

1. Raise and safely support the front of the vehicle.

2. Slide the boot from the rear of torsion bar spring, then paint an alignment mark from the torsion bar spring onto the anchor arm and the torque arm. There are right and left identification marks on the rear end of the torsion bar springs.

3. On the rear torsion bar spring holder, there is a long bolt that passes through the arm of the holder and up through the frame crossmember. Remove the locking nut only from this bolt.

4. Using a small ruler, measure the length from the bottom of the remaining nut to the threaded tip of the bolt and record this measurement.

5. Place a jack under the rear torsion bar spring holder arm and raise the arm to remove the spring pressure from the long bolt. Remove the adjusting nut from the long bolt.

6. Slowly lower jack.

7. Remove the long bolt, the spacers, the anchor arm and the torsion bar spring. The torsion bar should be easily pulled out of the anchor and the torque arms.

NOTE: Inspect all parts for

wear damage or cracks. Check the boots for rips and wear. Inspect the splined ends of the torsion bar spring and the splined holes in the rear holder and the front torque arm for damage. Replace as necessary.

To install:

8. Coat the splined ends of the torsion bar with multi-purpose grease.

9. If installing the old torsion bars, perform the following:

a. Slide the front of the torsion bar spring into the torque arm, making sure the alignment marks are matched.

b. Slide the anchor arm onto the rear of the torsion bar spring, making sure the alignment marks are matched. Install the long bolt and it's spacers.

c. Tighten the adjusting nut so it is the same length as it was before removal.

NOTE: Do not install the locknut.

10. When installing a new torsion bar spring, perform the following:

a. Raise the front of the vehicle, replace the wheel and tire assembly, place a wooden block (7½ in. high) under the front tire. Lower the jack until the clearance between the spring bumper (on the lower control arm) and the frame is ½ in.

b. Slide the front of the torsion bar spring into the torque arm.

c. Install the anchor arm into the rear of the torsion bar spring, then the long bolt and the spacers. the distance from the top of the upper spacer to the tip of the threaded end of bolt is 0.310–1.100 in. for ½ ton vehicles or 0.430–1.220 in. for ¾ ton vehicles.

NOTE: Make sure the bolt and bottom spacer are snuggly in the holder arm while measuring.

d. Remove the wooden block and lower the vehicle until it rests on the jackstands.

e. Install and tighten the adjusting nut until the distance from the bottom of the nut to the tip of the threaded end of the bolt is 2.7–3.5 in.

NOTE: Do not install the locknut.

11. Apply multi-purpose grease to the boot lips, then refit the boots to the torque and the anchor arms.

12. Lower the vehicle to the floor and bounce it several times to settle the suspension. With the wheels on the ground, measure the distance from the ground to the center of the lower control arm-to-frame shaft. Adjust the vehicle height using the adjusting nut on

the anchor arm. The height should be approximately 10.31 in.

NOTE: If, after achieving the correct vehicle height, the distance from the bottom of the adjusting nut to the top of the threaded end of the long bolt is not within 2.7–3.5 in., change the position of the anchor arm-to-tension bar spring spline and reassemble.

13. Install and torque the locknut on the long bolt to 61 ft. lbs. (83 Nm).

NOTE: Make sure the adjusting nut does not move when tightening locknut.

4WD Van

NOTE: Great care must be taken to make sure springs are not mixed after removal. It is strongly suggested that before removal, each spring be marked with paint, showing front and rear of spring and from which side of the vehicle it was taken. If the springs are installed backwards or on the wrong sides of the vehicle, they could fracture. If replacing the springs, it is not necessary to mark them.

1. Raise and safely support the vehicle.

2. Using a piece of chalk, remove the boots, then, matchmark the torsion bar spring, the anchor arm and the torque arm.

3. Remove the locknut.

4. Measure the protruding length of the adjusting arm bolt, from the nut to the end of the bolt.

NOTE: The adjusting arm bolt measurement is used as a reference to establish the chassis ground clearance.

5. Remove the adjusting nut, the anchor arm and the torsion bar spring.

NOTE: When installing the torsion bar springs, be sure to check the left/right indicating marks on the rear end of the springs; be careful not to interchange the springs.

To install:

6. Using molybdenum disulphide lithium base grease, apply a coat to the torsion bar spring splines.

7. If installing a used torsion bar spring, perform the following procedures:

a. Align the matchmarks and install the torsion bar spring to the torque arm.

b. Align the matchmarks and install the anchor arm to the torsion bar spring.

c. Tighten the adjusting nut until the bolt protrusion is the same as it was before.

8. If installing a new torsion bar spring, perform the following procedures:

a. Make sure the upper and lower arms rebound.

b. Install the boots onto the torsion bar spring.

c. Install one end of the torsion bar spring to the torque arm.

d. Install the torsion bar spring onto the opposite end of the anchor arm.

e. Finger tighten the adjusting nut until the adjusting bolt protrudes about 1.570 in.

f. Tighten the adjusting nut until the adjusting bolt protrudes about 2.480 in. for 4WD Van/Wagon or 2.400 in., for 4WD Van or 2.76 in. for 2WD Van.

g. Install the wheel(s) and lower the vehicle. Bounce the front of the vehicle to stabilize the suspension.

9. To adjust the ground clearance, turn the adjusting nut until the center of the cam plate nut, located of the front end of the lower suspension arm, is 11.17 in. for 4WD Van/Wagon, 11.99 in. for 4WD Van or 10.00 for 2WD Van above the ground.

10. After adjusting the ground clearance, torque the locknut to 58 ft. lbs. (78 Nm), then, install the boots.

4WD Pick-Up and 4Runner

These vehicles are equipped with torsion bar front springs.

NOTE: Great care must be taken to make sure springs are not mixed after removal. It is strongly suggested that before removal, each spring be marked with paint, showing front and rear of spring and from which side of the vehicle it was taken. If the springs are installed backwards or on the wrong sides of the vehicle, they could fracture. If replacing the springs, it is not necessary to mark them.

1. Raise and safely support the vehicle.

2. Using a piece of chalk, remove the boots. Matchmark the torsion bar spring, the anchor arm and the torque arm.

3. Remove the locknut.

4. Measure the protruding length of the adjusting arm bolt, from the nut to the end of the bolt.

NOTE: The adjusting arm bolt measurement is used as a reference to establish the chassis ground clearance.

5. Remove the adjusting nut, the anchor arm and the torsion bar spring.

NOTE: When installing the torsion bar springs, be sure to check the left/right indicating marks on the rear end of the springs; be careful not to interchange the springs.

To install:

6. Using molybdenum disulphide lithium base grease, apply a coat to the torsion bar spring splines.

7. If installing a used torsion bar spring, perform the following procedures:

    a. Align the matchmarks, install the torsion bar spring to the torque arm.

    b. Align the matchmarks and install the anchor arm to the torsion bar spring.

    c. Tighten the adjusting nut until the bolt protrusion is the same as it was before.

8. If installing a new torsion bar spring, perform the following procedures:

    a. Make sure the upper and lower arms rebound.

    b. Install the boots onto the torsion bar spring.

    c. Install one end of the torsion bar spring to the torque arm.

    d. Install the torsion bar spring onto the opposite end of the anchor arm.

    e. Finger tighten the adjusting nut until the adjusting bolt protrudes about 1.570 in.

    f. Tighten the adjusting nut until the adjusting bolt protrudes about 3.430 in.

    g. Install the wheel(s) and remove the jackstands. Bounce the front of the vehicle to stablize the suspension.

9. To adjust the ground clearance, turn the adjusting nut until the center of the cam plate nut, located of the front end of the lower suspension arm, about 11.220 in. above the ground.

10. After adjusting the ground clearance, torque the locknut to 61 ft. lbs. (83 Nm), then, install the boots.

Upper Ball Joints

INSPECTION

1. Raise the vehicle and place jackstands under the lower control arms and check for excess play.

2. Move the suspension arm up and down. Maximum vertical play should be 0.

3. If the ball joints are within specifications and a looseness problem still exists, check the other suspension parts for wheel bearings, tie rods and etc.

4. The bottom of the tire should not move more than 0.200 in. when the tire is pushed and pulled inward and outward. The tire should not move more than 0.090 in. up and down.

5. If the play is greater than these figures, replace the ball joint.

REMOVAL & INSTALLATION

1. Raise and safely support the vehicle. Remove the wheel and tire assembly.

2. Support the lower control arm with a floor jack.

3. Remove the brake caliper and support it aside, with a wire.

4. Using a ball joint removal tool, separate the tie rod end from the knuckle arm.

5. Remove the ball joint-to-control arm mounting bolts and separate the joint from the arm.

6. To install, reverse the removal procedures. Torque the ball joint-to-upper control arm bolts 20 ft. lbs. (27 Nm) for 2WD Pick-Up, 25 ft. lbs. (34 Nm) for 4WD Pick-Up or 22 ft. lbs. (30 Nm) for 2WD Van, the ball joint-to-lower control arm bolts to 51 ft. lbs. (69 Nm) for Pick-Up or 49 ft. lbs. (68 Nm) for Van, and the lower ball joint-to steering knuckle nut to 25 ft. lbs. (34 Nm) for 4WD Pick-Up, 76 ft. lbs. (103 Nm) for 2WD Van or 83 ft. lbs. (115 Nm) for 4WD Van.

NOTE: Be sure to grease the ball joints before moving the vehicle.

Lower Ball Joint

INSPECTION

1. Raise the vehicle and place jackstands under the lower control arms and check for excess play.

2. Move the suspension arm up and down. Maximum vertical play should be 0.

3. If the ball joints are within specifications and a looseness problem still exists, check the other suspension parts for wheel bearings, tie rods and etc.

4. The bottom of the tire should not move more than 0.200 in. when the tire is pushed and pulled inward and outward. The tire should not move more than 0.090 in. up and down.

5. If the play is greater than these figures, replace the ball joint.

REMOVAL & INSTALLATION

1. Raise and safely support the vehicle. Remove the wheel and tire assembly.

2. Support the lower control arm with a floor jack.

3. Remove the brake caliper and support it aside, with a wire.

4. Using a ball joint removal tool, separate the tie rod end from the knuckle arm.

5. Using a ball joint removal tool, separate the upper ball joint from the steering knuckle.

6. Remove the ball joint-to-control arm mounting bolts and separate the joint from the arm.

7. To install, reverse the removal procedures. Torque the ball joint-to-upper control arm bolts 20 ft. lbs. (27 Nm) for 2WD Pick-Up, 25 ft. lbs. (34 Nm) for 4WD Pick-Up or 22 ft. lbs. (30 Nm) for 2WD Van, the upper ball joint-to-steering knuckle nut to 80 ft. lbs. (122 Nm) for 2WD Pick-Up, 105 ft. lbs. (140 Nm) for 4WD Pick-Up, 58 ft. lbs. (79 Nm) for 2WD Van or 83 ft. lbs. (115 Nm) for 4WD Van.

NOTE: Be sure to grease the ball joints before moving the vehicle.

Knuckle Joint Bearings

REMOVAL & INSTALLATION

Land Cruiser

1. Raise the vehicle and support safely. Remove the front wheels.

2. Remove the caliper, axle hub and rotor.

3. Remove the spindle mounting bolts, dust seal and cover.

4. Using a brass drift, tap the spindle off of the steering knuckle. Tap around the side to dislodge.

5. Position 1 flat part of the outer shaft upward and pull out the axle shaft.

6. Using a tie rod remover 0961122012 or equivalent, remove the tie rod end.

7. Remove the oil seal and retainer from the rear of the knuckle.

8. Remove the upper and lower knuckle arm and bearing caps. Tap with a brass drift on the cone washers and remove them from the arm or bearing cap.

9. Using a bearing remover tool 0960660020 or equivalent, push the bearing cap and shims from the steering knuckle. Label each bearing set upper and lower for installation.

10. Remove the outer race using a brass drift.

To install:

11. Using a race installer tool 0960560010 or equivalent, drive in the new bearing outer races.

12. Coat the knuckle bearings with molybdenum lithium grease.

13. Mount the bearings on tool 0963460013 or equivalent.

14. Add preload to the bearing by

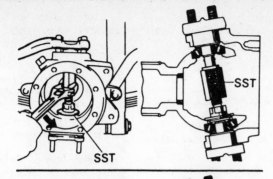

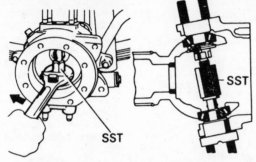

Knuckle joint bearings—Land Cruiser

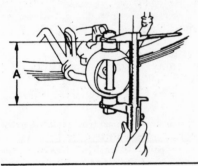

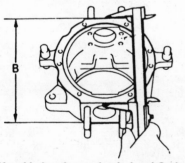

Knuckle bearing preload—Land Cruiser

tightening the upper nut on the bearings installation tool. Preload should be 6.6–13.2 lbs. (29–59 N).

15. Measure distance **A** and **B**. The difference between the 2 measurements is required to maintain the correct bearing preload (shim thickness).

16. Install the oil seal into the axle housing using tool 0951860010 or equivalent.

17. Install the felt dust seal, rubber seal and steel ring to the knuckle.

18. Pack the bearings with molybdenum lithium grease.

19. Install the bearings into position on the knuckle and axle housing. Insert the knuckle on the axle housing.

20. Using a support tool 0960660020 or equivalent, to support the bearings. Install the knuckle over the shims that were originally used or selected in the adjustment operation.

21. Using a hammer, tap the knuckle arm into the bearing inner races.

22. Install the bearing cap over the shims and tap with a hammer.

23. Install the cone washers and torque the nuts to 71 ft. lbs. (96 Nm).

24. Install the tie rod end and torque to 67 ft. lbs. (91 Nm). Install a new cotter pin.

25. Install the seal set to the knuckle.

26. Install the axle shaft and pack with grease.

27. Install the spindle dust cover and axle hub.

28. Install the remaining components and check operation.

Upper Control Arm

REMOVAL & INSTALLATION

Pick-Up and 4Runner

1. Raise and safely support the vehicle. Remove the wheel and tire assembly.

2. Using a floor jack, support the lower control arm.

3. Remove the upper ball joint-to-upper control arm nuts/bolts, then, disconnect the upper control arm.

4. Remove the upper control arm-to-chassis bolts and camber adjusting shims and the upper control arm.

NOTE: When removing the camber adjusting shims, be sure to record their location and thickness of shims, so they may be reinstalled in their original positions.

5. To install, reverse the removal procedures. Torque the upper control arm-to-chassis bolts to 72 ft. lbs. (96 Nm) and the upper control arm-to-upper ball joint nuts/bolts to 20 ft. lbs. (27 Nm). Check and/or adjust the front wheel alignment.

2WD Van

1. Raise and safely support the vehicle. Remove the torsion bar spring.

2. Remove the cool air intake duct.

3. Remove the upper control arm-to-upper ball joint nuts/bolts and separate the upper control arm from the ball joint.

4. Remove the upper control arm-to-chassis bolts and the control arm from the vehicle.

5. To install, reverse the removal procedures. Torque the upper control arm-to-chassis bolts to 65 ft. lbs. (88 Nm) for front and 112 ft. lbs. (150 Nm) for rear, then, the upper control arm-to-ball joint nuts/bolts to 22 ft. lbs. (29 Nm). Check and/or adjust the front end alignment.

4WD Van

1. Raise and safely support the vehicle. Remove the torsion bar spring. Lower the vehicle.

2. Remove the front-right seat and the console box. Disconnect the control and shift cables from the shift levers, then, remove the transmission/transfer shifting levers with retainer.

3. Disconnect the parking brake cable from the brake lever, then, remove the parking brake lever assembly from the vehicle.

4. Disconnect the parking brake cable from the intermediate lever and remove it. Disconnect the shift cable from the transmission and remove it.

5. Remove the seat floor panel.

6. Remove the fan shroud, the radiator mounting bolts/nuts and move it aside; do not drain the coolant. Raise and safely support the vehicle.

7. Remove the shock absorber-to-frame nuts and disconnect the shock absorber from the frame.

8. From the upper ball joint, remove the cotter pin and the nut. Using a ball joint removal tool, press the ball joint from the steering knuckle.

9. Remove the upper control arm-to-chassis bolts and the arm from the vehicle.

10. To install, reverse the removal procedures. Torque the upper control arm-to-chassis bolts to 112 ft. lbs. (152 Nm), the upper ball joint-to-steering knuckle nut to 83 ft. lbs. (122 Nm).

Lower Control Arm

REMOVAL & INSTALLATION

2WD Pick-Up

1. Raise and safely support the vehicle. Remove the torsion bar spring.
2. Remove the shock absorber, the stabilizer bar and the strut bar from the lower arm.
3. Remove the shock absorber from the lower arm.
4. From the lower ball joint, remove the cotter pin and the nut. Using a ball joint removal tool, press the ball joint from the lower control arm.

NOTE: If the lower ball joint is not to be replaced, simply unbolt it from the lower control arm. It is not necessary to separate the ball joint from the steering knuckle.

5. Remove the lower control arm shaft nut. Remove the spring torque arm from the other side of the lower control arm, then, remove the lower arm shaft bolt and the lower arm.
6. To install, reverse the removal procedures.
7. Tighten the bolt(s) holding the lower control arm to the frame but do not torque them until the vehicle is on the ground.
8. Torque the ball joint-to-lower control arm nuts/bolts to 51 ft. lbs. (69 Nm), the strut bar-to-lower control arm bolts to 70 ft. lbs. (95 Nm), the stabilizer bar-to-lower control arm bolts to 9 ft. lbs. (12 Nm), the lower shock absorber bolt to 13 ft. lbs. (18 Nm), upper shock absorber bolt to 18 ft. lbs. (25 Nm) and the lower arm mounting nuts to 166 ft. lbs. (220 Nm).
9. Check and/or adjust the front end alignment.

NOTE: Do not torque the control arm bolts fully until the vehicle is lowered and bounced several times; if the bolts are tightened with the control arm(s) hanging, excessive bushing wear will result.

4WD Pick-Up and 4Runner

1. Raise and safely support the vehicle. Remove the shock absorber.
2. Disconnect the stabilizer bar from the lower suspension arm.
3. Remove the lower ball joint-to-lower control arm bolts, then, separate the control arm from the ball joint.
4. Using a piece of chalk, place matchmarks on the front/rear adjusting cams.
5. Remove the nuts and adjusting cams and the lower control arms.
6. To install, reverse the removal procedures. Torque the lower ball joint-to-lower control arm bolts to 20 ft. lbs. (27 Nm), the stabilizer bar-to-lower control arm bolts to 19 ft. lbs. (26 Nm), the shock absorber-to-lower control arm nut/bolt to 101 ft. lbs. (137 Nm).
7. Lower the vehicle to the ground, bounce it a few times, align the matchmarks and torque the adjusting cam nuts to 203 ft. lbs. (270 Nm). Check and/or adjust the front wheel alignment.

2WD Van

1. Raise and safely support the vehicle.
2. Remove the stabilizer bar and the strut bar from the lower arm.
3. Remove the shock absorber from the lower arm. If necessary, disconnect the tie rod end from the steering knuckle.
4. From the lower ball joint, remove the cotter pin and the nut. Using a ball joint removal tool, press the ball joint from the lower control arm.

NOTE: If the lower ball joint is not to be replaced, simply unbolt it from the lower control arm. It is not necessary to separate the ball joint from the steering knuckle.

5. Using a piece of chalk, matchmark the adjusting cam of the lower control arm.
6. Remove the adjusting cam, the nut and the lower control arm.
7. To install, reverse the removal procedures. Align the cam matchmarks and finger-tighten the nut:
Torque the ball joint-to-lower control arm nuts/bolts to 49 ft. lbs. (68 Nm), the lower ball joint-to-steering knuckle nut to 76 ft. lbs. (103 Nm).
Torque the strut bar-to-lower control arm bolts to 49 ft. lbs. (68 Nm).
Torque the stabilizer bar-to-lower control arm bolts to 9 ft. lbs. (12 Nm).
Torque the tie rod end-to-steering knuckle nut to 43 ft. lbs. (58 Nm).
Torque the lower shock absorber bolt to 13 ft. lbs. (17 Nm).
Torque the upper shock absorber bolt to 19 ft. lbs. (25 Nm).
Torque the adjusting cam unit to 152 ft. lbs. (205 Nm).
8. Check and/or adjust the front end alignment.

NOTE: Do not torque the control arm bolts fully until the vehicle is lowered and bounced several times.

4WD Van

1. Raise and safely support the vehicle.

2. Remove the stabilizer bar from the lower control arm.
3. Remove the shock absorber from the lower control arm.
4. From the lower ball joint, remove the cotter pin and the nut. Using a ball joint removal tool, press the ball joint from the lower control arm.

NOTE: If the lower ball joint is not to be replaced, simply unbolt it from the lower control arm. It is not necessary to separate the ball joint from the steering knuckle.

5. Using a piece of chalk, matchmark the adjusting cam of the lower control arm.
6. Remove the adjusting cam, the nut and the lower control arm.
7. To install, reverse the removal procedures. Align the cam matchmarks and finger-tighten the nut:
Torque the ball joint-to-lower control arm nuts/bolts to 83 ft. lbs. (110 Nm).
Torque the stabilizer bar-to-lower control arm bolts to 14 ft. lbs. (17 Nm).
Torque the tie rod end-to-steering knuckle nut to 43 ft. lbs. (57 Nm).
Torque the lower shock absorber bolt to 70 ft. lbs. (95 Nm).
Torque the adjusting cam nut to 152 ft. lbs. (205 Nm).
8. Check and/or adjust the front end alignment.

NOTE: Do not torque the control arm bolts fully until the vehicle is lowered and bounced several times.

1991–92 Land Cruiser

1. Raise the vehicle and support safely.
2. Support the frame and body. Support the axle assembly with a floor jack.
3. Remove the lower control arm-to-frame bolt.
4. Remove the 2 bolts and nuts from the axle housing and remove the control arm.
To install:
5. Install the control and bolts loosely. Do not tighten at this time.
6. Install the front wheels and lower the vehicle. Bounce the front end to stabilize the front suspension.
7. Torque the retaining bolts to 127 ft. lbs. (171 Nm).

Previa

1. Raise the vehicle and support safely. Remove the front wheel.
2. Remove the under cover and disconnect the lower ball joint from the knuckle.
3. Remove the 2 bolts and lower

arm bracket, nut and arm shaft and lower control arm with ball joint.

To install:

4. Install the lower arm and retainers. Do not tighten at this time.

5. Connect the lower ball joint and torque to 94 ft. lbs. (127 Nm).

6. Install the front wheels and lower the vehicle.

7. Bounce the suspension up and down several times.

8. Raise the vehicle and support under the control arms with jackstands.

9. Torque the lower arm retainers to 121 ft. lbs. (164 Nm).

10. Install the remaining components and lower the vehicle.

Sway Bar

REMOVAL & INSTALLATION

1. Raise the vehicle and support safely.

2. Remove the under covers.

3. Disconnect the sway bar links-to-control arms.

4. Disconnect the sway bar-to-frame bolts and remove the sway bar.

5. Installation is the reverse of removal. torque the frame bracket bolts to 14 ft. lbs. (19 Nm) and the link nuts to 76 ft. lbs. (103 Nm).

Front Wheel Bearings

REMOVAL & INSTALLATION

RWD vehicles only, refer to drive axle for 4WD vehicles.

Pick-Up and Van

1. Raise the vehicle and support safely. Remove the front wheel.

2. Remove the brake caliper and suspend with a wire.

3. Remove the bearing cap, cotter pin, nut and outer bearing. Do not drop the bearing.

4. Remove the rotor/hub assembly.

5. Pry the inner grease seal from the hub and remove the inner bearing.

6. Clean the grease out of the hub using shop rags and compressed air. Do not use solvent to remove grease.

7. Using a brass drift and hammer, drive out the bearing outer races. Do not use old races with new bearings.

NOTE: Grind the outer circumference of the old bearing race with a grinder. Use this race to press in the new race.

8. Install the outer race into the hub using a press. If the new race gets damaged, replace with a new one.

9. Pack the bearings with high temperature bearing grease and install inner bearing.

10. Install the grease seal with tool 0960804100 or equivalent.

11. Install the rotor, outer bearing, washer, nut and nut lock.

12. Torque the nut to 25 ft. lbs. (34 Nm). Loosen the nut until it can be turned by hand.

13. Using a spring tension gauge attached to a lug bolt.

14. Tighten the nut until the bearing preload is 0.9–2.2 lbs. (3.9–9.8 N) for dual wheel vehicles or 1.3–4.0 lbs. (5.9–17.7 N) for single wheel vehicles.

15. Install nut lock, new cotter pin and grease cap.

16. Install the remaining components and check operation.

2WD Previa

1. Remove the steering knuckle from the vehicle. Place the assembly in a vise.

2. Remove the grease cap and lock nut. Remove the spacer or speed sensor rotor.

3. Using a hub remover tool (slide hammer) 0952000031 or equivalent, remove the wheel hub from the knuckle.

4. Remove the bearing from the hub using a press and arbor tool 0955010012 or equivalent.

5. Remove the dust protector and oil seal.

6. Remove the bearing snapring from the knuckle and press the inner bearing from the knuckle.

To install:

7. Using a press and arbor tool 0960810010 or equivalent, press the bearing into the knuckle.

8. Install the outer oil seal and dust cover.

9. Press the axle hub onto the knuckle.

10. Install the new nut and torque to 147 ft. lbs. (199 Nm). Stack the nut using a chisel. Install the grease cap.

11. Install the knuckle assembly onto the vehicle.

REAR SUSPENSION

Shock Absorbers

REMOVAL & INSTALLATION

Pick-Up, 4Runner and Land Cruiser

1. Raise and safely support the vehicle.

3. Remove the upper shock absorber retaining bolts from the upper frame member.

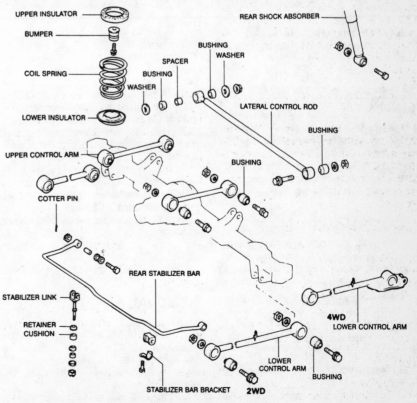

Rear suspension—Van and Previa

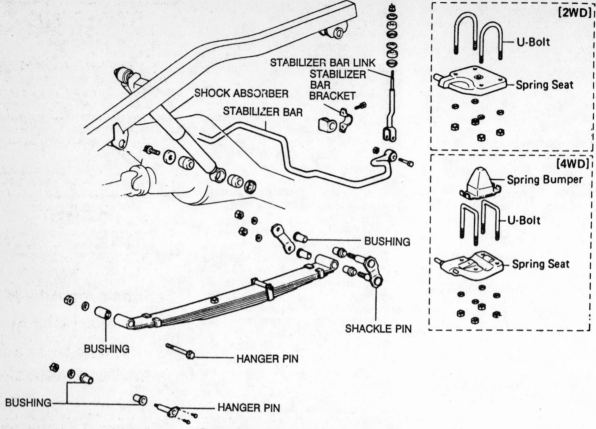

SHOCK ABSORBER

STABILIZER BAR

STABILIZER BAR LINK
STABILIZER BAR BRACKET

BUSHING

SHACKLE PIN

BUSHING — HANGER PIN

BUSHING — HANGER PIN

[2WD]
U-Bolt
Spring Seat

[4WD]
Spring Bumper
U-Bolt
Spring Seat

4. Remove the lower end bolt of the shock absorber from the spring seat.

5. Remove the shock absorber from the vehicle.

NOTE: Inspect the shock for wear, leaks or other signs of damage.

6. To install, reverse the removal procedures. Torque the upper bolt to 19 ft. lbs. (25 Nm) for 2WD vehicle or 47 ft. lbs. (63 Nm) for 4WD vehicle and the lower bolt to 19 ft. lbs. (25 Nm) for 2WD vehicle or 47 ft. lbs. (63 Nm) for 4WD vehicle.

Van

1. Raise and safely support the vehicle.

2. Remove the shock absorber-to-axle housing bolt.

3. Working inside the vehicle, remove the locknut, the retaining nut, the retainers and the rubber bushings from the top of the shock absorber.

NOTE: When removing the retaining nut, from the top of the shock absorber, it may be necessary to hold the top of the shock to keep it from turning.

4. Remove the shock absorber from the vehicle.

5. To install, reverse the removal procedures. Torque the shock absorb-

er-to-body nut to 16–24 ft. lbs. (22–33 Nm) and the shock absorber-to-axle housing bolt to 27 ft. lbs. (38 Nm).

Previa

1. Raise the vehicle and support safely.

2. Support the axle housing with a jack.

3. Disconnect the shock from the lower control arm. Hold the shaft with a suitable tool and remove the nut.

4. Remove the upper mount bolt and shock absorber.

To install:

5. Install the shock and torque the upper mount to 27 ft. lbs. (37 Nm) and tighten the lower mount nut until the shaft protrudes 0.0059 in. (1.5mm).

6. Lower the vehicle and check operation.

Coil Spring

REMOVAL & INSTALLATION

Van and Previa

1. Raise and safely support the vehicle. Support the axle housing with a floor jack. Remove the wheel and tire assembly.

2. Remove the shock absorber-to-axle housing bolt.

3. Remove the stabilizer-to-axle housing bar bushing bracket bolts.

4. Remove the lateral control arm-to-axle housing nut and disconnect the lateral control arm.

5. Lower the floor jack, then remove the coil spring(s) and the insulators.

NOTE: While lowering the axle housing, be careful not to snag the brake line of the parking brake cable.

6. To install, reverse the removal procedures. Torque the shock absorber bolt to 27 ft. lbs. (34 Nm), the lateral control arm-to-axle housing nut to 43 ft. lbs. (60 Nm) and the stabilizer-to-axle housing bolts to 27 ft. lbs. (34 Nm).

NOTE: Before tightening the lateral control arm and the stabilizer nuts/bolts, bounce the vehicle to stabilize the suspension.

Leaf Springs

REMOVAL & INSTALLATION

1. Raise and safely support the vehicle. Support the axle housing with a floor jack. Remove the wheel and tire assembly.

2. Lower the floor jack to take the

tension off of the spring. Remove the shock absorber mounting nuts/bolts and the shock absorber.

3. Remove the cotter pins and the nuts from the lower end of the stabilizer link. Detach the link from the axle housing.

4. Remove the spring-to-axle housing U-bolt nuts, the spring bumper and the U-bolt.

5. At the front of the spring, remove the hanger pin bolt. Disconnect the spring from the bracket.

6. Remove the spring shackle retaining nuts and the spring shackle inner plate, then carefully pry out the spring shackle with a pry bar.

7. Remove the spring from the vehicle.

To install:

8. To install, perform the following procedure:

a. Install the rubber bushings in the eye of the spring.

b. Align the eye of the spring with the spring hanger bracket and drive the pin through the bracket holes and rubber bushings.

NOTE: Use soapy water as lubricant, if necessary, to aid in pin installation. Never use oil or grease.

c. Finger-tighten the spring hanger nuts/bolts.

d. Install the rubber bushings in the spring eye at the opposite end of the spring.

e. Raise the free end of the spring. Install the spring shackle through the bushings and the bracket.

f. Install the shackle inner plate and finger-tighten the retaining nuts.

g. Center the bolt head in the hole which is provided in the spring seat on the axle housing.

h. Fit the U-bolts over the axle housing. Install the lower spring seat for 2WD vehicle or spring bumper for 4WD vehicle and the nuts.

9. To complete the installation, reverse the removal procedures. Torque the U-bolt nuts to 90 ft. lbs. (122 Nm), the hanger pin-to-frame nut to 67 ft. lbs. (90 Nm), the shackle pin nuts to 67 ft. lbs. (90 Nm), the shock absorber bolts to 47 ft. lbs. (65 Nm) for 4WD vehicle.

NOTE: When installing the U-bolts, tighten the nuts so the length of the bolts are equal.

Rear Control Arms

REMOVAL & INSTALLATION

Van and Previa

1. Raise and safely support the ve-

hicle. Place a floor jack under the axle housing to support it.

2. Remove the upper control arm-to-body bolt, the upper control arm-to-axle housing bolt and the upper control arm from the vehicle.

3. Disconnect the brake line from the lower control arm.

4. Remove the lower control arm-to-body bolt, the lower control arm-to-axle housing bolt and the lower control arm from the vehicle.

To install:

5. Install the upper control arm to the body and to the axle housing with the nuts. Do not tighten the nuts.

6. Install the lower control arm to the body and to the axle housing with the nuts. Do not tighten the nuts.

7. Remove the jack and the supports from under the vehicle. Bounce the vehicle to stabilize the suspension.

8. Using the floor jack under the axle housing, raise the vehicle. Place jackstands under the frame but do not let them touch the frame.

9. To complete the installation, torque the upper control arm-to-body bolt to 105 ft. lbs. (140 Nm), the upper control arm-to-axle housing bolt to 105 ft. lbs. (140 Nm), the lower control arm-to-body bolt to 130 ft. lbs. (176 Nm) and the lower control arm-to-axle housing bolt to 105 ft. lbs. (140 Nm).

Lateral Control Rod

REMOVAL & INSTALLATION

Van and Previa

1. Raise and safely support the vehicle. Place a floor jack under the axle housing and support it.

2. Remove the lateral control rod-to-axle housing nut.

3. Remove the lateral control rod-to-body nut and the control rod from the vehicle.

To install:

4. Raise the axle housing until the frame is just free of the jack.

5. Install the lateral control rod-to-body with the nut. Do not tighten the nut.

6. Install the lateral control rod-to-axle housing in the following order: washer, bushing, spacer, lateral control rod, bushing, washer and nut. Do not tighten the nut.

7. Remove the jack, lower the vehicle to the floor and bounce it to stabilize the suspension.

8. Using the floor jack under the axle housing, raise the vehicle. Torque the lateral control rod-to-body nut to 81 ft. lbs. (110 Nm) and the lateral control rod-to-axle housing nut to 43 ft. lbs. (58 Nm).

STEERING

Steering Wheel

REMOVAL & INSTALLATION

1. Disconnect the negative battery cable.

2. Position the wheels in a straight ahead position.

3. Remove the steering wheel center cover, some vehicles use a screw to retain the cover.

4. Disconnect the horn wire. Matchmark the wheel and the shaft.

5. Using an appropriate wheel puller tool 0960920011 or equivalent, remove the steering wheel.

6. Installation is the reverse of the removal procedure. Tighten the steering wheel nut to 25 ft. lbs. (34 Nm).

Steering Column

REMOVAL & INSTALLATION

1. Disconnect the negative battery cable.

2. Remove the instrument panel and steering column finish panels.

3. Disconnect all electrical connectors from the column.

4. Remove the lower joint protectors, if equipped.

5. Remove the pinch bolt and nut from the intermediate shaft or flex joint. Disconnect the intermediate shaft from the column shaft or disconnect the flex joint from the steering gear.

6. Remove the column bracket-to-instrument panel nuts.

7. Remove the floor boot retainers and pull the boot away from the floor.

8. Remove the column from the vehicle. Do not hammer on the shaft.

To install:

9. Install the column into the vehicle. Make sure the plastic retainers are properly aligned. Torque the column-to-instrument panel nuts to 18 ft. lbs. (25 Nm).

10. Install the floor boot and retainers. Make sure the boot is sealed from water. Use sealer between the floor and boot.

11. Connect the intermediate shaft to the column shaft or connect the flex joint to the steering gear. Install the pinch bolt and nut to the intermediate shaft or flex joint. Torque the bolt to 26 ft. lbs. (35 Nm).

12. Install the lower joint protectors, if equipped.

13. Connect all electrical connectors to the column.

14. Install the instrument panel and steering column finish panels.

15. Connect the negative battery cable and check operation.

Manual Steering Gear

ADJUSTMENT

Pick-Up and 4Runner

1. Install a special socket 0961600010 or equivalent, and inch lbs. torque wrench on the worm shaft.
2. Turn the adjusting screw while measuring the preload. It should be 6.9–9.5 inch lbs. (0.8–1.1 Nm).
3. Install the lock nut and torque to 34 ft. lbs. (46 Nm).

VAN

1. Loosen the lock nut and torque the rack guide spring cap to 18 ft. lbs. (25 Nm).
2. Return the cap 30 degrees.
3. Turn the control valve shaft right and left 1 or 2 times. Loosen the spring cap until the rack guide compression spring is not functioning.
4. Using an inch lbs. torque wrench on the worm shaft, tighten the rack guide cap until the preload is within 4.3–11.3 inch lbs. (0.7–1.3 Nm).
5. Apply sealer to the lock nut threads and torque the locknut to 41 ft. lbs. (56 Nm).

REMOVAL & INSTALLATION

2WD Pick-Up

1. Raise and safely support the vehicle. Remove the pitman arm-to-relay rod cotter pin and nut. Separate the relay rod from the pitman arm.
2. Matchmark the flexible steering coupling-to-steering gear, then remove the lock bolt and separate the steering coupling from the steering gear.
3. Remove the steering gear housing mounting bolts and the gear housing.
4. To install, reverse the removal procedures. Torque the housing-to-frame bolts to 48 ft. lbs. (62 Nm), the pitman arm-to-relay rod nut 67 ft. lbs. (93 Nm) and the steering gear-to-coupling yoke to 15–20 ft. lbs. (20–27 Nm).

4WD Pick-Up and 4Runner

1. Raise and safely support the vehicle. Remove the stone shield from the gear housing, if equipped.
2. Matchmark the intermediate shaft-to-steering gear and disconnect them.
3. Remove the cotter pin and plug from the drag link.
4. Disconnect the drag link from the pitman arm.

5. Remove the pitman arm nut. Using a puller tool, separate the pitman arm from the steering gear.
6. Remove the steering gear housing-to-frame bolts and the gear housing.
7. To install, reverse the removal procedures. Torque the steering gear-to-frame bolts to 42 ft. lbs. (56 Nm), the steering gear-to-intermediate bolts to 29 ft. lbs. (40 Nm), the pitman arm-to-steering gear nut to 127 ft. lbs. (170 Nm).

NOTE: When installing the drag link to the pitman arm, tighten the plug completely and loosen it 1⅓ turns.

Van

1. Raise and safely support the vehicle.
2. Remove the wheel and tire assemblies.
3. Disconnect the tie rod ends from the steering knuckle.
4. Matchmark and disconnect the intermediate shaft from the gear.
5. Remove the retaining bolts from the 2 brackets and slide the rack out from under the vehicle.
6. Installation is the reverse of the removal procedure. Torque the bracket bolts to 56 ft. lbs (76 Nm), the coupling bolts to 26 ft. lbs. (35 Nm) and the tie rod bolts to 43 ft. lbs. (59 Nm).

Land Cruiser

1. Raise and safely support the vehicle. Remove the worm yokes from the worm and the main shaft.
2. Remove the intermediate shaft assembly.
3. Remove the pitman arm from the sector shaft.
4. Remove the steering gear-to-frame bolts and the steering gear from the vehicle.
5. To install, reverse the removal procedures. Torque the pitman arm to 119–141 ft. lbs. (163–190 Nm).

NOTE: The intermediate shaft must be installed with the wheels in a straight-ahead position and the steering wheel straight-ahead.

Power Steering Gear

ADJUSTMENT

Pick-Up and 4Runner

1. Install a special socket 0961600010 or equivalent, and inch lbs. torque wrench on the worm shaft.
2. Turn the adjusting screw while measuring the preload. It should be 2.6–4.8 inch lbs. (0.3–0.5 Nm).

3. Install the lock nut and torque to 34 ft. lbs. (46 Nm).

Van and Previa

1. Loosen the lock nut and torque the rack guide spring cap to 18 ft. lbs. (25 Nm).
2. Return the cap 30 degrees.
3. Turn the control valve shaft right and left 1 or 2 times. Loosen the spring cap until the rack guide compression spring is not functioning.
4. Using an inch lbs. torque wrench on the worm shaft, tighten the rack guide cap until the preload is within 7.8–11.3 inch lbs. (0.9–1.3 Nm) for the Van or 6.1–11.3 inch lbs. (0.7–1.3 Nm) for the Previa.
5. Torque the lock nut to 41 ft. lbs. (56 Nm).

Land Cruiser

1. Turn the worm shaft in both directions to determine the exact center.
2. Install special socket 0961600010 or equivalent, and an inch lbs. torque wrench to the worm shaft.
3. Turn the adjusting screw while measuring the preload. It should be 6.5–9.5 inch lbs. (0.7–1.1 Nm).
4. Install a new seal and torque the locknut to 34 ft. lbs. (46 Nm).

REMOVAL & INSTALLATION

2WD Pick-Up

1. Raise and safely support the vehicle. Disconnect and plug the pressure line clamp bolts at the steering gear.
2. Matchmark the intermediate shaft-to-steering gear, then, remove the coupling bolt and separate the intermediate shaft from the steering gear.
3. Remove the pitman arm-to-steering gear and the pitman arm-to-relay rod nuts.
4. Using a puller tool, separate the

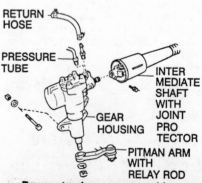

Power steering gear assembly— all except Van

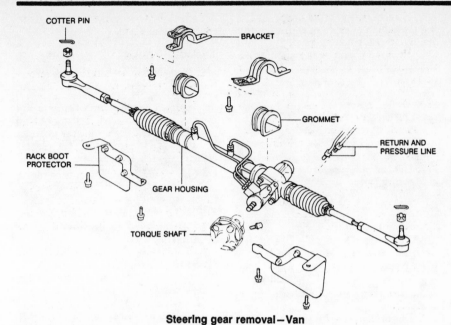

COTTER PIN

BRACKET

GROMMET

RETURN AND
PRESSURE LINE

RACK BOOT
PROTECTOR

GEAR HOUSING

TORQUE SHAFT

Steering gear removal—Van

pitman arm from the relay rod and the pitman arm from the steering gear.

5. Remove the steering gear-to-frame bolts and the steering gear from the vehicle.

6. To install, reverse the removal procedures. Torque the steering gear-to-frame bolts to 48 ft. lbs. (66 Nm), the pitman arm-to-steering gear nut to 90 ft. lbs. (122 Nm), the pitman arm-to-relay rod nut to 67 ft. lbs. (92 Nm), the intermediate shaft-to-steering gear bolt to 19 ft. lbs. and the pressure line nuts to 33 ft. lbs. (45 Nm). Bleed the power steering system.

4WD Pick-Up and 4Runner

1. Remove the battery and the engine lower gravel shield.

2. Raise and safely support the vehicle. Disconnect and plug the pressure lines at the steering gear.

3. Remove the steering gear stone shield.

4. Matchmark the intermediate shaft-to-steering gear, then remove coupling bolt and the intermediate shaft from the steering gear.

5. Remove the pitman arm-to-steering gear nut. Using a puller tool, separate the pitman arm from the steering gear.

6. Remove the gear housing-to-frame bolts and the steering gear from the vehicle.

7. To install, reverse the removal procedures. Torque the steering gear-to-frame bolts to 42 ft. lbs. (57 Nm), the pitman arm-to-steering gear nut to 127 ft. lbs. (172 Nm), the intermediate shaft-to-steering gear bolt to 29 ft. lbs. (41 Nm) and the pressure line union nuts to 33 ft. lbs. (44 Nm). Bleed the power steering system.

Van

1. Raise and safely support the vehicle.

2. Remove the wheel and tire assemblies.

3. Disconnect and plug the power steering fluid lines from the gear.

4. Disconnect the tie rod ends from the steering knuckle.

5. Matchmark and disconnect the intermediate shaft from the gear.

6. Remove the retaining bolts from the 2 brackets and slide the rack out from under the vehicle.

7. Installation is the reverse of the removal procedure. Torque the bracket bolts to 56 ft. lbs. (76 Nm), the coupling bolts to 26 ft. lbs. (35 Nm) and the tie rod bolts to 43 ft. lbs. (59 Nm).

Land Cruiser

1. Raise and safely support the vehicle. Disconnect the pressure lines from the steering gear.

2. Remove the intermediate shaft-to-steering gear bolt and the steering column-to-firewall bolts.

3. Loosen the steering column-to-dash bolts. Remove the pitman arm-to-steering gear nut.

4. Using a puller, separate the relay rod from the pitman shaft and the pitman arm from the steering gear.

5. Pull the steering column towards the passenger compartment to uncouple the steering shaft from the steering gear.

6. Remove the steering gear-to-frame bolts and the steering gear from the vehicle.

7. To install, reverse the removal procedures. Torque the steering gear-to-frame bolts to 40–63 ft. lbs. (54–86

Nm), the pitman arm-to-steering gear nut to 120–141 ft. lbs. (163–190 Nm), the intermediate shaft-to-steering gear bolt to 22–32 ft. lbs. (29–43 Nm), the pressure hose fitting to 29–36 ft. lbs. (40–48 Nm) and the return hose fitting to 24–30 ft. lbs. (35–41 Nm). Bleed the power steering system.

NOTE: During installation of the hydraulic lines, position each line clear of any surrounding components, then tighten the fittings.

Previa

1. Raise the vehicle and support safely.

2. Disconnect the tie rod ends using a separator tool 0961112010 or equivalent.

3. Place matchmarks on the universal joint and shaft. Remove the lower and upper joint bolts and slide the joint upward to disconnect.

4. Disconnect the pressure and return pipes using flarenut wrenches.

5. 4WD only: remove the front differential assembly, equipment driveshaft housing insulator and drive housing stay.

6. Remove the bracket bolts and grommets. Turn the housing toward the back side and slide the housing to the right side. Put the left tie rod end in the body panel. Pull the housing out through the opening in the left lower side of the body.

To install:

7. Install the gear housing and torque the mounting bolts to 56 ft. lbs. (76 Nm).

8. 4WD only: install the front differential assembly, equipment driveshaft housing insulator and drive housing stay.

9. Connect the pressure and return pipes using flarenut wrenches. Torque the fittings to 33 ft. lbs. (44 Nm).

10. Install the lower and upper joint bolts. Torque the bolts to 26 ft. lbs. (35 Nm).

11. Connect the tie rod ends, torque to 36 ft. lbs. (49 Nm) and install a new cotter pin.

12. Align the front end and lower the vehicle.

Power Steering Pump

REMOVAL & INSTALLATION

Pick-Up, Land Cruiser and 4Runner

NOTE: Disconnect the air hoses from the air control valve and the high tension wires from the distributor.

1. Disconnect the negative battery

cable. Loosen the power steering pump pulley nut.

NOTE: Use the drive belt as a brake to keep the pulley from rotating.

2. Place a container under the pump. Disconnect the return line and the pressure tube, then drain the fluid into the container.

3. Loosen the idler pulley nut and the adjusting bolt, then remove the drive belt.

4. Remove the drive pulley and the Woodruff key from the pump shaft.

5. Remove the mounting bolts and the power steering pump from the vehicle.

6. To install, reverse the removal procedures. Torque the pump pulley mounting bolt to 29 ft. lbs. (40 Nm), the pump pulley nut to 32 ft. lbs. (42 Nm) and the pressure hoses to 33 ft. lbs. (45 Nm). Adjust the drive belt tension. Bleed the power steering system.

Van

1. Disconnect the negative battery cable. Disconnect the air hoses from the air control valve of the power steering pump.

2. Drain the fluid from the power steering reservoir tank.

3. At the power steering pump, disconnect the return hose and the pressure tube.

4. Loosen the power steering pump adjusting bolt, then, remove the drive belt, the pulley and the Woodruff key.

5. Remove the mounting bolts, the power steering pump and the bracket from the vehicle.

6. To install, reverse the removal procedures. Torque the power steering pump-to-engine bolts to 29 ft. lbs. (40 Nm), the pulley set nut to 32 ft. lbs. (42 Nm) and the pressure tube to 33 ft. lbs. (45 Nm).

BELT ADJUSTMENT

Use a belt tension gauge and make

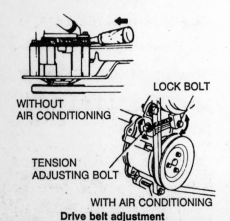

WITHOUT
AIR CONDITIONING

LOCK BOLT

TENSION
ADJUSTING BOLT

WITH AIR CONDITIONING
Drive belt adjustment

sure the tension is 105 lbs. (140 N) for new or 85 lbs. (115 N) for and old belt. Loosen the air pump mounting and adjust the pump until the proper tension is obtained. tighten the mounting brackets and check operation.

SYSTEM BLEEDING

1. Raise and safely support vehicle.

2. Fill the pump reservoir with power steering fluid.

3. With the engine running, rotate the steering wheel from lock to lock several times. Add fluid as necessary.

NOTE: Perform the bleeding procedure until all of the air is bled from the system.

4. The fluid level should not have risen more than $\frac{2}{10}$ in.; if it does, check the pump.

Tie Rod Ends

REMOVAL & INSTALLATION

1. Raise and safely support the vehicle.

2. Remove the wheel and tire assembly. Remove the cotter pin and nut.

3. Using a tie rod end puller, disconnect the tie rod from the relay rod.

4. Using a tie rod end puller remove the tie rod from the steering knuckle.

5. Remove the tie rod end from the vehicle.

6. Installation is the reverse of the removal procedure. Tighten the clamp nuts to 19 ft. lbs. (25 Nm) and the knuckle-to-arm nuts to 67 ft. lbs. (90 Nm). Always install a new cotter pin.

BRAKES

For all brake system repair and service procedures not detailed below, please refer to "Brakes" in the Unit Repair section.

Master Cylinder

REMOVAL & INSTALLATION

Pick-Up, 4Runner, Land Cruiser and Previa

1. Disconnect the negative battery cable. Using a syringe, remove the brake fluid from the master cylinder.

2. Disconnect and plug the hydraulic lines at the master cylinder.

3. If equipped, disconnect the level

warning switch connector from the master cylinder.

4. Remove the master cylinder-to-power booster nuts and the master cylinder assembly from the power brake unit.

5. To install, reverse the removal procedures. Torque the master cylinder mounting bolts to 9 ft. lbs. (12 Nm) and the brake lines-to-master cylinder to 11 ft. lbs. (15 Nm). Refill the master cylinder with new brake fluid and bleed the brake system.

Van

1. Disconnect the negative battery terminal.

2. To expose the master cylinder, perform the following:

   a. Remove the master cylinder reservoir cap, located at the left-side of the instrument panel.

   b. Remove the instrument cluster finish panel and the lower cluster finish panel. Disconnect the electrical connectors and the speedometer cable from the instrument panel, then remove the instrument panel.

   c. Remove the No. 1, 2 and 3 air ducts.

3. Using a syringe, remove the brake fluid from the master cylinder reservoir.

4. Remove the reservoir hoses from the master cylinder. Disconnect and plug the brake lines at the master cylinder.

5. Remove the master cylinder mounting nuts, the vacuum check valve bracket and the master cylinder from the vehicle.

6. To install, reverse the removal procedures. Torque the master cylinder mounting nuts to 9 ft. lbs. (12 Nm) and the brake lines-to-master cylinder to 11 ft. lbs. (15 Nm). Refill the master cylinder with new brake fluid and bleed the brake system.

Load Sensing Proportioning Valve

REMOVAL & INSTALLATION

1. Raise and safely support the vehicle, so it is level.

2. Disconnect the No. 2 shackle from the bracket.

3. Disconnect and plug the brake lines from the load sensing valve.

4. Remove the load sensing valve bracket from the frame.

5. To install, reverse the removal procedures. Torque the load sensing valve-to-frame bolts to 14 ft. lbs. (18 Nm) and the brake tubes to 11 ft. lbs. (15 Nm). Bleed the brake system.

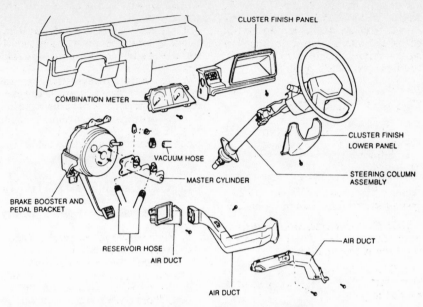

Removing the brake booster and master cylinder—Van

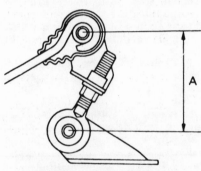

Load sensing proportioning valve adjustment—Except Previa

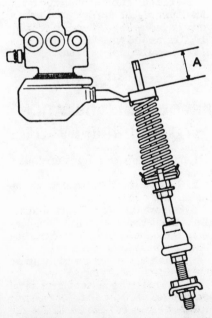

Load sensing proportioning valve adjustment—Previa

ADJUSTMENT

Except Previa

1. Adjust the load sensing valve and the rear axle load.
2. Check and/or adjust the length of the No. 2 shackle for distance from the center of the No. 2 shackle-to-shackle bracket bolt to the center of the No. 1 shackle-to-spring bolt, 3.07 in. for 2WD Pick-Up, Van and 1989–92 Land Cruiser, 4.72 in. for 4WD Pick-Up, 4Runner and 1988 Land Cruiser or 1.18 in. for Previa.
3. Shortening the **A** length raises the pressure to the rear brakes and lengthening lowers the pressure. 1 turn will adjust the pressure to the rear brakes 28–51 psi (196–353 kPa).

Power Brake Booster

REMOVAL & INSTALLATION

Except Van

1. Disconnect the negative battery cable. Separate the master cylinder from the power brake booster.
2. Remove the vacuum hose form the power brake booster.
3. Working under the instrument panel, remove the brake pedal-to-brake booster rod clevis pin. Remove the power brake booster mounting bolts and the booster from the vehicle.
4. To install, reverse the removal procedures. Torque the power brake booster nuts to 9 ft. lbs. (12 Nm).
5. Adjust the length of the booster pushrod. Install the master cylinder gasket on the booster. Set the measurement tool 0973700010 or equiva-lent, so the pin slightly touches the piston.
6. Turn the tool upside down and measure the clearance between the booster pushrod and pin head. The clearance should be **0**.

NOTE: When installing a new booster, make sure there is a little clearance between the pushrod end and the master cylinder piston.

Van

1. Disconnect the negative battery terminal.
2. To expose the master cylinder, perform the following:
   a. Remove the master cylinder reservoir cap, located at the left-side of the instrument panel.
   b. Remove the instrument cluster finish panel and the lower cluster finish panel. Disconnect the electrical connectors and the speedometer cable from the instrument panel, then remove the instrument panel.
   c. Remove the No. 1, 2 and 3 air ducts.
3. Using a syringe, remove the brake fluid from the master cylinder reservoir.
4. Remove the reservoir hoses from the master cylinder. Disconnect and plug the brake lines at the master cylinder.
5. Remove the master cylinder mounting nuts, the vacuum check valve bracket and the master cylinder from the vehicle.
6. Remove the bolts retaining the brake pedal and booster bracket. Remove the bracket from the vehicle.
7. To install, reverse the removal procedures. Torque the master cylinder mounting nuts to 9 ft. lbs. (12 Nm) and the brake lines-to-master cylinder to 11 ft. lbs. (15 Nm). Refill the master cylinder with new brake fluid and bleed the brake system.

Brake Caliper

REMOVAL & INSTALLATION

1. Raise ands safely support the vehicle.
2. Remove the wheel and tire assembly.
3. Remove the 2 wire clips at the ends of the brake pad pins.
4. Pull out the pads and anti-rattle springs.
5. Lift out the anti squeal shims.
6. Plug the vent hole on the master cylinder cap, to prevent fluid leakage. Disconnect and plug the brake line from the caliper.
7. Remove the 2 caliper mounting bolts and remove the caliper.

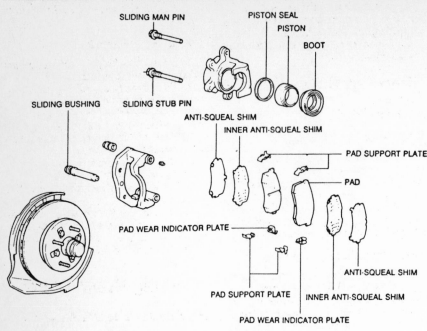

Front brake assembly—typical 2WD vehicle

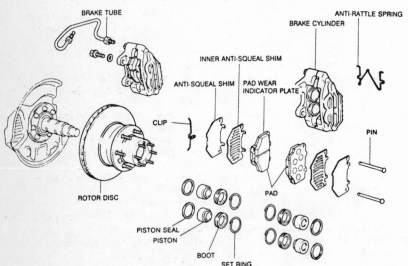

Front brake assembly—typcal 4WD vehicle

To install:

8. Position the caliper and install the mounting bolts. Tighten the caliper mounting bolts to 18 ft. lbs. (25 Nm) on Previa, 29 ft. lbs. (41 Nm) on 2WD vehicles or to 90 ft. lbs. (122 Nm) on 4WD vehicles.

9. Install the brake pads. Install the wheel and tire assembly and lower the vehicle.

10. Road test the vehicle.

Disc Brake Pads

REMOVAL & INSTALLATION

2WD Vehicle

1. Raise and safely support the vehicle.

2. Remove the wheel and tire assembly.

3. Remove the bottom caliper retaining bolt and loosen the top bolt.

4. Pivot the caliper upward and suspend it with wire, do not disconnect the brake line.

5. Remove the anti-squeal springs, brake pads, anti-squeal shims, wear indicator plates and the 4 pad support plates.

6. Installation is the reverse of the removal procedure. Compress the caliper piston into the bore using a C-clamp. Torque the caliper bolts to 29 ft. lbs. (39 Nm).

4WD Vehicle

1. Raise and safely support the vehicle.

2. Remove the wheel and tire assembly.

3. Remove the brake pad retaining clips, 2 locating pins, anti-rattle spring, brake pads and anti-squeal shims.

4. Installation is the reverse of the removal procedure. Compress the caliper piston into the bore using a C-clamp. Torque the caliper bolt to 27 ft. lbs. (36 Nm).

Brake Rotor

REMOVAL & INSTALLATION

1. Raise and safely support the vehicle.

2. Remove the wheels and tires.

3. Remove the caliper and brake pads. Hange from the suspension with a wire.

4. Remove the brake torque plate.

5. On 2WD vehicles, remove the grease cap, cotter pin and nut from the hub, except Previa.

6. On 4WD vehicles, remove the bolt retaining the locking hub assembly and lift it off the hub. Remove the nuts and washer from the inside of the hub.

7. Remove the rotor from the vehicle. Be careful not to drop the bearings from the hub.

To install:

8. Install the rotor. Be careful not to drop the bearings from the hub.

9. On 4WD vehicles, install the hub bolts.

10. On 2WD vehicles, install the grease cap, cotter pin and nut to the hub. Adjust the bearing preload.

11. Install the brake torque plate and torque the bolts to 77 ft. lbs. (104 Nm).

12. Install the caliper and brake pads.

13. Install the wheels and tires.

14. Bleed the brake system.

Brake Drums

REMOVAL & INSTALLATION

1. Raise and safely support the vehicle.

2. Remove the wheel and tire assemblies.

3. Remove the brake drum retaining screws.

4. Remove the drum from the vehicle. If the drum is difficult to remove, release the brake adjusters from behind the drum.

5. Install the drum on the vehicle and install the retaining screws.

6. Install the wheel and tire assemblies.

7. Lower the vehicle. Check the operation of the brakes and adjust as needed.

Brake Shoes

REMOVAL & INSTALLATION

1. Raise and safely support the vehicle.
2. Remove the wheel and tire assemblies.
3. Remove the brake drum retaining screws and remove the brake drum.
4. Disconnect the return spring from the rear shoe.
5. Remove the rear shoe hold-down spring, cups and pin. Disconnect the anchor spring from the shoe and remove the shoe.
6. Remove the front shoe hold-down spring, cups and pin. Disconnect the parking brake cable from the lever and remove the front shoe with the adjuster.
7. Remove the adjuster from the front shoe.
8. Installation is the reverse of removal procedure.
9. Adjust the brake shoes after the drum is installed. Once the vehicle is lowered, pump the brakes several times to seat the shoes.

Wheel Cylinder

REMOVAL & INSTALLATION

1. Raise and safely support the vehicle.
2. Remove the wheel and tire assemblies.
3. Remove the brake drum and the brake shoes.
4. Disconnect and plug the brake line from the wheel cylinder.
5. Remove the wheel cylinder mounting bolts and remove the wheel cylinder from the vehicle.

6. Installation is the reverse of the removal procedure. Torque the mounting bolts to 7 ft. lbs. (10 Nm).
7. Bleed the brake system after installation.

Parking Brake Cable

ADJUSTMENT

2WD Pick-Up

1. Working under the vehicle, tighten the adjusting nut at the equalizer until the travel is within limits and there is no drag at the rear shoes.
2. Apply the parking brake several times and again check that there is no drag with the brake released.

4WD Pick-Up and 4Runner

1. Working under the vehicle, tighten the bellcrank stopper screw until the play at the rear brake links is gone, then loosen the nut one full turn. Tighten the locknut.
2. Tighten one of the adjusting nuts on the intermediate lever while loosening the other, until the travel is correct. Tighten the locknuts.
3. Confirm that the bellcrank is in contact with the backing plate.

Van and Previa With Rear Drums

1. Raise and safely support the rear of the vehicle.
2. Remove the shift knob and the console box.
3. At the parking brake handle, loosen the cable locknut. Pull the hand brake upward about 7–9 clicks for Van or 4–5 clicks for Previa.
4. Turn the adjust nut until the rear wheels can no longer be turned, then, tighten the locknut.

5. Install the console and the shift knob.

Previa With Rear Discs

1. Raise the vehicle and support safely. Remove the rear wheel.
2. Temporarily install lug nuts and remove the plug hole in the rotor.
3. Turn the adjuster and expand the shoes until the rotor disc locks.
4. Back off the adjuster 8 notches and install the hole plug.
5. Make sure there is no excessive drag on the parking brake.
6. Install the wheel and lower the vehicle.
7. Loosen the 2 adjusting nuts under the vehicle and adjust the parking brake No. 1 cable until the parking brake lever clicks 4–5 times.

Parking Brake Shoes

REMOVAL & INSTALLATION

Previa With Rear Discs

1. Raise the vehicle and support safely. Remove the rear wheels.
2. Remove the shoe return spring using a remover tool 0971720010 or equivalent.
3. Remove the shoe strut and spring. Remove the shoe hold-down cups.
4. Remove the front shoe, adjuster and tension spring.
5. Remove the rear shoe and disconnect the parking brake cable.
6. Remove the parking brake shoe lever by spreading the clip with a prybar.

To install:

7. Lubricate all moving parts with high temperature lithium grease. Install the parking brake lever and new

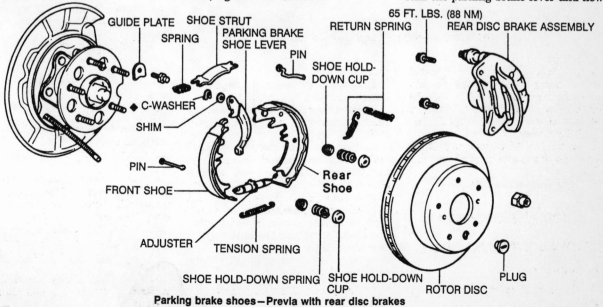

Parking brake shoes—Previa with rear disc brakes

clip. Bend the clip around the shaft with a pliers.

8. Connect the cable to the parking brake lever.

9. Install the rear shoe and hold-down.

10. Install the front shoe, tension spring and adjuster.

11. Install the strut with spring and shoe return springs.

12. Install the rotor and remaining components.

13. Adjust the parking brake and lower the vehicle.

Bleeding the Brake System

1. Fill reservoir to the maximum fill mark.

2. Bleed the brakes in the following sequence: right rear, left rear, right front, left front and load sensing proportioning valve.

3. Attach a bleed hose to the caliper or wheel cylinder being bled, immerse the end of the bleed hose in a glass container partially filled with brake fluid.

4. Have a helper apply brake pedal to pressurize the system. Open the bleed screw ½ turn. Close the bleed screw when the fluid entering the glass container is free of bubbles.

5. Check the reservoir fluid level and add fluid to the **MAX** fill mark.

6. Repeat bleeding operations at remaining wheels.

NOTE: Do not allow the master cylinder reservoir to run dry while bleeding the brakes. Running dry will allow air to re-enter the system making a second bleeding operation necessary.

Anti-lock Brake System Service

PRECAUTIONS

● When welding with an electric welding unit, unplug the electric control unit.

● During paint jobs, the electronic control unit may be exposed to a maximum of 203°F (95°C) for up to 2 hours or 185°F (85°C) if more time is needed.

● When removing the rear axle centerpiece, make sure the correct toothed wheel with the correct ratio for the wheel speed sensor is installed. If a wheel with the wrong number of teeth is installed, this fault will not show up when checking the system with the ABS tester. The stopping distance, however, will be increased during controlled braking.

● If work was done to non-ABS brake components, a simple operation-al test will be sufficient. This means that after driving about 5 mph, the warning light on the instrument panel should go out if the ABS system is intact.

● If ABS components have been replaced, the entire system should be checked using the appropriate tester in combination with brake test bench or an adaptor in combination with a multimeter.

RELIEVING ANTI-LOCK BRAKE SYSTEM PRESSURE

Pump the brake pedal at least 20 times with the ignition key in the OFF position. Place a shop rag around the hydraulic line fitting and wear safety glasses when disconnecting the hydraulic lines.

Anti-Lock Brake Actuator

REMOVAL & INSTALLATION

Previa

1. Disconnect the negative battery cable.

2. Remove the brake fluid from the master cylinder with a syringe.

3. Remove the plastic cover from the actuator.

4. Disconnect the hydraulic lines from the actuator. Plug the lines to prevent loss of fluid.

5. Disconnect the electrical connectors from the actuator.

6. Remove the actuator from the bracket.

7. Installation is the reverse of the removal procedure. Fill and bleed the system.

8. Connect the battery cable, start the engine and check brake system operation before driving the vehicle.

Pick-Up and 4Runner

1. Disconnect the negative battery cable.

2. Remove the brake fluid from the master cylinder with a syringe.

3. Remove the battery and tray.

4. Disconnect the electrical connectors from the actuator.

5. Using a flarenut wrench, disconnect the hydraulic lines from the actuator.

6. Turn the steering wheel clockwise until it locks before disconnecting the power steering lines at the actuator. Using a flarenut wrench, disconnect the power steering lines from the actuator.

7. Remove the actuator and bracket.

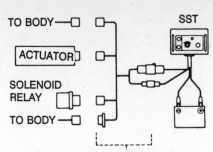

SUB WIRE HARNESS (SST)

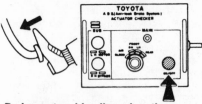

Brake system bleeding when the power steering system at the actuator is disconnected—Pick-Up and 4Runner with ABS

To install:

8. Install the actuator and torque the bolts to 21 ft. lbs. (28 Nm).

9. Connect the hydraulic lines and torque to 11 ft. lbs. (15 Nm) and the power steering lines to 34 ft. lbs. (47 Nm).

10. Connect the electrical connectors and install the remaining components.

11. Bleed the brake system.

ABS Electronic Control Unit (ECU)

REMOVAL & INSTALLATION

1. Disconnect the negative battery cable.

2. On Pick-up and 4Runner, remove the glove compartment assembly, disconnect the ECU and remove. The unit is located above the engine ECU.

3. On Previa, remove the audio power amplifier, if equipped.

4. Remove the ABS ECU and disconnect the wiring.

5. Installation is the reverse of removal.

ANTI-LOCK BRAKE SYSTEM BLEEDING

Previa

Use the conventional brake system bleeding procedure.

Pick-Up and 4Runner

Use the conventional brake system bleeding procedure unless the power steering hoses are disconnected from the brake actuator.

1. Bleed the power steering system using the conventional method.

2. Bleed the brake system with the engine running.

3. Bleed the brake system with the engine not running.

4. Disconnect the connector from the actuator and solenoid relay.

5. Connect an actuator tester tool 0999000150 and 0999099205 or equivalent, to the actuator solenoid relay and body side of the wiring harness through a sub-wire.

6. Connect the red cable of the tester to the battery positive terminal and the black wire to the negative terminal.

7. Start the engine and run at idle.

8. Turn the selector switch on the tester to **AIR BLEED**.

9. Strongly depress the brake pedal and hold.

10. Push **ON** and release the ON/OFF switch, 3 seconds each for 5 times.

11. Release the switch and brake pedal.

12. Check the power steering fluid level and add Dexron®II fluid.

13. Remove the tester and reconnect the actuator and solenoid.

CHASSIS ELECTRICAL

Heater Blower Motor

REMOVAL & INSTALLATION

Pick-Up, Van and 4Runner

1. Disconnect the negative battery cable.

2. Disconnect the electrical connector from motor.

3. Remove the blower motor-to-case screws and lift the motor from the case.

4. To install, reverse the removal procedures. Make sure the seal around the motor flange is in good condition.

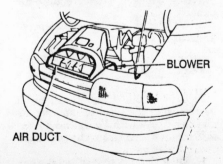

Blower motor location—Previa

Land Cruiser

1. Disconnect the negative battery cable. Disconnect the electrical connector from the blower motor.

2. Disconnect the flexible tube from the side of the blower motor.

3. Remove the blower motor fasteners and lower the blower motor out of the air inlet duct.

4. To install, reverse the removal procedures. During installation, be sure to position the motor so the flexible tube can be attached to the motor.

Previa

1. Disconnect the negative battery cable and open the hood.

2. Remove the air duct and disconnect the motor connectors.

3. Remove the blower motor from the housing.

4. Installation is the reverse of removal.

Windshield Wiper Motor

REMOVAL & INSTALLATION

Pick-Up, Van and 4Runner
FRONT

1. Disconnect the negative battery cable. Disconnect the wiring from the wiper motor. Remove the motor from the fire wall.

2. Remove the nut, then, pry the wiper link from the crank arm.

3. Remove the motor.

4. To install, reverse the removal procedures and inspect he operation.

REAR

1. Disconnect the negative battery cable. At the rear of the vehicle, remove the wiper motor cover panel.

2. Remove the wiper arm from the wiper motor.

3. Disconnect the electrical connector from the wiper motor.

4. Remove the wiper motor-to-door bolts and the motor from the vehicle.

5. To install, reverse the removal procedures and inspect the operation.

Land Cruiser and Previa

NOTE: On these vehicles, the wiper motor is removed with the linkage assembly.

1. Disconnect the negative battery cable. Remove the wiper arm retaining nuts, then, the wiper arm/blade assemblies.

2. Remove both wiper arm pivot covers and the pivot-to-cowl attaching screws.

3. Remove the service hole covers from the cowl area of the engine compartment.

4. Disconnect the wiring from the wiper motor.

5. From the engine compartment, remove the wiper motor plate-to-cowl screws. Withdraw the wiper motor and the linkage from the cowl panel as an assembly.

6. Pry the linkage from of the wiper motor.

7. To install, reverse the removal procedures.

Windshield Wiper Linkage

REMOVAL & INSTALLATION

Pick-Up, Van, 4Runner and Previa

1. Disconnect the negative battery cable. Remove the wiper motor.

2. Remove the wiper arms by removing their retaining nuts and working them off their shafts.

3. Remove the wiper shafts nuts/spacers and push the shafts down into the body cavity. Pull the linkage out of the cavity through the wiper motor hole.

4. To install, reverse the removal procedures.

Land Cruiser

1. Disconnect the negative battery cable. Remove the wiper arm assemblies.

2. Remove the end plate from the pivot housing.

3. Remove the wiper motor with the linkage cable.

4. Separate the wiper motor and the transmission.

5. Remove the linkage cable.

6. To install, reverse the removal procedures.

Windshield Wiper Switch

REMOVAL & INSTALLATION

Front

1. Disconnect negative battery terminal from the battery.

2. Remove the upper and the lower steering column shrouds.

3. Disconnect the combination switch electrical connector.

4. Remove the terminal from the horn contact.

5. To remove the windshield/wiper switch wires from the electrical connector, place a small prybar into the end of the connector, pry up on the retaining tab and pull the wire(s) from the connector.

6. Remove the windshield/wiper

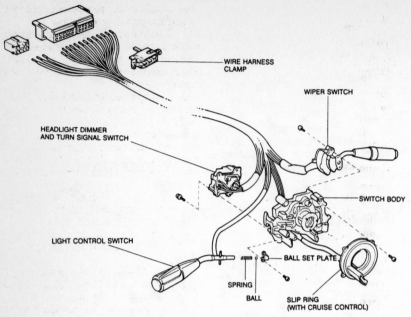

Typical combination switch assembly

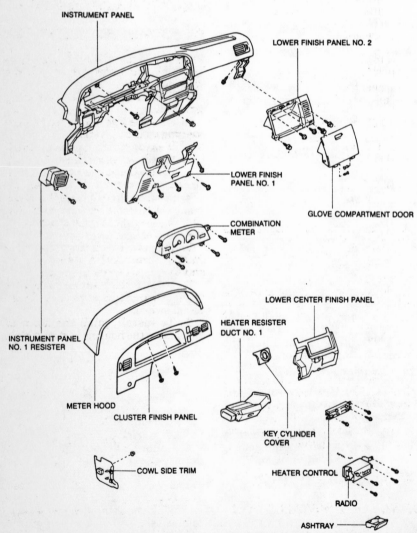

Typical Instrument cluster and panel assembly—Pick-Up and 4Runner

switch-to-combination switch screw and the switch.

7. Installation is the reverse of the removal procedure. To install, place the wire(s) into the electrical connector's slots, place a suitable tool behind the wire terminal and push the wire into the connector until the retaining tab locks it into place.

Rear

If equipped with a rear wiper switch, it will be located in the center of the dash.

1. Disconnect the negative battery cable. Using a small prybar, pry the rear wiper switch from the center of the dash.

2. Disconnect the electrical connector from the rear of the switch.

3. To install, reverse the removal procedures.

Instrument Cluster

REMOVAL & INSTALLATION

Pick-Up, Van and 4Runner

1. Disconnect the negative battery terminal from the battery.

2. Remove the upper and lower steering column covers.

3. Remove the instrument trim panel screws and the panel.

4. Disconnect the speedometer cable from the speedometer.

5. Remove the instrument panel screws and pull the panel forward. Disconnect the electrical connectors from the back of the panel and remove the panel.

6. To install, reverse the removal procedures.

Land Cruiser

1. Disconnect the negative battery terminal from the battery.

2. Disconnect the speedometer cable. Remove the instrument panel screws.

3. Loosen the steering column clamp by removing the attaching bolts.

4. Pull out the instrument panel and the speedometer, disconnect the electrical connectors and remove the panel.

5. To install, reverse the removal procedures.

PREVIA

1. Disconnect the negative battery cable.

2. Remove the cluster finish panel by removing the 2 screws and prying on the left side with a prybar. Be careful not to damage the plastic components.

3. Remove the 4 screws and disconnect the cable from the automatic

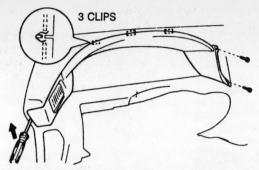

3 CLIPS

Instrument cluster finish panel removal—Previa

transmission indicator. Remove the cable from the roller and cluster.

4. Disconnect the electrical connectors and remove from the vehicle.

5. To install, reverse the removal procedures.

Speedometer Cable

REMOVAL & INSTALLATION

1. Disconnect the negative battery cable. Remove the instrument cluster and disconnect the cable from the speedometer.

2. Disconnect the other end of the speedometer cable from the transmission extension housing and pull the cable from its jacket at the transmission end.

NOTE: If the cable is being replace because it is broken, be sure to remove both pieces of the broken cable.

3. Using graphite, lubricate the new speedometer cable and insert it into the cable jacket at the lower end.

4. Connect the speedometer cable to the transmission, then, to the instrument cluster.

5. To complete the installation, reverse the removal procedures.

Radio

REMOVAL & INSTALLATION

1. Disconnect the negative battery cable.

2. Remove the screws retaining the lower center finish panel to the dash board.

3. Remove the radio retaining screws.

4. Remove the radio by pulling it out of the instrument panel.

5. Disconnect the antenna wire and the multi-plug connector.

6. Remove it from the vehicle.

7. Installation is the reverse of the removal procedure.

Headlight/Dimmer Switch

REMOVAL & INSTALLATION

1. Disconnect the negative battery terminal from the battery.

2. Remove the upper and lower steering column covers.

3. Disconnect the electrical connector from the combination switch.

4. Remove the headlight switch-to-combination switch screws and the headlight switch from the combination switch.

5. To install, reverse the removal procedures.

Turn Signal Switch

REMOVAL & INSTALLATION

1. Disconnect negative battery terminal.

2. Remove the upper and the lower steering column shrouds.

3. Disconnect the combination switch electrical connector.

4. At the left-rear of the combination switch, remove the mounting screws and the turn signal switch.

5. If necessary to the remove the turn signal switch wires from the electrical connector, place a small prybar into the end of the connector, pry up on the retaining tab and pull the wire(s) from the connector.

6. To install, place the wire(s) into the electrical connector's slots, place a prybar behind the wire terminal and push the wire into the connector until the retaining tab locks it into place.

7. To complete the installation, reverse the removal procedures.

Combination Switch

REMOVAL & INSTALLATION

The combination switch is composed of the turn signal, the headlight con-

trol, the dimmer, the hazard, the wiper and the washer switches.

1. Disconnect the negative battery cable. Remove the steering wheel.

2. Remove the upper and lower steering column shroud screws and the shrouds.

3. Remove the combination switch screws and the switch from the column.

4. Disconnect the electrical connector from the combination switch. To remove the wires from the electrical connector, perform the following procedures.

   a. Using a small prybar, insert it into the open end between the locking lugs and the terminal.

   b. Pry the locking lugs upward and pull the terminal out from the rear.

   c. To install the terminals, simply push them into the connector until they lock securely in place.

5. To complete the installation, reverse the removal procedures.

Ignition Lock/Switch

REMOVAL & INSTALLATION

The ignition lock/switch is located behind the combination switch on the steering column.

1. Disconnect the negative battery terminal.

2. Remove the upper and lower steering column covers.

3. Disconnect the ignition switch from the electrical connector.

4. Using the key in the ignition switch, turn it to the **ACC** position.

5. Using a thin rod, place it into the hole of the cylinder lock housing. Pushing down on the thin rod, pull out the cylinder lock.

6. Remove the unlock warning switch-to-combination switch screws and the unlock warning switch.

7. Remove the ignition switch-to-combination switch screw and the ignition switch.

8. To install, push the ignition switch into the housing and install the

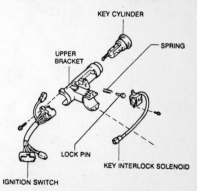

KEY CYLINDER

SPRING

UPPER BRACKET

LOCK PIN

KEY INTERLOCK SOLENOID

IGNITION SWITCH

Ignition lock assembly

screw. Using the key, install cylinder lock into the housing until the retaining tab locks it in place.

9. To complete the installation, reverse the removal procedures.

Stoplight Switch

REMOVAL & INSTALLATION

1. Disconnect the negative battery cable.
2. Remove the electrical connector from the switch at the brake pedal.
3. Remove the mounting nut and remove the switch from the bracket.
4. Installation is the reverse of the removal procedure.

Clutch Switch

REMOVAL & INSTALLATION

1. Disconnect the negative battery cable.
2. Remove the electrical connector from the switch at the clutch pedal.
3. Remove the mounting nut and remove the switch from the bracket.
4. Installation is the reverse of the removal procedure.

Fuses, Circuit Breakers and Relays

LOCATION

There are 3 fuse boxes on the Pick-Up, 4Runner and Land Cruiser. One is located in the engine compartment, 1 at the drivers side kick panel and 1 behind the glove box. The Van has a combination fuse box and relay panel, located behind a panel on the passengers side of the instrument panel. The junction/relay and fuse block is located behind the heater controls, under the center instrument panel hood.

Each fuse box has the fuse numbers and circuits protected on the lid of the box.

Computer

LOCATION

The ECU is located at the right kick panel for the Pick-Up and 4Runner, behind the glove compartment for the Land Cruiser, at the left door pillar for the Van, behind the trim panel or under the driver's seat for the Previa.

Flashers

The turn signal/hazard flasher is located under the instrument panel near the steering column for the Pick-Up, 4Runner and Previa. The flasher unit is located behind the heater controls for the Van and in the relay block at the left kick panel for the Land Cruiser.

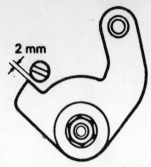

Cruise control actuator adjustment—Van

Cruise Control

ADJUSTMENT

Pick-Up, 4Runner and Land Cruiser

1. Inspect that the control cable freeplay is less than 0.39 in. (10mm).
2. Connect the positive lead from the battery to terminals **1** and **2** and negative lead to terminal **3** of the actuator.
3. Slowly apply vacuum from 0–11.81 in. Hg. (0–300mm Hg), check that the control cable can be pulled smoothly and the cable stroke is at least 1.42 in. (36mm).

Van

Inspect that the control cable freeplay is less than 0.08 in. (2mm).